Science Networks . Historical Studies
Founded by Erwin Hiebert and Hans Wußing
Volume 34

Edited by Eberhard Knobloch and Erhard Scholz

Jens Høyrup

Jacopo da Firenze's *Tractatus Algorismi* and Early Italian Abbacus Culture

Birkhäuser
Basel · Boston · Berlin

Jens Høyrup
Section for Philosophy and Science Studies
Roskilde University
Box 260
4000 Roskilde
Denmark
e-mail: jensh@ruc.dk

2000 Mathematical Subject Classification: Primary 01A35; Secondary 01A30, 01A32, 01A40, 01A74, 01A80

Library of Congress Control Number: 2007935140

Bibliographic information published by Die Deutsche Bibliothek
Die Deutsche Bibliothek lists this publication in the Deutsche Nationalbibliografie;
detailed bibliographic data is available in the Internet at http://dnb.ddb.de

ISBN 978-3-7643-8390-9 Birkhäuser Verlag AG, Basel - Boston - Berlin

© 2007 Birkhäuser Verlag AG
Basel · Boston · Berlin
P.O. Box 133, CH-4010 Basel, Switzerland
Part of Springer Science+Business Media
Printed on acid-free paper produced from chlorine-free pulp. TCF∞
Printed in Germany

ISBN 978-3-7643-8390-9   e-ISBN 978-3-7643-8391-6

9 8 7 6 5 4 3 2 1   www.birkhauser.ch

A honore di tucti magistri et scolari de questa scienza. Et de qualunqua altra bona persona vedesse et legesse questo tractato devoto e ragionevolmente

and in particular in memoriam

Joseph Needham
Nikolay Bukharin
Edgar Zilsel
Kurt Vogel
Gino Arrighi

Preface

In 1984, I was present when Joseph Needham gave a dinner talk at the Sarton Centennial Conference. During or just after the talk I whispered to the colleague sitting next to me "He does not know, but he was my only teacher in the history of science". Back whispered Samuel Edgerton – he, indeed, was the neighbour, and I hope he will forgive me for telling – "He was mine, too".

One of the things I learned from Joseph Needham's *Grand Titration* [1969] – recently published at the time my serious interest in the history of science began, and the first thing from his hand I read – was to take the relation between "clerks and craftsmen" seriously as a complicated interplay. Needham, of course, made no secret of having received inspiration both from Edgar Zilsel and from the Soviet contributions to the legendary 1931-conference on the history of science. When reading these, what impressed me as intellectually most sophisticated was Nikolay Bukharin's London-paper [1931/1971].[1]

My receptivity to Needham's and Bukharin's interest in the intricate relations between theoreticians' and practitioners' knowledge was certainly enhanced by my own experience as a physics teacher in an engineering school, where I was confronted time and again with the inability of university-trained physicists and mathematicians to create a bridge between the shape of their own knowledge and the interests, orientation and knowledge of students who thought about building houses and bridges.

As I was also engaged at the times in the Danish debate about the "new math" reform, my general interest in the history of science concentrated naturally on the history of

---

[1] I thus agree with I. Bernard Cohen's words as quoted by Loren Graham [1993: 241] that "Bukharin's piece remains impressive today to a degree that Hessen's is not".

mathematics as soon as I started doing my own work in the field – and, as it turned out, mostly on the history of pre-Modern mathematics. Along with other topics I maintained my interest in practitioners' knowledge as it could be encountered in the sources – which, when we speak of pre-Modern mathematical practice, was often only possible through teaching material directed at future practitioners or otherwise reflecting practitioners' knowledge. Though for a long time I did not work directly on late medieval European vernacular mathematics, I collected material when I encountered it – in particular source editions. Much I found in Baldassare Boncompagni's marvellous *Bullettino*, much I owe to Moritz Cantor and the circle around him. I also had the good luck to come in personal touch with Kurt Vogel and Gino Arrighi and to receive directly from them publications which even my excellent interlibrary service would have been unable to get hold of.

Two accidents pushed me to capitalize on this material. First, in November 1996, at a meeting at the Mathematisches Forschungsinstitut Oberwolfach, Henk Bos asked me to improvise on 43 hours notice a presentation of what had happened in the historiography of European medieval mathematics during the last forty years and to summarize the picture which now emerged. I had no support at hand beyond my memory, my personal catalogue and my own writings from the last decade, catalogue and writings as present on my laptop. None the less I accepted, maybe because of my personal affection for the organizers, maybe because I felt as absurdly flattered as the father asked by his kid to repeat a magnificent sunset – maybe for still other reasons. While working on my talk I discovered structures in the story which I had not been conscious of knowing about; during the next months I therefore set myself the goal to do the work as it *should* have been done, looking once more at relevant sources and publications (which for a project of this kind were also primary sources), sifting and ordering the material, etc.

The next accident was that I reread for this purpose an article which Louis Karpinski had published in 1929 about "The Italian Arithmetic and Algebra of Master Jacob of Florence, 1307". Knowing now Arabic algebra much better than I had when first reading the paper in 1977, I came to suspect that Jacopo's treatise might have astounding implications for our understanding of the origins of European vernacular algebra (the reasons for this are explained in detail in the Introduction, p. 3). The suspicion had to be verified, and I shelved my medieval survey (it remains shelved) and got hold of the manuscript Karpinski had described. Working first on the algebra section alone, afterwards on the rest of the treatise and other early abbacus writings – that is, mathematical writings made in the context of the late medieval and Renaissance *scuole d'abbaco* for merchant youth – I discovered that much more had to be changed in the conventional picture than I had at first suspected.

Much of abbacus mathematics was still around when I went to primary and early secondary school in the 1950s (that was before the arrival of the "new math" and the reactions to it, which to some extent changed the situation). Abbacus mathematics itself, however, can be known only by those who read Italian, since no editions of the manuscripts are accompanied by translations. I therefore realized that any challenge to the conventional picture had to be based on a text edition accompanied by a translation and copious references to other parts of the record.

I also thought that the challenge *had* to be made. According to the conventional picture, abbacus mathematics is a simplified version of a *great book* written in Latin (Leonardo Fibonacci's *Liber abbaci*), adapted by less able teachers to a school with rather modest pretensions. Close scrutiny of the sources shows that interactions between different cultures at the practitioners' own level was much more important than the "great book". In a certain way, the present book thus pays a debt to the manes of Joseph Needham, Nikolay Bukharin and Edgar Zilsel.

So much about what induced me to work on the history of abbacus mathematics and to write the present book. About the book itself I shall only say that it falls into two parts of roughly equal length, one examining Jacopo's treatise together with the whole context of early abbacus mathematics, the other containing an edition of the treatise itself (made from the Vatican manuscript Vat. Lat. 4826) accompanied by a very literal translation; an appendix contains a "semi-critical" edition of a revised version of the treatise, made from Milan, Trivulziana MS 90, collated with Annalisa Simi's transcription of Florence, Riccardiana MS 2236. Details are better read from the table of contents. Indexes of names, subjects and sources referred to should facilitate the use of the book.

References are made according to the author-date system (in the case of text editions, the editor takes the place of the author). All translations into English are mine if nothing else is stated.

For many years I have been in helpful contact with Menso Folkerts, Ivo Schneider, Ulrich Rebstock, Jacques Sesiano, Barnabas Hughes, Tony Lévy and Ahmed Djebbar. Specifically for my work on abbacus matters this last decade I have also benefitted from exchanges with Warren Van Egmond, Mahdi Abdeljaouad, Wolfgang Kaunzner, Charles Burnett, Raffaella Franci, Laura Toti Rigatelli, Annalisa Simi, Enrico Giusti, Lucia Travaini, Maryvonne Spiesser, Stéphane Lamassé, Betsabé Caunedo del Potro and Maria do Céu Silva, and occasionally with many other colleagues. It is a pleasant duty to thank all of them for assistance, inspiration and challenges.

PICARDY

Paris
CHAMPAGNE

Troyes

Tours

Milan
LOMBARDY   Venice

PROVENCE
Toulouse   Avignon   Genova   Bologna
Aigues-Mortes   Nîmes   Lucca   Florence
Montpellier   Pisa TUSCANY
CASTILIA   Perugia
                 CATALONIA   UMBRIA
PORTUGAL   Barcelona
                                                Rome

AL-ANDALUS   Naples
Seville   APULIA

MOROCCO   Bejaia   SICILY

MAGHREB

Table of Contents

JACOPO, HIS TREATISE, AND ABBACUS CULTURE

JACOPO, HIS TREATISE, AND ABBACUS CULTURE

Introduction

In [1929a], Louis Karpinski published a short description of "The Italian Arithmetic and Algebra of Master Jacob of Florence, 1307". Around one third of the note describes the algebra chapter of the treatise, pointing out among other things that Jacopo presents the algebraic "cases" in a different order than al-Khwārizmī, Abū Kāmil, and Leonardo Fibonacci, and that the examples that follow the rules also differ from those of the same predecessors. Karpinski did not mention explicitly the absence of geometric proofs of the rules, nor that the examples differ from those of the other authors already in general style, not only in detailed contents; but attentive reading of Karpinski's text and excerpts from the manuscript leave little doubt on either account.

In retrospect, Karpinski's succinct article should therefore have been a challenge to conventional thinking, both about the history of pre-Renaissance algebra and about applied arithmetic in general from the same period. Nonetheless, I have not been able to discover any echo whatsoever of his publication. This may have at least three reasons.

Firstly, 1929 fell in a period where the interest in European medieval mathematics was at a low ebb – probably the lowest since the Middle Ages, at least since 1840. From 1920 to c. 1948 (from Moritz Cantor's death to the beginning of Marshall Clagett's work in the field), the total number of scholarly publications dealing with European medieval mathematics (Latin as well as vernacular) does not go much beyond the dozen.

Secondly, the existence of the distinct *abbaco*[2] mathematical tradition was not recognized, although Karpinski had already described another abbacus treatise in [1910], and Italian nineteenth-century local history had dealt with the *scuole d'abbaco* of many localities. As early as [1900: 166], it is true, Cantor had spoken of the existence throughout the fourteenth century of two coexisting "schools" of mathematics, one "geistlich" ("clerical", that is, universitarian), the other "weltlich oder kaufmännisch" ("secular or commercial", supposedly derived from Leonardo Fibonacci's work). Part of Cantor's basis for this was Guglielmo Libri's edition [1838: III, 302–349] of a major section of what has now been identified beyond reasonable doubt as Piero della Francesca's *Trattato*

---

[2] The late medieval sources alternate between the spellings *abaco* and *abbaco*. In order to avoid confusion with the calculation-board or -frame I shall stick to the spelling *abbaco* and the Anglo-Latin analogue *abbacus* (except of course in quotations, which follow the source that is quoted).

As formulated by Richard Goldthwaite [1972: 419], "Originally the term *abbaco* referred to a device for calculating by means of disks, beads, counters, etc.; but in Italy after the introduction of the use of the Arabic numerals [...], the term *abbaco* came to be used in a general sense for instruments, methods, manuals, schools, teachers or anything else related to the skill of doing computations, especially with reference to practical applications in the mercantile world".

d'abaco[3] (which Cantor, accepting Libri's wrong dating, had located in the fourteenth century); also known to Cantor was the material collected and published by Baldassare Boncompagni, for instance in [1854], which, as a matter of fact, was already a sufficient basis for the claim. Georg Eneström [1906] had done what he could to make a fool of Cantor by twisting his words.[4] Sensitive reading should have exposed Eneström's arrogant fraud; but the kind of knowledge that was required for that had come to be deemed irrelevant for historians of mathematics and hence forgotten, and George Sarton [1931: 612*f*] not only cites Eneström's article but embraces the whole thesis uncritically.

Thirdly, like Cantor, Karpinski took the continuity from Fibonacci onward for granted, and concluded [1929a: 177] that the

> treatise by Jacob of Florence, like the similar arithmetic of Calandri, marks little advance on the arithmetic and algebra of Leonard of Pisa. The work indicates the type of problems which continued current in Italy during the thirteenth to the fifteenth and even sixteenth centuries, stimulating abler students than this Jacob to researches which bore fruit in the sixteenth century in the achievements of Scipione del Ferro, Ferrari, Tartaglia, Cardan and Bombelli.

Only those interested in manifestations of mathematical stagnation – thus Karpinski's suggestion – would gain anything from looking deeper into Jacopo's treatise.

---

[3] On the identification of Libri's manuscript with the very manuscript from which Arrighi [1970a] made his edition, see [Davis 1977: 22*f*]. Margaret Davis also gives the argument that the anonymous treatise must indeed be ascribed to Piero.

[4] Arguing from his own blunt ignorance of the institution within which university mathematicians moved, Eneström rejected the epithet "clerical" as absurd ("Sacrobosco und Dominicus Clavasio waren meines Wissens nicht Geistliche"; actually, all university scholars were at least in lower holy orders, as evident from the familiar fact that they were submitted to canonical jurisdiction). He was further convinced that merchants' mathematics teaching would never treat useless problems like the "100 fowls", and claimed that Fibonacci was only spoken of as a merchant in late and unreliable sources; therefore no "commercial" school could have been inspired by Fibonacci and teach such useless problems. Actually, Cantor had not argued from Fibonacci's profession, although he does refer to him elsewhere in pseudo-poetical allusions as the "learned merchant" – pp. 85*f*, 154; yet in the very preface to the *Liber abbaci* Fibonacci refers to what can hardly be anything but business travelling – in Boncompagni's edition [1857: 1] Fibonacci speaks of his travels to "places of business". What Eneström could not know is that all other manuscripts speak of travelling "for reasons of business" [Grimm 1976: 101*f*].

Three Manuscripts

Whatever the reason, nobody seems to have taken interest in the treatise before Warren Van Egmond inspected it in the mid-seventies during the preparation of his global survey of Italian Renaissance manuscripts concerned with practical mathematics [1976; 1980]. By then, the autonomous existence of the abbacus tradition in the fourteenth and fifteenth centuries was well-established; but Van Egmond noticed that the manuscript which Karpinski had examined (Vatican MS Vat. Lat. 4826, henceforth **V**) could be dated by watermarks to the mid-fifteenth century, and that the algebra chapter (and certain other matters) were missing from two other manuscripts which also claim to contain Jacopo's *Tractatus algorismi* (Florence, Riccardiana MS 2236, undated;[5] henceforth **F**; and Milan, Trivulziana MS 90, c. 1410; henceforth **M**).[6] Because **M** can be dated by watermarks to c. 1410, some 40 years before **V** (yet still a whole century after 1307), and since **V** contains rules for the fourth degree not present in the algebra of Paolo Gherardi's *Libro di ragioni* from 1328, Van Egmond decided (personal communication) "that the algebra section of Vat.Lat. 4826 [was] a late 14th-century algebra text that [had] been inserted into a copy of Jacopo's early 14th-century algorism by a mid-15th-century copyist".

Prima facie, this might seem to be a reasonable conclusion – at least if one disregards the fact that reducible fourth-degree equations were solved routinely in Arabic algebra at least since al-Karajī's time and therefore were no innovation, neither in 1307 nor in the late fourteenth century. As we shall see, however, close textual analysis of **V** shows that this manuscript is very coherent in style as well as regarding the presence of various idiosyncratic features both in the chapters that are shared with **M** and **F** and in those that are not; **F** and **M**, on the other hand, are less coherent. Van Egmond's explanation of the differences between the two versions must therefore be turned around: **V** is a quite faithful descendant of Jacopo's original (or at least of the common archetype for all three

---

[5] Van Egmond's dating [1980: 148] is misleading, since it is merely the date of Jacopo's original (which is given in all three manuscripts), not that of the manuscript.

[6] A transcription of **F** was made by Annalisa Simi in [1995]. A critical edition of **F** and **M** by the late Jean Cassinet and Annalisa Simi was almost finished in 1999, but it got stuck with the publisher and is not going to appear (Maryvonne Spiesser, personal communication), for which reason I give a transcription of **M** with indication of all not merely orthographic variants with respect to **F** in the Appendix.

As pointed out by Karpinski [1929a: 170], **F** had already been mentioned by Boncompagni in 1883 and by B. Lami, librarian of the Biblioteca Riccardiana, in 1754; however, Karpinski had not seen **F** and therefore could not know that it differs from **V** on important points.

manuscripts), whereas the closely related **F** and **M** are the outcome of a process of rewriting and abridgement, an adapted version – well adapted indeed to the curriculum of the abbacus school known from a couple of documents to which we shall return (below, p. 28). To this version I shall henceforth refer as **M+F** when there is no need to distinguish the single manuscripts.

Internal evidence shows that **V** is a meticulous (yet not blameless) library or bookseller's copy made from another meticulous copy – for details, see p. 8. Seeming set-offs from Provençal orthography (and further internal textual evidence) suggest that preceding steps in the copying process (if any there are) can have been no less meticulous. All in all it is thus legitimate to treat **V** as identical with Jacopo's treatise from 1307 apart from a few errors and omissions.

The Vatican Manuscript

Van Egmond [1980: 224] describes **V**, the Vatican manuscript, in these terms:

> s. XV (c. 1450, *w*[atermark]), Holograph *libreria* treatise
> Paper, 4° gr[ande], 286×203 mm., 72 [leaves] num[bered] orig[inally] 1–59, 59–66, 69–73, plus 2 enclosing guard sheets. [...]. Single hand, a neat semi-cursive Gothic bookhand in 1 col[umn] press-ruled 194–205×118 mm., 32 regularly-spaced lines. Dark brown ink with alternating red and blue initials, some decorated; capitals shaded in yellow on 1r, 33v–41r, lined in red on 41v–42r; some titles in red. Tables on 2v, 3r, 4v–14v outlined in red; many colored diagrams and drawings in the margins.

Time, Place and Author

The incipit runs as follows:

> Incipit tractatus algorismi, huius autem artis novem sunt speties, silicet, numeratio, addictio, subtractio, ⟨mediatio,⟩[7] duplatio, multiplicatio, divixio, progrexio, et radicum extractio. Conpilatus a magistro Jacobo de Florentia apud Montem Phesulanum, anno domini Mᵒ CCC° VII° in mensis septenbris.

The treatise was thus written in Montpellier in September 1307 by one Jacobus or Jacopo coming from Florence. The Latin incipit, the Latinizing spellings and hypercompensations (see imminently) and a reference to Boethius's *Arithmetic* in the introduction should probably not be taken as indications that Jacopo was a university scholar, only that he moved in an environment where scholars and mathematical practitioners were in contact (other indications suggest that this was indeed the situation in Montpellier – cf. [Hahn 1982: xxi*ff*]);[8] later on he demonstrates repeatedly not to know the difference between

---

[7] Inserted in agreement with **M+F**, and needed in order to fill out the number of nine species. In all quotations, ⟨ ⟩ is used to indicate scribal omissions, and { } to mark superfluous passages (e.g., dittographies). Editorial explanations and observations are in [ ].

[8] Also in other respects, Montpellier was the site of cultural contact; in 1271, Moïse Ibn Tibbon

proportio and *propositio*.[9] The religious invocations of the introduction correspond well to the style of other abbacus writers (and of Arabic writings and Catalan-Provençal vernacular culture – cf. below, note 119) but only enter the style of more scholarly work in the form of gentle parody.[10]

This is as close as we can come to the author for the moment. He may possibly be identical with a certain "ser Jacopo dell'abacho" who taught in 1334 in the "Scuola verso Piazza Peruzzi" in Florence (where he would obviously not be identified as "from Florence") – with one Jacopo da Firenze who taught the abbacus in Lucca and Pisa from 1345 to 1347; with the very first attested abbacus teacher from Florence, who taught in 1283 – with two of them (all three is extremely implausible, given the life expectancy of the age) – or with none of them.[11] As we shall see, he is likely to have left his native town in search for that new knowledge which the emerging abbacus school seemed to be in need of. Such a choice is most likely to have been made by a rather young man, which should rule out that he had begun his career in 1283, but would just allow him to have been still alive and active though around sixty years of age in 1347.

Language, Orthography, Copying Quality and Writing

After the incipit, the rest of the manuscript is in Tuscan with a somewhat Latinizing orthography (*dicto/decto, facto, septimo, scripto, octavo, exemplo, tractato, sanctissima,* etc. – none of them used quite systematically)[12]. A number of non-standard spellings might reflect the Provençal linguistic environment of Montpellier, the place where the incipit says the treatise was made, or some other northern region: *el* or *lo* almost

---

had translated in the same town al-Ḥaṣṣār's treatise on *ḥisāb* into Hebrew [Lévy 2003: 282] – seemingly the very first Hebrew reference to *al-jabr* (though no systematic presentation), and the first diffusion of the Maghreb mathematical tradition in Christian Europe (though apparently not further diffused to its Christian segment).

[9] The attempted fidelity of the copyists and the consistency of the confusion allows us to conclude with fair (not full) certainty that Jacopo himself is responsible for the mistake.

[10] I know of two such examples: *Liber Jordani de triangulis*, a student *reportatio* of a series of lectures probably held by Jordanus de Nemore himself [Høyrup 1988: 347–351], and Chuquet's trinitarian argument for the title of his *Triparty* [ed. Marre 1880: 593] – a work situated in the borderland between the abbacus-type and the scholarly tradition.

[11] See [Van Egmond 1976: 383], and [Ulivi 2002b: 196]. Ulivi, after a tentative identification with the ones from 1283 and 1334, admits in a note that "we observe that the corresponding documents that we have found might refer to different abbacus masters named Jacopo. One master Jacopo da Firenze was employed as an abbacus teacher by the Commune of Lucca in the years 1345–1347, and then taught in Pisa".

[12] We also encounter Latinizing hypercompensations: thus for instance *librectine* (fol. 4ᵛ), *cictadino* (fol. 36ʳ), *soctilita* (fol. 1ʳ), *tucto, rocto, socto* (all three regularly).

consistently instead of *il*,[13] consistently *sera* or *serra* instead of *sara* (i.e., *sarà*) and almost consistently *mesura* instead of *misura*; mostly *de* instead of *di*, and occasionally *que* instead of *che* (both as an interrogative and as a relative pronoun); mostly *remanere* (with declinated forms) instead of *rimanere*[14], and mostly *vene* instead of *viene*; mostly also *doi* or *doy* instead of *due*, occasionally *dui* or *duy*; mostly *amendori*, or occasionally *amendoi/-dui*, instead of *amendue*;[15] ten, when not written with numerals, is almost always *dece*; *-ximo* is used instead of *-simo* as ordinal suffix (and *-x-* also elsewhere where we might expect *-ss-*); *lira/lire* are *libra/libre*. Also possibly influenced by the Provençal (or French or Iberian) environment is the use of *ha* (written *a*) or *si ha* instead of *c'è*.[16]

The copyist seems to have aimed at fidelity toward his model both concerning its orthography and in other respects: at times he corrects a spelling, even though both the new and the old spelling are present elsewhere in the text, which suggests an aspiration for orthographic accuracy (but also shows that he did commit errors on this account, some of which he will probably not have noticed); when leaving a sequence of open spaces of c. 2 cm instead of numbers on fol. 39^r, he makes a note in the margin, "così stava nel'originale spatii" ("in this way there were spaces in the original"). The lacuna should have contained the transformation of 4√54 into a pure square root; obviously, the original author did not want to compute 16×54 mentally but postponed – and forgot; and all intermediate copyists conserved the blank. Similarly, fols 46^v–47^r present us with firm evidence that not only the ultimate but also the penultimate copyist tried to be faithful: fol. 46^v starts by telling us that a section on silver coins has been omitted by error and

---

[13] *Lo* (occasionally *lu*) appears in all cases where modern Italian requires it except before *zero*, and is used inconsistently before *s*+vowel, *d*, *m*, *p*, *q*, *r*, and *t*. *El*, on the other hand, is used exclusively when *following* a vowel, thus allowing contraction (its use after the word *et* indicates that this was meant to be pronounced *e*). This habit agrees with a custom that can be seen both in the *Divina Commedia* (with spelling *il*), cf. [Rohlfs 1966: II, 99*f*] and in contemporary writings of the Provençal troubadour n'At de Mons [ed. Bernhard 1887] (with consistent contraction, which explains the habit).

[14] At one occasion *romane*, which looks Catalan but could also be north eastern Italian, see [Rohlfs 1966: I, 169; II, 363].

[15] *Amendore* also occurs – as *ame(n)doro* – in a probably Umbrian abbacus (*Livero de l'abbecho*) from not much later than 1290, to which we shall return repeatedly (Florence, Riccardiana MS 2404, fols 1^v, 69^r, 69^v, 81^r, 113^r, ed. [Arrighi 1989: 9*f*, 87, 119]). According to [Rohlfs 1966: III, 319], these and related forms (from Umbria, Marche, Naples, Lucca) result from interaction of *amendue* and its cognates with old French *ambure*.

[16] The Florence manuscript of Paolo Gherardi's algebra [ed. Van Egmond 1978], also originally written in Montpellier, uses *el* (not *il*) and *lo* as does **V**, but has *rimanere* and *di*. It uses *v'à* for "there is". Two and ten are written with numerals and are hence uninformative. It has some Latinizing spellings, genuine as well as hypercompensations, which may thus be taken to characterize the environment rather than individual propensities of the writer (however, it can also be found elsewhere).

is inserted "de rimpecto nel sequento foglio", "opposite on the next sheet" (it follows indeed on fol. 47^r) – but the organization of the page shows that this passage was not inserted after the writing of the following section on "le leghe de monete picciole", "the alloys of small coins", and thus that it was present (together with a mark ⸸ indicating the location of the omitted section) in the original used by the ultimate copyist, who followed this original rather than running the risk of accumulating extra errors in an attempt to repair the mistake. The fact that whole pages are interchanged shows that **V** had approximately the same number of lines to a page as the two latest precursor manuscripts.

The insertion of forgotten words above the line makes it plausible that the finished copy was collated with the original. It will come as no surprise that the elaborate initials were also inserted after the completion of the text – as demonstrated by the omission of one on fol. 17^r and of another one on fol. 42^r. The insertion of a missing passage in the margin of fol. 48^r may be due to the same hand as the initials.

Evidently, many abbreviations are made use of in the manuscript, including abbreviations for *libre, soldi, denari, fiorini, oncie* and *braccia*. What is remarkable in comparison with other abbacus manuscripts is that, with one sole exception, key terms for a number of mathematical operations – essential in particular in the algebra – are *never* abbreviated: neither *più* nor *meno*, nor *radice, cosa, censo* or *chubo*. The exception is a single appearance of the standard abbreviation for *radice* in V.19.2 – obviously a slip only demonstrating that either Jacopo or one of the copyists knew it and used it this single time by mistake, since *radice* occurs 141 times in the manuscript, of which 86 are in the algebraic chapters V.16–19. In contrast, in the tabulation of degrees of fineness of gold coins on fol. 46^r, *meno* is abbreviated: "charati 24 ⓜ $^1/_5$ per oncia", etc.[17] The absence even of as simple an abbreviation as *cēso* for *censo* can only be the consequence of a deliberate choice.

Remarkable is also a less standardized technical vocabulary than in other treatises – *el diametro*, for instance, may also be *el dericto de mezzo, el mino longho*[18] or *el dericto de mino longho*. That a number *m* falls short of another number *n* by an amount *p* may be expressed not only in the phrase "da 7 infino in 3 menoma 4" (V.21.8, fol. 50^r) – the standard expression of **M+F** – but also (in parallel passages) "da 7 infino in 4 mancha 3" (*ibid.*) or "da 9½ infino in 8½ à uno" (V.21.6, fol. 49^r), with the variant "da 4 a 7 si à 3" (21.4, fol. 48^r).

Structure, Contents and Character

The structure of the Vatican text is as follows:

---

[17] This abbreviation appears to have belonged specifically to the domain in question, cf. [Vogel 1977: 11].

[18] Most likely a misunderstood expansion of *miluogo*, cf. French *milieu* – see below, p. 135.

In the following, paragraphs in **V** will be referred to as V.*a*.*n* (or, if there is no doubt that **V** is spoken about, simply as *a*.*n*), *a* being the chapter number in Arabic numerals and *n* the section number within the chapter (counted in agreement with the initials). The

---

[19] A row containing the numbers 2000–5000 is evidently omitted by error. These numbers (though distributed over two rows) are present in **F** as well as **M**.

[20] E.g., 600 written as c above vi.

[21] Seeing which of two is greater, and finding the excess; kept apart from subtraction though evidently solved in the same way.

corresponding paragraphs in **M** are designated M.*a.n*, since my numbering of **M** in the Appendix keeps as close to that of **V** as possible. References to **F** will have the form F.*r.s*, *r* being the chapter number in Roman minuscule numerals and *s* the section number within the chapter (both counted as in [Simi 1995]). None of the three manuscripts contains any of these numbers. References like M.14.11=F.v.12 will be used when there is no reason to distinguish between equivalent problems in **M+F**. For further details regarding the numbering of paragraphs in **M** and **F**, see below, p. 380.

The text of the Vatican manuscript is characterized by interspersed personal and colloquial-pedagogical remarks – for instance:

(a) Abiamo dicto de rotti abastanza, però che dele simili ragioni de rotti tucte se fanno a uno modo e per una regola. E però non diremo più al punte.[22] (11.1, fol. 17ʳ).
(b) Et se non te paresse tanto chiara questa ragione, sì te dico che ogni volta che te fosse data simile ragione, sappi primamente ... .[23] (14.19, fol. 26ʳ).
(c) Una torre ... sicomo tu vedi designata de rinpetto.[24] (15.9, fol. 31ᵛ; similarly *passim*).
(d) Ora non si vole agiungere inseme come tu facesti quella de sopra.[25] (15.10, fol. 32ʳ).
(e) Fa così, io t'ò anche dicto de sopra, ogni tondo, a volere sapere quanto è el suo diametro, si vole partire per 3 e ¹⁄₇.[26] (15.11, fol. 32ʳ).
(f) Et abi a mente questa regola. In bona verità vorrebbe una grande despositione; ma non mi distendo troppo però che me pare stendere et scrivere in vile cosa; ma questo baste qui et in più dire sopra ciò non mi vo' stendere.[27] (16.14, fol. 40ᵛ).
(g) Fa così, et questa se fa propriamente come quella che tu ài nanzi a quella ragione de sopra a questa, et in questa forma. Et però non me stendarò in si longho dire como feci in quella.[28] (21.6, fol. 49ʳ).
(h) Et però ho facta questa al lato a quella, che tu intende bene l'una et l'altra, et che l'una et l'altra è bona reghola. Et stanno bene. Et così se fanno le simili ragioni.[29] (22.6, fol. 52ʳ).

---

[22] "We have said enough about fractions, because of the similar computations with fractions all are done in one and the same way and by one and the same rule. And therefore we shall say no more about them here".

[23] "And if this computation should not seem too clear to you, then I say to you that every time that a similar computation should be given to us, know firstly ...".

[24] "A tower ... as you see drawn opposite".

[25] "Now one shall not join together as you made the one above".

[26] "Do thus, I have also said it to you above, (for) every round, if one wants to know how much is its diameter, one shall divide by 3 and ¹⁄₇".

[27] "And keep in mind this rule. Verily, a vast exposition would be needed. But I will not enlarge too much, because I seem to expand and write about a base thing. But this will be enough here, and I will not expand to say more about it".

[28] "Do thus, and this is properly done as the one you have before the computation above the present one, and in this (same) diagram. And therefore I shall not expand in as long speech as I did in that one".

[29] "And therefore I have made this beside that, so that you understand well one as well as the other,

A mathematical particularity of the text in **V** is the use of the partnership as a functionally abstract representation of proportional sharing. In 14.9–10 (fols 23ᵛ–24ʳ) it is used to determine the shares of a heritage; in the algebra section, a sub-problem of the partnership problem 16.3 (fol. 37ʳ) is represented by means of *a different* partnership; 21.4 (fol. 48ʳ) introduces it as a basic tool for computations concerned with alloying, which is referred to in 21.6 (fol. 49ʳ) and used again explicitly in 21.8 (fol. 50ᵛ); 22.1 (fols 50ᵛ–51ʳ), a partnership problem where the participants in the *compagnia* do not enter at the same time, introduces a different, fictitious partnership where the shares are the respective interests which the investments would have earned (and thus proportional to the product of investment and time). The latter has a few parallels, mostly in writings linked to the Provençal environment – cf. p. 130; but I have not observed any other treatise which used the model as broadly as **V**.

V also exhibits a rhetorical particularity. When explaining the reason for a particular step, it regularly ascribes to the *tu*/"you" of the text such knowledge or conditions that were originally stated by its *io*/"I" ("perché tu di' che ..."/"because you say that ...", etc). Examples of this are found in all chapters which offer the occasion, that is, 14–16, 18–19, 21 and 22 (of which 16, 18–19 and 22 have no counterpart in **M+F**). In general, **M+F** has nothing similar except in Chapter M.21=V.vii, but some traces remain – see below, p. 20.

The manuscript is nicely illustrated. Some of the illustrations are neutral and to the point (circles with indications of the measures of diameter and perimeter, etc.), but many are not (most aberrant is a beautiful plant with flowers, stalk and *root* drawn along the rule for approximating square roots).

The Florence and Milan Manuscripts[30]

Van Egmond [1980: 148] describes the Florence manuscript as follows:

> s. XIV (1307, *d[ated]*), Holograph *libreria* treatise, Montpellier
> Vellum, 264 × 183 mm., 48 [leaves] numb[ered] mod[ern] type 1–48 at bottom right, orig[inal] 1–44, 44–47, plus 3 ant[ique] and 4 mod[ern] vellum guard sheets [...]. Single hand, a very neat *corsiva gothica cancelleresca* in 1 col[umn] dry-ruled 12 cm. wide 25 dry-ruled lines in 182 mm., text on second. Brown ink with alternating red and blue initials and ¶, title on

and that one as well as the other is a valid rule. And they go well. And thus the similar computations are done". In the preceding problem, the area of the circle is found as $1-\frac{1}{7}-\frac{1}{2}\cdot\frac{1}{7}$ times the square on the diameter; here, it is found instead as $\frac{1}{4}$ of the product of diameter and circumference.

[30] The following discussion of the Florence and Milan manuscripts is very technical and overloaded by oppressive detail; the reader who is willing to take for granted the conclusion – that **M+F** is the outcome of a process of rewriting of an original to which **V** is much closer – can skim it without bothering about details.

1r in red, 3-line decorated red on blue initials on 1r and 17r. Tables and calculations outlined in red on 2v–3r, 5r–16v; geometrical diagrams and illustrations in margins in red on 37r–43r. Wooden boards with leather backing labeled: Algorismo / di Maestro / Jacopo / da Firenze.

[. . .]

Con[tents]: Intro[duction], mult[iplication] tables, "numeri rotti" (17r), bus[iness] prob[lems] (19v), rec[reational] prob[lems] (27r), "tutte maniere di misure" (37r), "leghe di monete" (43r)

[. . .]

The Milan manuscript, in which the *Tractatus algorismi* in found on fols 1^r–46^v, is described in these terms by Van Egmond [1980: 166]:

s. XV (c. 1410, *w*), Holograph treatise, Genova

Paper, 4°, 227×164 mm, 64 cc. num. mod[ern] [1 unnumbered] 1–47, 49–64 at bottom right, orig[inal] 1–9, 12–59 [1 unnumbered] 61, 60, 73–75. [...] Single hand, a neat *semi-corsiva gothica italiana* in 1 col[umn] lead-ruled 104 mm. wide, 27 ink-ruled lines in 147 mm., text on first. Brown ink with initials and ¶ in red, title on 1r in red. Tables on 4v–9v, outlined in red, calculations on 10r–14r, geometrical diagrams and illustrations in margins on 34r–39v.

[. . .]

Con[tents]: Intro[duction], operations, bus[iness] prob[lem]s, "tute maniere di misure" (34r), "Tute maniere di leghe di monete" (39v).

[. . .]

to which can be added, firstly, that the two missing original folios (10^r–11^v) correspond to four pages of tables in **F**; it is thus obvious that two sheets have been lost, and that the original foliation was continuous (this is therefore the one I shall use in my transcription). Secondly, that the contents of **F** and **M** are strictly identical on other accounts; the different descriptions in [Van Egmond 1980] thus only illustrate that the same contents can be described in different ways.

In her edition, Annalisa Simi [1995] divides **F** as follows:[31]

		Pp./fol	Corresp. to
	Incipit, Prologue	7–9/1^r–4^v	V/M.1–2
Table 1–2	Tables about the writing of numbers	38–39/2^v–3^r	V/M.3–4
Chapter I	Tables of multiplication and squares	40–58/5^r–1^r	V/M.6–7
Chapter II	Division tables	59–62/14^v–16^r	V/M.8–9
Chapter III	Fractions	63,9–11/16^v–19^r	V/M.10
Chapter IV	Rules (of three, interest and metrology)	12–17/19^v–25^r	V/M.11–13

---

31 Pp. 7–37 contains her transcription of the text, pp. 38–63 facsimile reproductions of the various tables.

Chapter V	Mixed problems	17–26/25^r–37^r	V/M.14
Chapter VI	Geometrical problems	26–32/37^r–43^r	V/M.15
Chapter VII	Fineness of coins, alloying	32–37/43^r–48^v	V/M.20–21

As we see, **F** (agreeing with **M**) contains no counterparts of V.5 (the introduction to the multiplication tables), V.16–19 (mostly algebraic matters) and V.22 (supplementary mixed problems).

The language and orthography of **F** appears to be an even purer Tuscan than that of **V**. There are in total five instances of *el* in Simi's transcription, but three or four of them were probably meant as *e'l*, leaving one or at most two instances proper; the 273 *il* and 113 *lo* are distributed according to the same rule as in **V** (*il* exclusively after a vowel, cf. note 13). The 40 instances of *de* are all in expressions that in modern Italian would be *del'*, *della*, *degli* and *delle*, and thus quite regular Tuscan. The spelling *que* is not found, and the distribution of the 11 *ke* (against 541 *che*) appears to be independent of the 33 *que* (against 1901 *che*) in **V**.

In parallel passages, *remanere* (etc.), *vene* and *mesura* in **V** correspond to *rimanere* (etc.), *viene* and *misura* in **F**, *doi/doy* in **V** to *due* in **F**, *amendore* in **V** to *ambendue* in **F**. **F** has *diece* for ten, and speaks of *lire* (and *danaio* for *denaro*; but singular *libra* or *livra*).

Latinizing spellings are rarer than in **V**, and concern only the words *facto* (with declinated forms), *octavo* (etc.), *septe* (etc.) and *nocte*. There are 14 instances of *fact* (against 61 of *fatt*); of these, 7 are in passages that have no exact counterpart in **V**,[32] and all the others correspond to the same spelling in **V** (which has 95 instances of *fact*, 15 of *fatt*), which suggests one of the two manuscripts to be derived from an original which the other one follows quite precisely on this orthographic account.[33] The

---

[32] In five cases, the phrase "et è facta" does not appear in the corresponding place in **V** (F.iii.9, F.v.12, F.v.23, F.v.37, F.vi.14). One occurrence is in a proof that is absent from **V** (F.v.12), the last (in F.iv.24) is in the explanation "sì come avemo facto il quintale di libre 104 et la caricha di libre 312, a quella medesima ragione, possiamo fare di quantunque libre fosse la caricha e 'l quintale", which in V.13.10 appears in the shorter form "così facesti de quantunqua libra fosse el quintale".

[33] Of the 61 *fatt* in **F**, 33 have no exact counterpart in **V** (in 24 cases because **F** has a phrase "et è fatta" where **V** has none, in 6 because the corresponding problem is absent from **V**, in three because of slight reformulations). Of the remaining 28, 8 correspond to *fatt* in **V**, 20 to *fact*.

The distribution of the spellings in the shared occurrences can be summarized in the following scheme:

distribution of *octav/ottav*, *sept/sett* and *nocte/notte* present similar pictures but are not statistically significant, the numbers being much smaller. There are 58 shared occurrences of "said", but since all in **V** are *dicto/dicta/decta* and all are *detto/detta* in **F**, this only tells us that both scribes were consistent on this account. Similarly, **V** has *dericto/diricto* everywhere, while **F** has *diritto* (11 shared cases in all).

The language of **M** suggests a Tuscan base filtrated through the Genovese environment and a tendency to keep closer to pronunciations than the original: in **M**, 54% of all "and" are *e*, in **F** only 28% (in **V** only c. 5%)/; geminations are much fewer, as is to be expected in the north – in **M**, 82% of all feminine "done" are *fata*, 18% are *fatta*; in **F**, there are no *fata*, 23% are *facta*, 77% are *fatta* (**V** has 84% *facta*, 16% *fatta*). There are only two occurrences of *el* (disregarding *e'l*), against 269 of *il*; *il* and *lo* are distributed in the same way as in **V**. 12 instances of *de* are not in expressions that in modern Italian would be *del'*, *della*, *degli* and *delle*; only half of these are in passages which follow **V**; two are in a section which is added in **M+F**, namely M.14.27A=F.v.30, two are in the additional coin lists which are only in **M** (and one is an abbreviated *del*). They are thus spontaneous choices of the scribe, not carried over from his model, and thus evidence of his linguistic environment. The same may be the case for the spelling *vene*, which occurs 7 times (against 119 occurrences of *viene*): 4 correspond to **V**, 2 are simply *passato remoto* written without gemination (thus *venne* in **V** and **F**). The environment may also explain the constant spelling *ponto* (with *aponto*), where **F** has an equally constant *punto* and **V** has 11 *ponto* against 41 *punto*. The forms *que* and *ke* are absent, instead **M** uses *che* consistently.

In parallel passages, *remanere* (with derived forms), *mesura*, *doi/doy*, *amendore* and *dece* are as absent from **M** as from **F**. Not counting the abbreviations there are 116 instances of the forms *libra/libre*, against 21 of *livra/livre* and 2 of *lira/lire* (not counting the 237 occurrences of the abbreviation, in which the *b* is also visible).

Latinizing spellings and hypercompensations (*facto*, *octavo*, *septe*, *nocte* etc.) are totally absent from **M**.

	F		
	ct	*tt*	*ct+tt*
V *ct*	7	20	27
V *tt*	0	8	8
V *ct+tt*	7	28	35

If the spellings of both manuscripts had resulted from independent variation with respect to the archetype (the scribe of **V** mostly preferring *ct*, the one of **F** mostly *tt*), the 7 *fact* of **F** would have been distributed randomly over the relevant 35 *fact+fatt* of **V** (or, reciprocally, the 8 *fatt* of **V** randomly over the relevant 35 *fatt+fact* of **F**). In this situation odds are around 13.2% that no *fatt* in **V** will correspond to *fact* in **F** – namely $\frac{28!\cdot27!}{20!\cdot35!}$.

The conclusion that was suggested by the distribution of *fact/fatt* in **V** and **F** is confirmed, and immensely strengthened, if we investigate the distributions of certain ways to speak of operations (the countings in the following are evidently restricted to the truly shared occurrences). Firstly, there is the way of "taking" a part of something, which can be expressed as *pigliare* or *prendere* (in **V** also *togliere* a few times); alternatively it may simply be stated that, e.g., $\frac{2}{3}$ of 12 *è/sono* ("is"/"are") 8. All 20 occurrences of *prendere* in **V** correspond to *prendere* in **M+F**. Of the 23 *pigliare* in **V**, 3 correspond to *pigliare* in **M+F**, 17 to *prendere*,[34] three to *è*. Chances that all 3 *pigliare* occurring in **M+F** which correspond to either *pigliare* or *prendere* would correspond by mere accident to pigliare *in V* is c. 16% (precisely, $\frac{23!\cdot40!}{20!\cdot43!}$), and the odds that the 3 *è* in **M+F** that correspond to either verb in **V** should all correspond to *pigliare* by mere accident is the same. The odds for both to happen at the same time (the two events are not stochastically independent) are 1.66% (precisely, $\frac{23!\cdot37!}{17!\cdot43!}$).

Next there is the evidence offered by the way divisions are expressed. Division can be *partire in* or *partire per*[35] (and is always one of the two, which simplifies matters). Restricting ourselves again to occurrences belonging to parallel passages in the two manuscripts, **V** contains 67 cases of division *in*, of which 9 correspond to division *in* in **M+F** and 58 to divisions *per* in **M+F**. All 21 divisions *per* in **V** correspond to division *per* in **M+F**. The odds for this distribution to have arisen by accident is 7.48% ($\frac{67!\cdot79!}{58!\cdot88!}$). Combined with the *pigliare/prendere/è*-distribution, the odds are 0.124%.

If the evidence coming from the *fact/fatt*-distribution is also taken into account we may therefore safely conclude that one of these three things happened:

(1) Somebody, using an original very close to **V** in its distribution of the expressions discussed, disliked the use of *pigliare* and mostly replaced it by *prendere*, sometimes by an even less metaphorical term, finding it also more close at hand to divide *per* and having a tendency to eliminate Latinizing spellings; this produced something close to **F**. The scribe of **M** then eliminated the latter altogether, but conserved the same choices as **F** on other accounts.

(2) Somebody, starting from something close to **M**, had a certain predilection for Latinizing spellings and inserted them occasionally, thus producing something close to **F**. Further, somebody with more pronounced predilection for *pigliare*, for division *in* and for Latinizing spellings inserted more of each, thus producing a precursor of **V** (or **V** itself).

(3) Somebody, starting from something close to **F**, disliked Latinizing spellings and eliminated them altogether, producing thus a precursor of **M** (of **M** itself). Somebody

---

[34] 16 in **F**, since one instance is in a passage in M.10.12 which the copyist has forgotten in F.iii.11.

[35] The former expression corresponds, we may say, to the idea that a quantity is divided *into* a number of parts, the latter to the execution of the operation *by means of* the divisor.

else being fond of Latinizing spellings, *pigliare* and of divisions *in* inserted more of them, thus creating a precursor of **V** or **V** itself.

Three additional observations on the divisions speak against the assumption that **M** or **F** should be the faithful representative of Jacopo's original. Firstly, in all cases where a number has been explained to be the *partitore* (e.g., in proportional sharing), all divisions by this number are *in* in **V**; in **M+F**, all may be *per*, or the first may be *in*, and the following *per*.[36] Since the etymological reason for this choice is not clear to the writer of **V** (several times, a number is divided *per* so and so many "parts"), there is no reason that he should have introduced a system for whose purpose he had no feeling while copying a non-systematic source.[37] But it is quite likely that a copyist on the way toward **M** and **F**, not aware of the reason for the system of an original close to **V**, might emulate it badly.[38]

Secondly, when formulating the *rules* of three (there are several, depending on the presence or absence of fractions), **M+F** divides *in*, but in the corresponding examples *per*. **V** uses *in* consistently – as does Paolo Gherardi [ed. Arrighi 1987: 15*f*] when writing in Montpellier in 1328; also in partnership divisions his choice agrees with **V**.

Thirdly, inspection of V.14.5 and its counterpart M.14.5=F.v.5 is revealing. In **V**, a solution starting from one possible initial guess is shown first, then comes a proof, and finally the text proposes how to solve the problem starting from a different initial guess. Both solutions contain a division *in*. **M+F** starts by the same initial guess, agreeing with **V** in the choice of a division *in* – but then postpones the proof and goes on immediately to the alternative solution, now dividing *per*. The two divisions are so close to each other in **M+F** that the parallel calculations would have called for the same formulation in both cases if the writer had chosen freely. But if he followed a model similar to **V** for the first solution and then formulated the second solution on his own, the structure is understandable.

Comparison of chapters which are shared by **V** and **M+F** shows them to be far from identical even at higher textual levels. They largely contain the same problems, even though some problems from **M+F** have no counterpart in **V** (on the other hand, a number of introductory statements from **V** are absent from **M+F**); but the formulations often exhibit differences showing that at least one of the two is the outcome of a genuine rewriting of the common archetype going beyond the choice of one rather than the other synonym

---

[36] There is strict agreement between **F** and **M** on this account, apart from two cases where one manuscript has omitted a passage which is in the other (M.14.23=F.v.25 and M.21.3=F.vii.3) and a genuine disagreement in M.21.3=F.vii.3 in a passage which has no counterpart in **V**.

[37] If he had done so because of mere infatuation with *in* he should also have replaced some of the divisions *per* $3\frac{1}{7}$ (giving the circle diameter) by *in*. But that does not happen.

[38] In M.10.3 the scribe has first written *per* and then deleted it and replaced it by *in* (which is also in F.iii.2 and V.10.3). There thus seems to have been a spontaneous drift toward *per*.

or spelling. As examples we may look at the above-mentioned didactical remarks in **V** and their counterparts in **M+F** (omitting those that can have no counterpart because they appear in chapters that have none, and following the spellings of **M**):[39]

V	M
Abiamo dicto de rotti abastanza, però che dele simili ragioni de rotti tucte se fanno a uno modo e per una regola. E però non diremo più al punte. Et incomincia-remo ad fare et ad mostrare alcune ragioni secondo che appresso diremo. ⟨S⟩e ci fosse data alcuna ragione nela quale se propo-nesse tre cose, ... (11.1–2, fol. 17ʳ)	Se ci fosse deta alchuna ragione nela quale si pro-ponese in tre cosse ... (M.11.2)
Et se non te paresse tanto chiara questa ragione, sì te dico che ogni volta che te fosse data simile ragione, sappi primamente ... . (14.19, fol. 26ʳ)	(Omitted from M.14.19)
Una torre è alta 50 braccia sicomo tu vedi designata de rinpetto. Et a'ppe dela dicta torre è uno fosso, el quale è largho 30 braccia. (15.9, fol. 31ᵛ)	Una tore, la qual'è alta 50 bracia. E al piede de que-sta tore si ae un fosso, lo qual'è ampio 30 bracia.. (M.15.9)
Ora non si vole agiungere inseme come tu facesti ⟨in⟩ quella de sopra. Ancho se vole trare l'una multiplicatione dell'altra, cioè, 1600 de 2500, ... (15.10, fol. 32ʳ)	Ora tray 1600 di 2500 ... . (M.15.10)

---

[39] I expand some of the quotations as needed for the confrontation, and omit translation for brevity.

Uno tundo ad conpasso como tu vidi designato de rimpetto gira dintorno 100 braccia. Vo' sapere quanto serà el diametro suo, cioè el dericto de mezzo. Fa così, io t'ò anche dicto de sopra, ogni tondo, a volere sapere quanto è el suo diametro, si vole partire per 3 e $\frac{1}{7}$. Et però parti 100 per 3 e $\frac{1}{7}$, et quello che ne vene, tanto è'l suo diametro, cioè, tante braccia. Et se non sapissi partire per 3 e $\frac{1}{7}$, sì te insegnarò. Fo così, ogni volta che tu averai a partire per numero sano et rotto, si vole arrechare a rotti tucto el numero, cioè, quelli rotti che tu ài a partire. Et similemente anchora quello numero che tu ài a partire. Et fa così per meglio intendare. Io te dicho che tu parte 100 braccia per 3 et $\frac{1}{7}$. Et però arrecha a settimi tucto el partitore. Et di' così, 7 via 3 e $\frac{1}{7}$ fa 22, ... (15.11, fol. 32^r)	Uno ritondo a conpasso, lo quale gira d'intorno 100 bracia, dimi quant'è il suo diamitro, cioè il drito di mezo. Fa cosie, e quest'è la sua propia regola, parti 100 per 3 et $\frac{1}{7}$ in questo modo, die, 7 via 3 et $\frac{1}{7}$ fano 22 ... (M.15.11)
Fa così, et questa se fa propriamente come quella che tu ài nanzi a quella ragione de sopra a questa, et in questa forma. Et però non me stendarò in si longho dire como feci in quella. Et però di' così ... (21.6, fol. 49^r)	Fa cosie, die, ... (M.21.6)

This does not mean that there are no colloquial remarks in **M+F**, but they are fewer and usually quite short – for instance, after a proof in M.21.4 (=F.vii.4), the terse remark "Dunque avemo bene alegato, però ch'avemo aponto ritrovati i deti 700 danari. S'avessimo trovato piue o meno starebbe male"[40] where **V** just takes note of the correctness of the result, states that the rule is general, and refers to a diagram that is used in the calculation but which is absent from **M+F**.

Yet another instructive difference between the manuscripts is the way illustrations are treated. **V** habitually refers to the illustrations and diagrams contained in the margin in words similar to those of the third quotation in the scheme above (cf. note 24). **M+F** has most of the same illustrations (somewhat fewer, **F** two less than **M**, and none which are not in **V**), but does not always give a reference in the text; when it does, the reference almost always comes after the solution of the problem, whereas **V** as a rule gives it immediately after the statement. All this excludes simple expansion or abbreviation and leaves us with two possibilities: either **V** is a rewriting aiming at greater completeness and consistency, or **M+F** is a rewriting aiming at conciseness.

Once again, some details imply that the manner and the illustrations of **V** correspond to those of the original. Firstly, we may notice the following passage in M.21.7(=F.vii.7)

---

[40] "Thus we have alloyed well, since we have found again precisely the 700 *denari*. If we had found more or less it would have been a shame".

(for once after the statement, not after the solution): "simigliantemente il mostriamo materialmente per figure come si fae lo deto alegamento". In the corresponding passage in V.21.7 (fol. 49^v), we read "Et simile⟨mente⟩ porremo la figura. Nel modo si fa como abbiamo facto de sopra nell'altra ragione". In **V**, the promised diagram is present – but it is absent from **M+F**.[41] Secondly, both V.15.21 (fol. 34^v) and the counterpart M.15.21= F.vi.17 contain a drawing of a pavilion, and both refer to it in the text; but whereas the illustration of **V** has the conic shape presupposed by the problem, that of **M+F** is so glaringly discordant with the description in the text that Annalisa Simi inserts a "(sic!)".

Many other features of the manuscripts also point to the conclusion that **V** is close to an original of which **M+F** is an adaptation.

Most striking is the shift from "io" to "tu" when reasons for a step are given. As related above, p. 12, it characterizes all chapters of **V** which allow it (14–16,18–19, 21, 22). Of these, V.16, V.18–19 and V.22 have no counterpart in **M+F**. In the counterparts of V.14 and V.15 (M.14–15=F.v–VI), only one such shift occurs (in F.v.8=M.14.8~ V.14.8) – but in M.21=F.vii(~V.21) it turns up regularly in places where it is also found in **V**, though less often and never in places where **V** does not make use of it (except an occurrence in M.21.2A=F.vii.2, which has no counterpart in **V**, perhaps because of an omission, see p. 126).[42]

At times the two manuscripts differ in the way a problem is dressed. On at least one occasion, however, **M+F** betrays descent from the formulation which we find in **V**. Indeed, V.14.30 starts "Egli è uno muro, el quale è lungho 12 braccia e alto sette. Et grosso uno et $\frac{1}{4}$".[43] The counterpart M.14.30(=F.v.34) has instead "Egli è uno tereno lo qual'è ampio 12 bracia, cio un muro, ed è alto bracia 7 ed è grosso braccia j et $\frac{1}{4}$".[44] Obviously, the compiler has started by changing the original wall into a "terrain" without having read the whole problem, and then discovers that the presence of a height makes the change impossible, and corrects himself.

---

[41] A related phenomenon is discussed below, note 202.

[42] Similar shifts seem to be very rare elsewhere. I have indeed only noticed one instance, namely in the *Libro di conti e mercatanzie* (Parma, Biblioteca Palatina, MS Pal. 312, [ed. Gregori & Grugnetti 1998: 50]) – according to problems dealing with partnership and interest in the years immediately after 1389–94. In most cases, however, this treatise (whose pedagogical pretensions are comparable to those of **V**) refers such explanatory information to "noi" and not to "tu" (a possibility which is also used regularly in **V**).

It may be worth noticing that this treatise divides consistently *in* when defining and applying the rule of three and in partnership divisions, as does **V**. As we shall see (below, note 154), at least its treatment of the rule of three and its alternatives is derived faithfully from a source which it shares with a treatise from c. 1330.

[43] "There is a wall, which is 12 *braccia* long and seven high. And thick one and $\frac{1}{4}$".

[44] "There is a terrain, which is 12 *braccia* large, that is, a wall, and it is high *braccia* 7 and it is thick *braccia* 1 and $\frac{1}{4}$".

Particularly striking is a passage where all three manuscripts differ. V.15.21 has "parti per mezzo el dericto de mezzo del padiglione, cioè 60, che è 30".[45] In F.vi.17 we find the meaningless "parti il diritto del mezzo, cioè 60, che nne viene 30",[46] obviously caused by the repeated *mezzo*; M.15.21 gives "parti il diritto del mezzo de⟨l⟩ padigio⟨n⟩e per mezo, cioè 60, che nne viene 30",[47] – thus agreeing with **F** in the reformulation of the way the result is stated, and with a word order suggesting that its scribe was either on the point of committing the same error or repaired a manuscript which contained it; in contrast to **F**, **M** also mentions the pavilion. There can be no doubt that **V** gives the original phrase.[48]

The treatment of approximate square roots ($\sqrt{a} \approx n+\frac{d}{2n}$ if $a = n^2+d$) is informative on several levels. Least decisive but still in accordance with the way illustrations are referred to is the contrast between the rather systematic habit of **V** of pointing out that this is *not* precise ("non è appunto") and the much more scattered observations of the same kind in **M+F**. Moreover, after having presented the rule, V.15.13 states that its outcome will be "la più pressa radice"/"the closest root", while M.15.13(=F.vi.10) believes it to be "radice vera o più pressa"/"true or closest root", as if the occasional reference to its only approximate character were mere lip service not supported by full understanding.

But there is more to square roots. **V** not only mentions repeatedly that $n+\frac{d}{2n}$ is merely "the closest root" but also shows time and again (in actual numbers) that $(n+\frac{d}{2n})^2$ exceeds a by $d^2/(4n^2)$. V.15.22, however, does none of this; but instead it gives the "improved" value $\sqrt{108} = 10\frac{2}{5}-\frac{4}{25}$; similarly, V.15.25 gives $\sqrt{569} = 23\frac{20}{23}-\frac{400}{529}$ – and in V.15.20 we are told that $\sqrt{33\frac{1}{3}} = 5\frac{7}{9}-\frac{4}{18}$.[49]

The counterpart of V.15.22 (M.15.22=F.vi.19) gives the simple first approximation $10\frac{2}{5}$, and that of V.15.25 (M.15.25=F.vi.22) similarly gives $23\frac{20}{23}$; in neither problem is it pointed out that the value is approximate. Once more, this might mean that the fallacy of **V** has been added onto a sounder stem, or that it has been eliminated in **M+F**. However, M.15.21A(20)=F.vi.1, the counterpart of V.15.20, tells $\sqrt{33\frac{1}{3}}$ to be "5 et $\frac{5}{6}$ meno $\frac{17}{54}$ non aponto". $5\frac{5}{6}$ is obviously found in the usual way, as $5+(33\frac{1}{3}-25)\div(2\cdot5)$. If the fallacious

---

[45] "Divide in half the straight in middle of the pavilion, that is, 60, which is 30".

[46] "Divide the straight in middle, that is 60, from which comes 30".

[47] "Divide the straight in middle of the pavilion in half, that is, 60, from which results 30".

[48] Slightly later, **M** also agrees with **V** when referring to "il deto padiglione", "the said pavilion", where **F** has only "il padiglione".

[49] Obviously determined in the following way: $\sqrt{33\frac{1}{3}}$ is found as $\frac{\sqrt{300}}{3}$, where $\sqrt{300}$ is approximated from $n = 18$, which gives $5\frac{7}{9}$ as the first approximation. Since $(5\frac{7}{9})^2 = 33\frac{1}{3}+\frac{4}{81}$, the fallacious rule of 15.22 and 15.25 would hence give the value $5\frac{7}{9}-\frac{4}{81}$, which is miswritten $5\frac{7}{9}-\frac{4}{18}$ (a similar inversion of digits is found in 14.3, fol. 22ʳ).

It is noteworthy that none of the manuscripts explains that square roots can be approximated from above as done implicitly here.

procedure of **V** had been used to find the correction, it would have been $(5^2)/(6^2)$. The actual correction instead is $(2 \cdot 5\frac{5}{6})/(6^2)$, which appears to mix the formula of **V** with the meaningful correction $(\frac{5}{6})^2 \div (2 \cdot 5\frac{5}{6})$.[50] The only explanation seems to be that the compiler of **M+F** has seen that something is wrong in the correction of **V**, and that he has tried at one point to repair it by having recourse to another formula – but unfortunately remembering or applying it wrongly.

These observations should already suffice to show that **M+F** is a remake, and **V** much closer to what Jacopo wrote in Montpellier in 1307 (as also suggested by the orthography of **V** and **F**, with all the provisos that are needed when medieval orthography is used as an argument – yet one of the manuscripts, as we have seen, must necessarily have conserved certain orthographic features of the common archetype very faithfully). However, they do not exhaust the list of characteristic differences between the **V** and **M+F** that point in the same direction.

The use of the partnership as a functionally abstract representation of proportional sharing was mentioned above, p. 12. The same idea turns up in M.14.9–10=F.v.10, the counterpart of V.14.9–10, but nowhere else in **M+F** – in particular not in the alloying problems of Chapter VII, where **V** uses it consistently.

V contains a general introduction to the multiplication tables (V.5), a specific introduction to the "librettine magiori" (V.6.2) and an introduction to the divisions (V.8.1); all of these are absent from **M+F**. Absent from **M+F** are finally a number of metatheoretical explanations – for instance the explanation which V.15.2 gives after telling how to find a circular diameter from the perimeter, and vice versa: "Et se volissi sapere per che cagione parti et moltiplichi per 3 e $\frac{1}{7}$, sì te dico che la ragione è perché ogni tundo de qualunqua mesura se sia è intorno 3 volte et $\frac{1}{7}$ quanto è el suo diametro, cioè el diricto de mezzo. Et per questa cagione ày a moltiplichare et partire como io t'ò dicto de sopra".[51]

All this does not necessarily mean that nothing in **M+F** apart from what is also found in **V** goes back to Jacopo. We cannot exclude that Jacopo revised his manuscript over time, either adding new material (which might imply that the earliest archetype of **M+F** precedes that of **V**) or preparing an abridged version corresponding to what he used in teaching (in which case the archetype of **V** might come first).[52] Only two things can

---

[50] This formula is described by al-Qalaṣādī [ed., trans. Souissi 1988: 61] – and, in ambiguous terms, by Ibn al-Bannāʾ [ed., trans. Souissi 1969: 79].

[51] "And if you should want to know for which cause you divide and multiply by 3 and $\frac{1}{7}$, then I say to you that the reason is that each every round of whatever measure it might be is around 3 times and $\frac{1}{7}$ as much as is its diameter, that is, the straight in middle. And for this cause you have to multiply and divide as I have said to you above".

[52] Since the dating to September 1307 is the same in both versions, this would have required that he left that information untouched when making a revised (either expanded or reduced) version.

be said with fair certainty: firstly, a large part of the deviations of **M+F** from **V** must be due to a later compiler, since they betray failing understanding and/or different stylistic preferences[53] – which means that the latest common archetype for **F** and **M** cannot be due to Jacopo; secondly, as we shall se below (p. 54), its listing of specious higher products (11×22, 12×32, ...) arranged in schemes ordered from right to left must almost certainly reflect the Provençal environment in which Jacopo worked; if they do not go back to his Jacopo's own hand (which ultimately seems implausible) they must go back to a revision still made in Provence or using material from Provence. This would also explain the presence of the preferred orthography and style of **V** in some of the problems that are present in **M+F** but absent from **V** (cf. notes 32 and 195).

The Vatican Chapters with No Counterpart

Evidently, all these arguments (and still others of the same kind) only show that **M+F** is the outcome of a rewriting of an original to which **V** is much more faithful[54] *in the chapters which are present in both versions*. They do not exclude the possibility that the

---

This is perhaps not very likely.

[53] In particular, M.14.27A=F.v.30 and M.14.31A=F.v.36 can hardly be from his hand, and they are thus likely to be due to the same compiler as the stylistic changes.

[54] If **V** were also the outcome of a process of rewriting, if would certainly contain passages in which inconsistencies betrayed dependency on something closer to **M+F**. I have noticed none (apart from the corrupt V.14.32, which appears to mix up two numerically different versions of the same problem).

This does not mean that no inconsistencies can be found in **V** – but the only ones I have observed are shared with **M+F** or equally problematic there. One of them (which points to the Arabic world) is a distinction seemingly made in the introduction (V.1.3) between "rocti sani e rocti in rocti", where "rocti in rocti" could refer to the *partes-de-partibus* usage of Arabic mathematics; but such "parts of parts" are never used afterwards (although they do turn up in a few other early abbacus treatises, and play a major role in the *Liber abbaci*). **M** and **F** seem to contain two slightly different versions of a misshaped attempt to repair something which the compiler did not understand – cf. the detailed discussion on p. 128.

Another possible inconsistency is the claim of Chapter V.8 that the prime numbers are the "most necessary" divisors. This claim is not born out by any calculation in the rest of the treatise. However, it corresponds to a method that was apparently well known (see below, p. 57); since Jacopo never explains how to perform the divisions that are required by his calculations (nor any other merely numerical algorithm) it may be unjust to consider it a real inconsistency. **M+F** omits this introduction.

A third apparent inconsistency is the discrepancy between the list of topics given in the Latin *incipit* and what is actually done afterwards. This list (shared with **M+F**) is borrowed from the Latin algorisms. It thus describes what is done in the (Latin) art of algorism, and may not at all be intended to announce what is to come – in which case the inconsistency may be said to a mere product of our modern expectations.

chapters on algebra (V.16–19) and the final sequence of mixed problems (V.22) might be additions.

However, other arguments speak strongly against this possibility. At first we may look at the contents of Chapter 22, which from the above general description seems to overlap Chapters 14–15. At closer inspection, however, the apparent overlap turns out to consist of duly cross-referenced variations and supplements; no single genuine repetition can be found. This would hardly have been the case if a later hand had glued another problem collection onto an originally shorter treatise, given the frequency of whole-sale borrowings of problems between different abbacus writings.

To this comes the homogeneity of **V**. On all levels – orthography, vocabulary, discourse, pedagogical style – the treatise is a seamless whole, also on points where it differs from **M+F** or other abbacus writings. Instances of such idiosyncrasies that were already referred to are the preferential use of the spelling *fact* and the consistent spellings *dicto*/*dicta* and *dericto*/*diricto*, the use of *pigliare* when fractions (or square roots) are taken, the *io*/*tu* shift and the way diagrams are referred to. Another example is the copious use of the phrase *et poi*, "and then", in sequences of numerical operations in all chapters of *V* where the occasion presents itself (including Chapters 16, 18, 19 and 22); it is completely absent from **M+F** (whereas this version employs it occasionally in the stepwise description of general procedures). Further, the phrase *como t'ò dicto de sopra* ("as I have said to you above") turns up with variants time and again not only in Chapters 4, 10, 11, 13, 15, 20 and 21 but also in Chapters 17 and 22. It is absent from **M+F**. The corresponding plural construction, which is less common in **V** (10 occurrences against 22 for the singular, found in Chapters 1, 4, 10, 11, 12, 13 and 19) does turn up a few times in **M+F**.[55]

A particular case is the phrase that marks the question. Almost all explicit questions in **V** are marked either "say me" (*dime* with orthographic variants) or "I want to know" (*vo' sapere* with orthographic variants). **M+F** has no single instance of "I want to know" but 32 "say me" that correspond to the same phrase in **V** and 57 that correspond to "I want to know" in **V**.[56] In the chapters with no counterpart in **M+F**, **V** has no single instance of "say me". The explanation for this is that **V** has a strong tendency to use "say me" when the question is quite elementary, and to use "I want to know" when calculations are more complex.[57] The chapters with no counterpart in **M+F** contain nothing which

---

[55] Namely in Chapters M.1 and M.4 (the "Prologue" in [Simi 1995]) and (without "above") in M.13.7=F.iv.24, M.13.10=F.iv.p, M.13.11=F.iv.25 and M.13.12=F.iv.26; the first two of these are in passages which are absent from **V**, the others correspond to what we find in **V**.

[56] In contrast, phrases "if you want to know" or "which you/we wanted to know" are fairly copious in **M+F**.

[57] The first occurrences of "I want to know" are in Chapter 11, when broken numbers enter the rule of three, and in Chapter 12, when the interest in a single day and the doubling time for a capital

is quite elementary.

The homogeneity also holds for the computational techniques when there is a choice, and for the mathematical approach (for instance the use of the partnership model, and the ever-recurrent emphasis that the approximate values for square roots *are* approximate). All this in itself does not necessarily imply that everything was written originally by the same author; but if it was not, it would have required a strong harmonization and reformulation by a later hand, as has happened to the text as we find it in **M+F**. But a harmonization of this kind would also have affected the chapters that have a counterpart in **M+F**, and at some points we would certainly have found incongruities that betrayed the departure of **V** from an original stem which in these passages was closer to **M+F**. We must therefore conclude that whatever harmonization occurred to the text we find in **V** occurred *before* the text developing into **M+F** split off from the stem, and that Chapters 16–19 and 22 are absent from **M+F** because they were eliminated (possibly but hardly by Jacopo already, cf. p. 22).

This still does not prove definitively that these chapters were part of the treatise written by Jacopo in 1307 – but their absence from **M+F** provides no evidence that they were not. As we shall see, at least the algebra must be dated well before Paolo Gherardi's work from 1328, which leaves very little time for insertion of extra material into the original treatise and for a thorough reworking by an independent hand. All in all, the most reasonable assumption is thus that the algebra (and the supplementary mixed problems of Chapter 22) belong to the original treatise, and that **V** as a whole reproduces all major and most minor features of Jacopo's treatise faithfully.

---

are asked for. The last instances of "say me" are in Chapter 15 (one is about the area of a rectangle with known sides, the other asks for $\sqrt{2}$. The same chapter contains 17 instances of "I want to know".

The Abbacus Tradition

In a famous passage from Giovanni Villani's Florentine *Cronica* [1823: VI, 184*f*] we read concerning the year 1338 that

> we find from the priest who baptized the children (since, for keeping track of their number, for each boy that was baptized in San Giovanni a black bean was put aside, and for each girl a white bean) that they numbered every year by then 5500 to 6000, the boys exceeding the girls by three to five hundred per year. We find that the boys and girls that were learning to read numbered from eight to ten thousand. The boys that were learning the abbacus and the algorism in six schools, from 1000 to 1200. And those who were learning grammar and logic in four higher schools, from 550 to 600.[58]

– all for a town population which Villani estimates to be around 90000.

Accepting Villani's numbers at face value and taking infant mortality into account we would find that well above half of all children learned to read; that at least 10% of all boys frequented the mercantile "abbacus school"; and that another 5% or so attended the Latin-based grammar school. According to Paul Grendler [1989: 72], each abbacus school is likely to have had at most c. 40 students, which suggests real numbers to have have been appreciably lower; still, data from 1480 (where nothing suggest percentages to have changed very much) indicate that more than one third of all Florentine boys between ages 6 and 14 went to school [Grendler 1989: 77*f*]. Some 5% could have passed through an abbacus school.[59] Other sources show that this institution shaped not only future merchant's clerks and artisans but also sons of the patriciate – including some who, like Machiavelli, were going to pursue a legal and not a commercial career. All in all the abbacus school was thus certainly important for the economic as well as the cultural life of the city.

---

[58] "Troviamo dal piovano che battezzava i fanciulli (imperocchè ogni maschio che si battezzava in san Giovanni, per averne il novero metteva una fava nera, e per ogni femmina una fava bianca) che erano l'anno in questi tempi dalle cinquantacinque alla sessanta centinaia, avanzando più il sesso masculino che'l femminino da trecento in cinquecento per l'anno. Troviamo, ch'e'fanciulli e fanciulle che stanno a leggere, da otto a dieci mila. I fanciulli che stanno a imparare l'abbaco e algorismo in sei scuole, da mille in milledugento. E quegli che stanno ad apprendere la grammatica e loica in quattro grandi scuole, da cinquecentocinquanta in seicento".

[59] It should be added that other statistical information given by Villani turn out at closer analysis to be better than at first assumed by critics – see [Sapori 1955: 25–33]. In particular, Grendler supposes (from what he admits is "limited evidence") each school to have had only one teacher. As we shall see imminently (p. 28), paid assistants might be employed, whether regularly or rarely we do not know. However, see imminently on unpaid assistants.

During the last three to four decades, our knowledge of this school and the masters who taught at it has increased immensely.[60] The earliest attested abbacus master taught in Bologna in 1265, probably on a private basis; within the next four decades, abbacus masters turn up in numerous other towns from Umbria and Tuscany in the south to Genoa, Lombardy and Venice in the north. Masters paid by city communes turn up in the sources from the 1280s onward[61] – mainly in smaller communes, Venice and Florence appear to have felt no need for a public undertaking. The Florentine schools were soon considered the best, and many Florentines went to teach in other places. In university towns – Bologna, Florence, Perugia – abbacus masters often also taught mathematical matters at the arts or the propedeutical level.

The normal entrance age in the abbacus school was ten to eleven years, and the normal duration of the training was around two years – sometimes slightly less, sometimes up to two and a half;[62] we must presume that future abbacus teachers were taught on an individual basis in the same institution for longer, but known sources are mute on this account. We may also assume without having positive evidence that such apprentices worked as unpaid assistants while being taught themselves – so did apprentices in general.

Two documents exist which inform us about the teaching in the school. One, claiming to describe "'l modo a insegniare l'anbaco al modo di Pisa" ("the Pisa way to teach the abbacus") and written by one Cristofano di Gherardo di Dino who flourished in 1428–29, was published by Gino Arrighi in [1967b]. The second is a Florentine contract between the abbacus master Francesco di Leonardo Ghaligai with an assistant, drawn up in 1519 and describing what the assistant is paid per student for each section of the syllabus; it was published by Richard Goldthwaite in [1972: 421–425]. The two documents supplement

---

[60] Important publications, among others, are [Goldthwaite 1972] – the first to emphasize the cultural importance of the school and to regret how little was known about it at the time; [Van Egmond 1976] – the first full-scale survey of abbacus treatises as well as teachers; Gino Arrighi's many publications with scattered information, and his summary of much of this in [1986]; and, from the last years, Elisabetta Ulivi's penetrating study [2002b] of all abbacus masters and schools attested in Florence between the mid-thirteenth and the mid-sixteenth century and her global survey of the state of the art as of [2002a]. When no specified source is given for years and cities in the present paragraph I build on [Ulivi 2002a].

[61] It is quite likely that such schools are older, but sources for their existence do not exist. Already in 1241, it is true, Leonardo Fibonacci received a yearly pension from the Commune of Pisa, but the document – quoted in [Bonaini 1858: 5] – speaks of *abbacandi* services as a consultant for the commune and its "officials *in abbacation*", not of teaching.

[62] Armando Sapori [1955: I, 67*f*] draws the following from a recording of the 15 children born between 1323 and 1343 to the merchant Giovanni di Durante: 5 died in the months of the Black Death, 6 in pre-school age (one of these during the Plague). For two of the boys, their stay in the abbacus school is known; one stayed 29 and one 31 months. The father being wealthy, we may guess that average infant mortality was higher than 35% even before the Plague, and that the normal stay in abbacus school was shorter than 2½ years.

each other and suggest a pattern of general validity though with some variation – probably depending on epoch and place as well as on the style of the single master. Beyond that, we have some scattered remarks about what was habitual. One is Pacioli's explanation of how division *a regolo* is usually taught – see below, p. 56. Another one is in Benedetto da Firenze's *Tractato d'abbacho* [ed. Arrighi 1974: 172][63]; it states that *recare a termine*[64] is usually the last topic dealt with in the school.

At first – this can be concluded from the two complete documents – students were taught to write numbers in the Arabic number system. Then followed the *libbrettine*, the multiplication tables (until a certain limit also learned by heart) and their application; addition and subtraction go unmentioned in both sources, and may be presupposed under the learning of the number system. Next follows division, in Pisa by the numbers 1 through 19 and then 23 and 29, and after some further training in division by the *danda* method[65] for divisors with up to 7 digits; in Ghaligai's contract first by one-, then by two-digit numbers, and finally by numbers with three or more digits, obviously again by the *danda* or some similar method. The Pisa document then lists simple and composite interest and reduction to interest per day; mercantile calculation making use of the rule of three and metrological conversions, including also partnership problems; area measurement; simple and composite discount, again with reduction to one day "in several ways"; coins; alloying; and the use of a single false position. Ghaligai's contract, after the section on fractions, has one on the rule of three, and one on monetary systems. The Pisa document also states that all teaching, from the *librettine* onward, is accompanied by *ragioni*, problems, that are to be done as homework.

The numerous *libri d'abbaco* written by the teachers can be used to give full body to these short descriptions; but one *caveat* is urgent: many of them contain considerably more than listed in the two explicit sources. Part of the excess may just be *ragioni* meant to represent advanced uses of the techniques that were taught, to serve the training of the intellect, or to procure some variation in the tediousness of the daily mercantile training,[66] and therefore not worth mentioning in a list of the important matters dealt

---

[63] For the correct identification of the author (whom Arrighi believes to be Pier Maria Calandri), see [Van Egmond 1980: 96] and [Ulivi 2002b: 54–56]).

[64] Normally, this should mean to find the value of a loan or debt including interest or with interest discounted at a certain term. Benedetto, however, is referring specifically to the settling of composite loans at the same day; this is a topic which often recurs in abbacus treatises, based on real or (probably sham) loan contracts – cf. an early example below, p. 34.

[65] See, the explanations of this method below, p. 412.

[66] As expressed by Benedetto da Firenze in his *Tractato d'Abbacho* from c. 1480 (ed. [Arrighi 1974: 86]:

Io mi credo certo per lo chapitolo passato l'appetito è infastidito di tanto avere scripto di mercatantia, onde in questo presente chapitolo intendo di scrivere alchuno chaso di dilecto

with.[67] But much of the excess cannot be explained in this way – not least everything that has to do with algebra, which goes much beyond anything that can have been reasonably taught during the two-years course. On p. 162 we shall return to the question of the function of the chapters on algebra that are found in numerous abbacus treatises and for the moment only observe that they seem to have been aimed at other abbacus authors and perhaps a few advanced *dilettanti*.[68] The same may be said about the occasional treatment of matters that belonged to the liberal-arts and university tradition – though with the difference that those abbacus masters who were also connected to the arts teaching at universities would have more direct reasons to integrate them in the expositions of their knowledge.

A "Fibonacci Tradition"?

As long as the existence of the abbacus tradition has been at least implicitly recognized – that is, at least since [Boncompagni 1854] – it has been taken for granted by almost everybody that it descended from Fibonacci's writings, perhaps with more or less marginal additions.[69]

> acciò che l'animo sia alquanto plachato

> I am convinced that the appetite has been disturbed by the previous chapter, writing so much about trading, and therefore I intend to bring here some pleasant problems in order to soothe the mind somehow.

and a few chapters later [ed. Arrighi 1974: 105]

> Io credo certo che lo'ntelletto humano usando sempre una medesima cosa qualche volta, benché dilectevole fusse, gli verrebbe in fastidio et per non occhorrere in questo inconveniente, nel principio della nostra hopera, abbiamo detto nel presente capitolo trattare di qualche chaso piacevole, e' quali sono absoluti per varie reghole chome per li exempli si manifesterà.

> I believe for certain that the human intellect, if it always repeats the same thing, however pleasant it be, will sometimes be irritated, and in order not to run into this mishap we have said in the beginning of our work to treat in this chapter some enjoyable cases, which can be resolved by various rules, as the examples will show.

[67] Or they are just mentioned in a generic way. The Pisa curriculum, as we remember, ends by the "single false position" ("la prima oppositione"); Benedetto da Firenze, when promising to "scrivere alchuno chaso di dilecto acciò che l'animo sia alquanto plachato" (see note 66), goes on indeed to explain that these cases are to be solved by the simple method of (false) position.

[68] Benedetto da Firenze's *Tractato* (see note 63) is in fact stated to be "fatto a un suo amicho" ("made for a friend of his") [ed. Arrighi 1974: 29], and thus not for his school teaching.

[69] Thus (to mention the latest phrasings of the view), according to [Ulivi 2002b: 10], the *libri d'abbaco* "were written in the vernaculars of the various regions, often in Tuscan vernacular, taking as their models the two important works of Leonardo Pisano, the *Liber abaci* and the *Practica geometriae*" – whereas, according to Enrico Giusti [2002], some of the abbacus writings

Apart from the conviction that every intellectual current has to descend from *a great book that is known to us* – a conviction that comes too easily to every scholar – this belief in descent from Fibonacci has its root in certain abbacus writings. The *Livero de l'abbecho*, the Umbrian abbacus treatise supposedly from c. 1288–90 which was mentioned in note 15, presents itself as being "secondo la oppenione de maiestro Leonardo de la chasa degli figluogle Bonaçie da Pisa", "according to the opinion of master Leonardo Fibonacci" [ed. Arrighi 1989: 9].[70] Later abbacus writers, if referring at all to intellectual ancestors (which is not done too often), also sometimes mention Fibonacci, sometimes together with more recent *maestri d'abbaco*.[71]

As we shall see, these references should often not be taken *ad litteram*. Being known

---

were genuine and proper vernacular versions of [Fibonacci's] works, made easier by elimination of the most abstract and theoretical parts; in other cases the author limits himself to dig in the mine of examples and problems from the *Liber abaci*, in order to find material he could insert in his own treatise.

Even stronger were the statement of Van Egmond [1980: 7], according to whom all abbacus writings "can be regarded as [...] direct descendants of Leonardo's book", and of Raffaella Franci and Laura Toti Rigatelli [1985: 28], that "the abacus schools had risen to vulgarize, among the merchants, Leonardo's mathematical works". The latter claim was mitigated, however, by the observation (p. 45) that

in Florence, in the 14th century, at least two algebraic traditions coexisted. One of them was inspired by Leonardo of Pisa and was improved by Biagio the Old and Antonio de' Mazzinghi, the other, the beginning of which is unknown until now, has Gerardi [i.e., the above-mentioned Paolo Gherardi / JH] as its first exponent.

In a similar vein, Gino Arrighi [1987: 10] suspects Paolo Gherardi's *Libro di ragioni* (the one referred to by Franci and Toti Rigatelli) and a *Liber habaci* (see below, note 114) to be either re-elaborations or translations of French writings – stating, on the other hand (p. 5), that these treatises are the only witnesses we have of important mathematical exchanges between Italy and France (i.e., the Provençal area).

[70] The *Livero del l'abbecho* has been believed to be the earliest extant abbacus treatise; if so, its claimed derivation from Fibonacci might be supposed to hold also for all those treatises that were written later and which shared at least in a general way much of its material. At closer inspection, however, the date 1288–90 turns out to be merely a *post quem* dating, see below, note 77. Moreover, a recent reexamination [Travaini 2003: 88–92] of the coin list in the "Columbia Algorism" (New York, Columbia University, MS X 511 A13, [ed. Vogel 1977]) shows it to have been prepared between 1278 and 1284. Even though the author may have borrowed his information from a money dealer or other specialist in the matter, the date of the archetype of the manuscript (itself a copy in a fourteenth-century hand) can therefore hardly be later than 1285–1290 – not least since the shape of the numerals agrees with this early date, though with a few slips where the scribe has used the shape of his own times [Vogel 1977: 12]. Although the *Livero de l'abbecho* may still be the earliest extant *manuscript* of its kind, the Columbia Algorism is thus likely to be a copy of a still earlier *treatise*.

[71] For the utter scarcity of such later references to Fibonacci, see [Franci 2003: 50*f*].

by name, and as the author of a famous though (until the later fourteenth century) rare book, Fibonacci very soon acquired the role of a *culture hero*.

The *Livero de l'abbecho*

However, let us first turn to the possibly earliest extant abbacus manuscript, the Umbrian *Livero de l'abbecho* (Florence, Riccardiana MS 2404, ed. [Arrighi 1989]), supposedly dated by internal evidence to c. 1288–1290.

As just quoted, the treatise claims to be written "according to the opinion" of Fibonacci. Van Egmond [1980: 156] describes it as follows:

> S. XIII (c. 1290, *i[nternal evidence]*, Holograph *libreria* treatise.
> Vellum, irreg. approx. 215–222 × 147 mm., 178 [leaves] num[erated] mod[ern] type 1–178 [...]. Single hand, a very neat Gothic bookhand in 1 col[umn] ink-ruled 10 cm. wide, 31 ink-ruled lines in 16–16.5 cm., text on second. Black ink with titles and numerals in red, alternating red/blue decorated initials in margins, initials of each chapter illuminated. Geometrical diagrams and drawings in margins in red on 139r–160v, 175r–178v.[72]

Fols 1^r–136^v contains the *Livero del abbecho* (the second part is an *Arte dela geometria*). The abbacus treatise consists of 31 chapters:[73]

Ch. 1, "de le regole de le tre chose" (fols 1^r–1^v).
> Introduces the rule of three for integers and the tricks to use when one or more of the given numbers contains fractions.

Ch. 2, "de le chose che se vendono a centonaio" (fols 2^r–3^r).
> Gives rules of the type "if something is sold in batches of a hundred pounds, then for each *libra* that the hundred are worth, the pound is worth 2$\frac{2}{5}$ d., and the ounce is worth $\frac{1}{5}$ d.". Nothing is borrowed from Fibonacci, but from the end of the chapter (and for the rest of the treatise) the writing of mixed numbers follows Fibonacci's system, the fraction being written to the left of the integral part; until then, the integral part stands to the left.

Ch. 3, "de le regole de pepe che senno" (fols 3^r–4^v).
> Problems about pepper and other spices, some of them involving reduced weight due to refining. Nothing borrowed from Fibonacci.

Ch. 4, "de le regole degle drappe che se vendono a channa e a br[accia]" (fols 4^v–6^r).
> Problems depending on the metrology for cloth. Nothing borrowed from Fibonacci.

Ch. 5, "de regole de chanbio" (fols 6^r–13^r).
> Mostly on change of one coin against another – but also of coin against weighed bullion, silk or fish, and of combination of coins, depending mostly on the rule of three and involving the

---

[72] Actually, there are also some marginal diagrams in the preceding part.

[73] The following pages contain a quite detailed and fairly pedantic examination of the manuscript. The impatient reader who is willing to accept the conclusions without the details may go directly to p. 40.

subdivisions of the *libra*. Nothing is borrowed from Fibonacci. From fol. 7^r onward, many results are given in the awkward form "d. $^{17}/_{49}$ 7 de denaio" and similarly, "*denari* $^{17}/_{49}$, 7 of denari" – obviously arising from infelicitous mixing of Fibonacci's notation for (pure) mixed numbers with the standard expression "*denari* 7, $^{17}/_{49}$ of *denaro*".[74] The implication is that the compiler has copied this section from another pre-existing written source and tried to adapt its notation to the one he is using himself (which is hardly unexpected).[75] The same construction turns up again in various later chapters, but *never* in problems taken over from Fibonacci.

Ch. 6, "de baracta de monete e denari" (fols 13^r–15^r).
More complex problems about exchange of coin (and merchandise), involving the (unnamed) rule of five. Nothing borrowed from Fibonacci.

Ch. 7, "de le regole de marche Tresce [from Troyes / JH] e de svariate ragione de lib" (fols 15^r–16^v).
Similar to Chapter 6, but even more complex. Nothing borrowed from Fibonacci.

Ch. 8, "da sapere quante d[enari] de chantra e charrubbe e grana è l'onzia" (fols 16^v–17^v).
On the subdivision of the ounce, and on the refining of alloyed bullion. Nothing borrowed from Fibonacci.

Ch. 9, "de conparare bolçone a numero de denare ed a peso de libr[e]" (fols 17^v–20^v).
Problems about the purchase of alloyed bullion and its evaluation in value of pure metal. Nothing borrowed from Fibonacci.

Ch. 10, "de regole de consolare ed alegare monete" (fols 20^v–29^v).
Problems about alloying. After ten simple problems that are independent of Fibonacci follow eighteen, some of them more complex, that are borrowed from the *Liber abbaci* [ed. Boncompagni 1857a: 144–158] – in part whole sequences of consecutive problems. At times the copying is so close that Fibonacci's cross-references are borrowed even though they are invalid in the actual context; at times minor variations are introduced, e.g. the conversion of $^{101}/_{163}$ ounce into $7^{71}/_{163}$ *denari*.

Ch. 11, "de svariate regole che s'apartengono al consolare de le monete" (fols 29^v–32^v).
Six rather artificial problems of alloying type, five of which are from the *Liber abbaci* [ed. Boncompagni 1857a: 159–164].

Ch. 12, "de regole de merto o vero d'usura" (fols 32^v–42^v).
Problems about loans and interest, first 24 on simple interest, then one (counting a numerical variant, two) problems about composite interest over full years and one on a decrease in geometrical progression; none of these come from Fibonacci (the last problem is structurally analogous to one found in the *Liber abbaci* [ed. Boncompagni 1857a: 313], but the solution

---

[74] This is the notation that is used until the Fibonacci notation for mixed numbers is adopted on fol. 2^v – e.g., "denare 19, $^8/_{13}$ de denaio", fol. 1^r.

[75] Indeed, fol. 7^r has an isolated "d. 10, $^6/_7$ de denaio" betraying the original, following a few lines after "d. $^8/_{11}$ 8 de denaio". Similar slips on fol. 45^r, "dr. 1, $^{21}/_{50}$ de denaio", and fol. 134^r, "d. 3, $^{15}/_{19}$ de denaio", "d. 8, $^4/_{19}$ de denaio". On fol. 57^r and again on fol. 121^r, whole schemes are organized accordingly. All of these instances are in problems not borrowed from Fibonacci.

runs along different lines).[76] In the end comes a section "De sutile regole de prestiare lib. quante tu vuogle ad usura sopre alchuna chosa", about giving a loan in a house which the creditor rents (the rent exceeding the interest on the loan and the difference being discounted from the capital); all 12 problems belonging to this section are borrowed from the *Liber abbaci* [ed. Boncompagni 1857a: 267–273].

Ch. 13, "de regole che s'apartengono a quille de la usura" (fols 42^v–44^r).

Eight problems somehow involving interest (combined with partnership, discounting, etc.), not derived from Fibonacci. There are a few instances of constructions like "d. $^{14}/_{17}$ 2 de denaio", and also one "dìne $^6/_{97}$ 13 de dìne" ("days $^6/_{97}$ 13 of day").

Ch. 14, "de regole de saldare ragione" (fols 44^r–51^r).

Loan contracts containing invocation of God, names and dates, thus real or pretendedly real, leading to the problem of settling the accounts for several loans made within a single year, that is, bringing all to the same term of repayment; only simple interest is involved. Everything is independent of the *Liber abbaci*. There are copious instances of expressions of the type "d. $^{11}/_{12}$ 6 de denaio" in all those problems that permit it, with the implication that this section is copied from a written source, either real contracts or another abbacus treatise.[77]

Ch. 15, "de svariate regole de conpagnie" (fols 51^r–58^v).

Various partnership problems, none of which come from Fibonacci. Most of them contain constructions of the type "staia $^5/_8$ 90 de staio" (the *staio* is a measure of capacity).

Ch. 16, "de chonpare de chavagle" (fols 58^v–65^r).

Ten variations of the "purchase of a horse": N men go to the market in order to buy a horse; each one needs a given fraction of what the others possess in order to have enough to buy the horse. The first two are independent of Fibonacci (the second is indeed instead of "partnership" type), the following eight are taken over from the *Liber abbaci* [ed. Boncompagni 1857a: 228–235, 253*f*]. The first two problems contain numerous constructions of the type "d. 4/7 8 de denaio", the others none.

Ch. 17, "de huomene che demandavano d[enaro] l'uno a l'altro" (fols 65^r–74^r).

Variations (with changing number of men and conditions) of the problem type "Two men have *denari;* if the first gets a of what the second has, he shall have p; if the second gets b of what the first has, he shall have q", where a, b, p and q may be given absolutely or relatively to what the other has. The first six are independent of Fibonacci, then come nine that are taken

---

[76] In this and in a slightly earlier problem, we also find constructions of the type "d. $^{1747}/_{2561}$ 1 de denare".

[77] Since the claimed "internal evidence" for the dating of the treatise is constituted by nothing but these problems, the fact that they are copied undermines the dating of the treatise to the years 1288–90; all they tell is that the treatise was written at some later date (it would be much less compelling for a compiler to use quite recent examples than to use an updated coin list if he included one – cf. also note 255 on the observations made by a compiler who did use a dated list). Other evidence, however, supports an early date (see, e.g., below, p. 39); so does, to the modest extent I can judge it, the apparently archaic orthography. It might be observed, on the other hand, that the vellum manuscript is so beautiful that it is likely to be a *de luxe* copy and not the original and hence even later than the text.

from the *Liber abbaci* [ed. Boncompagni 1857a: 189f, 198–202], then finally one that is not borrowed from that work.[78]

Ch. 18, "de huomene che trovaro borsce" (fols 74ʳ–79ᵛ).

Seven variations on the theme "*N* men find a purse with *denari*; the first says, 'If I get what is in the purse (with or without what I already have) I shall have *p*'; the second says ...", where *N* is 2, 3, 4 or 5, and *p* is given relatively to the possession of the other(s). All are from the *Liber abbaci* [ed. Boncompagni 1857a: 212–214, 220, 223, 227].

Ch. 19, "de huomene che cholsero denare emsiememente" (fols 79ᵛ–82ʳ).

Five variations on the problem type "*N* men find *denari* which they divide in such a way that ...", *N* being 2, 3, 5 or 6. In several cases "in such a way" regards the products between the shares two by two. All are borrowed from the *Liber abbaci* [ed. Boncompagni 1857a: 204–207, 330, 281, this order]. In the end comes a single problem of the type treated in Chapter 18, but which is independent of Fibonacci.

Ch. 20, "de regole de prochacio overo de viage" (fols 82ʳ–86ᵛ).

Fifteen problems "on gain and travelling", about a merchant visiting three or more markets, gaining every time a profit that is defined relatively to what he brought and having expenses that are defined absolutely; the initial capital is found from what he has in the end. All come from the *Liber abbaci* [ed. Boncompagni 1857a: 258–262, 266].

Ch. 21, "de huomene ch'andaro a guadagnare agl merchate" (fols 86ᵛ–91ʳ).

Nine problems concerning trade or markets – not all, in spite of the title, about gains (two, indeed, are of the type "a hundred fowls"[79]), but all are from the *Liber abbaci* [ed. Boncompagni 1857a: 399, 298, 160, 165f, 179, this order]; in some cases it is obvious that the compiler does not understand what he copies (see below, p. 39).

Ch. 22, "de choppe e del suo fondo" (fols 91ʳ–92ʳ).

Three problems about a goblet consisting of a cover, a foot, and "el meço" ("the middle", i.e., the cup), one part being given absolutely, the others relatively. The problems correspond to a sequence of consecutive problems in the *Liber abbaci* [ed. Boncompagni 1857a: 188f];[80] the second contains a cross-reference to the use of the "rule of the tree" (the single false position) even though this rule, earlier in the *Liber abbaci*, only comes later here. The same second problem is corrupt, seemingly because the manuscript that is used has employed "$\frac{1}{2}$" as a word sign for *medium* or *meço*, which the present writer repeats but understands as a

---

[78] Or which at least is not in the 1228 edition as published by Boncompagni; it *could* in principle be one of those problems from the 1202-edition which Fibonacci [ed. Boncompagni 1857a: 1] says he has eliminated as superfluous; indeed, no obvious stylistic features distinguish it from the *Liber-abbaci* problems that precede it.

[79] To buy for *M* monetary units *N* fowls of three different kinds, the price for each kind being given (as a rule $M = N = 100$, whence the name) – for instance, a rooster costing 5 units, a hen 3 units, chicks being sold 3 for a unit.

[80] In order to see that the second problem has a counterpart in the *Liber abbaci* one has to discover (from the subsequent calculation, or from the *Livero de l'abbecho*) that the words "ponderet quantum medii" [ed. Boncompagni 1857a: 188, line 5 from bottom] should be "ponderet quartum medii".

number.

Ch. 23, "d'arbore o vogle de legne" (fols 92^r–93^r).

Four problems about a tree, a certain fraction of which is either hidden underground, or added to the tree, the remainder or total being given absolutely. They correspond to a sequence in the *Liber abbaci* [ed. Boncompagni 1857a: 174*f*], but the wording of the first problem deviates so much from the *Liber-abbaci* counterpart (and corresponds so well to what is found in other abbacus writings, for instance Jacopo's treatise) that one may assume the writer to have rewritten this problem from Fibonacci in a familiar style.[81]

Ch. 24, "de vasa" (fols 93^v–95^r).

Two problems about three respectively four vases, relative relations between whose contents are given (e.g., that the first holds $\frac{1}{18}$ of what the second holds, plus $\frac{1}{3}$ of what the third holds). Both are from the *Liber abbaci* [ed. Boncompagni 1857a: 286].

Ch. 25, "de huomene che vonno per via chumunalemente ensieme" (fols 95^r–96^r).

Two problems about men putting part of their possessions or the total of these in a common fund, redistributing part of the fund arbitrarily and the rest according to given proportions, finding thus their original possessions. Both are from the *Liber abbaci* [ed. Boncompagni 1857a: 293, 297], but the anecdote in the first one differs from Fibonacci's version (while probably coinciding with the typical tale belonging with the problem).[82]

Ch. 26, "de huomene che portaro margarite a vendere em Costantinuopole" (fols 96^r–97^v).

First two problems about carrying pearls to Constantinople and paying the customs, both taken over from the *Liber abbaci* [ed. Boncompagni 1857a: 203*f*]; next one which combines a dress about precious stones and Constantinople with the mathematical structure of a problem dealing with fishes and commercial duty, neighbouring problems in the *Liber abbaci* [ed. Boncompagni 1857a: 276*f*].[83] Finally an independent problem about trade in pearls.

Ch. 27, "de tine e de botte cho' n'esce el vino per gle foramene cho sonno el fondo" (fols 98^r–101^r).

Six problems about perforated tuns and casks, all from the *Liber abbaci* [ed. Boncompagni 1857a: 183–186]. In the first problem, the compiler misrepresents and obviously does not understand the explanation of the procedure given by Fibonacci.

Ch. 28, "d'uno che manda el figlo en Alixandria" (fols 101^r–102^r).

---

[81] Fibonacci asks for the whole tree to be divided into 12 parts, instead of putting the whole tree equal to a convenient number, *viz* 12. That this latter formulation was what he knew as the habitual one he states explicitly [ed. Boncompagni 1857a: 174]: "consueuit dicere: pro 12, que pono, ueniunt 7; quid ponam, ut ueniant 21" / "it is habitual to say, for 12 which I posit, 7 result; what shall I posit so that 21 result?". The *Livero de l'abbecho* returns to the customary formulation; it also inserts the standard phrase "quista è la sua regola, che ...". It is of course possible that the 1202 edition used the customary phrasing, and that this was the edition the compiler was using.

[82] Once again, the formulation in the 1202 edition may have been traditional, and this could be the one used by the compiler.

[83] Given the modest competence of our compiler one may ask whether these two problems were one in the version of the *Liber abbaci* which he had at his disposal.

Four problems about the purchase of pepper and saffron (and in the fourth also sugar and cinnamon) for a given total, at given prices and at given weight proportions. Borrowed from the *Liber abbaci* [ed. Boncompagni 1857a: 180].

Ch. 29, "d'uno lavoratore che lavorava enn una uopra" (fols 102^r–104^v).

First two identical problems about a worker who is paid for the days he works and pays a fine for the days he does not work, solved with different methods; the two versions are taken from widely scattered places in the *Liber abbaci* [ed. Boncompagni 1857a: 323, 160] – the first of them appears to be badly understood by the compiler. Next comes a problem about a complex mode of wage payment, again in two versions resolved by different procedures, both from the *Liber abbaci* [ed. Boncompagni 1857a: 186, 324].

Ch. 30, "de huomene ch'andano l'uno po' l'altro" (fols 104^v–105^r).

Two problems about two travellers, one going with a constant speed, the other pursuing him with a speed that increases arithmetically. Both are from the *Liber abbaci* [ed. Boncompagni 1857a: 168].

Ch. 31, "de regole per molte guise forte e ligiere de molte contintione" (fols 105^r–136^v).

A mixed collection of mainly recreational problems, some from the *Liber abbaci*, others not; several of the latter are simpler versions of problems borrowed from Fibonacci that appear in the preceding chapters. With reference to the pagination in [Boncompagni 1857a], the distribution is as follows; "indep." means that the problem has no counterpart in the *Liber abbaci* (a consecutive "≠" indicates that superficial similarity suggests a borrowing but closer inspection shows instead that both writers draw on a common fund of basic problems and variations):

p. 273, p. 273, p. 274, p. 297, p. 298, p. 298, p. 283,[84] p. 329, p. 312, p. 182, p. 182, indep., indep., indep. (≠p. 307), indep. (≠p. 323, ≠p. 160),[85] indep., indep.,[86] indep. (≠p. 179), indep., indep., indep., indep.,[87] p. 403, indep.,[88] indep., indep., indep., indep., indep., indep. (≠176*f*), p. 181, indep. (≠172),[89] indep. (≠174), indep.,[90] indep., indep.,[91] in-

---

[84] The famous "rabbit problem", transformed into a "pigeon problem" with no other change, and with a reference to a marginal diagram that is indeed found in the *Liber abbaci* but not in the present treatise.

[85] Apparently based on a source which is understood and copied badly.

[86] Meaningless as it stands, probably resulting from defective copying of a source. To be solved "sença regola [...] a palpagione e per apositione falsa" ("without a rule [...] by feeling one's way and by false position").

[87] This problem (about two kinds of wool that shrink at different rates when washed) seems to be borrowed from the Columbia Algorism [ed. Vogel 1977: 83*f*] or a very close relative. Not only the story and the rather particular parameters are shared; so are also the roundings of results and explanatory offhand remarks.

[88] A problem about a pavilion that also turns up in Jacopo's *Tractatus* (V.15.21=F.vi.17).

[89] Apparently based on a source that is copied thoughtlessly.

[90] Solved wrongly.

[91] A question touching at a real-life problem for long-distance trade which is rarely mentioned in

dep.,[92] indep.,[93] p. 177, p. 182,[94] p. 274, p. 311, p. 316,[95] p. 313, p. 309,[96] p. 311, indep., indep., indep., indep., indep., indep., indep., indep., indep., p. 132,[97] p. 133, indep.,[98] (≠167),[99] (≠167),[100] (≠166),[101] indep., indep., indep., indep. (≠p. 179), indep., indep., p. 283, indep., indep., indep.

Some of the independent problems contain expressions like "d. $\frac{2}{7}$ 4 de denaio", "d. $\frac{1}{3}$ 9 de denaio"; "dì $\frac{11}{30}$ 354 de d.".

To the (non-exhaustive!) observations made here on editorial changes, blind cross-references and misunderstandings of Fibonacci's text two of a more general character

---

abbacus treatises: a ship beating up against the wind.

[92] A perpetual calendar.

[93] This problem type is often found in *al-jabr* treatises: to divide a given number (mostly 10, here $16\frac{1}{4}$) into two parts with a given ratio.

[94] About ships that encounter each other. The compiler has added names to the points of departure and destination (Genoa and Pisa).

[95] Contains a cross-reference to the problem that precedes – in the *Liber abbaci*. Here it follows.

[96] The chess-board problem. The beginning copies Fibonacci in a way that suggest failing understanding.

[97] Makes use of the "rule of five" but without explaining what goes on (Fibonacci explains).

[98] Another instance of sloppy copying from a source – the problems starts, in word-for-word translation, "There is a well and a serpent deep 90 palms, by day $\frac{3}{3}$ palms and ascends and by night descends the fourth". Apart from the displaced and superfluous words, "$\frac{3}{3}$" should be "$\frac{2}{3}$".

[99] The rule for the summation of square numbers from 1^2 to 10^2, found as $10 \cdot (10+1) \cdot (10+[10+1])/6$. The same computation is found in the *Liber abbaci* [ed. Boncompagni 1857a: 167], but the formulations are too different to make a borrowing plausible. The general case (with the corresponding formula) is proved in Fibonacci's *Liber quadratorum* [ed. Boncompagni 1862: 262], but nothing in the formulations suggest the compiler to have known that work.

[100] The rule for the summation of odd square numbers from 1^2 to 11^2, found as $11 \cdot (11+2) \cdot (11+[11+2])/(2 \cdot 6)$. The *Liber abbaci* [ed. Boncompagni 1857a: 167] finds the sum $1^2+3^2+\cdots+9^2$ according to the same formula, but explaining that the factor $9+2$ is the following member in the sequence of odd numbers, and that the divisor 2 is the distance between the squared numbers. Once again, a general proof is found in the *Liber quadratorum* [ed. Boncompagni 1862: 263], but nothing in the formulations suggests that the compiler knew this advanced treatise.

[101] The rule for the computation of $1+2+\cdots+99$, found as the product of the last member by its half rounded upwards! The *Liber abbaci* [ed. Boncompagni 1857a: 166] gives two general formulae, either half the number of terms multiplied by the sum of the extremes, or half this sum multiplied by the number of terms, and one less general for sums of the type $p+2p+\cdots+np$; all Fibonacci's numerical examples differ from the present one. The *Liber quadratorum* [ed. Boncompagni 1862: 265] contains the rule that the sum of a number n and other numbers pairwise equidistant from it (i.e., $n+(n+d_1)+(n-d_1)+\cdots+(n+d_p)+(n-d_p)$) equals the product of the number (n) and the number of terms ($2p+1$). It is hardly necessary to argue that this generalization of the summation of an arithmetical series with an odd number of terms was not used by our compiler.

should be added. Firstly, the compiler does not understand Fibonacci's notations for composite fractions, e.g., for ascending continued fractions, and reads Fibonacci's $\frac{33\ \ 6\ \ 42\ \ 46}{53\ \ 53\ \ 53\ \ 53}$ [ed. Boncompagni 1857a: 273], standing for

$$\cfrac{46-\cfrac{42-\cfrac{6-\cfrac{33}{53}}{53}}{53}}{53},$$

as if it meant simply $\frac{3364246}{53535353}$ [ed. Arrighi 1989: 112].[102] The implication is that the compiler never performed these computations and probably did not intend to do so in his teaching – that is, that Fibonacci's more sophisticated problems (at least those where such fractions occur, but probably many others too[103]) are taken over as external embellishment.

Secondly, the compiler does not understand Fibonacci's *regula recta*, the application of first-degree *res*-algebra (see further explanation below, p. 105). Mostly, Fibonacci's alternative solutions by means of *regula recta* are simply skipped, but in one place (fol. 83^r) the compiler takes over a *regula-recta* solution from Fibonacci [ed. Boncompagni 1857a: 258], promises to teach the solution "per regola chorrecta", omits the first *res* from Fibonacci's text (the position) while conserving some of the following as *cosa*, obviously without noticing that this *thing* serves as an algebraic representative for the unknown number. Beyond elucidating once again the function of Fibonacci's sophisticated problems in the treatise,[104] this shows that the compiler worked at a moment when even the

---

[102] It can be excluded that it was just the copyist who committed the error. Firstly, the manuscript *never* explains these composite fractions as does Fibonacci; secondly, they are never translated into other fractions, as happens when Fibonacci's composite fractions of *solidi* are occasionally transformed into *denarii* and fractions of these. Actually, the manuscript regularly inserts small spaces between the digits of the enumerator when this number is shorter than the denominator; but in such cases there are no spaces between the digits of the denominator – the purpose is nothing but graphical regularity.

[103] Two obvious examples are found on fols 86^v–87^r. The first corresponds to a problem which Fibonacci [ed. Boncompagni 1857a: 399] solves by means of his letter formalism ("Somebody has 100 *libras*, on which he earned in some place; then he earned proportionally in another place, as he had earned before, and had in total 200 *libras*"). The compiler speaks of two different persons; does not tell us that the second goes on with what the first has in total; and eliminates the letters from the text when translating. The outcome is evidently pure nonsense.

The second corresponds to a mixed second-degree problem ("Somebody had 100 *libras*, with which he made a travel, and earned I do not know what; and then he received 100 *libras* more from a partnership, and with all this he earned in the same proportion as in the first travel, and thus had 299 *libras*"), which Fibonacci [ed. Boncompagni 1857a: 399] first transforms by means of continued proportions into a rectangle problem and next solves by means of *Elements* II.6. All letters and lines have disappeared in the translation, as has the Euclidean reference.

[104] Indeed, the method in question is well explained by Fibonacci in the *Liber abbaci* [ed. Boncompagni 1857a: 191] and regularly used after that in Chapter 12 [ed. Boncompagni 1857a:

most elementary level of algebra was still unknown in his environment – thus hardly much later than 1300–1310.

Let us then turn our attention to those chapters which teach matters of real commercial use – that is, to Chapters 1–10 and 12–15. As we see, only Chapters 10 and 12 contain problems taken over from Fibonacci; moreover, those which are taken over all belong to the most sophisticated and often rather artificial category.

The claim that the treatise is shaped "according to the opinion of master Leonardo Fibonacci" is thus in itself an instance of embellishment. The treatise is certainly no "genuine and proper vernacular [version of one of Fibonacci's] works, made easier by elimination of the most abstract and theoretical parts" (see note 69), nor is it written in order "to vulgarize, among the merchants, [one of] Leonardo's mathematical works". The supposedly earliest extant abbacus manuscript thus contains a treatise in which, in Giusti's words, "the author limits himself to dig in the mine of examples and problems from the *Liber abaci*, in order to find material he could insert in his own treatise" – not so much for providing understanding as for showing off. In the outgoing thirteenth century, Fibonacci had already acquired the status of the culture hero of abbacus culture; the rare "genuine and proper vernacular versions" came later, when a few abbacus masters felt the ambition to trace the sources of their field.[105]

198, 203*f*, 207, 213, 258, 260, 264, 280].

[105] A couple of translations from the *Liber abaci*, one of Chapters 14–15, another one of most of Chapter 12 and a little of Chapter 13, go back to c. 1350; another translation of Chapters 14–15 can be dated c. 1400, as can a translation of the *Liber quadratorum*; a translation of the *Pratica geometrie* is dated 1442; see [Van Egmond 1980: 363]. This can be contrasted with the total number of extant vernacular mathematical writings made within consecutive 25-year periods according to Van Egmond [1980: 407–414] (with the proviso that some datings are approximate, and others too early because copied internal evidence has suggested an early date for an actually later treatise):

1276–1300	1301–1325	1326–1350	1351–1375	1376–1400	1401–1425	1426–1450	1451–1475	1476–1500
1	8	10	6	19	16	39	56	66

The age distribution of surviving *Liber-abbaci* manuscripts is not very different from that of the translations: 3 appear to be from the later 13th century, 4 from the 14th, 2 or 3 from the 15th, 3 or 2 from the 16th (Menso Folkerts, private communication, emphasizing that some of the dates are uncertain or disputed).

It is interesting that the vernacular versions of al-Khwārizmī's algebra are slightly more copious than those of Fibonacci – see [Franci & Toti Rigatelli 1985: 28–30] and [Van Egmond 1980: 361]. One is from c. 1390, one from c. 1400, and several from the fifteenth century. In some cases, interest in al-Khwārizmī goes together with interest in Fibonacci – obviously, both play the role of (mythical) fathers, those *fontes* which it was not uncommon to look for in the Italian fourteenth and fifteenth centuries. In this connection it is suggestive that Benedetto da Firenze, one of those who actually

Our compiler certainly *could* have found even the material for his basic chapters in the *Liber abbaci* – all of it is copiously there, with the exception of simple interest. But he may have preferred to use examples referring to the metrologies and exchange rates of his own times and area; alternatively, he may already have had a treatise which was ready for all practical purposes and then have decided to insert into it the embellishments borrowed from the hero[106] (and for the last chapter probably also from other sources that went beyond his mathematical wits). We cannot know. What we *can* know from the analysis is that the abbacus tradition of the outgoing thirteenth century was *no Fibonacci tradition*, even though it was already *a tradition*.[107]

Fibonacci and the Abbaco

Fibonacci took his inspiration from many sources, some of which can be identified – the notation for ascending continued fractions emulates that of the Maghreb school, the algebra section copies creatively but unmistakeably from Gherardo da Cremona's translations of al-Khwārizmī's *Algebra* [Miura 1981], the *Pratica geometrie* from the same translator's version of Abū Bakr's *Liber mensurationum*.[108] Most of his sources, however, are

---

mention Fibonacci, follows al-Khwārizmī's ordering of the algebraic cases, and neither Fibonacci nor his abbacus predecessors [ed. Salomone 1982: 3–20]; he also borrows al-Khwārizmī's examples and his geometric proofs. See below, p. 101.

[106] The model for this possibility is Bombelli, whose *L'algebra* was already finished in a first version when he discovered Diophantos – see [Jayawardene 1970: 280].

[107] Raffaella Franci [2003: 40], admitting that the stem of the *Livero de l'abbecho* has nothing to do with the *Liber abbaci* but still wanting this treatise as well as the later abbacus tradition to descend somehow from Fibonacci, suggests that the inspiration was a *liber minoris guise*, "book in a smaller manner", which Fibonacci mentions in one place in the *Liber abbaci*. However, all conserved treatises of Leonardo use the same writing of mixed numbers, for which reason we must assume even the *liber minoris guise* to have done so, while – as we have just seen – the contributions coming from the *Liber abbaci* and those used for the stem of the *Livero de l'abbecho* differ on this account. In the first instance this only disqualifies the lost work as a source for the *Livero de l'abbecho*. However, Franci's argument for the general importance of the "book in a smaller manner" is the similarity of many later treatises to the Umbrian compilation on various accounts, e.g. in the presentation of the rule of three – which means that the argument breaks down generally.

This observation of course does not invalidate the reasonable assumption that the "book in a smaller manner" treated all or some of the same matters as later abbacus books. All we know about it is that Fibonacci states that he has borrowed from it an alternative method to treat the alloying of three kinds of bullion for the *Liber abbaci*, and that an anonymous abbacus writer knew it by name and as a *libro de merchaanti*, "book for merchants" in c. 1460 (Biblioteca Nazionale di Firenze, Pal. 573, fol. 433ᵛ, see [Franci 2003: 37]).

[108] Ed. [Busard 1968]. This treatise is indeed the source for most (if not all) of what Fibonacci is normally taken to have borrowed from Savasorda, as becomes evident as soon as the three texts are compared.

unidentified. If the abbacus tradition does not descend from Fibonacci, could then Fibonacci also have been inspired by an already emerging abbacus tradition?

As pointed out by Boncompagni [1854: 88–94], Fibonacci mostly refers to what we know as the *Liber abbaci* as his *liber numerorum* or, in the dedicatory letter of the *Flos*, as his *liber maior de numero*.[109] In one place, however – namely the *Pratica geometrie* [ed. Boncompagni 1862: 148] he speaks of it as his *liber abbaci*.[110] Such changing references suggest that at least the ones that only occur once were not thought of as titles (to the limited extent this concept was at all valid at the time) but rather as descriptions. Our *Liber abbaci* was thus thought of by Fibonacci as a "book about abbacus matters". But what did *abbaco* mean to him?

The word *abbaco* does appear at least thrice (as a Latinized genitive *abaci/abbaci*) in the treatise: in the prologue, where Fibonacci says [ed. Boncompagni 1857a: 1] that his father had wanted him to pursue *studio abaci* for "some days" in Bejaïa in present-day Algeria;[111] when Chapter 12 is said to treat of *questionibus abbaci* [ed. Boncompagni 1857a: 166]; and when the computation of the approximate square root of 743 is said to be done *secundum abaci materiam* [ed. Boncompagni 1857a: 353]. At least the latter two occurrences do sound as if something more precisely delimited is meant than mere references to "instruments, methods, manuals, schools, teachers or anything else related to the skill of doing computations", and could well refer to such things as we find in the earliest abbacus treatises; but it hardly proves anything.

Indirect supplementary information can be derived from certain key phrases that abound in the Umbrian *Livero de l'abbecho* as well as later abbacus treatises. Very often, problems start with the phrase *famme quista ragione* ("make this problem for me" – as we remember, the Pisa curriculum explains that *ragioni*, problems, were to be done as homework) or *se ci fosse dicto* ("if it was said to us, ..."). Very often, the procedure description ends with a phrase like *e chusì fa' le semeglante ragioni* ("and make similar

---

[109] Since the *Flos* was written in 1225, the 1202 edition must be meant. The expression "book in a smaller manner" can therefore not be meant as a reference to this, presumably shorter edition, as opposed to the major edition from 1228, however tempting such a hypothesis might seem. Nor would there be any reason to refer explicitly to the "book in a smaller manner" as the source for a particular method if it was identical with the very treatise which, with omissions and additions, was reshaped as the 1228-*Liber abbaci*.

[110] I thank Barnabas Hughes for directing me to this passage.

[111] These words are not changed in the critical edition of the passage in [Grimm 1976: 100]. As Grimm points out in a note, even a generous reading of "some days" can hardly correspond to more than a fortnight – but as he also observes, "some days" were what his father wanted, which does not strictly exclude that the son actually spent more time. What is clear is that Fibonacci wishes the reader to believe that his school days in Bejaïa were few.

Fibanacci says nothing about the kind of school he visited. Since he was no Muslim, both a mosque school and a madrasah can probably be excluded.

problems in this way"). Often, the procedure description also starts by the declaration that *quista è la sua regola* ("this is its rule").

In the *Livero de l'abbecho*, such phrases are particularly copious in problems that are not taken from the *Liber abbaci*, but many are also glued onto *Liber-abbaci* problems. What is more interesting is that Fibonacci has scattered but rarer instances of the "make similar problems in this way", as if somewhat influenced by the style of an environment where this usage was pervasive. We also find copious references to "the rule of [e.g.] the tree", meaning the rule introduced by means of a problem about a tree.

Similar evidence comes from the particular way in which many of the first alloying problems of the *Livero de l'abbecho* (but none of its other problems) begin (the initial problems of Chapter 10, which are not derived from the *Liber abbaci*), namely in the first person singular, "I have silver which contains *n* ounces per pound"; the later problems, those taken from Fibonacci, start in different ways, and so do the alloying problems in the *Liber abbaci* itself – but in one place, in a general explanation [ed. Boncompagni 1857a: 143], we find *cum dicimus: habeo monetam ad uncias quantaslibet, ut dicamus ad 2, intelligimus quod in libra ipsius monete habeantur uncie 2 argenti* ("when we say, I have bullion at some ounces, say at 2, we understand that one pound of it contains 2 ounces of silver"). It is not credible that the later abbacus tradition should have grasped this hint and generalized it;[112] instead, Fibonacci must be citing – and the only place where such a standard beginning is possible is in *problems* about alloying (the construction "we say, I have" shows that the choice of the grammatical person *I* belongs within the citation). Further evidence that the formulation was already standard for this problem type is offered by a problem about the payment of a worker [ed. Boncompagni 1857a: 160]. Fibonacci reduces it to an alloying problem, stated in the words "you say, 'I have coin at 26 and at 31; and I want to make of it 30 pounds [...] at 31'". This must be the formula in which he expects his reader to recognize the procedure.

There is thus no doubt that Fibonacci, though mostly trying to be neutral and to emulate scholarly style, was familiar with a tradition that influenced the style of later abbacus writings heavily.

Where can he have encountered this tradition and the environment which carried it? Italy is not excluded, even though he had had to go to Bejaïa in order to learn about the Arabic numerals. Various apparent Italianisms that creep into his text (e.g., *viadium/viagium* for travel, from *viaggio*, *avere* as an occasional translation of Arabic *māl* instead of *census*) might suggest that Italy played a role without excluding that the environment ranged more

---

[112] It is found not only in Jacopo's treatise and other abbacus writings but also in Pegolotti's early-fourteenth–century *Pratica di mercatura* [ed. Evans 1936: 342–357], not suspect of dependency of Fibonacci. The formula also turns up in the words "E si te dixeren, yo tengo de tres suertes plata, la una suerte es de 1 marco, 7 onças ⅓ de onça de plata fina ..." in the Castilian manuscript *De arismetica*, fol. 151ʳ [Caunedo del Potro 2004: 45].

widely; so does the observation that Italian merchants must already have had an urgent need for such things as are taught in the first 15 chapters of the *Livero de l'abbecho*.

We should take note of exactly what Fibonacci says in the prologue of the *Liber abbaci:* that his father brought him to Bejaïa, where his (supposedly brief) *studio abbaci* introduced him to the "nine figures of the Indians", that is, to the use of the Hindu-Arabic numerals; nothing is said about methods like the rule of three, partnerships, or alloying.[113] Latin culture, it is true, had already been introduced to Hindu-Arabic numerals in the early twelfth century; none the less it is more than likely that whatever commercial teaching went on in Italy during Fibonacci's youth was still based on Roman numerals[114], and that the consistent application of Arabic numerals to otherwise familiar matters is what makes his treatise really new (apart of course from its exorbitant scope and its integration of algebra and Euclidean material and of numerous sophisticated variations of many recreational problems); such an interpretation would fit his words better than the belief that *everything* in the book was new to his world. We may also take note of his reference to the "habitual" way to formulate the single false position (see note 81); since nothing is said about the formulation being habitual in some other place, he must refer to a habit which the reader can be supposed to know about.

However, the apparent Italianisms could also have been inspired for instance from the Catalan-Provençal (*viaje*, written *viatge* in modern Catalan; and *aver*, modern Catalan *haver*). As we shall see, Italian abbacus *algebra*, when it emerged, received its inspiration not from Fibonacci but from some non-Italian (probably Ibero-Provençal) environment, and algebra can hardly have thrived there without being part of a broader mathematical culture of abbacus type flourishing in the same place. Once again we may return to Fibonacci's own words. After his boyhood visit to Bejaïa, he continued his study of the "nine Indian figures" (and, certainly to be tacitly understood, matters that belong with them) on his travels to Egypt, Syria, Greece, Sicily and Provence.

---

[113] Later, of course, the *regula recta* and the *elchatayn* rules are ascribed to the Arabs [ed. Boncompagni 1857a: 191, 318]; but these are higher-level matters that go beyond basic abbacus teaching as delineated by the Pisa curriculum and Chapters 1–15 of the *Livero de l'abbecho*.

[114] Indeed, a *Liber habaci* (Florence, Magl. XI, 88, fols 1ʳ–40ᵛ, ed. [Arrighi 1987: 109–166], dated by Van Egmond [1980: 115] on the basis of internal evidence to 1310, gives all integers in Roman numerals – also those in the brief exposition of the place-value system (p. 109) – and all fraction denominations in words. Comparison of its introduction of the rule of three with what we find in the *Livero de l'abbecho* shows close affinity between the two. Various Franco-Provençal terms and in particular a lapse into Occitan orthography [ed. Arrighi 1987: 96] shows that the treatise was written in Provence.

It is true that a few notaries from Genoa and Perugia used some Arabic numerals between the mid-twelfth- and the early thirteenth century for numbers that had no juridical importance [Langeli 2000] – but they are so much exceptions that they prove that the new numerals were *not* broadly known in the commercially active environment rather than the opposite.

The Contents of Jacopo's *Tractatus*

We shall return to the identification of the environment (or, the various environments) which inspired Fibonacci, the compiler of the *Livero de l'abbecho*, Jacopo, etc., but first we shall have a closer look at the contents of Jacopo's treatise.

Ch. 1. Incipit and General Introduction

The treatise begins with the Latin *incipit* that was quoted on p. 6, in which Jacopo introduces the topic and identifies himself as well as the place and the time of the composition. Apart from a forgotten word (*mediatio*) in **V**, it is shared by **V** and **M+F**. Among the 250–300 items catalogued in [Van Egmond 1980] that are mathematical and *might* have started in a similar way,[115] most have nothing, but many do contain an identification of author or copyist and place and time.

An opening in Latin followed by a vernacular text is, on the other hand, utterly rare (not counting short phrases like *Laus Deo*). Jacopo's opening is also found in a late ps.-Jacopo treatise,[116] while a brief phrase that *could* be an abbreviated distortion of the same opening is found in a treatise from before c. 1350.[117] Beyond that, I have noticed one Latin presentation of author, time and place from 1545, one simple Latin fifteenth-century incipit and three Latin dedications from the decades around 1500.[118]

---

[115] Many of the codices listed contain several works, not all of which are vernacular mathematical writings (the normal condition for inclusion of a codex being that at least *one* piece is both vernacular and mathematical). Some pieces are not mathematical at all (a "general medication" ascribed to Arnaldo di Villanova turns up time and again), some are Latin works (various writings from Fibonacci's hand, various translations of al-Khwārizmī's *Algebra*, etc.), and some are fragments or collections of scattered notes. All in all, some 150 items (out of c. 430) can be discarded on these accounts as being irrelevant for comparison.

[116] *Tratatus algorissimi* copied in 1513 by one Andrea di Salvestro di Piero Nardi (Florence, Palat. 1162, see [Van Egmond 1980: 130]), which however contains nothing but the *incipit*, the introduction, numeration, multiplication tables and a collection of miscellaneous problems.

[117] "Incipit libre algorissimi et dicitur a magistro de [blank space] Florentia", in a *Libre algorissimi* from the first half of the fourteenth century (Florence, Magl. Cl. XI, 73, see [Van Egmond 1980: 107]). It is just possible but rather unlikely that the master whose name the compiler did not remember was meant to be Jacopo.

[118] München, Bayerische Staatsbibliothek, Cod. ital. 335, "Anno 1545 factus est liber hic Venetiis a Bortholamio Rote ..."; Bergamo, Biblioteca Civica MA 220 (Bartholomeo de Porta); Vatican, Vat. Lat. 3129 (Luca Pacioli); Florence, Magl. CL. XI, 15; Oxford, Bodleian Canon. Ital. 197 [Van

Initial dedications to God and his Holy Mother are widespread in abbacus writings, alone or in combination with the presentation of author, time and place.[119] Long discursive or philosophical introductions like V.1.2–4, on the other hand, are rare; most of those that can be found, moreover, copy one of a few standard specimens. The most widespread of these is the one from Jacopo's hand; apart from **V**, **F** and **M**, it is present in no less than 10 items listed by Van Egmond, written between c. 1370 and c. 1513, 6 of which present themselves as *Algorism treatises*.[120]

Another standard introduction is the following:[121]

Concio sia cosa che tutti li animagli rationali viventi nella presente vita abino disposto per

---

Egmond 1980: 243, 47, 106, 247].

[119] Two instances of some interest are Paolo Gherardi's *Libro di Ragioni*, written in Montpellier in 1328 "al nome di Dio e della sua madre santissima e di tucta la corte celestiale [...] a honore di tuct'i buoni ragionieri", and a *Trattato di tutta l'arte dell'abbacho* (Avignon, c. 1334) written in honour of the Lord, his mother "vergine santa Maria et del beato Giovani Batista et di tutta la corte celestiale" and of Pope Benedict XII (**T**$_\text{F}$, fol. 17$^\text{r}$, with omission of John the Baptist once again on fol. 125$^\text{v}$); on the latter treatise, see note 144 and surrounding text. They are too different from Jacopo and from each other to indicate any direct influence, but the otherwise rare reference to "the whole celestial court" (beyond these I have observed eight instances in Van Egmond's total corpus, all but one postdating 1490) and the reference to the colleagues in two of them (Gherardi's "in honour of all good accountants", Jacopo's "with the assistance of our predecessors, and in honour of all masters and scholars of this science") suggest a local Provençal abbacus style.

The initial part of the dedication points more broadly toward Catalan or Catalan-Provençal vernacular culture. It is indeed quite similar in several details to Joanot Martorell's dedication from 1460 of *Tirant lo Blanch* "a honor, lahor e gloria de nostre senyor Déu Jesu Crist, e de la gloriosa sacratissima verge Maria, mare sua, senyora nostra" [ed. Aguiló y Fuster 1905: I, 3].

Introductions to fourteenth- and fifteenth-century writings from other parts of the Romance world also often contain religious dedications or (more often) invocations, but the ones I have inspected are definitely different in style.

[120]   Giovanni de Danti d'Arezzo, *Tractato del'algorisimo*, c. 1370 (Florence, Plut. 3026);

–   *Libro delle ragioni d'abacho*, c. 1395 (Florence, Conv. sopp. G: 7. 1137);

–   Bernardino de Faliva, *Opera de far de rasone*, early 15th c. (Bologna, MS 2780);

–   Tomaso de Jachomo Lione, *Libro da razioni*, 1430 (Vatican, Vat. lat. 4825);

–   *Tractato d'alghorissimus*, c. 1473 (Florence, Magl. XI. 75/I). Contains only a counterpart of V.1.4;

–   *Tractado algorismo*, c. 1475 (Florence, Ash. 1168);

–   Giovanni di Lucha, *Libro sopra arismetricha*, 1478 (New York, Plimpton 195/I);

–   Pietro Paolo Muscharello, *Algorismus* 1478 (private possession, ed. [Chiarini et al 1972]);

–   *Tractatus algorisimus*, c. 1485 (Florence, Ricc. 2991);

–   Ps.-Jacopo, *Tratatus algorissimi*, 1513 (Florence, Palat. 1162).

[121] "Admittedly, all the rational animals living in the present life are disposed by nature to wish to distribute their life in several different activities [...]". I quote the *Arte d'arismetrica* (c. 1452; Florence, Ricc. 2369) from [Van Egmond 1980: 155]. The other manuscripts present us with minor variations.

l'ordinamento della natura di volere distribuire la loro vita in alcuni exercitri diversi [...]

which is found in eight treatises written between 1452 and 1571. Another group of eight writings written from between c. 1428 and c. 1463 begin with a philosophical reference:[122]

Tutti gli uomini secondo che dicie Aristotile nel principio della metafisica naturalement desiderano di sapere. [...]

None of the 16 treatises carrying one or the other of these alternative introductions characterizes itself as an *Algorism*; the large majority refer to the notion of *Arismetrica*. This link between title and choice of introduction can hardly avoid being statistically significant – but the significance may hinge simply on the repeated occurrence of the word "algorism" in the Jacopo-introduction.

Jacopo's introduction starts by a high-flown praise of knowledge – emphasizing also its ongoing increase much more strongly than Bernard of Chartres' oft-quoted dictum about the dwarfs on the shoulders of giants. His witness as to the importance of knowledge is not Aristotle but Solomo,[123] but in spite of the predominantly pious tenor the exposition betrays influence from the Aristotelian mood of the scholarly culture of the time – namely in the distinction between *senno*[124] which is *natural* and *scienza* which is *accidental*. The proper Aristotelian distinction would seem to be between that which is *by nature* and that which is *by art*, but at the time it was not unusual to conflate this dichotomy with that between the *necessary* and the *accidental*.[125]

This, and the Latin start, are not the only suggestions of influence from scholarly culture. As mentioned in note 54, the listing of the nine species of algorism in the *incipit* has little to do with what follows in the treatise. It belongs with that genre that originally carried the name *algorism*, the Latin introductions to the computation with Arabic numerals, and comes *verbatim* from Sacrobosco's *Algorismus vulgaris* [ed. Pedersen 1983:

---

[122] "All men, according to what Aristotle says in the beginning of the *Metaphysics*, naturally want to know". I quote the precise wording and the spelling of *Ragioni apartenente all'arismetricha* (Florence, Magl. Cl. XI, 119) from [Van Egmond 1980: 119].

[123] The story that is told is based on but far from identical with the one which (with slight variation) is told in 1 Kgs. 3:5–14 and 2 Chr. 1:7–12. I have not been able to find Jacopo's story in the Bible, including the "Wisdom" books that were eliminated by Luther, nor in the Qur'ān or in the Latin Fathers; I suspect its origin to be in the pious vernacular literature.

[124] In 2 Chr. 1:11–12, Solomo as well as the Lord speak of "wisdom and knowledge"; however, since *senno* is a *capacity*, the explanation in 1 Kgs. 3:11, "understanding to discern judgment" is a better guide to what Jacopo intends. For brevity, I translate "discernment".

[125] The same conflation makes Albert the Great state in *De mineralibus* I.i.4 [ed., trans. Wyckoff 1967: 20] that some alchemists claim minerals to come to be *by accident*, explaining afterwards that what they really say is that the processes that engender them are identical with the artificial production of bricks.

174f].[126] Sacrobosco is also the probable source for the information that the root of this art is Arabic, and that this is the reason why the numbers are written from right to left (beginning with the order of units) and for the idea that *algo* stands for *art* and *rismus* for *number* – but not for claiming that these etymologies are Arabic (evidently Jacopo or some direct source of his extrapolated from the fact that the art itself had this origin). The reference to Boethius in V.1.3 is pure decoration and not to be traced to any Latin source, since the five chapters into which Boethius is said to have divided his *Arithmetic* are those of an abbacus treatise.

The passage "El dividere[127] si è in rocti sani e rocti in rocti. Sonno moltiplicare, dividere, giungere, sobtrare, e dire quale è più l'uno rocto che l'altro, overo quanto meno"[128] looks strange. The beginning of the second period "they are", obviously referring to fractions, suggests that the subject of the preceding period might be *rocti/* "fractions" – that is, that a passage has dropped out, perhaps something like "El dividere ⟨si è in uno o in doy o in più figure. I rocti⟩ si è in rocti sani e rocti in rocti".[129]

"Fractions in fractions" is no familiar concept from other abbacus writings (how unfamiliar we shall see imminently). The most obvious association is to the composite fractions known from Arabic mathematics, of the type "one half of one seventh". A Castilian *Libro de arismética que es dicho alguarismo* from 1393 is an even better clue. A whole sequence of its problems (ed. Caunedo del Potro, in [Caunedo del Potro & Córdoba de la Llave 2000: 182–186, 190]) treat of the multiplication or division by "sano e roto e roto de roto" ("integer and fraction and fraction of fraction"), for instance "5 quintales e 3 arrovas e 6 libras" ("5 quintals and 3 *arrobas* and 6 pounds"), 1 *arroba* being $\frac{1}{4}$ of a quintal and 1 pound being $\frac{1}{25}$ of an *arroba*. "Fractions and fractions of fractions" thus coincide with the ascending continued fractions (see p. 39) which Fibonacci had borrowed from Maghreb mathematics, applying them to structures of sub- and sub-sub-units. Since fractions of fractions turn up nowhere in the following, it is obvious that this passage is taken over from elsewhere, just as the list of the nine species.

In **F** we simply find "Il dividere si è di numeri sani e rotti. I rotti sono multiplicare,

---

[126] Not necessarily directly, to be sure. But Leonardo da Pistoia's roughly contemporary *Arithmetrica* [ed. Arrighi 1977: 20], though in Latin and rooted in the scholarly tradition, is less precise when copying the same passage.

[127] The term *divixio* in the Latin incipit seems to cause the appearance of *dividere* instead of the regular *partire*. This happens thrice in 1.3 and in no other place in the manuscript. Elsewhere *dividere* appears only as bisection, namely in the locution *dividere per mezzo*. In 10.1 there is a single instance of the noun *divisione*. All of this also holds for **M+F**.

[128] "Division falls in integer fractions and fractions in fractions. They are to multiply, to divide, to join, to subtract, and to say which fraction is greater that the other, or how much smaller".

[129] "Division ⟨is in one or in two or in more figures. Fractions⟩ fall in integer fractions and fractions in fractions".

dividere, ...";[130] **M** has "Il dividere si è i numeri sanny et roti. I rotti sono multipricare, dividere, ...".[131] None of these are very meaningful.

In Giovanni de' Danti d'Arezzo's *Tractato de algorisimo* [ed. Arrighi 1985: 10] from 1370 , which copies Jacopo's whole introduction wholesale, the same passage becomes

> E i quagli capitoli ⟨e⟩ spetie àno i'lloro molti divisioni e molti membri sì come è de multipricare d'una o di due o di tre o di quactro e d'infinite figure e del partire dei numeri sani e rocti. E i rocti ànno i'lloro per sè regola cioè di multipricare, dividere,[132] giongnere et soctrare e dire quanto è più o meno 'uno che l'altro vedendoli figurati,

in translation

> And these chapters ⟨and⟩ species have many subdivisions and many sections, such as multiplying by one or two or three or four (digits) or digits without limit, and the division of whole numbers and fractions. And the fractions have rule(s) by themselves, that is, to multiply, to divide, to join and to subtract and say how much one is more or less than the other seeing them in figures.

Neither from this nor from the formulations of **F** or **M** would it be possible to get the idea of fractions of fractions if one did not know it from elsewhere. Instead, Giovanni de' Danti and the compiler of **M+F** must have tried to make sense of an enigmatic and probably also corrupt passage in their original (enigmatic indeed for anyone who had not heard of *rocti in rocti*, which we may then conclude they had not).

Except for this passage, **M+F** contains the whole of V.1.2–4 with minor deviations in the wording.

Ch. 2. Introduction of the Numerals and the Role of Zero

In principle, section V.2.1 (common to **V** and **M+F**) already belongs to the Latin algorism genre as this genre had looked since its very beginning with *Dixit algorismus* [ed. Folkerts 1997: 28] (which does not necessarily imply a borrowing – it is difficult to present the use of Arabic numerals without telling an audience that is supposed to be unfamiliar with them how they look). But the presentation of the old and the new way of writing them has no parallel in the Latin algorisms. The "old art" seems to be of Iberian origin: the writing of 4 as X is indeed an Iberian modification of the Roman numeral; for the rest, the shapes are derived from the "Western" Arabic form. The 4 of the "new art" is a simplification of the "Western" form, and the new 7 also comes from there (without simplification); for the rest, "old" and "new art" do not differ.[133] The habit

---

[130] "Division is of whole numbers and fractions. Fractions are multiplying, dividing, ...".

[131] "Division is the whole numbers and fractions. Fractions are multiplying, dividing, ...".

[132] Giovanni de' Danti gets rid of the first two instances of the atypical *dividere* from **V** (and **M+F**), but forgets this one. His reliance on Jacopo and not just of a common unknown source is obvious.

[133] Both the "old" and the "new art" agree with "Western" numerals as these look in Arabic rather

of presenting together an old and a new style for the numerals goes back at least to the Maghreb/al-Andalus mathematician Ibn al-Yāsamīn from c. 1200, who confronts the "Eastern" and the "Western" style [Burnett 2002: 269 pl. 1; Kunitzsch 2005: 24, 17]. Quite new is apparently the inclusion of 10.[134]

The same combination of tradition and innovation is found in V.2.2. The happy formulation that zero does not signify in itself but "has the power to make signify" is borrowed from Sacrobosco [ed. Pedersen 1983: 176], as it was borrowed by many other writers on the subject.[135] But the observation that this does not happen "always but according to where it is put, either before or behind" (absent from **M+F**, which breaks off after "make signify") appears to be Jacopo's own contribution (the talkative style of the explanation fits what he presents us with elsewhere).

We notice an early instance of the word *zero* (if **V** follows the original on this account; **F** and **M** have the spelling *zevero*) and the information that the treatise is meant to serve (also) for self-tuition.

Ch. 3. Tabulated Writing of Numbers

The pedagogical examples of numbers written "by figures, and also by letters" – that is, by Arabic as well as Roman numerals – which are promised in V.2.2 follow in Chapter V.3. **V** lists all integers from 1 through 50 (apart from 47 and 48, which were probably skipped during copying by error), then continues by tens to 100, by hundreds to 1000, etc. until 1000000 – repeating 500 and 600 after a change of page, probably because the original was organized with different line breaks, omitting 2000 through 5000, probably a whole line in the original, and apparently forgetting 600000. From 500 onward the Roman writing becomes multiplicative, the "denominator" being written above the "numer-

---

than in scientific Latin manuscripts – see the tabulations in [Burnett 2002: 265*f*]. Only a early fourteenth-century Latin manuscript of the *Liber abbaci* (Florence, Biblioteca Nazionale Centrale, Conv. Sopp. C.1.2626) comes fairly close to Jacopo's "new art". Whether this means that its numerals rendered those used by Fibonacci himself or simply that the copyist was rooted in the abbacus environment is difficult to decide.

[134] Originally, only nine figures had been listed, zero being explained later. Sacrobosco, followed by the Italian Dominican Leonardo da Pistoia [ed. Arrighi 1977: 20] has a list of ten symbols, but ending with 0 and not with 10; so do both lists in **F**, whereas **M** ends the old art with 10 (written 01) and the new with 0.

Without explanation, the two sequences of numerals also turn up in the *Trattato di tutta l'arte dell'abacho*, written in Avignon in 1334 (see note 144). In the author's own draft manuscript (**T**$_F$, fol. 23^r) the numerals in both sequences have almost the same shape as in **F**.

[135] It is also in one of the four manuscripts on which Steele [1922: 72] based his edition of Alexandre de Villedieu's *Carmen de algorismo*; its absence from the others suggests that it was inserted under inspiration from Sacrobosco.

ator" (500 thus appearing as $\underset{v}{\varsigma}$). At 500000 this system becomes problematic, and instead of $\overset{m}{\underset{v}{\varsigma}}$ we simply get ς. In the next steps "v" stands for 500, as can be seen from the writing of 700000 as $\underset{vcc}{m}$. The numbers are written in increasing order from right to left, that is, in the Arabic way.

In **M+F**, the counting by tens starts after 30, there are no repetitions or omissions, and the line breaks are different. In the end, the numbers 234, 345651121 and 135512 are listed together with Roman equivalents (using again the multiplicative writings); the numbers 500000, 600000, 700000, 800000 and 900000 are written with "numerators" d, dc, dcc, dccc and dcccc. Apart from that, everything is identical with what we find in **V**, including the right-to-left orientation – except that **M**, like **V**, forgets 600000. This is therefore likely to be the way of **V′**, the common archetype for all three manuscripts (see the stemma on p. 145), and the way of **F** the outcome of a repair.

Ch. 4. Explanation and Exemplification of the Place-value Principle

Comparing Chapter V.4 with the corresponding section in the Latin introductions to the place value system, from *Dixit algorismus* to Leonardo di Pistoia, we notice that these offer little or nothing beyond the contents of sections V.4.1–2. This is one reason to guess that sections V.4.3–6 represent Jacopo's independent elaboration. Another reason for this assumption is the shift in what is considered the "first" digit – section 2 reads "in the forward way", counting 2 as "first" in 21, whereas sections 3–6 sees it to be placed in "the second rank". A third reason is that Jacopo goes wrong when trying to explain the insertion of points to distinguish successive levels of thousands after the level of millions, mixing up inclusive and exclusive counting. The effect is seen in section V.4.6, when 23456789.8765432 is understood as 2345,000,000,000+6789,000,000+8765,000+432.[136] The analogous example in V.4.5 is dealt with correctly.[137]

Just as interesting is a fairly deep observation made in section V.4.7: That the displacement of a number one place to the left corresponds to a decupling, *irrespective of whether this number has one or more digits* – that is, that 240 can be read as "24 in the ten's place". This is the great advantage of "pure" place value system as against "mixed" systems like the Babylonian so-called sexagesimal system, which was rather treated as a mixed decimal-seximal system, with levels 10, 60, 600, 3600.[138] The latter

---

[136] **M** has the same interpretation, whereas **F** replaces *oto* by *otanta*, which gives the interpretation 2345,000,000,000+6789,000,000+80765,000+432.

[137] Since **M+F** has 987654321 and a corresponding interpretation in words, one must presume that the original had this number, which a copying error changed into 987644321 in the branch of the stemma leading toward **V**, and that another copyist (or the same, after having written a wrong number) changed the words correspondingly.

[138] See [Høyrup 2002a: 15 n. 19] for an analysis of revelatory errors of Old Babylonian date; further,

do not permit computation by means of uniform place-independent algorithms[139] –
in particular by those paper-based algorithms that seem to have been created by the abbacus
masters of the fourteenth century [Van Egmond 1986: 56–60; 2001]. In spite of the
importance of this characteristic of the pure system I have never noticed that it was pointed
out in any other discussion of the place value system.

The point is not made in **M+F**, which instead repeats preceding matters in slightly
different and fewer words – as if its compiler rewrote something which he did not grasp.

Ch. 5. Introduction to the Multiplication Tables

This small introduction, wholly in the intimate-talkative style of V.4.3-7, is absent
from **M+F**. From the mathematical argument it is clear that *raccoglere*, "to aggregate",
must refer to the addition of *fractions*.

Ch. 6. Multiplication Tables, Including Multiples of Soldi

The contents of this chapter speaks for itself. Noteworthy in the *librettine magiori*,
the "major booklets", is that all multiples until 20×20 are listed (the squares being omitted,
they come in Chapter 7), together with multiples until 10 of the primes until 47 (the
purpose of which is revealed in Chapter 8) and of products $p \times q$ soldi, $10 < p < q \leq 20$ *soldi* =
1 *libra*. The "simple higher squares" obtained from 2×2, ..., 10×10 by means of
decuplation[140] appear to belong with the *librectine magiori*, since they are written
in direct continuation within the same scheme.[141] Noteworthy in the multiplication
of *soldi* is that 0 appears a few times (but not systematically) as the *number nothing*
(15×16 ß = 12 £ 0 ß, etc.).[142] This function is not implied by its use as a digit that
does not signify in itself but "has the power to make signify", but only the two functions
together allow us to attribute to Jacopo the "concept zero" (that is, our concept). Whether
the incipient innovation is due to Jacopo or to an inadvertency of a later copyist cannot
be decided at this point of the manuscript; however, in the casting out nines (fols 9ᵛ–11ʳ,

---

four of the five presumed errors of "type II" in the Seleucid text AO 6456 listed in [Proust 2000:
299] are indeed not errors but indications of empty decimal places (as recognized by Christine Proust
but not discussed in the article).

[139] Just try to add two lengths consisting of yards, feet, inches and lines!

[140] The sequence 200×200, ..., 1000×1000, at first omitted in the proper place on fol. 9ʳ, comes on
9ᵛ.

[141] Possibly because this repaired (copyist's?) omission takes up one column in the upper scheme
of fol. 9ᵛ, the squares on half-integers is split into two, 1½ through 7½ being shown on fol. 9ᵛ,
8½ through 19½ following only on fol. 12ᵛ.

[142] £ renders the abbreviation for *libra*, ß that for *soldo*. Further on, δ shall render that for *denaro*.

also in **M+F**), the use of 0 in this function is indubitable.

The graphical organization of the corresponding tables is less clear in **M+F** than in **V**, making no distinction between major columns by means of empty space. As regards the contents, **M+F** contains the *librectine minori*, the "minor booklets", skipping $1 \times 1 =$ 1 and going thus from $2 \times 2 = 4$ until $10 \times 10 = 100$. The introduction to the *librettine magiori* is absent, and these are arranged in a different manner – starting with the table of simple higher squares $20 \times 20 = 400, ..., 100 \times 100 = 10000$ and what else can be obtained from the *librectine minori* by simple decuplation. Only then come the multiples of the number 11 through 19. The systematic listing of the multiples of primes until 47 is absent, as are the multiples of *soldi*.

Ch. 7. Tables of Higher Squares

The squarings in this chapter are organized in a different scheme, one factor being written on top of the other within a box, and the product above the box – at times with a symbol × between the factors, standing for the cross-multiplication of integers and fractions respectively of ones and tens (it is large and connects these). First the squares on the half-integers from 1½ through 7½ are given (2½ being forgotten), then the squares on all integers from 11 to 100 are listed. For each of the latter, the remainder of the product modulo 9 is indicated in the lower left corner (in a few cases enclosed within a curved line); since the casting out of nines is never explained in the treatise this could have been used by the author for control (or by somebody from whom Jacopo copied his schemes). Next, the squares $110 \times 110, 120 \times 120, ..., 990 \times 990$ are listed, without any indication of remainders (they were probably not calculated directly but from 11×11 etc.). In the end come the squares on the remaining half-integers until 20.

In **M+F**, we remember, higher squares obtained by decuplation preceded the multiples of the numbers 11 through 20. After these follow:

(a)   the squares 11×11 through 99×99 (17×17 being forgotten);

(b)   products $11 \times 22, ... 88 \times 99$ ($n \times (n+11)$);

(c)   products $75 \times 675, 76 \times 676, 77 \times 677, ..., 79 \times 679, 80 \times 780, ..., 89 \times 789, 90 \times 890, ...,$
        $99 + 899, 99 \times 999$;

(d)   squares $453 \times 453, ..., 459 \times 459, 560 \times 560, ..., 569 \times 569, 670 \times 670, ..., 679 \times 679,$
        $780 \times 780, ..., 789 \times 789, 890 \times 890, ..., 899 \times 899, 999 \times 999$;

(e)   squares $453 \times 453, ..., 469 \times 469, 670 \times 670, ..., 679 \times 679, 780 \times 780, ..., 789 \times 789,$
        $890 \times 890, ..., 899 \times 899, 999 \times 999$;

(d)   56 products of the type $n \times (n+111)$, between 343×464 and 780×891;

(g)   35 products produced according to other rules, $456 \times 4567, 457 \times 4568, 458 \times 4569,$
        $...786 \times 7897$;

(h)   20 scattered squares, between 6565×6565 and 8784×8784;

(i)   9 products $9 \times p$, p representing scattered numbers between 4669 and 67895;

(j)   42 squares on and products of broken numbers, starting with $11\frac{1}{2} \times 11\frac{1}{2}$ but
         containing much more difficult computations like $78\frac{1}{9} \times 89\frac{1}{8}$;
(k) $987654321 \times 987654321$ and 9×987654321.

Everything in (a)–(i) and (j) is written in a scheme similar to that of V.7, every time with indication of the remainder mod 9 (but with the product written inside the box and the remainder always enclosed). Apart from the initial squares 11×11 through 99×99, the accidental picking of what to calculate suggests the purpose to be mere training; the rules generating the factors are seemingly meant as a help for the invention of new examples. (j) uses a different scheme, explained on p. 403. We observe that it makes use of the casting out of sevens, more tedious than casting out nines; even this method is never explained in the treatise. The sub-scheme for multiplication of the numerators (247×375 in the example) is similar to that used in V.7, the result being written above the line.

In contrast to the scheme containing the *librectine*, all these schemes in **M+F** are arranged from right to left, thus once again in the Arabic way. The mixed numbers are also written with the fraction to the left, in the Arabic or Fibonacci way.[143] This calls for comparison with the *Trattato di tutta l'arte dell'abacho* (written in Avignon in c. 1334, as shown by Jean Cassinet [2001]).[144] This manuscript lists the squarings 11×11

---

[143] However, Fibonacci's *schemes* are all arranged left-to-right, at least in the manuscript on which Boncompagni's edition was based [ed. Boncompagni 1857a: 25*f*, 37, 54*f*]; it thus seems safe to disregard the possibility of inspiration from Fibonacci.

[144] I have consulted two early manuscripts (datings based on watermarks): Rome, Biblioteca dell'Accademia Nazionale dei Lincei, Cors. 1875 (henceforth T_R); and Florence, Biblioteca Nazionale Centrale, fond. prin. II,IX.57 (henceforth T_F). For the latter, I refer to the most recent foliation.

According to Van Egmond [1980: 140], T_F appears to be the author's own draft treatise, thus from c. 1334. In his words, this manuscript "is now considerably disordered but by paper and original numeration can be reorganized into the following original components":

    I.   *Trattato di tutta l'arte dell'abacho* (fols 17–26, 1–12, 27–81, 94–117, 82–93, 118–141).
    II.   Medical matters, same hand and format as I.
    III.   Miscellaneous problems and notes.
            1. Problems in the same hand and format as I (fols 157–168).
            2–6. Problems and other matters in different hands and/or on different paper (fol. 169–181).
    IV.   Astrological matters.

Since even watermarks show continuity between I and III.1 it seems safe to count III.1 as part of the *Trattato*, at least in the sense that it bears witness of the same intentions, environment and sources.

T_R may have been written a few years later but no more. It also contains many of the miscellaneous problems of III.1.

Van Egmond [1977: 19; 1980: 140, 179] ascribes the treatise to Paolo dell'Abbaco, but the only reason for doing so appears to be a guess by a fifteenth-century owner of a manuscript from c. 1340 (Florence, Ricc. 2511) seemingly based on the insertion of the "Regholuzze di Maestro Pagholo astrolagho" within the treatise (which is rather a proof that the one who inserted it, that is, the author of the treatise, was *not* that same Paolo). No more weight carries an ascription of

through 100×100 on $\mathbf{T_R}$ fols 5^v–6^r, $\mathbf{T_F}$ fols 2^v–3^v, written as in $\mathbf{V}$ and controlled (in $\mathbf{T_R}$ only until 46×46) by the casting out of nines, followed on fol. 6^r–8^r (as in $\mathbf{M+F}$) by higher products in the style of $\mathbf{M+F}$ but only initially the same.[145] With the exception of $\mathbf{T_R}$, fol. 5^v–6^r, these squares and products are arranged right-to-left – though with repeated errors in $\mathbf{T_R}$ showing that its scribe was not accustomed to working like this.

Also to be taken note of is the way mixed numbers are multiplied in this section of $\mathbf{M+F}$, *viz* by bringing "to fractions both parts". This is the method explained in V.11.14, whereas M.11.14=F.iv.d tells the reader to "multiply the [integer part of] the smaller number by the whole of the other number, and then the fraction of the smaller by the whole of the other number". Together with the similarity with the *Trattato di tutta l'arte dell'abacho* this suggests, firstly, that the compiler of $\mathbf{M+F}$ took this material wholesale and inserted it in the treatise without editing it or thinking about it, and secondly that this compiler, if not working in Provence, used material which was produced there.[146]

Ch. 8. Divisions *a regolo*

The opening of this chapter announces that it is to "teach to divide in the numbers that are most necessary". This it does only in an indirect way, by showing the outcome of a large number of divisions but not how they are obtained – neither by showing an algorithm (*a danda*, *a galera*, or whatever it might be) nor in any other way. The method, indeed, is what according to Luca Pacioli [1494: 32$^{r–v}$] is commonly called division "a regolo: o a tavoletta", which in his times was used in Florence for all divisors contained in the *librettine*, and in which all remainders that are carried to the subsequent level are kept in mind and not written. The Pisa curriculum (see p. 28) also explains [ed. Arrighi

---

a fragment of *Trattato di tutta l'arte* inserted after the same *Regoluzze* to the "libro di maestro Pagolo" in a manuscript from c. 1465 (Florence, Ricc. 1169); see [Van Egmond 1980: 146, 158; Cassinet 2001: 107]. The unusual inclusion of John the Baptist in the dedication (see note 119) might make us suspect that the author's own name was Giovanni; that he was a member of a *confraternità* dedicated to the Baptist; or that he felt personally obliged toward John in some other way.

[145] 11×22, 12×23, ..., 88×99; 66×266, 67×267, ..., 90×290; 91×391, ..., 99×399; 55×366, ..., 79×390; 55×366, ..., 79×390; 80×491, ..., 88×499; 231×231, ..., 290×290; 390×390, 351×351, ..., 390×390; 491×491, ..., 499×499; 599×599; 260×371, ..., 264×375; ...; then a number of products in the style of 572×4572, of 681×6792, and of 684×6295; finally some even larger products, ending with 123456×123456; 234567×234567; 703480×87935, 8888877×9876535 – the last two without indication of the result.

In the first 6 of these, $\mathbf{T_R}$ again gives what should be remainders; but it copies badly, all are wrong ($\mathbf{T_F}$ is correct).

[146] The stylistic differences between the schemes used for higher squares in $\mathbf{V}$ and $\mathbf{M+F}$ speaks against the hypothesis that what we find in the latter version alone should have been in Jacopo's original and then have been left out in the copying process leading toward $\mathbf{V}$.

1967b: 122] that divisions by 2, ..., 19, 23 and 29 are taught, and afterwards 10 divisions *per reghulo* of 30-digit numbers are made – presumably linked in the same way, even though this is not explained.

The examples that are offered in **V** are divisions of numbers with 15, 16 or 17 digits. For each of the divisors, a sequence of ten divisions is given. The divisors are all integers from 2 through 19, and all primes from 23 through 47 – which are then presumably those "that are most necessary"; and finally 48, filling out the last page. In each sequence, the first dividend contains the digits 12345678910, preceded by 4 more digits. The next dividend consists of the 15 digits of the outcome of this division (with initial zeroes, if necessary), preceded by the remainder it has left. Pacioli [1494: 32^v] explains the use and purpose of the seemingly strange procedure in the teaching of division *a regolo*: "When we teach this, we put a number with many digits on the tablet, and we tell the students to divide it by 2, then by 3, then by 4, etc. for all". Further, "in order that they do not lack division work", we ask them to "divide the first outcome. And this again we ask them to divide. Always with the same divisor. [...] And if anything remains we ask them to put it in front of this outcome, as we do it,[147] in order that they do not run out of numbers, as they would soon do if remainders were not picked up".

The calculations contain a fair number of errors. I must confess, however, that I made many more myself when trying to check the calculations manually, and only the access to a 32-digit electronic calculator allowed me to perform the controls and interpret the errors. Some of these, as could be expected, are mere copying errors, either made by the original author himself (whether Jacopo or somebody he borrows from) when transferring results from the medium of computation or by a later copyist; others, however, are genuine miscalculations.

Such sequential divisions are rare in the abbacus treatises, but Pacioli's words tell us that they reflect a common teaching practice. The idea that primes belong to the group of divisors "that are most necessary" corresponds to a division by composite numbers through successive division by prime factors. Jacopo does not say so; nor does anything else in the treatise suggest that he used the method, but perhaps (as suggested above, note 54) for the simple reason that he never explains how numerical operations are performed.[148] However, division by way of factorization is described by a variety of authors, from Fibonacci [ed. Boncompagni 1857a: 36*f*] and al-Qalaṣādī [ed. Souissi 1988: 42] to Pacioli [1494: 33^{r-v}]; it is also used in *Trattato di tutta l'arte dell'abbacho*.[149]

---

[147] In the margin, Pacioli gives five consecutive divisions of 234567 by 4, and six to seven of 987654321 by 5, 8 and 12.

[148] With one apparent exception: in 14.23, a division by 600 is said to be done as (successive) division by 20 and by 30 – but the subsequent explanation shows that what is meant is that the successive divisions by 20 and 30 (caused by metrology) amount to a division by 600.

[149] This treatise has sequential divisions **T$_R$** fols 9^r–10^r, **T$_F$** fols 9^v–12^v, the sequences containing

We may legitimately suppose that even this was a common practice.

M+F omits the initial explanatory passage. It brings the same type of divisions in sequence, but starting with dividends of 6–8 digits, with divisors 1 through 19 (omitting 10), and performing for each only 10 to 12 divisions. Instead of the sequential divisions by higher prime quotients it shows the division of 987654321 by 23, 29, 37, 43, 53, 79, 83, 97, 131, 311, 1234 and 9011 (1234 = 2×717, 717 and all the other divisors are prime) *a danda* (without giving this name); this method (a paper version of the delete-and-replace algorithms used on the dust board, see the explanation below, p. 412) is also taught, for instance, by Giovanni de' Danti d'Arezzo [ed. Arrighi 1985: 19–21] and in *Trattato di tutta l'arte dell'abacho*,[150] which means that the tables of multiples of primes above 20 (absent from **M+F**) are indeed not needed, as they are for the divisions *a regolo* which we find in **V**. In all examples, a control by means of the casting out of sevens is shown.[151] Since the way the remainders mod 7 are indicated is graphically identical with the one used for higher squares and products, one may assume the source for higher products and higher divisions to be the same – not least because the principle itself was rarely used.[152] In the whole section of **M+F** on divisions, however, tables are systematically arranged left-to-right (so they are in **T_F**, in the rare places where this manuscript has divisions in two columns).

---

8 to 12 divisions of numbers with 15 to 20 digits, with divisors 2, 3, 4, ..., 19, 23 and 29. Subsequent examples of division *a reghola* (as they are called here) make use of factorization when dividing by large numbers; thus, the division 7651 £ 12 ß 2 δ by 168 is performed as a division by 12 followed by a division by 14.

[150] The divisions here thus do not follow the "galley" method, *pace* Jean Cassinet [2001: 110]. Actually, the manuscript itself (**T_F** fol. 29ʳ) states that the divisions are "a'ddanda".

  Giovanni de' Danti shows the division of 987654321 by 23, 29, 37, 43, 131 and 713 (713 = 23×31, the other divisors are prime). In the third division there is a slight error in the numbers representing the last intermediate calculations ("02" instead of "20"), in the fourth these numbers are wholly wrong from a certain point onward. Giovanni appears to have calculated himself, miscalculated, and then have copied the results from somebody else (a nice example to his students!). But even though he borrows Jacopo's introduction, and some of the examples coincide with what we find in **M+F**, his source for the divisions is hardly an early representative of that version since other examples differ (prime divisors and sequences like 987654321 are pervasive in the abbacus material).

[151] Better, the control is prepared but not made. The principle is that divisor×quotient+remainder is equivalent to the dividend mod 7. The dividend being always the same (and equivalent to 3 mod 7), the text gives the remainders of the divisor and the quotient mod 7 and the product of these (again mod 7). Since the remainder of the division is also indicated, the control is straightforward, but it is never shown (nor is there any explanation of the numbers in question, they are just there). The last result is wrong, as would have been revealed if the check had been performed.

[152] It is not only used but also explained in Giovanni de' Danti d'Arezzo's *Tractato de l'algorisimo* [ed. Arrighi 1985: 20], but the notation is somewhat different.

Ch. 9. Graphic Schemes Illustrating the Arithmetic of Fractions

On the whole, these schemes speak for themselves. Noteworthy is the distinction between *comparison* (seeing which of two is greater) and *subtraction*. Indeed, in an arithmetic not dealing with negatives a subtraction only gives meaning if the subtrahend is the smaller number.

The third example of multiplication and the second of addition are absent from **M+F**. In return, a reversed comparison (how much one broken number is *less* than another one, exemplified by $\frac{1}{3}$ and $\frac{9}{10}$) is added in the end. In the addition of $\frac{7}{8}$ and $\frac{11}{12}$, the result is given in partially reduced form ($1\frac{38}{48}$).

Ch. 10. Examples Explaining the Arithmetic of Fractions

This chapter explains in words the methods that were set out in diagrams in Chapter 9 (omitting division, even though multiplication is introduced by words that suggest division); everything is still made on the basis of paradigmatic examples. Mathematically we notice that the common denominator is explained in the idiom of the single false position ("$\frac{1}{2}$ and $\frac{1}{3}$ one finds in six", V.10.2). This implies that the least common denominator is not always looked for.

There are some traces that Jacopo used a model which he tinkered with, at times not quite consistently (which of course is the condition that the use of a model can be detected): V.10.9 is identical with V.10.4 apart from its choice of common denominator – but V.10.10, using the phrase "this number is again 24", is formulated as if it stood immediately after V.10.4.

This inconsistency is absent from **M+F**, which also contains a few extra examples and sometimes have things in a slightly different order.

Ch. 11. The Rule of Three, with Examples

We have now arrived at the realm of concrete number, and at the point where many abbacus treatises start – for instance the Umbrian *Livero de l'abbecho* from not much later than 1290, the *Liber habaci* written in Provence in c. 1310 and Paolo Gherardi's *Libro di ragioni* from 1328 (also written in Provence), all of which were mentioned above and which together with Jacopo's treatise represent most of the extant full abbacus writings from before 1330 (the Columbia Algorism is an exception). So do also two treatises we shall encounter repeatedly below: a *Libro di molte ragioni d'abaco*, a conglomerate from Lucca from c. 1330,[153] and a *Libro di conti e mercatanzie* probably written around

---

[153] Biblioteca Statale di Lucca, MS 1754, ed. [Arrighi 1973].

1395 (see note 42).

Before discussing how the topic is dealt with by Jacopo, some general observations concerning the rule may be appropriate. Notwithstanding the way the matter is often handled in the historiography of mathematics, the rule of three should not be confounded with the treatment of problems of proportionality in general, and in particular not with their treatment in the Near Eastern Bronze Age cultures.

Jacopo's presentation of the rule in V.11.2 is quite orthodox in an abbacus context:

> If some computation should be given to us in which three things were proposed, then we should always multiply the thing that we want to know against that which is not similar, and divide in the other thing, that is, in the other that remains.

So is the ensuing example:

> VII *tornesi* are worth VIIII *parigini*. Say me, how much will 20 *tornesi* be worth. Do thus, the thing that you want to know is that which 20 *tornesi* will be worth. And the not similar (thing) is that which VII *tornesi* are worth, that is, they are worth 9 *parigini*. And therefore we should multiply 9 *parigini* times 20, they make 180 *parigini*, and divide in 7, which is the third thing. Divide 180, from which results 25 and $\frac{5}{7}$. And 25 *parigini* and $\frac{5}{7}$ will 20 *tornesi* be worth.

The reason that this works is not easily explained without paper to somebody who is not too well trained in mathematics. The reason for the difficulty is that the intermediate result 9 *parigini* × 20 *tornesi* has no concrete interpretation.

Babylonian, Egyptian and ancient Greek calculators would have proceeded differently. Their normal procedure would have been to divide first (by whatever method they would use for division) 9 *parigini* by 7 *tornesi*. The result has an obvious concrete interpretation, the value of 1 *tornese* in *parigini*. Next, this could be multiplied by 20 in order to find the value of 20 *tornesi*.

In [1716: 867], Christian Wolff observed in his *Mathematisches Lexikon*:

> It is true that performing mathematics can be learned without reasoning mathematics; but then one remains blind in all affairs, achieves nothing with suitable precision and in the best way, at times it may occur that one does not find one's way at all. Not to mention that it is easy to forget what one has learned, and that that which one has forgotten is not so easily retrieved, because everything depends only on memory.

– in other words, only procedures that are performed so often that one runs no risk of forgetting them (like changing gears in a car) can be safely taught as mere skills. Probably the scribes of Near Eastern Antiquity did not perform the kind of proportional operations we are speaking of here so often that the appeal to their understanding could be given up safely.

While being pedagogically inefficient, the rule of three has computational advantages, as inherent in the above phrase "by whatever method ...". Division is difficult, and often leads to rounding (either for reasons of principle, namely if one has to multiply by a reciprocal, or because it may lead to a very unhandy string of aliquot parts). Subsequent multiplication will also lead to multiplication of the rounding error, quite apart from the practical difficulty of multiplying an inconveniently composite numerical expression. Better therefore postpone the division and make it the last step.[154]

The first place where computational adequacy outweighed pedagogical efficiency was probably India. Already around 500 BCE if not before we find the rule that multiplication is to be performed first [Sarma 2002: 135]. Another possible origin is in China, where the rule is formulated in the *Nine Chapters on Arithmetic* half a millennium later or so [trans. Vogel 1968b: 17*f*; trans. Chemla & Guo 2004: 225] and already used in the *Suàn shù shū* from no later than c. 186 BCE [trans. Cullen 2004: 62*f* and *passim*]. Mathematics must have been taught in China before this time, but no sources antedating the first century CE offer any direct evidence (Christopher Cullen, personal communication). All in all, Indian priority is thus likely but not certain. More certain is that borrowing one or the other way has taken place.[155]

---

[154] Quite unusually, the *Libro di conti e mercatanzie* ([ed. Gregori & Grugnetti 1998: 3*f*], see note 42) combines the two ways. After resolving the initial example

7 cubits of cloth are worth 9 florins, how much will 89 cubits be worth?

by the rule of three it shows two comprehensible alternatives: finding out that the price of one cubit must be $9 \div 7 = 1\frac{2}{7}$ florin, and that 89 cubits is $89 \div 7 = 12\frac{5}{7}$ times as much as 7 cubits, a ratio that must also hold for the prices. Then comes a question

7 cubits of cloth are worth 9 florins, how much will one get for 89 florins?

solved by the first and the third method. Only then the abstract rule is formulated (even this is an exception, cf. imminently); these alternatives may therefore be meant as ways to make the rule comprehensible, even though that is not said.

The *Libro di molte ragioni d'abaco* ([ed. Arrighi 1973: 17*f*], see note 153), though earlier, presents us with a messy mix-up of this system with the usual one: it starts by giving the rule in abstract form, then presents the same five computations (three with numbers 8 cubits of cloth, 11 florins and 97 cubits, two with numbers 8 cubits, 11 florins and 97 florins) as if all solved the same problem, and finally gives the rule in abstract form again. Since the observation

and you see how I have showed it without fractions in order that it should be quicker to understand the principle, but as it will follow further on, the present book will show you similar computations with fractions and without fractions, as I shall find it needed

is found verbatim in both works, the derivations from a common source (antedating 1330) is indisputable.

Ibn Thabāt [ed., trans. Rebstock 1993: 43–45] writing in Baghdad somewhere around 1200 also presents the third method of the *Libro di conti e mercatanzie*, without explaining it but giving it a name ("by *nisba*", a word that means "relation", specifically "ratio"), making thus manifest that this alternative was regarded a standard method in the Islamic world. So does al-Karajī in the *Kāfī* [ed., trans. Hochheim 1878: II, 17] – preferring even the solution by ratio to the one by "multiplication and division".

[155] Cécile Michel [1998: 251*f*] suggests that the rule was used in the solution of an Assyrian problem from c. 1800 BCE. This would be interesting, not least because the problem comes from a school

West of India, the earliest extant source for the rule seems to be al-Khwārizmī's *Algebra*, which dedicates a short chapter to *al-muʿāmalāt* [ed. Rosen 1831: Arabic 48] – [the mathematics of] social life or, more specifically, of business relations. In Rosen's translation [1831: 68], the initial passage runs as follows:

> You know that all mercantile transactions of people, such as buying and selling, exchange and hire, comprehend always two notions and four numbers, which are stated by the enquirer; namely, measure and price, and quantity and sum.

Being a mathematical scholar, al-Khwārizmī goes on to explain the situation in terms of a proportion. He then continues,

> Three of these four numbers are always known, one is unknown, and this is implied when the person enquiring says *how much?* and it is the object of the question. The computation in such instances is this, that you try the three given numbers; two of them must necessarily be inversely proportionate the one to the other. Then you multiply these two proportionate numbers, and you divide the product by the third given number, the proportionate of which is unknown.

In this form, the rule reached Latin culture through the translations due to Gherardo da Cremona and Robert of Chester. Fibonacci [ed. Boncompagni 1857a: 83], also in possession of scholarly competence and pretensions, introduces the rule in a similar way in the *Liber abbaci* (possibly with an eye to al-Khwārizmī, but this is less certain here than elsewhere). However, the formulation of Jacopo and other abbacus writers, referring to that which is similar and that which is not, coincides with what can be found in Brahmagupta and the ninth-century Jaina mathematician Mahāvīra; like the compiler of the *Livero de l'abbecho* and many other abbacus writers (not Jacopo, nor the Columbia Algorism), Āryabhaṭa, Brahmagupta and Mahāvīra also speak of the "rule of three (things)".[156] Ibn al-Bannāʾ (*Talkhīṣ*, ed., trans. [Souissi 1969: 88, Arabic 69]), though speaking like al-Khwārizmī of proportional numbers, also refers to the number "that is not similar"; so does Ibn Thabāt [ed., trans. Rebstock 1993: 45, 47] around 1200, whereas this primarily legal scholar omits the explicit notions of proportion and proportionality.

From al-Khwārizmī's presentation of the topic and from the shared terminology of the abbacus writers, Ibn al-Bannāʾ, Ibn Thabāt, Brahmagupta and Mahāvīra it becomes evident that the rule was first of all the common property of a trading community ranging from India to the western Mediterranean – and perhaps also to China, as intimated but not said directly by LAM Lay Yong [1977: 329]. It was certainly written about by more

---

for future merchants (that is, a school of abbacus type), but seems to be founded on a pure guess – a guess which is furthermore made improbable by the metrology that is involved.

[156] For Āryabhaṭa, see [Elfering (ed., trans.) 1975: 140], for Brahmagupta [Sarma 2002: 135, 137] (quotation p. 137), cf. [Colebrooke (ed.) 1817: 283]. Regarding Mahāvīra, see [Raṅgācārya (ed., trans.) 1912: 86].

professional mathematical scholars in various parts of this region and connected to their general way to practice mathematics, but these scholars and their writings had no pivotal role in the diffusion of the rule.

Theoretically oriented authors, once they have explained elsewhere how to multiply and divide by mixed numbers, should see no reason to split the treatment of the rule of three according to whether the occurring three numbers are integer or mixed. That expectation is confirmed by Brahmagupta, Mahāvīra, Ibn al-Bannā᾿ and Fibonacci. Jacopo's way, on the other hand, corresponds to what is generally found in the abbacus treatises – for instance, citing only the specimens closest in time, the *Livero de l'abbecho*, the *Liber habaci*, and Paolo Gherardi.

Common to all the writings that have been cited in this connection from Brahmagupta to Jacopo and Paolo Gherardi (the *Nine Chapters* being different in this respect) is that the rule (the rules, if more are given) is/are stated in abstract form, as dealing with three numbers.[157] This is no wonder, e.g., in the Indian writings, where everything is first stated in compact abstract rules, paradigmatic examples coming only afterwards or in commentaries. Ibn al-Bannā᾿'s *Talkhīṣ* (which by its very title is an epitome) also gives abstract rules only. But in the *Liber abbaci* and in the abbacus writings, the abstract rule is an exception which is to be taken note of. Fibonacci [ed. Boncompagni 1857a: 177 and *passim*], for instance, speaks of the method of the single false position as the "rule of the tree" because the method *can* be abstracted from a paradigmatic example dealing with a tree (actually he does not do so, no rule in the proper sense is ever stated).

Jacopo's real-life examples are all concerned with the exchange of coins. This is not uncommon in abbacus writings – the situation is the same in the *Livero de l'abbecho*, in the *Liber habaci* and in Paolo Gherardi's *Libro di ragioni* (bullion against coin). Others (for instance, as we have seen in note 154, the *Libro di molte ragioni d'abaco* from Lucca and the *Libro di conti e mercatanzie*) agree with the Arabic authors and confront goods and price.[158] The *Liber abbaci* starts with such examples, dealing afterwards with

---

[157] Just to avoid a frequent conceptual confusion: "abstract" means that the references to the concrete have been removed – references to this or that particular number, to this or that metrological unit, to coins, pepper or cloth. It has nothing to do with the question whether arguments or proof are present. These might just as well be formulated with reference to paradigmatic examples – until the invention of symbolic algebra indeed better (a Greek geometric demonstration, we should remember, was formulated with reference to a *particular* diagram with paradigmatic status).

[158] Ibn al-Thabāt [ed., trans. Rebstock 1993: 50] treats the exchange of money in a separate sub-chapter.

In order to avoid mistaken generalizations: cloth versus money can also be found in treatises from the whole time-span of abbacus writing that do *not* (like the *Libro di molti ragioni* and the *Libro di conti*) give the intuitively comprehensible alternatives to the rule of three – thus in the *Livero de l'abbecho*; in Giovanni de' Danti's *Tractato de l'algorisimo* [ed. Arrighi 1985: 29], which first has one example of this type and then a number of problems about monetary exchange; and

exchange of money.

The end of Jacopo's chapter on the rule of three – a rule for the multiplication of mixed numbers – seems rather out of place. Comparison with some of early treatises that take the rule of three as their starting point explains why the rule is there: in the *Liber habaci*, in Paolo Gherardi's *Libro di ragioni* and in the *Libro di molte ragioni d'abaco*, the section on the rule of three is followed by a long section on computation with fractions and mixed numbers. In the *Liber habaci* and in the *Libro di ragioni*, this section starts precisely with the rule for multiplying mixed numbers, *numero sano et rocto contra a numero sano et rotto* (as **V** spells the words they all share). Jacopo was apparently inspired by a model of this kind but then stopped after having borrowed a single rule with appurtenant example, fractions having already been treated by him.

Beyond the shift from division *in* to division *per* (see p. 17), five differences between **V** and **M+F** can be observed. First, no transition passage similar to V.11.1 is found in **M+F**. Second, in the formulation of the rule, "quella che non è simegliante"/"that which is not similar" becomes "quella che non è di quela medessima"/"that which is not of the same [kind]", and "l'altra che remane"/"the other that remains" becomes "la terza cossa"/"the third thing". Third, after the first example **M+F** offers a variant asking for the value of 30 instead of 20 *parigini*. Fourth, as mentioned above (p. 55), the rule in **M+F** for multiplying mixed numbers prescribes to "multiply the [integer part of] the smaller number by the whole of the other number, and then the fraction of the smaller by the whole of the other number", whereas **V** wants us to "bring to fractions both parts".[159] Fifth and finally, M.11.4=F.iv.3 seems to announce an alternative way to solve the problem,[160] then apparently abandons this idea and recalculates instead according to the rule of three – and errs, converting a remainder of 2 £ into 60 ß instead of 40 ß.

In other respects, the two manuscripts deal with the rule of three in fairly identical ways.

---

in Piero della Francesca's *Trattato d'abaco* [ed. Arrighi 1970a: 42–46].

[159] The *Liber habaci* has the same rule as **M+F**. Since the preceding examples are wholly different from Jacopo's, this only shows that both used a fixed formulation that was in circulation.

[160] Since 7 (£) *tornesi* are worth 9 (£) *parigini*, 7 ß *tornesi* must be worth 9 ß *parigini*, and 7 δ *tornesi* must be worth 9 δ parigini. What is suggested by this observation is an instance of the so-called *welsche Praktik* (see below, p. 84): 120 £ *tornesi* are split into 119 £ and 40 ß. Since 119 = 17×7, 119 £ *tornesi* must be worth 17×9 £ *parigini*. Similarly, 20 ß *tornesi* are split into 14 ß and 72 δ, the former converted into 18 ß *parigini*, the latter split into 70+2 δ, etc.

The *welsche Praktik*, as we shall see, is used in several problems that are present in **M+F** but not in **V**.

An Aside on Counterfactual Mathematics

Along with examples drawn from commercial real life, Jacopo has a couple of strangely counterfactual calculations, "if 5 times 5 would make 26, say me how much would 7 times 7 make at this same rate?" and "If 3 times 4 would make 13, how much would 7 times 9 make at this same rate?". I do not remember having seen anything similar in Indian or Arabic material, but a few texts suggest that simple counterfactual *questions* were not fully unknown – see below, p. 67. Such counterfactual questions are rather common in Italian abbacus treatises, in the shape of rule-of-three problems of the types "if $\frac{2}{3}$ were $\frac{3}{4}$, what would $\frac{4}{5}$ be?" or "$7\frac{1}{2}$ is worth $9\frac{1}{3}$, what will $5\frac{3}{4}$ be worth?".[161]

Similar counterfactual questions are found in *all* Ibero-Provençal treatises I have had the opportunity to inspect, and in all of these they constitute the first introduction of the rule of three; in contrast, the Italian examples are either secondary illustrations of this rule or number problems located elsewhere in the treatises. In the Castilian *Libro de arismética que es dicho alguarismo* (ed. Caunedo del Potro, in [Caunedo del Potro & Córdoba de la Llave 2000: 148]) we find "If 3 were 4, what would 5 be?" and "If 15 were 22, what would 32 be?". In two Provençal treatises – the "Pamiers algorism" from c. 1430 [Sesiano 1984a: 45] and Francés Pellos's *Compendion de l'abaco* from 1492 [ed. Lafont & Tournerie 1967: 103–107] – the basic examples of the rule of three are all of the kind "if $4\frac{1}{2}$ are worth $7\frac{2}{3}$, what are $13\frac{3}{4}$ worth?".[162] This might be quite irrelevant to the Castilian example, if the latter treatise had not given this problem immediately after the ones just cited (since the text is corrupt I leave it untranslated):

> Sy quisieres saber sy el 3° de 3 menos terçio fuese el 4° de 4° menos 4° ¿que sería el 5° de 5 menos 5°?, deves primeramente saber qué es el 3° de 3 menos 3° e qué es el $\frac{1}{4}$ de 4 menos 4° e qué es el 5 de 5 menos 5 e sabe que $\frac{1}{3}$ de 3 menos $\frac{1}{3}$ que son $4\frac{4}{5}$ e agora moltiplica los $3\frac{4}{5}\frac{3}{4}$ por los $4\frac{4}{5}$ e fallarás que montan 18 e pártelo por los $2\frac{2}{3}$ e verná a la parte $6\frac{3}{4}$.[163]

---

[161] The former example is from a fifteenth-century anonymous *Arte giamata aresmetica*, Torino N.III.53 [ed. Rivolo 1983: 11*f*], the latter from the Columbia Algorism [ed. Vogel 1977: 54], cf. note 70. Elsewhere [ed. Vogel 1977: 32, 80] the latter uses the formulations "If 25 were 12, what ..." and "If $1\frac{3}{4}$ were $1\frac{7}{8}$, what ...?". The "Treviso-arithmetic", printed in 1478, has a third formulation, "If 8 should become 11, what would 12 become?" (trans. D. E. Smith [ed. Swetz 1987: 103]).

[162] Chuquet [ed. Marre 1880: 632*f*], whose inspiration from the Provençal tradition is well known, is similar – but in the likeness of Fibonacci (see imminently) he feels the need for a mathematically coherent interpretation, even though the format he has to explain is less stunning than Fibonacci's first example: "If 8 are worth 12, what will 14 be worth? Or, if 8 asks for 12 as its proportional, what will 14 ask for?"

[163] Taken phrase for phrase, the text of course has a meaning. Translating this and repairing what has probably been confused in copying gives the following (⟨...⟩ indicates what has to be inserted, {...} what has to be deleted):

Even with emendations this is only correct if we interpret "the third of 3 less one third" as $3-\frac{1}{3} = 2\frac{2}{3}$, etc. But that does not fit the words in any way. Instead, Pellos gives us what is almost by necessity the right interpretation (I translate, since this text is meaningful):

> If one third of 3 less one third is worth one fourth of 4 less one fourth, how much will a fifth of 5 less one fifth be worth? Do thus: first you observe what one third of 3 is, this is one, integer. Therefore remove one third from one, integer, and two thirds remain. Likewise next observe what one fourth of 4 is, it is one, integer, from which remove one fourth, and three fourths remain. And put for the second number three fourths. Likewise, next observe what one fifth of 5 is, and it is one, integer, from which remove one fifth, and four fifths remain. Therefore put four fifths for the third number, and order your computation in the way you know

$$\begin{array}{ccc} 2 & 3 \!\!\!\!\diagdown \!\!\!\!& 4 \\ & \times & \\ 3 & 4 \!\!\!\!\diagup\!\!\!\! & 5 \end{array}$$ The amount to divide is : 36

                     The divisor is   : 40

It seems certain that the Castilian corrupt problem descends from something very much like this, that somebody baffled by the absence of algebraic parentheses tried a different way, and that the copyist may have made things worse (the manuscript of the Castilian treatise is a copy). In consequence we may also assume that the idea of a number a "being" a different number b arose as another version of a "being worth" b, thereby changing what may be seen simply as an abstract formulation into indubitable counterfactuality (this is in any case a reasonable hypothesis). In the second instance, "5 times 5 making 26" (and similar fancies) became possible once mathematical validity seemed no longer to be a prerequisite.[164]

A few Italian treatises contain counterfactual *calculations* similar to those of Jacopo. Paolo Gherardi's *Libro di ragioni* [ed. Arrighi 1987: 17] presents us with the question "If 9 is the $\frac{1}{2}$ of 16, I ask you what part 12 will be of 25"; quite similar, the Columbia Algorism asks twice, "If 9 is the $\frac{1}{2}$ of 22, what part will 13 be of 26?" [ed. Vogel 1977: 101, 112].[165] The *Trattato di tutta l'arte dell'abacho* (see note 144) offers the following

---

> If you want to know, if the 3rd of three less a third were the 4th of 4 less a 4th, what would the 5th of 5 less a 5th be? You must first know what the 3rd of 3 less a 3rd is and what the $\frac{1}{4}$ of 4 less a 4th is and what the 5⟨th⟩ of 5 less a 5⟨th⟩ is, and know that $\frac{1}{3}$ of 3 less $\frac{1}{3}$ ⟨is 2$\frac{2}{3}$ and that $\frac{1}{4}$ of 4 less $\frac{1}{4}$ is 3$\frac{3}{4}$ and that $\frac{1}{5}$ of 5 less $\frac{1}{5}$⟩ is 4$\frac{4}{5}$, and now multiply the 3{$\frac{4}{5}$}$\frac{3}{4}$ by the 4$\frac{4}{5}$, which by necessity amounts to 18, and divide it by the 2$\frac{2}{3}$, and from the division will result 6$\frac{3}{4}$.

[164] We notice that Jacopo's solution in V.11.6 proceeds via a derived question, "if 25 is worth 26, what will 49 be worth?"

[165] Two more examples from the same treatise are "If the $\frac{1}{3}$ and the $\frac{1}{4}$ of 14 were the $\frac{1}{2}$ and the $\frac{1}{5}$ of 11, what would the $\frac{1}{3}$ and the $\frac{1}{5}$ of 23 be?" and "If the $\frac{1}{3}$ of 13 were the $\frac{1}{3}$ of 11, what would 7 be?" [ed. Vogel 1977: 110f].

illustration of the rule of three by what is presented as "a false position" (after five commercial examples):

> Let us posit that 3 times 7 would make 23, tell me how much 5 times 9 would make at that
> same rate. Do thus, say, 3 times 7 in truth makes 21, and we say it makes 23, and it is not
> true. And we want to know what 5 times 9 makes, which truly makes 45. Hence say, the thing
> that we want to know is 45, and that which is not of the same (kind) is 23, and therefore
> multiply by 23, it makes 1035, and divide this by the third thing, that is by 21 which is true,
> from which results $49\frac{2}{7}$. If you want to verify it, join $\frac{2}{21}$ to 45, in total it is $49\frac{2}{7}$, and it is
> verified.[166]

A third example ("if 4 times 4 would make 17, how much would 5 times 5 make at this same rate?"), found in the *Istratti di ragioni*, a collection of problems from c. 1440,[167] is throughout so close to Jacopo in its wording that it might well descend from him (but certainly not from *Trattato di tutta l'arte dell'abacho*).

Of particular interest is a passage in the Columbia Algorism [ed. Vogel 1977: 31*f*] which shows us a passage from abstraction to counterfactual calculation. A number problem with the structure

$$(n-\tfrac{1}{3}n-\tfrac{1}{4}n)\times(n-\tfrac{1}{3}n-\tfrac{1}{4}n) = n$$

is solved from the false position $n = 12$, whence $5\times5 = 25$ should be 12. The text runs "5 times 5 are 25; I want that this 25 should be 12, what would 12 be? Say, if 25 were 12, what would 12 be?"

Even in the *Liber abbaci* [ed. Boncompagni 1857a: 170] there are two instances of counterfactuality, long after the introduction of the rule of three – one a calculation, the other a simple question: "If 7 were the half of 12, what would be the half of 10?", and "If $\frac{1}{3}$ were $\frac{1}{4}$, what would $\frac{1}{5}$ be?" Fibonacci apparently does not like them, and explains that by the first can be understood "that the half of 12, which is 6, grows into 7; or 7 is diminished into the half of 12, which is 6"; about the second he tells the reader that it is as if one said, "$\frac{1}{3}$ of a *rotulo* [a weight unit, c. $2\frac{1}{2}$ *libre*] for $\frac{1}{4}$ of one bezant, how much is $\frac{1}{5}$ of one *rotulo* worth?". It is thus fairly obvious that Fibonacci had found these problems in circulation and added a mathematically satisfactory explanation when including them in his book.

---

[166] "Pogniamo che 3 via 2 faciessi 23, dimmi quanto farebbe 5 via 9 a quella medesima ragione. Fa cosi, di', 3 via 7 fa veramente 21, e noi diciamo che fa 23, e non è vero. E noi vogliamo sapere che fa 5 via 9, ch'è vero che fae 45. Dunque die, la cosa che noi vogliamo sapere si è 45, e quella che nonn'è di quella medesima si è 23, e però multiplica via 23, fa 1035, e questo parti per la terza cosa, cioè per 21 ch'è vero, che'nne viene $49\frac{2}{7}$. Se'llo vuoi provare, giugni $\frac{2}{21}$ sopra 45, sono'ttutto $49\frac{2}{7}$, ch'è provata" ($\mathbf{T_R}$, fol. 32$^\mathrm{v}$, similarly $\mathbf{T_F}$ fol. 95$^\mathrm{r}$).

[167] Florence, Magl. Cl. XI, 86, [ed. Arrighi 1964: 89]; the problems are claimed by the compiler to go back Paolo dell'Abbaco, active a century earlier, that is, about one generation after Jacopo. There seems to be no particular reason to doubt this claim, and in any case the dating of the problems to c. 1340 seems reasonable.

No Arabic work I have looked at contains counterfactual *calculations*. For linguistic reasons, it also seems improbable that works written in Arabic should contain counterfactual *questions* in "be"-formulation, since the verb "be" in copula function does not exist in Semitic languages. However, this argument does not rule out that the "be-worth" formulation might come from or through the Arabic world. Most of the Arabic introductions to the rule of three I have looked at do not use it. In one place, however, ʿAlī Ibn al-Khiḍr al-Qurašī's eleventh-century *Al-tadhkira bi-uṣūl al-hisāb wa l'farāʾiḍ* ("Book on the Principles of Arithmetic and Inheritance-Shares", probably written in Damascus) [ed., trans. Rebstock 2001: 64], explains that "[this computation is] as when you say, so much for so much, how much for so much". Strictly speaking, this has nothing of the counterfactual; but it is strikingly similar to a reference to the rule of three as "the rule that if so much was so much, what would so much be", found in the *Libro de arismética que es dicho alguarismo* (ed. Caunedo del Potro, in [Caunedo del Potro & Córdoba de la Llave 2000: 183]), and coupled there to the counterfactual questions. Even more remarkable is the way al-Baghdādī refers to the way profit and loss are calculated, namely by the Persian expressions *dah yazidah*, "ten is eleven", and *dah diyazidah*, "ten is twelve".[168]

Ch. 12. Computations of Non-Compound Interest

Interest rates were not stated in percent per annum, but normally as the number of *denari* earned by one *libra* in a month. 1 *libra*, we remember, is 20 *soldi*, and 1 *soldo* is 12 *denari*. 1 *denaro* per month thus amounts to 1 *soldo* = $\frac{1}{20}$ *libra* a year, i.e., 5% per annum.[169]

The chapter not only teaches how to solve the basic problem – given the rate of interest, how much will a certain capital yield in a given time – but also a number of shortcuts and rules depending on the metrology: the interest per day; the amount which yields 1 *denaro* a day at a given rate of interest; the time it takes to double a capital "without making (up accounts at the) end of year", that is, at non-composite interest; etc. Obviously, such rules were useful by reducing the need for multiplications and divisions.

Throughout the chapter, Jacopo makes computations in several steps instead of using the rules "of five" and "of seven".

Apart from a copying error in M.12.9=F.iv.9 (borrowing part of the wording from

---

[168] Thus according to A. S. Saidan (Mahdi Abdeljaouad, personal communication). Unfortunately, neither Abdeljaouad nor I have so far been able to get hold of Saidan's edition of al-Baghdādī.

[169] The interest rates of most examples are thus 12% or 15% per annum (one having 5%). This is in the upper end of what was regarded as acceptable in Florence in the earlier fourteenth century, but still within the limits of the acceptable and not usury – see [Sapori 1955: I, 236–240].

V.12.11) and slightly different formulations, the contents of **M+F** is identical with that of **V**.

Ch. 13. Problems Involving Metrological Shortcuts

Medieval metrology was a complex affair – much more complex than Jacopo's treatise forces us to realize.[170]

A first difficulty for the reader of **V** is that the *libra* occurs both as a monetary and as a weight unit. To make things a bit easier for the reader, my translation distinguishes between the weight unit pound and the monetary unit *libra*. For the latter I shall use the abbreviation £ in the commentary when it stands as a unit; in the same function, *soldo* and the plural *soldi* will appear as ß, and *denaro/denari* as δ. All problems in this chapter deal with the units pound, quintal and *carco*/"load", presupposing at first (V.13.1–V.13.7) that the quintal is 100 pounds, and the load 3 quintals or 300 pounds. These values, we should note, were not of general validity (quite apart from the fact that the absolute value of the pound was far from fixed). Already the Arabic *qinṭār*, from which the quintal descends, was not exactly 100 *raṭl* but somewhat more or less – how much more or less depending on place and type of goods [Rebstock 1992: 129]. The *Liber abbaci* [ed. Boncompagni 1857a: 91] speaks of the load of 300 pounds as *carica provincie*, "load of Provence" – in order to differentiate it from the Italian value of 4 quintals or 400 pounds [Zupko 1981: 83*f*], we may presume.[171] Actually, 3 quintals or 300 pounds was the usual value not only of the Provençal *carga* but also of the *charge* of France proper [Zupko 1978: 41*f*]; 3 quintals of 104 pounds each was the Barcelona value, as explained in Francesc Santcliment's *Summa de l'art d'aritmètica* from 1482 [ed. Malet 1998: 183].

Manipulation of these metrological ratios permits a simplification of the conversion between the price of the load in *libre* or *soldi* and that of the pound in *denari*, analogous to the conversions between interest by the year, the month and the day taught in the previous chapter. However, Jacopo did not need to take the idea from there, the strategy was familiar. A similar rule is taught in the *Liber abbaci* (that is where Fibonacci refers to the "load of Provence"), although Fibonacci clearly prefers the use of the rule of three.

The simple rules taught in V.13.1 fail when the quintal is not equal to 100 pounds. To circumvent this difficulty, Jacopo introduces in V.13.8–10 a fictitious unit of account,

---

[170] Two handbooks to be recommended are [Edler 1934] and [Zupko 1981]. [Ciano 1964: 133–254] is an "Indice di misure" to the *Pratica di mercatura* found in the Archivio Datini, Prato. Unusual weights and measures occurring in Pegolotti's *Pratica di mercatura* are indexed in [Evans 1936: 410].

Medieval Islamic metrology (sometimes useful for comparison) is presented in [Hinz 1970].

[171] *Trattato di tutta l'arte dell'abacho*, written in Avignon and dealing with quintal and *charcha* over several folios, presupposes throughout the load to be 3 quintals or 300 pounds.

a monetary "grain" defined as a *denaro* divided by one fourth of the number of pounds to a quintal.[172][173] There is nothing shocking in this, even the *libra* and the *soldo* are monies of account that were not minted.[174] The number of these grains to a *denaro* is thus $\frac{1}{4}$ of the number of pounds to a quintal. Then for each *soldo* which the quintal is worth the pound is worth 3 grains. The inspiration could be the Barcelona metrology, where a monetary grain of $\frac{1}{26}$ of a *denaro* was in use according to Francesc Santcliment's *Summa* [ed. Malet 1998: 183]. About Barcelona, where the quintal was 104 pounds (and the load or *càrrega* still 3 quintals, but that is not used in the present passage), he writes

> for each *lliura* in money which the quintal of 104 pounds costs, results for each pound of weight 2 *diners* 8 grains (counting 26 grains to a *diner*). Further, for each additional *sou*, 3 grains (at 26 grains the *diner*).

Santcliment knows of no general rule valid irrespective of the value of the quintal; he states indeed that no such rule can be given. Jacopo may not have invented the artifice, but it was certainly a calculator's trick (which did not spread), not a use of existing coinage (whether minted or of account).

In the end, the price in grains of course has to be converted into *denari* by means of a division. The advantage may not be arithmetically obvious to us. If the price of a quintal of N pounds is x soldi, then the price of 1 pound is $12x/N$ denari. Instead of multiplying by 12 and dividing by N, Jacopo suggests to multiply by 3 and to divide by $N/4$. But if many calculations have to be made with the same N, and if $N/4$ is an integer, then computations become somewhat simpler. More important, the introduction of the fictitious unit gives a concrete meaning both to the number $3x$ (the price of one pound in grains) and the division by $N/4$ (a conversion of the monetary unit); this would certainly be an advantage for merchants not too well trained in mathematics.[175]

M+F contains the same rules and examples as **V** in the same order. The most

---

[172] 24 weight *denari* amount to an ounce, 8 ounces to a mark, two marks to a "heavy pound" (*libra grossa*). The pound occurring in alloying problems equals 12 ounces (at times characterized as a *libra sottile*, "light pound", which however may be as low as 8 ounces). In Italy, the weight *denaro* was usually divided into 24 (weight) grains; the grain introduced by Jacopo, instead, would be $\frac{1}{25}$ of a monetary *denaro*, since the normal Italian quintal is 100 pounds. As we shall see, a grain equal to $\frac{1}{600}$ of an ounce – that is, $\frac{1}{25}$ of a weight *denaro* – is spoken of as a weight unit for gold in V.14.23 and as a unit of account used in Sicily and Puglia in V.14.27.

[173] A corresponding but different rule "engendered by the avoidance of multiplication and division of the same numbers" is found in the *Liber abbaci* [ed. Boncompagni 1857a: 101].

[174] When introduced in Carolingian times, the monetary pound was equal to a pound of silver; but this was five centuries of inflation ago.

[175] Similar rules are explained in Arabic *muʿāmalāt* writings [Rebstock 1992: 131*f*], in good agreement with the preference of several of these for the concretely meaningful alternatives to the rule of three.

conspicuous difference between the two manuscripts is that M.13.6=F.iv.o and M.13.7=
F.iv.24 repeat that the rule and the computation presuppose a load of 300 pounds, whereas
V is satisfied with having stated it in V.13.1 and repeated it in V.13.3.

Ch. 14. Mixed Problems, Including Partnership, Exchange and Genuine "Recreational" Problems

Since the problems in this chapter constitute a mixed collection with little inner
coherence, the best way to approach it is to discuss the problems one by one.

V.14.1. This problem introduces the partnership, but not the usual way to solve partnership
problems – the way Jacopo himself follows elsewhere. This usual way can be seen as
a generalization of the rule of three; in the actual situation, the first share would be found
as $(100 \cdot 150)/(150+230+420)$. Instead, the present solution starts by finding the gain per
libra, in agreement with the concretely meaningful alternative to the rule of three (cf.
above, p. 59).

M.14.1=F.v.1 adds a clause stating that thus is the "way and rule for every partnership
whatever the number of partners and whatever each partner has put into the principal of
the partnership, and you see what comes per *livra*" – but like **V**, **M+F** switches to the
habitual method in all partnership problems that follow. Apart from that, the two
manuscripts only differ in their detailed wording.

V.14.2. The way this discount problem is solved may not be very satisfactory from a
mathematical point of view, but it probably corresponds to what was often done in
commercial practice. The method can be described as informal iteration. In order to find
the value of 200 £ $2\frac{1}{2}$ months in advance at a rate of 2 δ the *libra* a month the text starts
tacitly from the fact that the interest of 200 £ at that rate (1000 δ) is close to 5 £ (= 1200
d; 4 £ would have been much better, but may have been discarded because it would lead
to an approximation from above). It therefore takes 195 £ as its first approximation (that
amount on which you "put yourself"), finding that this would develop into 199 £ 1 ß 3
δ, that is, 18 ß 9 δ less than required. Adding this deficit to the first approximation would
give an extra interest of (slightly less than) 5 δ, the interest on which in $2\frac{1}{2}$ months is
negligible. Therefore, the answer is 195 £ + (18 ß 9 δ) – 5 δ = 195 £ 18 ß 4 δ.

A solution of a similar problem by quite formal iteration in 7 steps is found in the
Libro di conti e mercatanzie [ed. Gregori & Grugnetti 1998: 95], which shows it to give
the same result as a solution by means of the rule of three but regards the iterative
procedure as simpler (the rule-of-three solution involves a division by $22\frac{7}{20}$).[176]

---

[176] Obviously, this corresponds to the development of $\frac{1}{1+\rho}$ as $1-\rho(1-\rho(1-\rho(1-\rho(1-\rho)))))$. It is
just as obvious that the text says nothing like this – but it constructs an ordered scheme in which
the alternating subtractions are performed. The iteration stops when the correction vanishes.

M.14.2=F.v.2 only differs slightly from **V** in the wording.

V.14.3. For a modern reader, this calculation with its *denari* of different kinds is probably opaque. Introducing the notations δ_B for "*denari* of *bolognini piccioli*" and δ_F for "*denari* of the coin of Florence", B_g for the *bolognino grosso* and f for the *fiorino*, and remembering that the relation between *libra*, *soldo* and *denaro* is the same in both places (and everywhere), we see that the computation in **V** is correct (apart from the writing error $^{15}/_{36}$).[177]

M.143=F.v.3, however, errs, apparently because its editor recalculates and trusts his own errors.[178] As one might perhaps expect in the case of a recalculation, the formulation of the last clause is independent of **V**.

M+F starts "Io". The prescription, however, speaks of the "he" who has to do the payment according to **V**. Perhaps the first person in the question "What is better for me" has incited the compiler to make this change, an ellipsis of the structure "If somebody says, I have ...", with subsequent explanation of what this asking *he* should do.

V.14.4. Once again, the modern reader may be baffled by the seeming confusion of units and dimensions. First of all, the "200 de *pisani*" should be understood to be *libre* (from *Pisa*), and these are to be converted into numerically equal amounts of *libre* from Provence and Imperial *libre*. It is not easy for us to emulate the ease with which Jacopo supposed his reader to be able to handle the units, but in our terms the argument can be explained approximately like this: one Imperial *soldo* and one Provençal *soldo* equal 40+32 Pisan *denari*, i.e., 6 Pisan *soldi*. 200 Pisan *soldi* are worth 200/6 = $33\frac{1}{3}$ times as much, i.e., $33\frac{1}{3}$ Imperial *soldi* and $33\frac{1}{3}$ Provençal *soldi*. Therefore, 200 Pisan *libre* are worth as much as $33\frac{1}{3}$ *libre* of each, i.e., 33 £ 6 ß 8 δ.

As we see, however, Jacopo skips approximately two lines of this, without committing any error. The explanation sets out the reasons that he is right, but hardly renders the thinking of a calculator trained in the handling of medieval currencies faithfully.

M.14.4=F.v.4 agrees with **V**, except for jumping directly to the result in the proof.

---

[177] 100 δ_B = 100/$13\frac{1}{3}$ B_g, 1 B_g = $15\frac{1}{4}$ δ_F, thus 100 δ_B = $\dfrac{100 \cdot 15\frac{1}{4}}{13\frac{1}{3}}$ δ_F, and similarly 100 £$_B$ = $\dfrac{100 \cdot 15\frac{1}{4}}{13\frac{1}{3}}$ £$_F$. Similarly for the calculation in *fiorini*.

[178] At first, **F** (but not **M**) miswrites 31 as 13 and 1525 as *millecinquecento venti*, but these are mere copying errors. Afterwards, however, when dividing 1525 both find 123 £ 12 ß $11^{25}/_{37}$ δ (**F** at first writes *soldi* 13, but has 12 in the repetition), which corresponds to a division by $12\frac{1}{3}$ instead of $13\frac{1}{3}$ (thus probably coming from a conversion of 1525/$13\frac{1}{3}$ into $^{4575}/_{37}$ instead of $^{4575}/_{40}$. Afterwards, the division of 3950 by $31\frac{1}{2}$ gives 123 £ 7 ß 9 δ – which is actually the outcome of the division 3950÷$32^{128}/_{9871}$. I am unable to explain this with certainty but presuppose that the calculator combined at least two errors – he may, for instance, have transformed 3950/$31\frac{1}{2}$ into 7900/64 instead of 7900/63, which *should* give 123 £ 8 ß 9 δ, and have found or written instead 123 £ 7 ß 9 δ.

V.14.5. This is a problem that would often be solved by means of the "rule of double false", "double false position" or *regula elchatayn*[179] – thus for instance a very similar problem in the *Liber augmenti et diminutionis* [ed. Libri 1838: I, 363–365].[180] In the present case, and using together the two false positions that occur as alternatives in the text, this would amount to using the following scheme:

```
        0 new   100 new

             X
        60 ß    140 ß

      sum of errors 200
```

On top we have the two positions, either that none of the *fiorini* are new[181] or that all 100 are new, below we have the respective errors resulting from these positions, and in bottom the sum of the errors. The calculation consists in a cross-multiplication, addition of the two products, and division of the sum by the sum of the errors. The result is that $\frac{0 \cdot 140 - 60 \cdot 100}{60 - 140} = 30$ of the *fiorini* have to be new (if both errors had been in deficit or both in excess, the sums should have been replaced by differences. The principle is the same as the one used in alloying calculations – namely an inversely weighted mean, $\frac{0 \cdot 140 - 60 \cdot 100}{60 - 140} = \frac{140}{60 - 140} \cdot 0 + \frac{60}{60 - 140} \cdot 100$.

In our symbols, this is easily understood. As far as I have noticed, explanations of why the principle works are never given in the abbacus texts,[182] nor is the analogy with alloying ever mentioned. For the masters and *a fortiori* for their students, the rule seems to be purely mechanical, and it must have been quite opaque. Jacopo never uses it; he may not have known it, or he may have avoided it for pedagogical or other reasons.

V.14.8. A goblet consisting of three parts was a favourite theme, occurring also (for instance) in the *Liber abbaci* [ed. Boncompagni 1857a: 188*f*] and borrowed by the compiler of the Umbrian *Livero de l'abbecho* (cf. above, p. 35). Normally (and thus in the *Liber abbaci*, and below in V.14.25), the weight of one part is given absolutely and that of the others relatively. Here, the dress is used for a different purpose – also found

---

[179] This is Fibonacci's spelling, certainly closer to the Arabic origin than what is found in many abbacus writings!

[180] One hundred gold coins, some of them *melichini* worth 5 *solidi* each and some of them *revelati* worth 3 *solidi* each, in total worth 460 *solidi*.

[181] A medieval Italian, Latin or Arabic calculator would never have taken zero as one of the positions. In the *Liber augmenti* ... the first position is that only one coin is a *melichinus*, the second that two are of this kind. But the principle does not change if we base the explanation on Jacopo's choices.

[182] Fibonacci [ed. Boncompagni 1857a: 320–322] does give a demonstration, but it is clearly of his own making; and its complexity seems to have prevented adoption.

in almost identical words in *Trattato di tutta l'arte dell'abbaco* ($\mathbf{T_F}$ fol. 122^r, $\mathbf{T_R}$ fol. 51^v); two fairly similar stories about a cup in three parts are told in the Columbia Algorism (see note 70) [ed. Vogel 1977: 69, 103].[183] For once Jacopo does not state it explicitly, but the structure is that of the partnership.

M.14.8=F.v.8 agrees with **V** in all essentials.

M.14.8A = F.v.9. In **M+F** a problem follows which has no counterpart in **V**. As I shall do in similar cases in the following, I insert the translation of this problem at the corresponding place in the translation of **V**, without intending this as a claim that the problem belonged necessarily in the original from which **V** descends.[184] What is certain in the present case is that the text is wholly in the style of the surrounding problems in **M+F**,[185] and thus that it is no secondary insertion made after the stylistic recasting of this version of the *Tractatus*.

The problem is one of of inverse proportionality, but the solution is phrased in terms that make it look like an application of the rule of three, without explanation. The equation

$$\text{old weight} \times \text{old price} = \text{new weight} \times \text{new price} ,$$

i.e.,

$$8 \times 66 = 7\tfrac{1}{4} \times u ,$$

is formulated as if the ratio 8:66 had to do with the matter. The calculation is wholly pertinent, however, and the result correct.[186]

V.14.9–10. This problem, including the ratios 2:1 and 1:2, goes back to ancient Roman jurisprudence,[187] and it is quite widespread in abbacus writings (but absent from the

---

[183] The Columbia Algorism uses the dress in the habitual way in a third problem [ed. Vogel 1977: 104].

[184] In some cases, as we shall see, there are good reasons to believe that problems in **M+F** that are absent from **V** were also absent from Jacopo's original – see p. 88. In one case it is at least possible that such a problem has disappeared from **V** because of faulty copying – namely M.21.2A= F.vii.2, see p. 126.

[185] Thus, to mention two features that are characteristic of **M+F** but do not recur in all abbacus writings: the marking of the question by the phrase "say me" (*dimmi*) and the repetition of the result after a specification of its meaning, marking the repetition by "that is" (*cioè*).

[186] [Simi 1995: 19 n.6] claims that the result should be ß 72 δ 5$^{23}/_{29}$. This, however, is the outcome of the division of £ 105 ß 2 by 29, not as required of 105 £ 12 ß.

[187] This was pointed out by Moritz Cantor [1875: 146–149]. The question – Dig.2.13 [*Ius romanum*] – stands as a quite hypothetical illustration ("If it be written thus, ...") of a legal principle formulated by the second-century jurist Salvius Julianus.

Actually, one may suspect Julianus to have made use of a pre-existing mathematical recreational problem: the earliest extant Chinese mathematical manuscript, the *Suàn shù shū* from no later than c. 186 BCE, contains a problem about a fox, a wild-cat and a dog going through a customs-post and sharing the tax according to the ratios between the prices of their skins, which are pairwise

Liber abbaci as well as the *Livero de l'abbecho*). What is interesting in Jacopo's approach is his explicit introduction of a partnership, and his splitting of the exposition of the general principle from the calculation regarding a specified amount. The only other places where I have noticed a similar invention of a partnership are in the *Liber habaci* [ed. Arrighi 1987: 145], in Paolo Gherardi's *Libro di ragioni* [ed. Arrighi 1987: 37], in the *Libro di molte ragioni* from c. 1330 (as a hint only), and in Pietro Paolo Muscharello's *Algorismus* from 1478 [ed. Chiarini et al 1972: 189–191]. The second treatise was written in Montpellier in 1328, the first one around 1310 and also in Provence, see note 114. The *Libro de molti ragioni* was written in Lucca but, as we shall see repeatedly, influenced by Jacopo's *Tractatus* or something quite close to it, while Muscharello's algorism is one of those which borrows Jacopo's introduction (cf. note 120).

The version in the *Liber habaci* appears to be unrelated to Jacopo's: instead of making a false position the text states that each time the girl gets 1 *libra* the mother gets 2 and the boy 4; the wording is also very different. The total inheritance is supposed to be 100 *libre*. Gherardi's words, in contrast, might suggest that he knew Jacopo's text (in version **V** rather than **M+F**) and then have changed the numbers. Indeed, his text begins with these words (emphasis added for identical words; **M+F** has *giudicamento* instead of *testamento*):

> Elli è 1 huomo lo quale è malato molto *gravemente* e fa suo *testamento* finalmente e *ha una sua* moglie *la qual'è* gravida .... Or viene *che'l buono huomo* muore e la donna damdo a uno tempo si parturiscie e fa uno figlio e una figlia. ....

However, Pellos's *Compendion de l'abaco* from 1492 [ed. Lafont & Tournerie 1967: 135] shares even more of Jacopo phrasing, but also part of that which Gherardi does not share with Jacopo:

> un home ven malaut, et ha de bens 6.000 fl., et ha una molher che es grossa d'enfant. Et lo home si sente fort agreujat de la malautia, et fa son testament ensins: si sa molher fa un masche, adunques es sa voluntat che lo enfant mascle prenga 2 pars, et la mayre una part de tots sos bens.

In *Trattato di tutta l'arte dell'abbaco* (**T$_R$** fol. 28^v , similarly **T$_F$** fol. 75^v) we find:

1:2 [trans. Cullen 2004: 45]. A similar story about animals (eating in the same proportions) is found in the *Nine Chapters on Arithmetic* [trans. Vogel 1968b: 28; trans. Chemla & Guo 2004: 285–287].

Although the problem (in inheritance shape) can be found in several cultural settings where it would be juridically impossible, all versions I have inspected but one respect the culturally shared presupposition that the brother gets more than the sister. The one, unexpected exception is in a Byzantine collection written after the Turkish conquest [ed. Hunger & Vogel 1963: 38]: here the sister gets twice the inheritance of the mother, and the mother twice that of the brother. Later in the same treatise, however, we find the problem with the traditional numbers but with a unique argument that allows the brother to get $^4/_9$, the mother $^3/_9$ and the sister $^2/_9$.

Un signore sta gravamente, e vuole fare suo testamento, ed a una sua donna la quale è grossa. Il signore fa testamento di 1000 ff. , ...

After all it thus seems more plausible that Gherardi was inspired, not directly by Jacopo's text but rather by a common source tradition (to which **V** should indeed be closer than **M+F** according to preceding arguments). *Trattato di tutta l'arte* states explicitly that the solution is made *al modo della compangnia*; Pellos does not refer explicitly to the partnership, but lists the problem as "the 14th example" of *la regola de companhia*. It hence seems likely that Jacopo, the author of the *Liber habaci* and Gherardi have borrowed the explicit reference to the partnership from the Provençal environment of Montpellier.[188]

Problem V.14.9 is not the first one in which Jacopo makes use of a single false position; but it is the first time the method is referred to by name. In view of Jacopo's general pedagogical care his reference to the name as something with which his reader is already familiar must be regarded as a slip, and as evidence that it was indeed fully current in the place where he found his material.

Grosso modo, M.14.9=F.v.10 follows the same pattern as **V**, including the reference to a "position" and the explicit "reduction to a partnership".

V.14.11. Even the fish consisting of head, tail and body – the weight of one part being given absolutely, the others relatively – turns up in various writings that do not depend on Jacopo (e.g., the Columbia Algorism [ed. Vogel 1977: 91, 105] – but not in the *Liber abbaci*, nor in the *Livero de l'abbecho*). An amusing variant is found in *El arte del alguarismo*: the fish is effectively cut into the pieces in question, but since it is an eel the parts are still able to tell each other about their respective weights.[189]

Structurally similar problems about collections of persons, heaps of apples or walnuts, amounts of money or precious metals are found in book XIV of the *Anthologia graeca* [ed. Paton 1979]. Further variation of the dress is found in Indian sources, which often change the mathematical structure, introducing square roots and products along with the aliquot parts.[190] In the Islamic world, as in abbacus culture, the favourite basic examples were the tree and the fish [Bagheri 1997: 3].[191]

---

[188] It may be noticed that *El arte del alguarismo* does *not* refer to the partnership in its version of the problem (ed. Caunedo del Potro, in [Caunedo del Potro & Córdoba de la Llave 2000: 154]).

[189] Ed. Caunedo del Potro, in [Caunedo del Potro & Córdoba de la Llave 2000: 172].

Readers who only know frozen eels from the supermarket should be told that fresh eels, even after having been skinned and cut into pieces, still behave very lively on the frying pan.

[190] E.g., in Mahāvīra's *Gaṇita-sāra-sangraha*, [ed., trans. Rangācārya 1912], the links of which to the mathematical practice of the Ancient Near East and Mediterranean region are striking – see [Høyrup 2004].

[191] Mahāvīra [ed., trans. Rangācārya 1912: 80, 82] also has a lotus and a pillar in water and air

M.14.14=F.v.12 only differs slightly from **V** in its wording.

V.14.12–13. Problems of meeting and pursuit are known from ancient China, medieval India and the medieval Arabic world, and in Europe since Fibonacci – see the overview in [Tropfke/Vogel et al 1980: 588–598]. One may even suspect Zeno's paradox about Achilles and the tortoise to be inspired from the arithmetical problem,[192] but we have no direct evidence for it from the classical world. In all cultures where it is present we find simple versions where both travellers go with constant speed as well as more intricate questions where one of them increases his speed in arithmetical (at times also geometrical) progression.

Jacopo only treats of the simple version, the one whose method of solution coincides with that used for combined works – for instance for vessels filled from or emptied through several taps. Here, as in the parallel cases V.14.18 and V.22.9 and the semi-parallel case M.14.17A=F.v.18, no explanation is given, only a numerical prescription which is concise enough to prevent the reader who has not already grasped the trick from reconstructing it from the computation; plausibly, the underlying idea is that the first man can cover the distance 9 times in 9×11 days, whereas the second could cover it 11 times; together they would thus cover it 9+11 times in 9×11 days; a similar explanation is given in the *Liber abbaci* [ed. Boncompagni 1857a: 183] in connection with a somewhat different problem, cf. below, p. 78. The circumstance that 9+11 is referred to as *"the* divisor" (*el partitore*) indicates that an already established standard procedure is referred to,[193] which corresponds well to the absence of explanation. There is no indication as to whether Jacopo understood why the formula works.

M.14.12=F.v.13 omits the phrase "join together 11 and 9, which make 20. And this is the divisor", for which reason it has to explain when dividing by 20 that this is the sum 11+9; but the next problem follows **V** exactly on this point.

V.14.14. I have not noticed this problem type in non-European sources, but it already occurs in the *Liber abbaci* [ed. Boncompagni 1857a: 276], with different numbers and words that also differ enough to make direct use by Jacopo quite implausible (quite apart

---

(the pillar also with a part in the mud), and Ananias of Širak [ed., trans. Kokian 1919: 114] has a sepia consisting of head, body and tail.

[192] Much as the Chinese fox–wild-cat–dog problem seem to have inspired the twin problem of Roman jurisprudents, see note 187. The appearance of several problems also known from China in classical ancient sources (together with the appearance of other problems with Chinese affinity in Diophantos's *Arithmetica* and the *Anthologia graeca*) strengthens the plausibility of borrowing or common origin – the a priori plausibility of the existence of a network that would allow diffusion may not be very high, but it may still outweigh the fourth power of the probability of independent invention of a rather characteristic dress.

[193] The same expression "and this is the divisor" is used in the contexts of partnership calculations and a false position.

from the general lack of evidence for use of the *Liber abbaci* in his treatise) but are similar enough to suggest both to have been inspired by a circulating standard version.

M.14.14=F.v.15 calculates the freight rate and the price paid by each merchant, but does not formulate a proof.

V.14.15. Problems of this type are widely spread; Fibonacci [ed. Boncompagni 1857a: 173] actually uses one of them to introduce the method of the single false position, which he therefore refers to afterwards as *regula arboris*, the "rule of the tree" (cf. above, pp. 35 and 62). Fibonacci has several examples where the fraction is $\frac{1}{3}\frac{1}{4}$; the same fraction occurs in the example found in the *Libro de arismética que es dicho alguarismo* (ed. Caunedo del Potro, in [Caunedo del Potro & Córdoba de la Llave 2000: 161]); the Columbia Algorism [ed. Vogel 1977: 79] has $\frac{1}{3}\frac{1}{5}$. We must assume that the circulating standard versions contained such composite fractions – Jacopo uses them nowhere else except in V.22.4, a problem about a goblet with three parts, again of the same kind as three problems in the *Liber abbaci* [ed. Boncompagni 1857a: 188*f*] which Fibonacci solves with a reference to the *regula arboris*; and we may guess that the use of these composite fractions was a means to achieve mathematical complexity at the same level as the fish consisting of three parts for the tree which consists of only two.

M.14.15=F.v.16 omits the proof.

V.14.16. One may suspect the *compagni* of this problem to be comrades rather than commercial partners, whence my aberrant translation in this place. Anyhow, the method used to solve the problem combines that of the partnership (used tacitly and in a slightly distorted way) with that of a false position.

One may also suspect the formulation of the problem, with specified fractions that add up to more than the whole and therefore have to be reduced proportionally, to be somehow in debt to Islamic inheritance computation, where this was a standard procedure; but it is hardly possible to proceed from suspicion to conviction (in any of the senses of that word). However that may be, the reference to "el partitore"/"the divisor" suggests once again that a well-defined procedure or rule is referred to. In the present case, this reference might be to the partnership rule.

M.14.16=F.v.17 specifies that the 10 *soldi* etc. are one half etc. of 20 *soldi*; it is also somewhat more explicit in the following – for example when pointing out that the final summation is a proof.

V.14.17. This problem type is familiar from Arabic sources – for instance, al-Karajī's *Fakhrī* [Woepcke 1853: 83]. It also occurs in the *Libro de arismética que es dicho alguarismo* (ed. Caunedo del Potro, in [Caunedo del Potro & Córdoba de la Llave 2000: 164]) and in the Columbia Algorism [ed. Vogel 1977: 85] as well as in many later abbacus writings. The *Libro de arismética que es dicho alguarismo* contains a marginal diagram suggesting an argument that is halfway spelled out in the Columbia Algorism: using the parameters of the actual problem, 7 days of work and 5 days of non-work give payment zero. The 30 days are then divided in the same proportion (using the structure of the partnership but without naming it). This may also be the reasoning behind Jacopo's procedure (irrespective of whether he understood it or not).

The *Liber abbaci* [ed. Boncompagni 1857a: 160*f*] contains a version with a net gain for the worker, which enforces the use of a different argument based explicitly on the alloying model (cf. above, p. 43).[194] Later, a similar problem [ed. Boncompagni 1857a: 323] is solved by a double false position.

M.14.10A(17)=F.v.11 separates the proof clearly from the solution of the problem, which means that the formulations after "12 days and $\frac{1}{2}$ he stayed away from working" are independent.[195] Until that point, the two formulations of the problem only differ somewhat in the wording, but it is noteworthy that they are in different locations – M.14.10A(17)=F.v.11 is located between the counterparts of V.14.10 and V.14.11.

M.14.17A=F.v.18. This problem type was already diffused from China and India to the Mediterranean in late Antiquity; it is mathematically analogous to the problem of combined works, which is also treated in Old Babylonian sources. It is mathematically closely related to the simple meeting problems V.14.12–13. As in these, no explanation for the prescribed steps is offered. We may assume that the underlying argument was the one that was suggested on p. 76 – namely that 30 is chosen as a number of days in which each one of the taps can empty the cask an integer number of times (the *Liber abbaci* [ed. Boncompagni 1857a: 183] indeed gives the argument mentioned on p. 76 in a problem about the emptying of a cask). However, in an analogous problem (about a ship with two sails) the *Liber habaci* [ed. Arrighi 1987: 144] uses the present formulation in an argument explicitly based on the fractions of the task (there, the voyage) each agent (there, each sail) would perform in a day, corresponding to a solution based on a transformation

$$\frac{1}{\frac{1}{2}+\frac{1}{3}+\frac{1}{5}} = \frac{30}{\frac{30}{2}+\frac{30}{3}+\frac{30}{5}} \; ,$$

and in agreement with the way the same manuscript teaches to divide by non-integer numbers. The particular formulation in M.14.17A=F.v.18 may therefore well correspond to a difference in underlying argument. Slightly later, indeed, the *Liber habaci* has an

---

[194] Since alloys cannot contain a negative share of silver, the daily gain and penalty are added to and subtracted from the 30 days. This looks bewildering, but just serves to displace the zero – any number larger than the daily penalty could have been used.

[195] In this passage, **F** has a spelling *facta*. This *could* be taken to suggest that both formulations go back to Jacopo; however, since Paolo Gherardi also uses this spelling (cf. note 16), the explanation is more likely to be use of material coming from the environment where both Jacopo and Gherardi worked – which fits what could be concluded on p. 55 about the origin of the tables of higher squares and products.

analogue of V.14.12 = M.14.12=F.v.13, which is formulated in the same terms as this latter problem and not like the preceding two-sail question.

The division *in* suggests that the problem goes back to a source which shared Jacopo's stylistic characteristics.[196] As M.14.8A=F.v.9, M.14.17A=F.v.18 is wholly in the rhetorical style of the surrounding problems; even this problem must therefore have been inserted in the treatise no later than the stylistic recasting – most likely in the very process of recasting.[197]

V.14.18. This variant of the "combined works" is found in many abbacus treatises, and also in the *Libro de arismética que es dicho alguarismo* (ed. Caunedo del Potro, in [Caunedo del Potro & Córdoba de la Llave 2000: 156*f*]), but not in the *Liber abbaci*. It seems likely that it was invented in the environment of maritime commerce during the thirteenth century.

The argument, still not explicit, is apparently the same as in V.14.12–13; in any case, the sequence of operations is the same in the two cases, and the wording of V.14.18 is a simple abbreviation of that of 14.12. In contrast, the wording in 14.17A is quite different.

M.14.18–F.v.19 calculates the sum as 7+9 instead of 9+7, and states the result as 4 less $\frac{1}{16}$. Apart from that, differences are slight.

V.14.19, like 14.5, could have been solved by means of a double false position. Once again, this opaque trick is avoided, and the same transparent method is applied.

M.14.19=F.v.20 omits the final explanation "And if this computation should not seem too clear to you ...". Apart from that, only details in the wording are different.

V.14.20. With varying numbers (but normally powers of 2, and normally both chests being cubic) and more or less explicit indication of the fraud that might be intended, this problem is found in many abbacus writings, and also in the *Liber abbaci* [ed. Boncompagni 1857a: 403]. Normally, the calculation is straightforward, consisting of a division of the larger cubic contents by the smaller.[198] Jacopo's computation is puzzling. Most likely he has used a source which he does not understand but tries to make sense of – perhaps imagining the large container as consisting of 16 columns, each 4 *braccia* high. Since each of the small containers holds 8 "square" braccia, that is, two of these columns, 16 has to be divided by 2. Alternatively, he has mixed up two different procedures found

---

[196] As pointed out on p. 17, all divisions in **V** by a number defined as *the divisor* are divisions *in*. But cf. note 195.

[197] As we shall see below (pp. 82 and 88), Jacopo himself can hardly be responsible for the insertion of the extra problems. On the other hand, several consecutive thorough recastings of a treatise which still conserves its purported identity are not very likely.

[198] This is also done first by Fibonacci; afterwards he suggests as an alternative to find the ratio for each dimension separately and to multiply these ratios afterwards. The *Livero de l'abbecho* [ed. Arrighi 1989: 120] omits the alternative when copying.

in his source.

That the cubic contents of the two chests are measured in "square *braccia*" seems strange to us but is quite regular in the abbacus context. It corresponds to a view of volumes as consisting of "thick surfaces", that is, slices 1 *braccio* high.

M.14.20=F.v.21 is solved in the orthodox way, by division of one volume by the other. Even before this disagreement its formulations differ somewhat from those of **V**.

V.14.21. A sequence of similar problems about rectangular pieces of cloth turn up in the third-century BCE Demotic Papyrus Cairo J.E. 89127–30, 89137-43 [ed. Parker 1972: 20–24]. None of them obviously makes use of the rule of three, which only reached the Mediterranean region much later.[199]

Neither the *Liber abbaci* nor the *Livero de l'abbecho* contains any similar problem. The Columbia Algorism, however, does [ed. Vogel 1977: 80]. Like **V** it makes use of the inverse rule of three without pointing out that an inversion is involved.[200]

The words of M.14.21=F.v.22 are somewhat different, and the solution only prescribes the multiplications and the division, without hinting at the rule of three.

V.14.22. An early variant of this problem where each dispossession amounts to one half and one fourth is given by Ananias of Širak [ed., trans. Kokian 1919: 115]. A variant where each doorkeeper gets one half of the apples and two more is found in the *Liber augmentis et diminutionis* [ed. Libri 1838: I, 335–339], which the compiler – one Abraham who is probably *not* to be identified with Abraham Ibn Ezra – ascribes to Indian sources.[201] It is possible that this ascription builds on mere hearsay: nothing in the

---

[199] In the last of them, the width is seen to be reduced to $\frac{1}{2}$ of its original value, and the length therefore found as the original length divided by $\frac{1}{2}$. In the others, the area of the piece that is removed from, e.g., the length, is calculated and then distributed along what remains of the length – arithmetically somewhat clumsy, but perhaps inspired by the actual physical process of cutting and sewing.

[200] However, since it has never introduced the rule of three explicitly it refers to it by repeating the counterfactual structure, "If $1\frac{3}{4}$ were $1\frac{7}{8}$, what would $23\frac{1}{2}$ *braccia* be".

[201] Cf. Barnabas Hughes' supplement [2001] to Libri's edition, which transforms the latter into a critical edition. The Latin text must be a translation, since it refers to an initial praise of God but does not bring it ("hic post laudem Dei inquit"). The same happens, e.g., in Gherardo da Cremona's translation of al-Khwārizmī's algebra [ed. Hughes 1986: 233].

Libri [1838: I, 304 n. 1] suggested "Abraham" to be identical with Abraham Ibn Ezra without feeling sure; many later writers have discounted his doubts – without ever advancing valid reasons for the identification, as far as I can judge. Comparison with Ibn Ezra's *Sefer ha-mispar* [ed., trans. Silberberg 1895] enforces my doubts. For instance, the initial dedication to God seems to be wholly external to the subject-matter, as usually in Islamic and Islamic-impregnated Christian and Jewish writings. Ibn Ezra's inception, in contrast, fuses theology, numerology and mathematics into an inextricable whole (another twelfth-century Abraham, namely Abraham bar Ḥiyya [ed. Guttmann, trans. Millàs i Vallicrosa 1931: 9], restricts himself to soliciting divine assistance, yet another way).

treatise recalls specifically any Indian source known to us. Alternatively, it might build on genuine Indian merchant mathematics, which must have existed but for which we have no direct sources. The main purpose of the treatise is to teach the double false position, also ascribed to the Indians in Arabic sources; this could be the foundation for Abraham's claim. It also shows solutions *per regulam*, that is, by means of *regula recta*, first-degree *res*-algebra (see pp. 39 and 105). In any case the problem must have been known in the Arabic world. In the eleventh century, it was known in Persia [Tropfke/Vogel 1980: 585].

The *Liber abbaci* [ed. Boncompagni 1857a: 278] has a variant with seven doorkeepers, each one getting half of the apples and one more; the *Liber habaci* [ed. Arrighi 1987: 144] has three persons receiving half and, respectively, one, two and three more. The *Libro de arismética que es dicho alguarismo* (ed. Caunedo del Potro, in [Caunedo del Potro & Córdoba de la Llave 2000: 161]) has three doorkeepers, receiving different fractions and different absolute numbers. The Columbia Algorism [ed. Vogel 1977: 93] has five doorkeepers, each receiving half and one more; as in Jacopo (but none of the others), its apples belong to *una donna*.

In M.14.22=F.v.23, some details of the formulation are changed. In the solution, moreover, doublings precede additions, for which reason 2 have to be added each time instead of 1. This can hardly have been done in order to ease understanding.

V.14.23. With this problem we leave the realm of recreational mathematics for a short while, entering that of really commercial calculation. First of all we notice that the grain used here for weighing gold ($\frac{1}{600}$ of an ounce) is not identical with the usual Italian grain of $\frac{1}{24}$ of a weight *denaro*, i.e., $\frac{1}{576}$ of an ounce, but corresponds instead to $\frac{1}{25}$ of a weight *denaro* (cf. note 172). We also observe that the formulations presuppose the carat to be $\frac{1}{24}$ of an ounce, that is, the same as a weight *denaro*. Actually, its traditional value was 4 grains [Zupko 1981: 79], $\frac{1}{24}$ of the weight of the Roman gold *solidus* – cf. below, p. 123.

There is a small error in the multiplication of 8 £ 15 ß $9\frac{3}{8}$ δ by 13, the result should be 114 £ 5 ß $1\frac{7}{8}$ δ. The outcome of the ensuing calculation does not decide whether the error is due to copying; division of the correct result should have the fractional part $(1\frac{7}{8})/30$, which might be rounded downwards to $\frac{1}{30}$, while division of the number found in the manuscript yields the fraction $(\frac{7}{8})/30$, which might be rounded upwards. However, since M.14.23=F.v.24 rounds the number to 114 £ 5 ß 2 δ we may perhaps assume the error to have arisen during copying – and the repetition of the error to be due to a second copyist who has reestablished consistency at the cost of correctness. The way the value of the 14 grains is established illustrates the observations made on p. 60 in connection

---

It may be added that Shlomo Sela [2001] does not mention the *Liber augmentis et diminutionis* in his discussion of Ibn Ezra's complete scientific corpus, which otherwise includes *dubia* which are rejected.

with the rule of three. Indeed, when finding the value of one grain, Jacopo rounds $3\frac{33}{64}$ to $3\frac{1}{2}$; multiplication of this rounding by 14 gives an error that exceeds the precision he has aimed at earlier – even though it still disappears in the rounding to integer *soldi* in the total (here the error exceeds a whole *denaro* slightly).

M.14.23=F.v.24 changes some of the formulations slightly and seems to have rounded the numbers of the original to integer *denari*.

V.14.24. Mathematically, this problem is wholly equivalent to the fish cut into three pieces (V.14.11), and only different from the tree in ground and air by having one component more but in return no composite fractions. I have not noticed the type outside the abbacus corpus, and the topic itself corresponds well to an invention within this environment.

M.14.24=F.v.25 changes the colour sequence into white, green and vermilion without changing the numbers. It also reduces the proof to a mere sketch.

M.14.24A=F.v.26. Cloth problems, as said above, were traditional, and the comparison of prices was certainly a recurrent problem for merchants and textile artisans. However, this particular problem type I have not noticed elsewhere.

Until the determination of the price difference 6 *denari* per square palm, the calculation is wholly reasonable. The meaningfulness of what follows next is questionable, but of course depends on the wish to know the price difference for a rod, where this rod corresponds to different widths and hence different areas. Given this choice, the calculation of the price difference for a rod with the average width is as good an answer as any.

Stylistically – for instance, in the repetition of a numerical result after its meaning has been explained, the present problem does not differ from those that surround it in **M+F**. It should be observed that the unit *canna* never occurs in **V** but only in this and another problem from **M+F** with no counterpart in **V** (namely M.14.31A=F.v.36, see below, p. 88). Since there is no reason that a careless copyist should forget precisely these two problems, we may be confident that they are later additions to the original stem.

V.14.25. This problem represents the habitual use of the goblet, which was mentioned in connection with problem V.14.8. Mathematically, it is completely analogous to the fish consisting of three parts and the purse in three colours. Most remarkable is perhaps that the number of parts remains the same as always – namely three: in order to allow the present goblet to have a stem, Jacopo eliminates the cover.

M.14.25=F.v.27 has an explicitation that the final addition constitutes a proof. The proof in **M+F** is also more detailed than in **V**. Other differences in the formulations are minor.

V.14.26. The problem and its solution speak for themselves; most puzzling is perhaps that it is not treated in the geometrical chapter (where, for instance, *Trattato di tutta l'arte dell'abbaco* actually places it, see imminently; without having a separate chapter on geometry, Paolo Gherardi [ed. Arrighi 1987: 60*f*] also treats a similar problem about a *piaçça* in the context of geometric problems). The final remark "you see that it results in the way that we have done" is likely to explain why the proof goes via the number of slabs in a stripe one *braccio* large: by making the proof structurally different from the solution not only the computation but also the *way* is checked. Jacopo may not have been an outstanding mathematician, but this at least is a good idea.

M.14.27(=F.v.28) starts "Una chiesa overo palazo, la qual'è longa bracia 120 ed è ampia bracia 36 né più né meno" ("A church or palace, which is long *braccia* 120 and wide *braccia* 36, neither more nor less"). Apart from a less meticulous proof, this is the major difference between the two manuscripts. It only becomes interesting when we notice that a problem of the same structure in *Trattato di tutta l'arte dell'abbaco* (T_F, fol. 132^r, T_R fol. 57^v) starts precisely with the words "Uno palagio overo una chiesa o qualunque chosa simiglante fosse la quale ..." ("A palace or a church, or whatever similar thing it might be, which ..."). After this point, the texts diverge strongly in all respects, in phrasing as well as metrology and numerical parameters – so strongly that there is no reason to suspect the author of *Trattato di tutta l'arte* to have copied from a precursor of **M+F**.[202] It is possible, however, that the hand that reshaped **M+F** was familiar with the treatise from Avignon (even plausible, since it is one of the abbacus treatises of which most copies survive), and borrowed the introduction from there while taking over the rest of the problem from Jacopo's original. It is at least as likely, however, that both were inspired independently by the traditional problem type, which appears to have been widespread.[203]

Problems about the pavement of a church were familiar in the Latin post-agrimensorial tradition. One is found in the Carolingian *Propositiones ad acuendos iuvenes* [ed. Folkerts 1978: 62], another one in *Geometria incerti auctoris* IV.38 [ed. Bubnov 1899: 355].

V.14.27. The prolegomenon to this problem belongs to a different genre than the abbacus

---

[202] The failing verb in the initial sentence of *Trattato di tutta l'arte* is explained by the text being an explanation written below a preceding drawing. This drawing is absent from **M+F**, where a fully grammatical sentence would therefore be required. In **V** a similar drawing is found in the margin, but the text does not refer to it; occasionally **V** uses the same ungrammatical introduction, but only when a diagram is present and referred to explicitly in the text. In **M+F**, as observed on p. 19, the text often does not refer explicitly to appurtenant marginal diagrams, and when it does the reference mostly comes in the end. None the less, many problems start with an ungrammatical phrase that strictly speaking presupposes a connection to a diagram without caring for its actual absence.

[203] Tommaso della Gazzaia's early fifteenth-century *Praticha di geometria e tutte le misure di terra* [ed. Arrighi & Nanni 1982: 19], has the mixture "Una sala ovvero chiesa" (and a different set of numerical parameters). Tommaso was a nobleman from Siena who made this collection of mixed abbacus material "for his pleasure, taking delight in the science of geometry" [Van Egmond 1980: 187]. We shall return to the similarity of some of this material to what we find in Jacopo (p. 96 and *passim*).

book – namely the so-called *pratica della mercatura*.[204] Treatises of this kind abound in similar information about local metrologies for goods of various kinds. Similar elaborate lists are also found in the *Liber habaci* [ed. Arrighi 1987: 149–152] (see note 114), whereas 65^v–72^r of the Lucca conglomerate *Libro de molte ragioni d'abacho* ([ed. Arrighi 1973: 153–175], see note 153) constitutes a proper example of the genre. In al-Quraši's *Book on the Foundation of Arithmetic and Inheritance* [ed., trans. Rebstock 2001] it is also prominent, and so similar in style that one may believe the combination of local metrological information with basic mathematics teaching to be a borrowing from the Arabic world.

Turning attention to the substance of the problem we notice that the *teri* and *grani* of V.14.23 turn up again, this time specified however to be monetary units of account, as are the *libre*, *soldi* and *denari*, not weight units (cf. p. 69). Apart from the copying error "3 grani" instead of "13 grani", there is a small error in the calculation, the value of 6$\frac{1}{2}$ ounces of silver being 29 *teri*, 16$\frac{7}{16}$ *grani*, not 29 *teri*, 16$\frac{9}{16}$ *grani*.

The separate calculation for marks and ounces (similar to what we find in V.14.23) certainly corresponds to what merchants would prefer to do. Mathematically, conversion to a single unit may seem more straight (and Jacopo's preferences often go in this direction), but the intuitive feeling of verisimilitude (which was certainly more likely to be relied upon in the marketplace than a genuine proof by reverse computation) is lost by this method.

M.14.23=F.v.29 is closer in its details to **V** than elsewhere, as if this factual information invited less rewriting.

M.14.27A=F.v.30–31. The fairs of Champagne were of supreme importance for European long-distance trade of the epoch. The present problem, found only in **M+F**, thus belongs to the same genre as the preceding one. The calculation reflects even more closely than that of V.14.27 the way a merchant is likely to have performed it, with minimal use of paper. Among later German *Rechenmeister*, it came to be known as *Welsche Praktik*, "Italian Practice" – see, e.g., [Tropfke/Vogel et al 1980: 364*f*].

"Strong" or *forti* was a generic term for many silver monies minted in France and neighbouring regions from the eleventh through the fourteenth century. *Provisini* were minted both in Rome and in Champagne [Travaini 2003: 267*f*, 295; *idem*, personal communication].

V.14.28. This returns us to the normal type of abbacus problems, formulated in a fairy-tale style that mostly belongs with recreational problems ("it occurred after a certain time"). A problem found in the Columbia Algorism [ed. Vogel 1977: 83*f*] and apparently borrowed from there into the *Livero de l'abbecho* [ed. Arrighi 1989: 119*f*] (see note 87) explains

---

[204] Two published specimens are the *«Pratica di mercatura» Datiniana* [ed. Ciano 1964] and Pegolotti's *Pratica della mercatura* [ed. Evans 1936]. Cf. also below, p. 121.

why it was next to inescapable that wool became wet and shrunk when dried: raw wool is dirty and has to be washed.

The Columbia problem is at most a vague inspiration for Jacopo, but there is no reason to believe even this little: nothing but wool and shrinkage is shared by the two problems, and wool and shrinkage were problems of the real merchants' craft. Mathematically, Jacopo's problem is one of inverse proportionality; as in V.14.7 and V.14.21, the solution is phrased in terms of the direct rule of three.

M.14.28=F.v.32 agrees closely with the text in **V**, adding however the final observation that "do in this way and by this rule irrespective of how much the wool lessens, by way of the rule of the three things" (an expression which is never used in **V**, cf. p. 61).

V.14.29. Purses containing money usually serve a different purpose in abbacus- and similar literature: a number of men finding a purse, each telling how much he would have (in the absolute, or relatively to the others) if he should get all or a specified part of what it contained. No less than seven variants are found in Mahāvīra's *Gaṇita-sāra-sangraha*, [ed., trans. Rangācārya 1912: 155–158]; it is present in various Arabic treatises [Tropfke/ Vogel et al 1980: 607], copiously represented in the *Liber abbaci* and borrowed from there into the *Livero de l'abbecho* (see above, p. 35). The solutions of these problems are not easy to follow unless one translates them into symbolic algebra.

What Jacopo gives looks like a "poor man's use" of the dress – mathematically a complete analogue of the tree problem V.14.15. As in the tree problem, complexity is obtained by having one part consist of a sum of two aliquot parts of the whole (though written this time as a sum and not as a composite fraction). Jacopo is not alone in having a poor man's version: in the Columbia Algorism [ed. Vogel 1977: 122], "Somebody had *denari* in the purse, and we do not know how many. He lost $\frac{1}{3}$ and $\frac{1}{5}$, and 10 *denari* remained for him". The same problem, only with the unlucky owner of the purse being "I" and the remaining *dineros* being only 5, is found in the *Libro de arismética que es dicho alguarismo* (ed. Caunedo del Potro, in [Caunedo del Potro & Córdoba de la Llave 2000: 167]). Both of these solve it by way of the counterfactual question, "If 7 were 10 [respectively 5], what would 15 be?". Since this is the usual way of the Columbia Algorism as well as the Castilian treatise but not of other Italian treatises, and since the Columbia Algorism appears not to have been widely known, the problem type is likely to have circulated in the Ibero-Provençal area. Whether Jacopo's particular version was of his own design we cannot know.

M.14.29=F.v.33 uses 6 as its false position instead of 12. In general, its formulations are somewhat different. Without being a mere sketch, the proof is considerably more concise than that of **V**.

V.14.30. Not only we but also Jacopo's near contemporaries would have placed this problem in the geometry section, as they would have done with V.14.26. The geometry section is indeed where the close analogue in the *Trattato di tutta l'arte dell'abbaco* can be found[205], and geometry is also the context in which Paolo Gherardi [ed. Arrighi 1987: 64*f*] presents a similar but somewhat more intricate problem about the four walls of a house. One may speculate whether the location of V.14.26 and V.14.30 is evidence of gradual growth of a treatise which was not initially intended to comprise a specific chapter on geometry. Alternatively, Jacopo may have used one model for the present chapter (a problem collection including a little geometrical computation) and another one (possibly a collection of nothing but geometric problems) for the next one. In the same way he seems to have culled material for his final collection of mixed problems from yet another treatise (if not more than one) – cf. below, p. 128.

As mentioned on p. 20, M.14.30=F.v.34 starts by changing the wall into an impossible *tereno*/"terrain", and then corrects itself, demonstrating thereby derivation from something close to **V**. Apart from that and an erroneous repetition in **F** alone, the formulations follow those of **V** rather closely.

V.14.31. Zendado is a very subtle fabric, normally of silk, often used for shawls. The general conditions of trade depicted in the problem, with multiple metrologies, currencies and exchange rates, is similar to what we know from problem V.14.3. The calculations may be easier to follow for a modern reader if we introduce the subscripts $_L$ for Lucca, $_M$ for Montpellier, $_F$ for Florentine money[206] and $_T$ for money from Tours. If the ounce is abbreviated δ^z we then have the relations:[207]

12 $\delta^z{}_L$ cost 6 £$_L$ 5 ß$_L$,
12 $\delta^z{}_L$ = 15¼ $\delta^z{}_M$,
1 *fiorino* = 13 ß$_T$ 4 δ_T = 29 ß$_F$.

The last relation is equivalent to

1 ß$_T$ = 2 ß$_F$ 2¹⁄₁₀ δ_F .

We are interested in the price of 12 $\delta^z{}_M$, which is found correctly by the rule of three to be 4 £$_F$ 18ß$_F$ 4²⁰⁄₆₁ δ_F.

This is now converted into a mixture of *fiorini* and Tours money – or actually, what is converted is the amount miswritten or misremembered as 5 £$_F$ 18ß$_F$ 4²⁰⁄₆₁ δ_F. The first step, too easy to be written down, is to convert it into 118ß$_F$ 4²⁰⁄₆₁ δ_F. 116 ß$_F$ are converted into 5 *fiorini*, and the remaining 2 ß$_F$ 4²⁰⁄₆₁ δ_F into (approximately – in Jacopo's words, "we can say") 13 δ_T, a fair approximation to 13¹²⁹⁄₅₃₀₇ δ_T.[208] The "proof" that follows

---

²⁰⁵ **T**$_F$, fol. 132ᵛ–133ʳ, **T**$_R$ fol. 58ʳ ("Uno muro overo parete ..."). The numerical parameters are different, and the unit is palms instead of braccia; but the way the solution is structured and the key terms are shared.

²⁰⁶ The *fiorino* was a famous coin which was emulated in many places, and it is thus for good reasons that Jacopo specifies the *fiorini* to be Florentine.

²⁰⁷ It goes by itself and therefore is not said that the Montpellier pound is 12 $\delta^z{}_M$.

²⁰⁸ Jacopo did not calculate this latter number; since he states that 1 ß$_T$ is worth 2 ß$_F$ 2¹⁄₁₀δ_F, he will

only verifies the equivalence between the two relations *fiorini/tornesi*. In the end, the price in *tornesi* is found by means of a conversion of the 4 *fiorini* into *tornesi* according to the given value 13 ß_T 4 δ_T.

M.14.31=F.v.35 explains how to make the division by 15 $\frac{1}{4}$, namely as a multiplication by 4 followed by a division by 61 (miswritten as 16 in **F**) – but the result it gets is the erroneous 5 $\pounds_F$ 18ß_F 4$^{20}/_{61}$ δ_F from **V**, which can hardly have arisen in any other way than copying freely from a source where this error was already present, that is, from an original similar to **V**. The amount is furthermore claimed to be what should be the price of "the Lucca pound, that is, 12 ounces, at the weight of Montpellier", whereas it should be that of "the pound, that is, 12 ounces, at the weight of Montpellier" or of "the Montpellier pound, that is, 12 ounces, at the weight of Montpellier". This mess already starts in the introduction, which in **M+F** explains that "each pound of Montpelliers is ounces 15 and $\frac{1}{4}$, that is, that each Lucca pound in Montpelliers becomes ounces 15 and $\frac{1}{4}$"; it continues in the question, which asks "for how much can I give the pound in Lucca weight in Montpelliers", which, if taken to the letter, would make the multiplication by 12/15 $\frac{1}{4}$ completely superfluous.[209] What follows after this multiplication also looks most of all as a mistaken condensation of what is found in **V** (the passage ⌈...⌉ is forgotten in **F**):

> Now we bring them to *tornesi*, and £ 5 ß 16 are 4 gold *fiorini*. ß 2 and δ 4 of *fiorini* are saved, ⌈which are worth in *tornesi* δ 13. In Montpellier the *libra* will hence be worth 4 gold *fiorini* and 13 *tornesi* δ, ⌉ which are worth in *tornesi soldi* 54, *denari* 5, and so much he ought to sell the Lucca pound at the weight of Montpelliers, that is, 12 Lucca ounces are worth at the weight of Montpelliers *soldi* 54 and *denari* 5 of *tornesi*. And it is done. Do all the similar thus.

M.14.31A=F.v.36. This problem should offer supplementary comfort to the modern reader who feels disoriented by medieval metrology, being no less confused than the preceding problem in **M+F**. The text is contradictory, but real metrology shows that the initial information is reliable. We are told at first that the Florence rod[210] is 8 $\frac{3}{4}$ Nîmes palms, whereas a Nîmes rod is only 8 palms (still Nîmes palms, we must presume). This fits real measures well, the Florence mercantile *canna* being c. 234 cm, and all the Provençal *cana* listed by Zupko [1978: 33] (each equal to 8 corresponding palms) being some 10 percent or more shorter. The later reference to "palms 8 and $\frac{3}{4}$ of Florence" must thus be read as an (illegitimate) ellipsis for "[Nîmes] palms 8 and $\frac{3}{4}$, [that is, one rod] of Florence".

---

have split 2 ß_F 4$^{20}/_{61}$ δ_F into (2 ß_F 2 $^{1}/_{10}\delta_F$)+(2+($^{20}/_{61}$–$^{1}/_{10}$)) δ_F. The first addend is worth 1 ß_T = 12 δ_T, the second is close to 1 δ_T.

[209] The modern reader who is confused by medieval metrology may perhaps find comfort in this: even medieval professionals might lose track.

[210] Identified as that of 4 *braccia*, that is, the "mercantile rod" – see [Zupko 1981: 65].

This means that the price for a Nîmes rod should be $8/(8\frac{3}{4})$ times the price for a Florence rod, not $(8\frac{3}{4})/8$ times that same price (i.e., 47 *soldi* $\frac{3}{8}$ *denaro*, in Florentine coin) as claimed in **M+F** – which thus confuses the direct and the inverse rule of three. The conversion into *tornesi* goes by *Welsche Praktik*, as used in *M.14.27A*=F.v.30–31: if 29 Florentine *soldi* are worth $13\frac{1}{3}$ *soldi tornesi* $= \frac{2}{3}\cdot20$ *soldi tornesi*, then $1\frac{1}{2}\cdot29 = 43\frac{1}{2}$ Florentine *soldi* must be worth 20 *soldi tornesi*. The remainder is rounded to $3\frac{1}{2}$ Florentine *soldi*, that is, 42 Florentine *denari*, which are roughly converted into 21 *denari tornesi* (instead of $19\frac{9}{29}$ *denari*), with a conversion factor $\frac{1}{2}$ probably derived from the observation that 20 *denari tornesi* are $43\frac{1}{2}$ Florentine *denari*, that is, roughly 40 (if the equivalence had been used directly and not through a conversion factor, 42 Florentine *denari* would have been seen to be less than 20 *denari tornesi* – roughly $19\frac{1}{2}$, indeed).

The sharing of the unit *canna* ("rod") with M.14.24A=F.v.26 and of the *welsche Praktik* with M.14.27A=F.v.30+31 (but not with any problem contained in **V**) strongly suggests that these three problems were added by a later hand. Jacopo, as we have seen, does not distinguish the inverse rule of three from the direct rule when speaking about them but keeps them well apart in his calculations. The later hand, as we see, seems to be trapped by Jacopo's carelessness. We may now be fairly confident that this later hand was not Jacopo preparing a revised version, even though we cannot exclude that it inserted the three problems in question in a short version already prepared by Jacopo. Most likely, as argued in note 197, these supplementary problems were inserted by the same compiler as was responsible for the stylistic recasting; in any case their actual stylistic similarity to the rest of **M+F** shows that the insertion did not postdate the restyling.

V.14.32. Like V.14.29, this is a "poor man's use" of the purse – this time rather a mathematical analogue of the "fish problem" than of the "tree problem" – in the sense that complexity is achieved by having three participants in the game, not by use of an additively composite fraction. Unintended complexity (so we may assume) comes from an error, the fraction $\frac{1}{4}$ of the statement being changed into $\frac{1}{2}$ in the calculation – salvation coming from mistaking the total possession of the former two fellows for the possession of the third, thus taking this sum to be one *denaro*. The blunder is so elementary that one may mistake it to have arisen from thoughtless combination of two different versions of the problem.

M.14.32=F.v.37 does not commit the error, which in principle might therefore have arisen in the copying process between Jacopo's archetype and **V**. It is much more likely, however, that the compiler of **M+F** has corrected an error present in his original. Since the proof is similar to the one we find in M.14.29=F.v.33 (the previous purse problem), whereas the proof in V.14.32 is strictly analogous to the one of V.14.29, it seems implausible that somebody later than Jacopo should have tinkered with the text of **V** while emulating his characteristic style ("Now because you say", etc.), and thus likely that Jacopo is responsible for the mess – *interdum dormit Homerus*.

Ch. 15. Practical Geometry, with Approximate Computation of Square Roots

Chapter 15 is more coherent than Chapter 14. None the less, it may be a good idea to discuss its problems (and other constituents) one by one.

V.15.1. To be noticed in this introduction is merely that it is there, and that it refers to the notion of *misura*, not to *geometry* – as it might have done if Latin traditions (either "Boethius", other agrimensor treaties or Fibonacci's *Pratica geometrie*) had provided the main inspiration. *Misura*, instead, agrees with (but does not prove) principal roots in the Arabic *misaḥah*-tradition.[211] M.15.1=F.unnumbered agrees with **V** except for minor details in the wording.

V.15.2. This is the first of three consecutive problems about the circle (in the first and third a circular *terreno*/"terrain"). We notice that Jacopo (as many other abbacus authors) makes use of everyday terms – how much the circle *gira dintorno* ("goes around") and the figure itself being a *tundo* ("round") – before introducing the Latinizing technical terms *circumferentia* and *circhio* (coming only in V.15.4); only *diametro* comes before being explained as *lo diricto de mezzo* (the "straight in middle"). We also observe that the "measure" of the circle is something that can really be measured (namely with a string), not the area (which is to be calculated).

The belief that the ratio between circumference and diameter is exactly $3\frac{1}{7}$ is current in abbacus writings, and shared with Fibonacci's *Pratica*, with the *Liber mensurationum* and with Savasorda. Rather unusual is instead the choice to start by calculating the diameter from the perimeter, not vice versa nor by finding the area from the diameter. I have only noticed it in *Trattato di tutta l'arte dell'abbacho* (**T$_F$** fol. 130^r, **T$_R$** fol. 56^v); in the *Liber habaci* [ed. Arrighi 1987: 130]; in Tommaso della Gazzaia's *Praticha* (see note 203) [ed. Arrighi & Nanni 1982: 31]; and in a *Praticha di gieometria* [ed. Arrighi 1970b: 30] written by the mid-fifteenth-century military engineer Francesco di Giorgio Martini. *Trattato di tutta l'arte* and Tommaso della Gazzaia both speak of a *ruota*/"wheel" and not of a circular field (yet with rather implausible dimensions, the perimeter being 40 respectively 100 *braccia*), *Liber habaci* initially of a *tondo*/"round", later of a *ruota*.

M.15.2=F.vi.1 omits the explanation "And if you should want to know [...] as I have said to you above", and refers to the drawing in the end (as *forma*, omitted in **M**) instead of the beginning. These are the only major differences between the two texts.

V.15.3. Even though this is not said explicitly, this problem illustrates the second part of the double rule enunciated in V.15.2

M.15.3=F.vi.2 instead makes this relation explicit, replacing the reference to the diagram by a second period "We have said about the round how much is its diameter, now we shall say its straight in middle and show how much will be the whole circumference".

---

[211] It may be significant that Paolo Gherardi [ed. Arrighi 1987: 60–75] distinguishes between rules "di missure" and "di giomatria". The latter are indeed (possibly with a few exceptions) such as could have been borrowed from the Latin post-agrimensorial tradition.

V.15.4. The circular area is determined in many ways in medieval Arabic and Christian-European practical geometries – all obviously equivalent, since all assume the ratio between circular perimeter and diameter to be $3\frac{1}{7}$. Al-Karajī gives four in his *Kāfī* [trans. Hochheim 1878: II, 23*f*]:

   (1)   Semidiameter × semiperimeter
   (2)   ($\frac{1}{4}$ of diameter) x perimeter
   (3)   diameter x ($\frac{1}{4}$ of perimeter)
   (4) $(1-\frac{1}{7}-\frac{1}{2}\cdot\frac{1}{7})\cdot$diameter^2
   (5)   perimeter$^2 \div 12\frac{4}{7}$

Ibn al-Thabāt [ed., trans. Rebstock 1993: 112] adds these:

   (6)   ($\frac{1}{4}$ of perimeter^2)$\div 3\frac{1}{7}$
   (7) $\frac{1}{4}$ of (diameter × perimeter)

In Latin and in European vernacular, (4) became

   (4*) $\frac{11}{14}$ of diameter^2

A Latin treatise from the 13th century or earlier, known as *Geometrie due sunt partes principales* [ed. Hahn 1982: 157], has

   (8) $\frac{1}{2}$ of ([$\frac{1}{2}$ of diameter] × perimeter)

perhaps inspired from the Archimedean *De mensura circuli* [ed. Clagett 1964: 20, 40, etc.], that is, from the scholarly translations.

   Abū Bakr [ed. Busard 1968: 118] gives (4) and, as an alternative, (1). For Savasorda [ed., trans. Curtze 1902: 97; Guttman & Millàs i Vallicrosa 1931: 72*f*], (1) is the principal rule and (4) the alternative. Fibonacci's first rule [ed. Boncompagni 1862: 86] is (1), after which follow (4), a variant of this which relies on a metrological shortcut, and a variant of (6). Most abbacus geometries restrict themselves to (1) and (4).

   Jacopo, as we see, uses (7). The only other abbacus writings where I have noticed it are *Trattato di tutta l'arte* (**T**$_F$, fol. 130^v, **T**$_R$ fol. 56^v), the *Liber habaci* [ed. Arrighi 1987: 130], Tommaso della Gazzaia's *Praticha* [ed. Arrighi & Nanni 1982: 32],[212] and Francesco di Giorgio Martini's *Praticha di gieometria* [ed. Arrighi 1970b: 30*f*] – exactly those, we remember, which start by determining the diameter from the perimeter. They all also use the area formula first in a problem where the perimeter is given.[213]

   It is a vague possibility that the formula points back to the habits of Mesopotamian Bronze Age surveyors. One Old Babylonian text[214] gives $\frac{1}{4}$ as the "constant of the crescent", that is, of the semicircle – meaning that its area is $\frac{1}{4}$ of the product of the diameter and the semiperimeter (for the full circle, (5) was used, only with divisor 12). Moreover, Ibn Thabāt wrote in Baghdad. Finally, the formula is given by Mahāvīra in his *Gaṇita-sāra-saṅgraha* [ed., trans. Raṅgācārya 1912: 200] in the section dealing with

---

[212] Even though Tommaso can be seen to depend on something very close to **M+F** in other places, his treatment of the circle is independent.

[213] Paolo Gherardi [ed. Arrighi 1987: 68*f*, 76] makes use of (3) in the rather disorderly collection of *misura*- and *giomatria*-problems (always under the *misura*-heading).

[214] TMS III, 7, ed. [Bruins & Rutten 1961: 25].

"minutely accurate calculation of the measure of areas", which (along with other material) presents methods almost certainly of Near Eastern origin [Høyrup 2004]. Yet however conservative the practitioners often were in their use of inherited formulae, a Babylonian background to Jacopo's formula is nothing but a vague possibility – it is not difficult to get the idea that (2) or (3) should be changed into (7) in order to avoid multiplication of rounding errors.

The geographical distribution of the texts taking the perimeter and not the diameter as the basic circular parameter and determining the circular area from formula (7) has an important implication. As we see, these characteristics are absent from all early Italian treatises except those written in Provence, but present in all but one of these (namely Gherardi, who has no ordered exposition). This excludes that the Tuscan authors working in Provence were bringing Tuscan knowledge to Provence (at least as far as the geometry of the circle is concerned). Instead they were learning from the environment where they were working; eventually, they (or the diffusion of their writings) brought this knowledge to Italy. As we shall see, this concerns more than the geometry of the circle.

M.15.4=F.vi.3 moves the reference to the diagram (still *forma*) to the end. Apart from that, it only differs from **V** on minor points in the wording.

V.15.5. This is the first problem making use of what we may call the "Pythagorean rule" (since Jacopo does not know it as a theorem). Most conspicuous are the terminological difficulties – here, *canto* (then replaced by *lato*, replaced again by *faccia*) is indubitably an "edge". So it is in V.26.12. However, in V.26.19 and V.26.24 it is no less certainly a corner.[215] In order to state that the two given sides are orthogonal, the notion of *diricto*/"straight" is used – but "not orthogonal" is *squadrato*, in agreement with the way orthogonality of a corner is expressed in V.26.27 (which however still speaks of the orthogonal sides as *diricte*).

M.15.5=F.vi.4 omits the general prescription "multiply firstly each face by itself, of these of which you know the measure". It also omits the explanation "because 50 times 50 makes 2500" and moves the reference to the diagram (*forma*) to the end. Terminologically, it distributes the terms *lato*, *faccia* and *canto* differently in the beginning (and has the plural *cantoni*/*cantora* instead of *canti* when all three are spoken of).

V.15.6. This second application of the Pythagorean rule is more homogeneous in its terminology – perhaps because a square field was a more familiar object. The statement of the rule itself is puzzling, however, with its replacement of the squares on two sides by the rectangles contained by two sides (probably intended to be opposite each other); the formulation may be a pedagogical trick, but it shows clearly that the rule is not thought of by Jacopo as being derived from any general theorem.[216]

---

[215] In *Liber habaci* [ed. Arrighi 1987:134*f*], the *faccia* of a square is its side, its *canto* the diagonal; the term certainly had no fixed interpretation.

[216] A similar multiplication of different but equal sides instead of a squaring turns up in problem V.15.19 in a (wrong) rule for the area of a regular pentagon.

The problem contains the first determination of the approximate square roots of a non-square number. The approximation

$$\sqrt{m^2+n} \approx m+\frac{n}{2m} \quad m+\frac{n}{2m}$$

is traditional. The earliest source for it is the Old Babylonian text VAT 6598 [ed. Neugebauer 1935: I, 279]; for some of the numerous later occurrences, see [Tropfke/Vogel et al 1980: 264–277].

M.15.6=F.vi.5 reduces the prescription to the numerical essentials, and does not point out that the solution is approximate:

> Do thus, multiply 10 times 10, they make 100. Now double 100, they are 200. Now find the root of this number, that is, of 200, which is 14 and $\frac{1}{7}$. And it is done.

The reference to the diagram (*forma*) is moved to the end.

V.15.7. The only reason for this problem to be found in the geometry chapter is that it speaks of a tower, otherwise the topic of the properly geometric problems V.15.9, V.15.10 and (two towers) V.15.24.[217] We observe that the dress is not taken very seriously: at the end of the 356th day, the serpent has reached the top, and we do not need to count the sliding-down in the subsequent night. This slip was not Jacopo's alone; it is found in numerous other treatises, including indeed the problem in the *Liber abbaci* [ed. Boncompagni 1857a: 177] which gave the type the name *Leo in puteo*.[218] Indeed, "correct" solutions – that is, solutions which take the dress seriously – are so rare that they are the ones that should be taken note of.[219]

M.15.7=F.vi.6 first teaches a different procedure, namely to choose 12 as a convenient number, to calculate 12×30 = 360, and to divide this by $\frac{1}{3}\cdot 12-\frac{1}{4}\cdot 12 = 4-3 = 1$; the intermediate 360 has no concrete interpretation, we observe, since the sub-unit $\frac{1}{12}$ of *braccio* is not introduced. Next **M+F** presents the procedure found in **V** as an alternative, concluding that the same outcome results "in one as well as the other way". The reference to the diagram (*forma*) comes after the description of the first method. The presence of *pigliare* might possibly suggest that the formulation goes back to an environment where this term was more acceptable than to the compiler who gave **M+F** its final shape.

V.15.8. Most noteworthy in this rule for the determination of a rectangular area is probably that nothing is said which hints at the rectangularity of the quadrangle. Obviously, fields

---

[217] However, Jacopo was not the only author who thought of this as geometry. The problem is found with the same parameters in Tommaso della Gazzaia's *Praticha* [ed. Arrighi & Nanni 1982: 36] in words which are close to those of Jacopo but too different both from **V** and from **M+F** to suggest direct or semi-direct copying.

[218] To keep the record straight one may observe that Fibonacci's method is so different that inspiration from him can be safely excluded. Only the error is shared – but it is widely shared.

[219] One is in Barthélémy de Romans' *Compendy* [ed. Spiesser 2003: 387].

are supposed by default to be rectangular if nothing else is said.

M.15.8=F.vi.7 is no more explicit about rectangularity. It makes the calculation before pointing out that it represents a general rule, but for once does not move the reference to the diagram to the end, retaining also the phrase of **V** "as you see drawn" instead of speaking of a *forma*. The remaining differences are minor.

V.15.9. This time, the tower serves to formulate a properly geometric problem. The type is quite common in abbacus geometries.

M.15.9=F.vi.8 leaves out the passage "that is, the closest, [...] and precisely it is not found", and moves the reference to the diagram (*forma*) to the end; apart from that, differences only concern the precise wording.

V.15.10. This is an obvious companion piece to V.15.8.

M.15.10=F.vi.9 leaves out both cross-references to the previous tower-moat-rope problem and moves the reference to the diagram (*forma*) to the end. Remaining differences only concern the precise formulations.

V.15.11. This return to the problem of finding the circular diameter from the perimeter serves the purpose of explaining in detail how to perform the division by $3\frac{1}{7}$ – and indeed by any mixed number. The explanation is certainly more transparent than a prescription to multiply by the inverse would have been. It does not coincide with what is done in rule-of-three problems when a mixed number turns up as divisor (in V.11.9 and V.11.12), but the spirit is the same.

M.15.11=F.vi.10 abbreviates the explanation (and **F** then makes it slightly longer because of a dittography), but follows the same pattern in a way which leaves little doubt that its starting point is indeed the text found in **V**.

V.15.12. Square roots are not needed in abbacus mathematics before it treats of geometry or algebra. It is thus for fair reasons that Jacopo (as many other abbacus writers) deals with the topic under the heading of geometry.

In the end of the list of perfect roots we find the unusual formulation "100 times 100 is the root of 10000". That it is intentional and no slip is shown by its reappearance in V.15.15–16. The formulation could perhaps be read as evidence that the technical terminology as encountered (or rather understood?) by Jacopo was not quite stabilized.

The introduction (M.15.12=F.unnumbered) to the corresponding section of **M+F** is strongly abbreviated and suggests that the compiler of **M+F** did not fully grasp the difference between the true root and "the closest root" (which, as we shall see, refers to the first approximation). The non-standard formulation of the root of 10000 is replaced by the standard phrase "the root of 10000 is 100 because 100 times 100 make 10000".

V.15.13. This is the rule which was announced in the previous section, and which was already used in problem V.15.6 – see above, p. 91. On one occasion, Jacopo finds $\sqrt{33\frac{1}{3}}$ as $\sqrt{300}/3$ (see above, note 49), but the corresponding rule is never enunciated explicitly, and he is likely to have borrowed the value without knowing how it was found. He appears indeed to have been unaware of the possibility to approximate from above, as done in this calculation – cf. below, p. 99. His mistaken second-order approximations were discussed on p. 21.

The corresponding section of M.15.13=F.unnumbered confirms that the compiler of **M+F** had problems with the difference between the true root and "the closest root" – or at least did not grasp that the procedure described can never yield the true root. **M+F** also fails to specify both which remainder is meant and that the quotient that is found is to be added to the number which was squared.

The structure of V.15.12–13 (a list of the square roots of perfect squares followed by the approximation rule) recurs in Tommaso della Gazzaia's *Praticha* [ed. Arrighi & Nanni 1982: 28*f*]. Descent from something close to **M+F** is possible but not certain; the first example is the root of 67, the same as V.15.15 and M.15.15=F.vi.12.

V.15.14. This example is combined with a direct demonstration that the approximation is not precise.

M.15.14=F.vi.11 contains the same example, but does not show that the "closest root" is not precise. It is obvious that the text in **V** cannot be derived from that of **M+F**, whereas the opposite process is quite possible.

V.15.15. This second example of the rule serves to specify that the approximation is supposed to be made by means of integers. We may take note of the additive interpretation of the doubling – which indeed was not unusual[220] – and of the reappearance of the certainly unusual formulation "that 8 and $^3/_{16}$ times 8 and $^3/_{16}$ is the root of 67".

M.15.15=F.vi.12 makes use of the routine formulation "we shall say that the root of 67 is 8 and $^3/_{16}$" and leaves out the observation that the root is only approximate. Apart from that it is close to **V**, taking over also the additive doubling.

V.15.16. This third and last example of the rule for the determination of approximate square roots repeats the unusual formulation "*r* times *r* is the root".

M.15.16=F.vi.13 leaves out the initial pedagogical declaration as well as the cross-reference "Do thus as I have said to you. Which is that number [...]", and uses the standard formulation "we shall say that the root of 82 is 9 and $^1/_{18}$". The final "And this will be enough about this matter. And we turn back to the measures" is replaced by a slightly corrupt repetition of part of the rule, "always divide the excess [*soperchio*] which the multiplied results for you by the double of that which you multiplied by itself".

V.15.17. As in the case of the serpent in the tower, we observe that the dress of this problem is not taken seriously: 567 is a multiple of 7, but neither 31 nor 567 is divisible

---

[220] It is also found in Tommaso della Gazzaia's *Praticha* [ed. Arrighi & Nanni 1982: 76] the second time the rule is given.

by 11. Since the dimensions of the houses are fixed and not only their area, the whole terrain cannot be filled up – 213 houses seems to be the maximum allowed, leaving (for instance) a strip long 567 *braccia* and large 2 *braccia*, and a rectangle of dimensions 6 *braccia* × 7 *braccia*.

The problem type comes from (and the disrespect for the implications of the dress is shared with) the Latin tradition. It occurs thrice in the Carolingian *Propositiones ad acuendos iuvenes* [ed. Folkerts 1978: 60*f*],[221] in *Geometria incerti auctoris* IV.35–37 [ed. Bubnov 1899: 354*f*], and again in the *Artis cuiuslibet consummatio* I.34,36 [ed. Victor 1979: 212, 218]. From the latter it was borrowed into the *Pratike de geometrie* I.34,36 [ed. Victor 1979: 504, 506].

M.15.17=F.vi.14 keeps the reference to the diagram in the beginning, but refers to it as *forma*. It replaces the initial statement that this is a rule by a closing phrase "do the similar in this way" and uses the second person already when the question is asked ("say me how many houses you can locate there"). Remaining differences are slight.

V.15.18. As in V.14.20 (see above, p. 79), volumes are measured in square *braccia* – and as in that case, Jacopo demonstrates that he possesses no intuition for volumes. The problem type, strange as it appears to us, is quite common in abbacus and related texts – see, for instance, *Liber abbaci* [ed. Boncompagni 1857a: 403*f*], where five problems about large stones or columns falling into cisterns follow after a problem of the same type as V.14.20.

The first of Fibonacci's problems appears to explain what has gone wrong for Jacopo. Fibonacci gives the dimensions in feet for the cistern and for the stone which falls into it, but also states that the cistern contains 1000 barrels (the ratio of the barrel to the "square foot" being treated as unknown). The ratio between the volumes of the stone and that of the cistern measured in "square feet" therefore determines which fraction of these 1000 barrels flows out.

M.15.18=F.vi.15 is somewhat more concise but repeats the blunder. The reference to the diagram is moved to the end (as *forma*). In the *Praticha*, Tommaso della Gazzaia [ed. Arrighi & Nanni 1982: 36*f*] gives the same problem in words that are so close to those of **M+F** that the texts must certainly be related. However, at the moment when the volume of the column is found Tommaso declares that this is the amount of water that flows out; but then he recycles the division 200÷25, namely in order to predict that the water level would fall 8 *braccia* if the column were removed.[222] It is thus obvious

---

[221] One of the fields is a trapezoid, one is triangular, and one is circular; in all cases, the houses are rectangular, that is, unable to fit precisely.

[222] The text then adds that this is not true, and gives the correct solution. Even though the manuscript we possess is a seventeenth- or eighteenth-century copy it seems from the language and orthography that the correction is due to Tommaso himself.

that Tommaso copies from something very close to **M+F**, something which on its part, as we see from other problems, is derived from something quite close to **V**.

Two problems about columns or stones falling into wells are found in *Trattato di tutta l'arte dell'abbacho* (T_F fol. 162^r, T_R fol. 79^v); in V.22.13–14 we shall encounter two about stones falling by chance into basins. Since "stones" and "columns" are treated as neighbours in the ps.-Heronian *Stereometrica* [ed. Heiberg 1914: 94, 96], it is not excluded that the problem type points back to ancient Greek practical geometry, although the story of objects falling into water is likely to have been invented later in the transmission chain. In any case, the presence of the problem type in the *Liber mahamaleth* [Sesiano 1988: 87] suggests that Fibonacci and the abbacus authors borrowed it from the Arabic world.[223]

V.15.19. The *n*th pentagonal number is $\frac{1}{2} \cdot (3n^2-n)$, which – by omission of the halving – explains Jacopo's astounding formula for the area of the pentagon. The explanation of the factor 3 as that of the remaining sides (presupposing that the square on the side is thought of as the product of two sides, as in V.15.6) could be a pedagogical or mnemotechnic trick due to Jacopo himself – no source except **V** seems to know it.

Ultimately, the formula certainly comes from the "sub-Euclidean" Boethian-agrimensorial tradition.[224] Paolo Gherardi [ed. Arrighi 1987]: 61 has the "correct" formula in a slightly corrupt phrasing, the side being 6. Tommaso della Gazzaia's *Praticha* [ed. Arrighi & Nanni 1982: 24*f*] applies Jacopo's rule (without "the other three", which indeed have already disappeared in **M+F**) to three examples, the first of which with side 8, in words which might descend from a close parent of **M+F**.[225] In *Trattato di tutta l'arte dell'abbacho* (T_F, fol. 137^{r-v}) we also find an example with side 8 (in the unit *canna*), but with two different procedures. One is presented as being done *per l'arte di rismetricha*, the other as done *per giometria*. By arithmetic, the side is first multiplied by itself, 8×8 giving 64. This is multiplied by 3, with result 192; from this is subtracted not a side but the square on a side, that is, 64, leaving 128 for the area. By geometry,

---

[223] A sophisticated variant is also present in Elia Misrachi's Byzantine arithmetic [Wertheim 1896: 63–65] – the stone falls so slowly into a cistern filled and emptied through taps of different capacity that the cistern remains exactly filled without spilling over. But this treatise was written around 1500, and none of the earlier two extant abbacus-type Byzantine writings know it.

[224] The "correct" formula is in the *Geometria incerti auctoris* IV.17, ed. [Bubnov 1899: 346] and in Epaphroditus & Vitrivius Rufus 21 [ed. Bubnov 1899: 534]. It is also found in the ps.-Boethian *Geometria altera* II.xxv [ed. Folkerts 1970], "correct" when stated abstractly but illustrated (for side 6) by the calculation (6×6)×3+6 – that is, with addition replacing subtraction and omitted halving. The "correct" formula is also given in the treatise known as *Artis cuiuslibet consummatio* I.15 [ed. Victor 1979: 158], written in 1193 or somewhat later, perhaps in Paris [Victor 1979: 24*f*; Knorr 1990: 318].

[225] Even here, Tommaso (?) adds that the rule which he relates is not true, cf. note 222.

half of the side has to be multiplied by the measured height; since this is claimed to be $6\frac{9}{22}$, the area of each of the constituent isosceles triangles is found to be $25\frac{7}{11}$, and the whole area to be $128\frac{2}{11}$, almost the same.[226] The arithmetical method must have resulted from a repair of Jacopo's formula (which need not have been Jacopo's alone): this is the only explanation that it is not calculated straightforwardly as 2 times 8×8 – direct repair of the formula for the pentagonal number can be excluded, since the factor $\frac{1}{2}$ is omitted. The geometric method coincides with what is found in *Pratike de geometrie* I.15 [ed. Victor 1979: 489], written in Picardy in the late thirteenth century: to divide the pentagon into five triangles, and to find the measure of the distance from the mid-point of one side to the vertex – and the idea to measure is so rare in "mensuration" treatises (hard to believe, perhaps, but true!) that a link is plausible even though the Picardian treatise gives no numbers[227] – cf. also below, p. 140.

V.15.20. Although the *scudo*/"shield" is explained to be simply a triangle, the term mostly refers to an equilateral triangle – depicted in agreement with the non-technical interpretation of the term.[228] Jacopo does not explain the reason that his computation works, but it is so simple that we may assume he knew. How much he understood of the extraction of the square root is a different question (cf. note 49). Firstly, he has not explained – nor does he explain here – that $\sqrt{p} = \sqrt{\frac{pn^2}{n}}$. This principle was of course not new – it is

---

[226] Actually, the height should be 5.5055..., and the whole area 110.11... . It is not clear how the value $6\frac{9}{22}$ was obtained; backward calculation from the "arithmetical area" gives the simpler $6\frac{2}{5}$ Possibly, the outcome of a flawed measurement has been reduced to a fraction by means of the "Euclidean algorithm" – as shown by the two transformations

$$\frac{9}{22} = \cfrac{1}{2-\cfrac{1}{2-\frac{1}{4}}} \quad \text{and} \quad \frac{6\frac{9}{22}}{8} = \cfrac{1}{1-\cfrac{1}{4-\frac{1}{35}}},$$

the value $6\frac{9}{22}$ can be reached in a few steps in several ways. I have difficulty in imagining how the denominator 22 could turn up in other ways – but the historian's lack of fantasy should never count as a serious argument if not supported by other evidence!

[227] Most of this vernacular treatise is translated from the *Artis cuiuslibet consummatio*, but not this rule. As mentioned in note 224, that Latin treatise gives the rule for the pentagonal number.

Since the idea of measuring is also in *Geometrie due sunt partes principales* [ed. Hahn 1982: 155], it may have circulated in the Franco-Provençal region.

[228] But not always: in the *Libro di molte ragioni d'abaco* from Lucca [ed. Arrighi 1973: 114] a triangle (*terno*) in the shape of a *scudo* has the sides 6, 8 and 9 *braccia*; the same configuration as well as the triangle with sides 5, 6 and 7 *braccia* are also spoken of in the *Liber habaci* [ed. Arrighi 1987: 133] simply as *scudi*. Slightly later in the same treatise (p. 142), however, a triangle with sides 5, 6 and 7 *braccia* is only supposed to be "made almost like a shield" (*fatto quasi chome uno schudo*), and on p. 133 the rule for finding the side of a shield presupposes it to be equilateral. Strange as it may perhaps seem to us, the "normal shield" is thus equilateral, but specification of other dimensions allows it to deviate somewhat from this ideal shape.

explained, for instance, by Fibonacci in the *Liber abbaci* [ed. Boncompagni 1857a: 355] with $n = 100$ (not to speak of its use by Arabic authors). Secondly, Jacopo does not specify the first approximation (his so-called "closest root"), nor the origin of the (wrong) second correction.

M.15.21A(20)=F.vi.18 at first follows **V** rather closely, apart from moving the reference to the diagram to the end (as *forma*) and abbreviating the passage "multiply the straight in middle by itself, that is, 5 *braccia* times 5 *braccia*" into "multiply 5 times 5" (in agreement with its stylistic choices elsewhere). But the compiler has understood that something is wrong in the calculation of the square root, and rewrites the final part while repairing the value of the root – unfortunately inconsistently (see p. 22), by mixing the wrong correction term given by **V** with the appropriate correction known to Ibn al-Bannā' and al-Qalaṣādī. We may conclude that this appropriate correction was around in the environment where **M+F** was made, even though we have no better traces of it.

To add to the confusion, **M+F** claims the solution to agree with the rule that was stated earlier, and for once remembers to say that it is only approximate. It also specifies that the rule is meant for shields which are equilateral.

V.15.21. This problem shows that not only volumes but also other spatial questions were beyond Jacopo's ability. His use of rule (1) for the circular area instead of his usual (7) suggests that he has borrowed the problem without thinking too much about it, but such lack of attention corroborates the hypothesis that he did not feel comfortable about the matter.

M.15.21=F.vi.17 moves the reference to the diagram (*forma*) to the end; the drawing is also changed, corresponding perhaps better to how real pavilions were made – but not at all to the conic shape presupposed by the text, cf. above, p. 20. Apart from that, differences are slight. In particular, **M+F** shares the use of rule (1) and the belief that the area of the canvas is that of a triangle with height equal to the side and base equal to the diameter of the tent.

The conic pavilion was a rather widespread topic, but other authors usually solved it better. In the Umbrian *Livero de l'abbecho* [ed. Arrighi 1989: 122], the identical problem (but with a different terminology) is solved correctly; in *Trattato di tutta l'arte dell' abbacho* (T_F fol. 158^r, T_R fol. 77^v) the parameters are still the same, but the question is slightly more sophisticated: which length of fabric is needed if its width is $\frac{2}{3}$ of a *braccio*? Even in this treatise, the solution is correct.

V.15.22. In this slightly more elaborate repetition of V.15.20 with new numbers we observe that even the fallacious attempt to improve on the usual "closest possible approximation" to the square root is repeated; strangely, however, the correction $\frac{4}{25}$ is forgotten when the side is multiplied by 3. This might seem to imply that a first version of the text did not contain it; however, the traces of a corresponding correction in M.15.21A(20)=F.vi.18 shows the wrong second-order correction to have been present already in the manuscript from which **M+F** descends.

M.15.22=F.vi.19 makes no attempt at improving the first approximation, nor does it point out that $10\frac{2}{5}$ *is* an approximation. The statement of the problem is more elaborate than that of **V**, and so is the closing passage (but it does not contain the reference to the diagram that was left out in the beginning).

V.15.23. This simple application of the Pythagorean rule represents another common problem type. This time no attempt is made to improve on the first approximation of the square root.

M.15.23=F.vi.20 moves the reference diagram (*forma*) to the end. It is slightly more concise than **V** but follows it rather closely.

V.15.24. Like V.14.29 (see above, p. 204), this is a "poor man's version" of a much more sophisticated problem which is known not only from many abbacus writings but also from Mahāvīra: On the ground between two towers of different height, a cup is placed at the point where the two doves are equally distant from it; the distance of this point from each tower is asked for. As Mahāvīra points out in the *Gaṇita-sāra-saṅgraha* (VII.201½–203½, ed., trans. [Raṅgācārya 1912: 249*f*]), the solution is based on the same trick as the one that gives the height of a scalene triangle (a quasi-algebraic combination of the application of the Pythagorean theorem to two right triangles). This version is found, for instance, in the Columbia Algorism [ed. Vogel 1977: 139*f*], in Paolo Gherardi's *Libro di ragioni* [ed. Arrighi 1987: 65–67] with a correct but unargued solution, and in the *Liber habaci* [ed. Arrighi 1987: 139] with a sham solution that only works for the specific parameters chosen.[229]

Once it is decided that time can be measured as distance, Jacopo's problem is much simpler. Only the numerical calculations may be in need of explanation. $\sqrt{3125}$ is approximated from below according to the usual rule, although an approximation from above would be much more precise – which we may therefore suppose that Jacopo did not know. According to the same rule, $\sqrt{2900}$ should become $53\frac{91}{106}$. Jacopo probably calculates $\sqrt{2900} = 53 + \frac{2900-2809}{2\cdot 53} = 53 + \frac{100-9}{2\cdot 53}$, and by error replaces this by $53\frac{9}{2\cdot 53}$. The calculation $55\frac{10}{11} - 53\frac{9}{106} = 2\frac{87}{106} + \varepsilon$ may come from the transformation $\frac{10}{11} = \frac{100}{110} = \frac{100-4}{110-4} + \varepsilon = \frac{96}{106} + \varepsilon$. That a precise result is said *at this point* to be unattainable seems strange – Jacopo should know how to subtract fractions since he has taught it. Maybe he was caught by his shortcut, knowing directly (and thus for rather intuitive reasons) that $\frac{a-d}{b-d}$ is smaller than $\frac{a}{b}$ (for true fractions, where $0<a<b$) but not knowing how much. Not very plausible is that the observation comes at this point because

---

[229] If the heights of the two towers are h and H and their distance D, the distances of the cups from the towers are $\frac{D}{2} \pm \frac{H^2-h^2}{2D}$. The *Liber habaci* instead finds the smaller distance as $(\frac{H-h}{2}\cdot h) / (\frac{D}{2})$, which is numerically correct, but only because $D^2 = (H+h)^2-4h^2$ – in the actual case $D = H = 30$, $h = \frac{2}{3}H = 20$.

it holds for both square roots: since both square root approximation are too large, Jacopo would hardly be able to know (even if he had made no errors) which error is larger, and by how much; taking this into account he would not know whether *pocho pocho*/"a little bit" should be *più* or *meno*, more or less.

M.15.24=F.vi.21 has a somewhat different text. It keeps the reference to the drawing in the beginning (which is indeed too artistic to be considered a diagram – but this might also be claimed about other drawings which *are* spoken of as *forma*); it even gives a separate reference to the drawing of the cup. However, it interchanges the order of the towers (both in the presentation and in the calculations), and it extracts both square roots anew – but not exactly with better success. The root of 3125 is stated to be $56\frac{1}{112}$; since $56^2 = 3136$, it should have been $56-\frac{11}{112}$ (with an approximation from above). The root of 2900 *is* found with an approximation from above, as $54-\frac{4}{77}$, but should have been $54-\frac{16}{2\cdot54} = 54-\frac{4}{27}$. In the subtraction, $(56+\frac{1}{112})-(54-\frac{4}{77})$ is found as $56-54-\frac{4}{77} = 1\frac{73}{77}$; nothing is said about the result being approximate.

V.15.25. The main conundrum in this problem is architectonic. It does not make much sense to place gutters from the roof-ridge to the wall, but no more adequate translation seems to be at hand, nor is the original formulation more technically meaningful. The end of the text speaks of them as "drains (*dicorrenti*) [made] from the top of the wall to the roof-ridge", which does not help for a roof which carries no flat terrace at its top.

The application of the Pythagorean rule is straightforward, although the argument for splitting 40 into halves – namely that "he wants to make two gutters" – seems off the point. The determination of the square root repeats the fallacious second-order approximation of V.15.20 and V.15.22.

M.15.25=F.vi.22 replaces the "townsman" by the subject "I" and is more concise; in particular it leaves out the argument about two gutters together with the fallacious second-order correction. Further, it moves the reference to a drawing (not *forma*) to the end. In other respects, differences are minor. After the problem, **M+F** has a section (M.15.25A=F.vi.23) which describes in the usual way and words how to determine $\sqrt{101}$ (namely as $10\frac{1}{20}$), without observing that the value is not exact.

Ch. 16–19. Algebra and Quasi-Algebra, with a Non-Algebraic Intruder

In contrast to the preceding two chapters, Chapters 16–17 constitute a systematic treatise on their subject – first- and second-degree algebra and problems of the third and fourth degree that are either homogeneous or can be reduced to the second or first degree. Chapter 18 – consisting of a single problem – is an intruder in the context but equally absent from **M+F**. Chapter 19 contains four problems about continued proportions solved without reference to algebra but presupposing familiarity with what *we* would describe as the sequence of algebraic powers. Many Arabic mathematicians from al-Karajī onward would also have seen the technique as fundamentally algebraic, which perhaps provides the ultimate explanation why Jacopo (and presumably a source of his) treated these problems in an algebraic context.

It may seem strange that the algebraic treatise begins so abruptly – other transitions (the start of Chapters 11, 15, 20 and 21) are pointed out in clear words. A short fragment quoted in a manuscript from 1424 (Vat. Lat. 10488, fol. 28ᵛ–31ʳ, original foliation) may provide the explanation. The fragment (to which we shall return below, p. 176) is a similar but briefer presentation of algebra, excerpted from a treatise written by Giovanni di Davizzo de l'abacho da Firenze in 1339. After an introduction follow exactly the rules offered in **V**; the introduction, on its part, gives the sign rules (plus times plus makes plus, plus times minus makes minus, etc.) and rules for the products of powers of the algebraic unknown and for operations with roots. A similar introduction might have taken up a complete sheet,[230] which could have been lost. In any case something *must* have been lost, since Chapter 16 closes with the words "Here I end the six rules combined with various examples", which ask for an introductory phrase announcing these rules.; it is unlikely that Jacopo would not himself have produced a preamble if he was using another treatise where this introduction was lost.[231]

The discussion of the general conclusions that can be drawn from the examination of Jacopo's algebra asks for a chapter on its own (belov, p. 147), but a commentary to the text itself should still be useful.

Since anything algebraic in the European Middle Ages was ultimately derived from Arabic *al-jabr*, an introduction to this theme will be adequate.

An Aside on Arabic Algebra and Its Mixed Origin

The earliest extant treatise on the topic *al-jabr wa'l-muqābalah* (which may well represent the very first description of the technique in a regular treatise) was written by al-Khwārizmī between 813 and 833 CE at the behest of the caliph al-Maʿmūn. Six standard problems ("cases") about an unknown "possession" (Arabic *māl*, becoming *census* in Latin), its square root (*jidhr*, *radix* in Latin) and a known number (of dirhams, *dragmae* in Latin) make up the core of the art:[232]

---

[230] The number of lines in the Davizzo-fragments corresponds indeed to that of a sheet in **V**, and evidence from the coin lists discussed above (p. 9) suggests that these correspond well to the sheets of the archetype. However, it is unlikely that the text that could have been lost from Jacopo's original corresponded in details to Davizzo's very idiosyncratic presentation.

[231] I shall not repeat the arguments that the algebra chapters as they stand have been shaped by Jacopo or were at least present in the early common archetype for **V** and **M+F** – but see pp. 23 and 57. This certainly does not preclude that Jacopo built on pre-existing material; we shall see (pp. 177–180) that he did so at least in the formulation of the rules.

[232] As I have argued in [Høyrup 1998a], the best text for the part of the work with which we are here concerned is Gherardo da Cremona's Latin translation [ed. Hughes 1986], from which I translate in the following (cf. note 305). For the purpose of the present discussion Frederic Rosen's edition

(a)   possession is made equal to roots
(b)   possession is made equal to number
(c)   roots are made equal to number

(d)   possession and roots are made equal to number
(e)   possession and number are made equal to roots
(f)   roots and number are made equal to possession

Of these, the former three are spoken of as "simple" and the latter three as "composite".
For each, a rule based on a paradigmatic normalized numerical example is given;
afterwards it is shown how to reduce non-normalized examples to the normalized case.
For instance, case (d) runs as follows:

> But possession and roots that are made equal to a number is as if you say, "A possession and
> ten roots are made equal to thirty-nine dragmas". The meaning of which is: from which
> possession, to which is added ten of its roots, is aggregated a total which is thirty-nine? The
> rule of which is that you halve the roots, which in this question are five. Then multiply them
> by themselves, and from them 25 are made. To which add thirty-nine, and they will be sixty-
> four. Whose roots you take, which is eight. Then subtract from it half of the roots, which is
> five. There thus remains three, which is the root of the possession. And the possession is nine.
> And if two possessions or three or more or fewer are mentioned, reduce them similarly to one
> possession. And what are with them of roots or numbers, reduce them similarly as you reduced
> the possession. Which is as if you say, "Two possessions and ten roots are made equal to forty-
> eight". The meaning of which is that when to any two possessions are added the equal of ten
> roots of one of them, forty-eight are aggregated from it. Two possessions must hence be reduced
> to one possession. Now we know however that one possession is the half of two possessions.
> Reduce therefore everything that is in question to its half. And it is as if one said: "A possession
> and five roots are equal to twenty-four". Which means that with any possession five of the
> roots of the same are added, from which twenty-four are aggregated. Halve the roots, and they
> are two and a half. Multiply them then by themselves, and they make six and a fourth. [...]

Translated into symbols, the basic example for (d) is thus $y+10\sqrt{y} = 39$ rather than
$x^2+10x = 39$; the fundamental unknown appears to be the possession, not its square root.
Correspondingly, once the root is found, al-Khwārizmī also determines the possession
(except in case (b), where it is already known). But this reflects the origin of the technique,
not necessarily how al-Khwārizmī saw it, and not how he used it. Already when
introducing the basic terms al-Khwārizmī refers to roots as numbers, and defines the
possession as a root multiplied by itself; when later applying the technique, he also
identifies the "root" with the "thing" (*šay'*, Latin *res*) and the "possession" with the
arithmetical square on this thing, as we shall see.

[1831] and English translation is also acceptable (provided one is aware that his "square" translates
māl; in fact, Rosen distinguishes between "square", exclusively used in this sense, and "quadrate",
the geometric configuration, rendering Arabic *murabba*ᶜ.

Problems about quantities of money and their square root are likely to be related to what we find in Indian medieval treatises, where we meet the square root of Arjuna's arrows, of a flock of elephants, etc. – related probably by common origin rather than by direct descent [Høyrup 2001: 124].[233] They have nothing to do with Greek algebra as we know it from Diophantos, and even though we cannot exclude a tortuous inspiration from Old Babylonian mathematics there is nothing positive at all to suggest such a connection.

Maybe because he worked in an environment where others were engaged in the adoption of Greek mathematics, al-Khwārizmī was not satisfied with a mere presentation of the *al-jabr* technique: mathematics had to be based on arguments. When explaining the addition of binomials he tells that he had produced geometric proofs on his own (and that he had failed when trying to make something comprehensible for the addition of three-term polynomials); when providing geometric demonstrations for the rules by means of which he solves the cases (d)–(f) he says nothing of the kind, and already for this reason we may assume that he borrowed them from elsewhere – and closer investigation shows indeed that they are taken over from a non-scholarly practical surveyors' tradition whose geometrical riddles had already served as the starting point for Old Babylonian "algebra" in the early second millennium BCE.[234] We may look at the proof for case (d):

The cause [of the halving of the roots, characteristic of the mixed cases] is as follows. A possession and ten roots are made equal to thirty-nine dragmas. Make therefore for it a quadratic surface with unknown sides, which is the possession which we want to know together with its sides. Let the surface be *AB*. But each of its sides is its root. And each of its sides, when multiplied by a number, then the number which is aggregated from that is the number of roots of which each is as the root of this surface. Since it was thus said that there were ten roots with the possession, let us take a fourth of ten, which is two and a half. And let us make for each fourth a surface together with one of the sides of the surface. With the first surface, which is the surface *AB*, there will thus be four equal surfaces, the length of each of which is equal to the root of *AB* and the width two and a half.

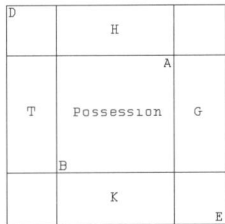

Al-Khwārizmī's first proof for case (d)

---

[233] Further evidence for a link to Indian mathematics is the use of the metaphor "root". In Sanskrit mathematics, the square root is understood as "that side on which a square rests" [Datta & Singh 1962: I, 169f], which makes the metaphor "root" (*mūla*) meaningful. In Arabic mathematics, the square root is understood in purely arithmetical terms, and the corresponding metaphor is therefore incomprehensible: it must be a calque, if not from Sanskrit then from another area where the square root was understood similarly. Since even a geometric conceptualization does not *per se* entail the idea of the root of a plant (the ancient Greeks spoke of "the side"), the Sanskrit term is likely to be inspired from the same place.

[234] The argument for this conclusion is far too extensive to be repeated here, but see [Høyrup 2001] or [Høyrup 2002a: 362–417].

Which are the surfaces *G*, *H*, *T* and *K*. From the root of a surface with equal and also unknown sides is lacking that which is diminished in the four corners, that is, from each of the corners is lacking the multiplication of two and a half by two and a half. What is needed in numbers for the quadratic surface to be completed is thus four times two and a half multiplied by itself. And from the sum of all this, twenty-five is aggregated.

Therefore, as the argument goes on, the completed square *DE* has the area 39+25 = 64, and its side must be 8. Subtracting 2·2½ = 5 = 10/2 we find the side of *AB* to be 8–5 = 3.

A second proof corresponds better to the details of the rule. It is based on a diagram where each of the rectangles *D* and *G* have an area (10/2)·*r* = 5*r*, and the lower left completing square an area (10/2)² = 5². Therefore, the total area of *AB* with rectangles *D* and *G* is *r*²+10*r* = 39, and the area of the large square is 39+25 = 64.

Al-Khwārizmī's second proof for case (d)

The proofs are followed by chapters on the arithmetic of monomials, roots and polynomials, after which follow a chapter containing six problems, one for each case, a chapter containing mixed questions, and a short chapter on commercial transactions (*mu'āmalāt*), actually on the rule of three (referred to above, p. 61). The original treatise goes on with a chapter on practical geometry and another one on the algebra of inheritance rules, but these were left out in the Latin translations.

The way the technique was applied can be illustrated by al-Khwārizmī's own illustration of case (e):

"Divide ten in two parts, and multiply each of them by itself, and aggregate them. And it amounts to fifty-eight". Whose rule is that you multiply ten minus a thing by itself, and hundred and a possession minus twenty things results. Then multiply a thing by itself, and it will be a possession. Then aggregate them, and they will be one hundred, known, and two possessions minus twenty things, which are made equal to fifty-eight. Restore then one hundred and two possessions with the things that were taken away, and add them to fifty-eight. And you say: "One hundred, and two possessions, are made equal to fifty-eight and twenty thing". Reduce it therefore to one possession. You therefore say: "Fifty and a possession are made equal to twenty-nine and ten things". Oppose hence by those, which means that you throw twenty-nine out from fifty. There thus remains twenty-one and a possession, which is made equal to ten things. Hence halve the roots, and five result. [...].

First a position is made (implicitly only, because the same position was made explicitly in the first three examples, all three similarly of the type "divided 10"): an entity linked to the original problem (here, one of the two parts) is represented by the *thing*, from which follows both what the other part will be and what results when each is multiplied by itself. Introducing the abbreviations that will serve in the following we have that one part is *t* (for *thing*), the other therefore 10–*t*. The product of the former with itself becomes *C* (for *census/censo*), that of the second with itself becomes 100+*C*–20*t*. We therefore have the equation

$$100+2C-20t = 58 .$$

The next operation is a "restoration" (*al-jabr* in Arabic): the lacking $20t$ are added, which gives us

$$100+2C = 58+20t ,$$

which by normalization becomes

$$50+C = 29+10t .$$

Now follows an operation called "opposition" (*al-muqābalah* in Arabic), which involves that the 29 which are present on both sides is removed. This could suggest that this removal is exactly what is meant by the term, and that is indeed what became the current interpretation; we shall see (p. 157), however, that the meaning may be different, namely the production of the simple three-term equation – an interpretation which is not excluded by al-Khwārizmī's words. In any case, we have now produced one of the six fundamental cases, and may proceed according to the rule. As we see, al-Khwārizmī makes no use of symbols (even his numbers are indicated by words, not by numerals); the piece is as pure an example as can be found of what since Nesselmann [1842: 302] has been known as "rhetorical algebra".

What is the reason that we have not one algebraic standard representation in terms of which "real" entities are expressed but a mixed system where the abstract representative *thing* is so to speak represented by the "root"? This may have to do with the *regula recta* mentioned above, p. 103 (which we know from the *Liber abbaci*, but which Fibonacci [ed. Boncompagni 1857a: 191] ascribes to the Arabs, and which also occurs systematically as *regula* in the twelfth-century translation *Liber augmentis et diminutionis* [ed. Libri 1838: I, 304–371].[235] Fibonacci presents it and uses it long before approaching *algebra et almuchabala*, and never connects the two methods; obviously, he sees them as distinct, even though the *regula recta* makes use of a *res*, a *thing*, just as we have seen with al-Khwārizmī.

Fibonacci introduces the *regula recta* as an alternative approach to a problem about two men who have money. The former, if receiving 7 *denari* from the latter, will have 5 times as much as remains to the giver. The latter, if receiving 5 *denari* from the former, will have 7 times as much as remains to the giver. The second man is posited to have *a thing* plus the 7 *denari* which the first man asks for.

This can be compared to Diophantos's *Aritmetica* I,xv [ed. Tannery 1893: I, 36], "to find two numbers such that each, after having received a given number from the other, will be in a given ratio to the remainder". Not only are the two problems strictly identical

---

[235] It turns up as *modo retto/repto/recto* in Benedetto da Firenze's *Tractato d'abbaco*, ed. [Arrighi 1974: 153, 168, 181] (see note 63). Benedetto speaks of the unknown as *una quantità*. None of the problems in question are taken over from Fibonacci, and Benedetto can hence be presumed to have other sources for the method and its name.

in mathematical structure – Diophantos also makes the position that the ἀριθμός is the second number diminished by that number which is to be transferred to the first, exactly as does Fibonacci. Since papyrological evidence shows that the ἀριθμός-technique was in wider use, at least for first-degree problems[236], this is strong evidence that the *regula recta* did not come from the *al-jabr wa'l-muqābalah* tradition but from that ἀριθμός-technique which Diophantos had known about, and that this tradition, originally carried by Greco-Hellenistic logistics (and perhaps by a wider group), had been taken over by Arabic-speaking practical arithmetic.[237]

Other evidence may point in the same direction. Al-Khwārizmī does not speak of higher powers, and within the framework of a possession and its roots there is not much space for them if we take the framework at its precise words.[238] However, already the next major source for Arabic algebra – Abū Kāmil's treatise from the late ninth or the earlier tenth century CE – makes ample use of *ka'b* (Latin *cubus*, the third power), *māl māl* (Latin *census census*, the fourth power), *māl māl māl* (Latin *census census census*, the sixth power), *ka'b ka'b* (Latin *cubis cubi*, also the sixth power) and *māl māl māl māl* (Latin *census census census census*, the eighth power).[239] The inspiration *could* come from Qusṭā Ibn Lūqā's translation of Diophantos, since Qusṭā had already fused Diophantine "theoretical arithmetic" and *al-jabr*, but it could also come from a broader inspiration, since the whole class of corresponding terms and not only the ἀριθμός were already in broader use when Diophantos wrote the *Arithmetica* – it "had been approved" (ἐδοκιμάσθη), as he says, that they were the elements (in the literal sense of "letters" or "building blocks") of arithmetical theory [ed. Tannery 1893: I, 4].[240] In any circum-

---

[236] See [Robbins 1929], [Karpinski & Robbins 1929b] and [Vogel 1930].

[237] It is true that Fibonacci says that the problem was posed to him by a master from Constantinople; but he does not say so about the solution, and the method, apart from being presented as an alternative, is ascribed to the Arabs.

[238] Not much, but still some. After his translation of al-Khwārizmī's algebra, Gherardo da Cremona adds a collection of 21 problems found "in another book"; though thus hardly from al-Khwārizmī's hand, they have crept into the fourteenth-century Arabic manuscript from which the two published editions were made [Rosen 1831; Mušarrafa and Aḥmad 1939]; four had already entered the text translated by Robert of Chester [ed. Hughes 1989]. Three of the additional problems are of the third degree but reducible; as an example we may look at this one [ed. Hughes 1986: 258]:

> And if [somebody] said: There is a possession, which I multiplied by its root, and the triple of the first possession resulted. Its treatment is: since, when you multiply the root of the possession by its third, the possession results, I say that the third of this possession is its root. And itself it is nine.

(that is, since $C \cdot t = 3C$, $(\frac{1}{3}C) \cdot t = C$ – but since $C = t \cdot t$, $\frac{1}{3}C = t$).

A complete concordance for the various versions of al-Khwārizmī's treatise is found in [Oaks & Alkhateeb 2005: 419].

[239] See the index of terms in [Sesiano 1993: 421–447].

[240] In fact, one agrimensor treatise (Epaphroditus and Vitruvius Rufus, ch. 40*f*, ed. [Bubnov 1899:

stance it is clear that Arabic algebra from Abū Kāmil onward was a melting pot, where the rules for solving the possession-root riddles, the surveyors' geometric proofs, and one or several strata of Greco-Hellenistic logistics were fused together.

Let us return for a moment to al Khwārizmī. If a mathematical author presenting a particular technique wants to prove its utility, it is not uncommon that he applies it to problems of a known type, demonstrating how much easier they are treated by means of the alternative technique he offers. Now it is striking that four of the first six examples – and seven of the twelve mixed problems – are of the type "divided ten". This might suggest that this type did *not* belong to traditional *al-jabr* but was used to illustrate the efficiency of this technique. Actually, seven of the 39 problems in Diophantos's *Arithmetica* I (which collects traditional "recreational" problems translated into the idiom of abstract numbers) ask for the division of a given number into two parts (two others ask for two or three different divisions of the same number). Even the problem type "divided ten" might thus represent an inheritance from ancient logistics.

Jacopo's First- and Second-Degree Algebra

In Chapter V.16 of Jacopo's *Tractatus* – the first part of his algebra – we find the same six fundamental cases as in al-Khwārizmī's treatise, though in a different order for the simple cases. Moreover, Jacopo's rules all start by a normalization, whereas those of al-Khwārizmī presuppose normalization.[241] We find nothing corresponding to al-Khwārizmī's paradigmatic examples in terms of possessions and roots, the examples given by Jacopo correspond rather to the six applications of the rules which al-Khwārizmī gives in a later chapter (but while al-Khwārizmī's examples are all in pure numbers, five of Jacopo's ten pretend to deal with commercial real life).

As we shall see, Jacopo's use of the term *al-jabr*, "to restore", is wider than what we find in al-Khwārizmī, and the term "to oppose" is absent (while a different translation of *al-muqābalah* may be obliquely present).

I shall make use of the symbols introduced above – *t* stands for *thing*, *C* for *censo*, the abbacus analogue of Latin *census*. Further on, *K* will stand for *cubo* and *CC* for *censo di censi* (Jacopo goes no further). In the translation, I have left *censo, cubo* and *censo de censi* untranslated in order to emphasize their role as technical terms within a particular algebraic representation that should not automatically be confounded with our unlimited

---

550] says that 100 is *decies in dynamum* and 1000 *decies in quibo*. According to a fragment that may go back to Varro (that is, to the first century BCE) [ed. Bubnov 1899: 511*f*], multiplying 16 *in dynamum* gives 256, multiplying 3 *in dynamodynamum* gives 81, and 4 *in dynamocubum* gives 1024. Cf. also [Folkerts 2003: II, 15–20].

Even if the exact ascriptions should be mistaken, this shows that the terms were around (though only as designations for powers) – the authors are not suspect of knowing Diophantos.

[241] This does not seem to be the case in the extant Arabic text as edited by Rosen [1831] and Mušarrafa & Aḥmad [1938] – but see notes 232 and 305.

sequence of powers. *Cosa*, whose everyday connotations could not be ignored by Jacopo and his readers, is translated accordingly as "thing".

V.16.1. In symbols,

$$\alpha t = \beta \implies t = \frac{\beta}{\alpha} .$$

V.16.2. In symbols,

$$10 = a+b , \quad a/b = 100 .$$

Positing $a := t$ we have $b = 10-t$, and thus

$$t/(10-t) = 100 , \quad t = 1000-100t , \quad 101t = 1000 , \quad t = 1000/101 = 9^{91}/_{101} .$$

The solution is thus

$$a = 9^{91}/_{101} , \quad b = 10-9^{91}/_{101} = {}^{10}/_{101} .$$

For his corresponding case (c), al-Khwārizmī has the problem

$$10 = a+b , \quad a/b = 4 .$$

However, his position is $b := t$, leading to $(10-t)/t = 4$.[242]

"To restore", we observe, is used in the familiar sense.

V.16.3. The investments of the three companions may be called a, b and c, and the profit rate p; the unit is the *libra*. Then

$$pa+pb+pc = 30 , \quad a = 10 , \quad b = 20 , \quad pc = 15 .$$

Positing $c := t$ we thus have

$$p\cdot(a+b+t) = 30 \quad \text{or} \quad p\cdot(30+t) = 30 .$$

By the partnership rule (cf. p. 70), the third companion gains

$$pc = \frac{t\cdot 30}{30+t} ,$$

and since this gain is 15,

$$15\cdot(30+t) = 30t \quad \text{or} \quad 450+15t = 30t \quad \text{or} \quad 15t = 450 .$$

Hence $t = c = 30$.

In consequence, the part of the gain which falls to the first two companions is $30-15 = 15$, while their investment is $10+20 = 30$; this is divided proportionally according to the partnership rule,

$$pa = \frac{15\cdot 10}{30} = 5 , \quad pb = \frac{15\cdot 20}{30} = 10 .$$

We observe that "restoration" is used in this problem in a non-standard way. We also observe the invention of a new, formal partnership for the purpose of dividing the profits between the first and the second partner, referred to above (p. 12 and elsewhere) as a characteristic of manuscript **V** as a whole.

---

[242] In the fifth of the "varied problems" [ed. Hughes 1986: 252], al-Khwārizmī does divide by "10 minus a *thing*".

As pointed out by Karpinski [1929a: 175], the same problem (with different parameters) is found in Mahāvīra's *Ganita-sāra-sangraha* (VI.223–225, ed., trans. [Rangācārya 1912: 151]). However, Mahāvīra's method is different, namely application of the "partnership" rule for proportionate distribution (which in the present case would tell us that the third partner must have made one half of the investment, since he gets half of the total profits); all that follows is therefore that the commercial mathematics of the abbacus school had its root in a mathematical culture reaching as far as the sea trade on the Indian Ocean and going back at least to the early Middle Ages.[243] That the riddles of so-called "recreational mathematics" were shared as widely has long been common knowledge[244]; it can come as no surprise that their diffusion was due to a professional culture engaged in commercial computation.[245]

V.16.4. In symbols,

$$\alpha C = \beta \implies t = \sqrt{\frac{\beta}{\alpha}} \ .$$

V.16.5. The problem can be expressed
$$a{:}b = 2{:}3 \ , \quad b^2{-}a^2 = 20 \ .$$
Positing $a := 2t$, $b := 3t$ we get
$$9C{-}4C = 5C = 20 \quad \text{or} \quad C = 4 \quad \text{or} \quad t = \sqrt{4} = 2 \ ,$$
and thus
$$a = 2t = 4 \ , \quad b = 3t = 6 \ .$$

V.16.6. In symbols,

$$\alpha C = \beta t \implies t = \frac{\beta}{\alpha} \ .$$

V.16.7. The problem can be expressed
$$a{:}b = 4{:}9 \ , \quad a{\cdot}b = a{+}b \ .$$
Positing $a := 4t$, $b := 9t$ we have
$$4t{\cdot}9t = 4t{+}9t \quad \text{or} \quad 36C = 13t \ ,$$
and hence $t = {}^{13}\!/_{36}$.

V.16.8. In symbols,

---

[243] Mahāvīra's *geometry*, on the other hand, integrated elements of Mediterranean and Near Eastern origin, some from the first and others from the second millennium BCE – see [Høyrup 2004].

[244] See, for instance, [Tropfke/Vogel et al 1980] and [Hermelink 1978].

[245] On the connection between recreational and practitioners' mathematics in the pre-Modern era, see for instance [Høyrup 1997].

$$\alpha C + \beta t = \gamma \;\Rightarrow\; t = \sqrt{\frac{\gamma}{\alpha} + (\frac{\beta}{\alpha}:2)^2} \; -(\frac{\beta}{\alpha}:2) \; .$$

V.16.9. As pointed out already, *fare capo d'anno*, "to make (up accounts at the) end of year" is the standard term for composite interest. If the *thing* had been taken to be for instance the debt after one year, the problem would have been of the simple type "the *censi* are equal to the number", $\alpha C = \beta$. Taking instead the rate of interest as the unknown, we get the composite problem type $\alpha C + \beta t = \gamma$. The calculations can then be summarized as follows:

The loan is 100 £, 1 £ = 20 ß, 1 ß = 12 δ. The monthly interest per £ is posited to be t δ, for which reason the yearly interest on 1 £ is $12t$ δ = $\frac{1}{20}t$ £. After 1 year, the 100 £ are therefore worth $(100 + \frac{100}{20}t)$ £ = $(100 + 5t)$ £, and the interest of the second year will be $(100 + 5t) \cdot \frac{1}{20}t$ £ = $(5t + \frac{1}{4}C)$ £. Therefore 150 £ = 100 £ + $10t + \frac{1}{4}C$. "Restoring", that is, *subtracting* 100 £, we get

$$10t + \tfrac{1}{4}C = 50 \quad \text{or} \quad C + 40t = 200 \; .$$

According to the rule, $t = \sqrt{20 \cdot 20 + 200} \; - 20 = \sqrt{600} \; - 20$, which will be the monthly interest on 1 £. As abbacus algebra in general, Jacopo stops here and does not calculate an approximate rational solution; obviously, abbacus algebra was not meant to serve in practice (as was abbacus geometry, where approximate square roots abound). Nor does he or other abbacus algebra authors transform $\sqrt{600}$ into $10\sqrt{6}$. If encountering the latter expression, they would indeed "reduce it to root", that is, replace it by $\sqrt{600}$.

We take note of another subtractive "restoration".

V.16.10. Superficially, this problem is similar to the one for which Fibonacci introduced the *regula recta* (above, p. 105), and thus also to Diophantos's *Arithmetica* I.xv. However, the appearance of the square root (a square root of real money, not of a mere formal *censo*) produces a wholly different mathematical structure.

If the two possessions are a and b, the problem is indeed

$$a + 14 = 4 \cdot (b - 14) \;, \quad b + \sqrt{a} = 30 \; .$$

Positing $a := C$, this gives

$$b - 14 = \tfrac{1}{4}C + 3\tfrac{1}{2} \quad \text{or} \quad b = 17\tfrac{1}{2} + \tfrac{1}{4}C \; .$$

Further, since $\sqrt{a} = \sqrt{C} = t$,

$$b + \sqrt{a} = 17\tfrac{1}{2} + \tfrac{1}{4}C + t = 30 \quad \text{or} \quad \tfrac{1}{4}C + t + 17\tfrac{1}{2} = 30 \; .$$

"Restoring" we get

$$\tfrac{1}{4}C + t = 12\tfrac{1}{2} \quad \text{or} \quad C + 4t = 50 \; ,$$

which according to the rule gives

$$t = \sqrt{2 \cdot 2 + 50} \; - 2 = \sqrt{54} \; - 2 \;, \quad a = C = t^2 = 58 - 2 \cdot 2\sqrt{54} \; = 58 - \sqrt{864} \;,$$

$$b = \tfrac{1}{4}C + 17\tfrac{1}{2} = 14\tfrac{1}{2} - \sqrt{54} \; + 17\tfrac{1}{2} = 32 - \sqrt{54} \; .$$

The text, we observe, contains yet another subtractive "restoration".

This is the problem where Jacopo left spaces instead of calculating that $4\sqrt{54} = \sqrt{864}$, and where all copyists conserved the spaces (above, p. 8).

V.16.11. In symbols,

$$\beta t = \alpha C + \gamma \Rightarrow$$

$$t = \sqrt{(\frac{\beta}{\alpha}:2)^2 - \frac{\gamma}{\alpha}} + (\frac{\beta}{\alpha}:2) \quad \text{or} \quad t = (\frac{\beta}{\alpha}:2) - \sqrt{(\frac{\beta}{\alpha}:2)^2 - \frac{\gamma}{\alpha}} .$$

V.16.12. The problem can be expressed:

$$10 = a+b , \quad a \cdot b = 20 .$$

Positing $a := t$ ($a<b$) we have $b = 10-t$ and thus

$$a \cdot b = t \cdot (10-t) = 10t - C = 20 .$$

"Restoring" we get

$$10t = C + 20$$

whence, according to the rule,

$$a = t = 5 - \sqrt{5 \cdot 5 - 20} = 5 - \sqrt{5} , \quad b = 10 - t = 5 + \sqrt{5} .$$

Jacopo does not say so, but the choice of the subtractive solution follows by necessity from the positing of t as "the lesser part". If t had been chosen just as one of the parts, both solutions would have been valid, and of course identical when it comes to the set of numbers.

V.16.13–14. If a is the initial capital, and p the rate of gain, the problem is

$$p \cdot a = 12 , \quad a + p \cdot a + p \cdot (a + p \cdot a) = 54 .$$

We posit $a := t$, and get $a + p \cdot a = t + 12$, whence

$$(t+12) \cdot (t+12)/t = (C + 24t + 144)/t = 54$$

whence

$$C + 24t + 144 = 54t \quad \text{or} \quad 30t = C + 144 .$$

According to the rule we therefore get an initial capital

$$a = 15 - \sqrt{15 \cdot 15 - 144} = 15 - \sqrt{81} = 6 ,$$

or, alternatively,

$$a = \sqrt{15 \cdot 15 - 144} + 15 = \sqrt{81} + 15 = 24 ,$$

both solutions being shown to be valid in this case. Yet another subtractive "restoration" occurs in section V.16.13.

The term for "equation" (*raoguaglamento*) probably appears in section V.16.14 for the first time in extant Latin and European-vernacular literature. Obviously Jacopo is not aware of contributing something new, the word must have been current where he took his inspiration.

"Two travels" was a common dress, often but not always leading to problems of the second degree. In note 103, two examples from the *Liber abbaci* were mentioned. None of them has the same mathematical structure as Jacopo's, even though both are of the second degree.

V.16.15. The problem may be expressed
$$10 = a+b \ , \ \ a \cdot b + (b-a) = 22 \ .$$
Positing $a := t$ we get $b = 10-t$, whence
$$t \cdot (10-t) + (10-2t) = 10+8t-C = 22 \ \ \text{or} \ \ 8t = 12+C \ .$$
The rule suggests the possibility
$$a = t = \sqrt{4 \cdot 4 - 12} \ + 4 = 2 + 4 = 6 \ , \ \ b = 10-t = 4 \ .$$
A proof shows that $a \cdot b + (b-a) = 26$, whence the alternative possibility has to be used,
$$a = t = 4 - \sqrt{4 \cdot 4 - 12} \ = 4-2 = 2 \ , \ \ b = 10-t = 8 \ .$$
Jacopo's intention (or that of his source) is to show that certain problems are only satisfied by one of the solutions. He could have done so on the example in *V.16.14*, but the source of the difficulty (the explicitation of t as the minor number) may have been too obvious. In the present case this choice has not been made explicit, instead it is implied by choosing the difference to be $10-2t$ not $2t-10$. The source of the difficulty was therefore better hidden for an audience which was not mentally trained on negative numbers and absolute values.

A subtractive "restoration" is followed by an addition of 1 *censo* to both terms which is spoken of as "giving to each part", *not* as a "restoration".

V.16.16. In symbols,

$$\alpha C = \beta t + \gamma \ \Rightarrow \ t = \sqrt{\frac{\gamma}{\alpha} + (\frac{\beta}{\alpha}:2)^2} \ + (\frac{\beta}{\alpha}:2) \ .$$

V.16.17. Positing t for the original rate of exchange of *fiorini* into *venetiani*, the quantity of *venetiani* first obtained is $40t$, of which 60 are changed back into $^{60}\!/_{t+1}$ *fiorini* and $40t-60$ remain as they are. Then

$$\frac{60}{t+1} + (40t - 60) = 100 \ \ \text{or} \ \ \frac{60}{t+1} = 100 + 60 - 40t = 160 - 40t \ ,$$

whence
$$(160-40t) \cdot (t+1) = 60 \ \ \text{or} \ \ 120t-40C+160 = 60 \ .$$
"Restoring" (in the additive as well as the subtractive sense) we get
$$40C = 120t+100 \ ,$$
whence
$$C = 3t + 2\tfrac{1}{2} \ .$$
According to the rule we therefore have
$$t = \sqrt{1\tfrac{1}{2} \cdot 1\tfrac{1}{2} + 2\tfrac{1}{2}} \ + 1\tfrac{1}{2} = \sqrt{4\tfrac{3}{4}} + 1\tfrac{1}{2} \ ,$$
which is the original value of the *fiorino* in *venetiani*.

Rules for the Third and Fourth Degree

In this chapter, only rules and no examples are given. All cases can either be reduced to the first- or second-degree cases dealt with in Chapter 16, or they are homogeneous and therefore solvable by means of cube roots or square roots of square roots.

V.17.1. Just as the second part of this section is reflected in section V.17.19, its first part must be a reflection of a missing introduction to the rules and examples of Chapter 16.

The ordering of the cases follows this system: first come cubic, then quartic cases; within each group, we first have homogeneous equations with increasing powers on the right-hand side, then mixed cases in the same order as for the second-degree equations – in the quartic case first those that can be reduced by means of division, next the biquadratics (of which only the first is given).

V.17.2. In symbols,

$$\alpha K = \beta \implies \sqrt[3]{\frac{\beta}{\alpha}} \ .$$

V.17.3. In symbols,

$$\alpha K = \beta t \implies \sqrt{\frac{\beta}{\alpha}} \ .$$

V.17.4. In symbols,

$$\alpha K = \beta C \implies \frac{\beta}{\alpha} \ .$$

V.17.5. In symbols,

$$\alpha K + \beta C = \gamma t \implies \sqrt{\frac{\gamma}{\alpha} + (\frac{\beta}{\alpha} : 2)^2} \ -(\frac{\beta}{\alpha} : 2) \ .$$

V.17.6. In symbols,

$$\beta C = \alpha K + \gamma t \implies$$

$$\sqrt{(\frac{\beta}{\alpha} : 2)^2 - \frac{\gamma}{\alpha} + (\frac{\beta}{\alpha} : 2)} \ \text{ or } \ t = (\frac{\beta}{\alpha} : 2) - \sqrt{(\frac{\beta}{\alpha} : 2)^2 - \frac{\gamma}{\alpha}} \ .$$

V.17.7. In symbols,

$$\alpha K = \beta C + \gamma t \implies \sqrt{\frac{\gamma}{\alpha} + (\frac{\beta}{\alpha} : 2)^2} \ + (\frac{\beta}{\alpha} : 2) \ .$$

V.17.8. In symbols,

$$\alpha CC = \beta \;\Rightarrow\; t = \sqrt{\sqrt{\frac{\beta}{\alpha}}} \;.$$

V.17.9. In symbols,

$$\alpha CC = \beta t \;\Rightarrow\; t = \sqrt[3]{\frac{\beta}{\alpha}} \;.$$

V.17.10. In symbols,

$$\alpha CC = \beta C \;\Rightarrow\; t = \sqrt{\frac{\beta}{\alpha}} \;.$$

V.17.11. In symbols,

$$\alpha CC = \beta K \;\Rightarrow\; t = \frac{\beta}{\alpha} \;.$$

V.17.12. In symbols,

$$\alpha CC + \beta K = \gamma C \;\Rightarrow\; \sqrt{\frac{\gamma}{\alpha} + (\frac{\beta}{\alpha}:2)^2} \; -(\frac{\beta}{\alpha}:2) \;.$$

V.17.13. In symbols,

$$\beta K = \alpha CC + \gamma C \;\Rightarrow\;$$

$$t = \sqrt{(\frac{\beta}{\alpha}:2)^2 - \frac{\gamma}{\alpha}} \; +(\frac{\beta}{\alpha}:2) \;\; \text{or} \;\; t = (\frac{\beta}{\alpha}:2) - \sqrt{(\frac{\beta}{\alpha}:2)^2 - \frac{\gamma}{\alpha}} \;.$$

V.17.14. In symbols,

$$\alpha CC = \beta K + \gamma C \;\Rightarrow\; t = \sqrt{\frac{\gamma}{\alpha} + (\frac{\beta}{\alpha}:2)^2} \; +(\frac{\beta}{\alpha}:2) \;.$$

V.17.15. In symbols,

$$\alpha CC + \beta C = \gamma \;\Rightarrow\; t = \sqrt{\sqrt{\frac{\gamma}{\alpha} + (\frac{\beta}{\alpha}:2)^2} \; -(\frac{\beta}{\alpha}:2)} \;.$$

V.17.16. Actually, the manuscript contains 14 rules only, not 15 as claimed in the manuscript. The omission will have regarded the biquadratic analogue of either the second or the third mixed second-degree case, respectively

Censi oguali a censi de censi et numero

or

Censi de censi oguali a censi et numero.

The formulation suggests that the lost introduction to the algebra specified that it would first give six rules with examples, and next 15 rules without examples that can be reduced to the initial six.[246] Elsewhere, Jacopo's cross references are precise, and the introduction to Chapter 17 says nothing about reducibility.

A Grain Problem of Alloying Type

V.18.1. The *staio* is a measure of capacity for liquids and dry products like grain (as other measures for grain, it may also serve as an area measure, but that is not the case here). The grain *staio* varies between c. 10 litre and c. 140 litre.

   The solution makes no use of algebra. Instead, if *a* is the quantity of grain of 12 *soldi* the *staio*, the solution builds on the alloying principle

$$\frac{a}{20 - 18} = \frac{100}{18 \quad 12}.$$

Similar diagrams (but without the circles) are found thrice on fols 48^v–50^r in problems concerned with alloying of bullion. Even more similar (namely with circles) are the corresponding diagrams found in *Trattato di tutta l'arte dell'abbacho* (T_F, fol. 140^v, T_R fol. 61^r).

   One may wonder why this problem is inserted between algebraic rules and problems of quasi-algebraic character. A guess near at hand, supported by what was just said about the character of the diagram, is that it was present in the same place in the model which from which Jacopo took his inspiration for the algebra (admittedly, this merely displaces the problem: why would the model have it here?).

Wages in Geometric Progression

The four problems in this chapter all deal with the wages of the manager of a *fondaco*, a warehouse (located abroad, seen from the owner's point of view; the word comes from Arabic *funduq*, and originally had a broader sense which does not concern us here); it is tacitly presupposed that these wages increase in geometric proportion. Before discussing the group as a whole, we may analyse its problems one by one.

V.19.1. If the salaries in the three years are *a*, *b* and *d*, we know that $a+d = 20$, and $b = 8$. Moreover, as it turns out, the salaries of the three years are presupposed to be in continued proportion, since $ad = b^2 = 64$. As in V.16.12 we thus have two numbers whose sum and product we know. However, the problem is not solved by means of *censo* and

---

[246] Of course this is not literally true for the cases $\alpha K = \beta$ and $\alpha CC = \beta t$; for Jacopo it seems to be sufficient that the former solved "in the same way" as $\alpha C = \beta$ and the latter "in the same way" after reduction.

thing, nor according to an algorithm derived from the solution of V.16.12; instead, the two numbers are found as

$$a = \frac{a+d}{2} - \sqrt{(\frac{a+d}{2})^2 - ad} \quad \text{and} \quad b = \frac{a+d}{2} + \sqrt{(\frac{a+d}{2})^2 - ad} \,.$$

This corresponds to the procedure used in Abū Bakr's *Liber mensurationum*, No. 25 [ed. Busard 1968: 25][247] and to Diophantos's *Arithmetica* I.27 [ed., trans. Tannery 1893: I, 60–62]. The same procedure turns up repeatedly in later abbacus writings, for instance in the *Trattato dell'alcibra amuchabile* from c. 1365 [ed. Simi 1994: 34].

V.19.2. If the respective salaries are a, b, d and e, we know that $a = 15$, $e = 60$, and that the salaries are in continued proportion. The solution is given as

$$b = \sqrt[3]{\frac{d}{a} \cdot a^3} \quad , \quad d = \sqrt[3]{(\frac{d}{a})^2 \cdot a^3} \,.$$

V.19.3. If the respective salaries are a, b, d and e, we know that $a+e = 90$, $b+d = 60$, and that the numbers are in continued proportion. The solution makes use of the formula

$$a \cdot e = b \cdot d = \frac{(b+d)^3}{3(b+d)+(a+e)} \,.$$

The formula is easily justified algebraically, if we introduce p as the ratio of subsequent salaries, whence $b = ap$, $d = ap^2$, $e = ap^3$, since then

$$\frac{(b+d)^3}{3(b+d)+(a+e)} = \frac{a^3 p^3 (1+p)^3}{a(3p+3p^2+1+p^3)} = a^2 p^3 = a \cdot ap^3 = ap \cdot ap^2 \,.$$

This, however, is not likely to be a faithful rendering of the underlying reasoning; in any case, the reference to "el partitore"/"the divisor" points to use of an established method or formula (as one might guess).

The determination, respectively, of a and e and b and d from their sum and product follows the pattern of V.19.1, the first *fondaco* problem.

V.19.4. If the respective salaries are a, b, d and e, we know that $a+d = 20$, $b+e = 30$, and that the numbers are in continued proportion. Without being so identified, the ratio p between successive members is found as $(b+e)/(a+d)$, whence

$$a = \frac{a+d}{1+p^2} \,, \quad d = (a+d)-a \,, \quad b = \frac{b+e}{1+p^2} \,, \quad e = (b+e)-b \,.$$

I have never noticed similar problems in earlier sources, whether Indian, Arabic or

---

[247] Abū Bakr's method goes back to the geometric solution used already in the Old Babylonian scribe school and by Near Eastern Bronze Age surveyors – see [Høyrup 2001]. The geometric procedure will be illustrated imminently.

Latin;[248] on the other hand, the circumstance that Jacopo supposes the reader to know in advance that the wages should be in geometric progression shows that the problem type was not his invention but current in the place where he worked or from where he took his inspiration.

The problem type did not catch on strongly in the Italian abbacus environment, nor was it however quite neglected. I have noticed five other texts where the salary either of the manager of a *fondaco* or of a servant is supposed to increase in the same way and two further texts borrowing the mathematical structure. In the *Istratti di ragioni* (see note 167 and preceding text) [ed. Arrighi 1964: 149], the increase of the salary is still taken for granted (which supports the claim that the contents of the *Istratti* goes back to Paolo dell'Abbaco or at least to his times). Benedetto da Firenze's selection from 1463 from maestro Biagio's collection of algebra problems [ed. Pieraccini 1983: 89–91] – the latter dated before 1340 – explains the presupposed increase meticulously. In an anonymous *Trattato d'algibra* from the 1390s [ed. Franci & Pancanti 1988: 80] the geometric progression is also explained in detail, in words so close that affinity between the two texts is evident. The fourth occurrence no longer deals with a *fondaco* and its manager but with the wages of a servant: Filippo Calandri's late-fifteenth century problem collection [ed. Santini 1982: 32*f*]; so does the occurrence in Pacioli's *Summa* [1494: 194ʳ].[249] Cardano presents some analogous pure-number problems and rules in the *Practica arithmetice, et mensurandi singularis* (see imminently) but considers Pacioli's acceptance of a geometrically increasing salary as a serious error which nobody of sane mind could accept (thus the words of the second edition [Cardano 1663: 170] – those of the first edition [1539: QQ viiᵛ] are less sharp). A pure-number problem belonging to the family is found in *Alchune ragione*,[250] fol. 78ʳ.

The *Istratti* contains a single problem with the structure of V.19.2 (but restricted to three years). Benedetto's extract from Biagio also contains a single representative of the family, which coincides with V.19.4 apart from a factor 2 but which is solved by means

----

[248] The closest we get is that proposition III.5 of Jordanus of Nemore's *De numeris datis* [ed. Hughes 1981: 88] is a pure-number version of V.19.1. Everything, however, speaks against a connection: Jordanus's solution is different (it goes via sum and difference, not half-sum and half-difference). Moreover, the analogues of V.19.3 and V.19.4 are absent from Jordanus's treatise, even though his III.18 and III.19 are their analogues for four numbers in non-continued proportion; if Jordanus had known Jacopo's problems, he would certainly have included them in his rather disparate collection of problems about numbers in proportion.

[249] By then, indeed, problems about *fondachi* and *fattori*, warehouses and their managers, usually had a different (and more realistic) mathematical structure – cf. also [Tropfke/Vogel et al 1980: 559*f*].

[250] Vatican, Vat. lat. 10488, an anonymous Venetian compilation of problems from 1424, see [Van Egmond 1980: 230]. In all references to this manuscript, I shall use the earliest foliation.

of algebra.[251] The problem in the *Trattato d'algibra* is of the same type, uses almost the same words in the statement and follows the same path in the solution, apart from positing the salaries of the first two years to be, respectively, 2 *censi* and 3 *censi*, not 2 *cose* and 3 *cose*.[252] Filippo Calandri's single specimen has the structure of Jacopo's V.19.3; it makes the type of increase explicit by means of the simple phrase "and pays him proportionally", suggesting that Calandri presupposed the type to be at least known. Calandri's numbers differ from Jacopo's ($93\frac{1}{3}$ *denari* in the first plus the fourth year, 40 *denari* in the second plus the third). Like Jacopo he formulates a rule in general terms, but in largely different words. It is noteworthy that another one of Filippo's problems (no 44) is a three-participant analogue of V.16.10, involving a square root in the data.

Pacioli [1494: 194ʳ] has a whole small section about "the wages of servants"; two of its six problems presuppose the wage to increase in geometric proportion. The first of them, an analogue of V.19.2, explains the salary of the fourth year to be "at the rate of the others". The second, an analogue of V.19.3, sharing even Jacopo's numbers and part of his phrasing, says nothing about the nature of the increase, even though the two problems are separated by one where the increase is of a different type; we may guess that Pacioli took over the problem wholesale from some older source, which *could* be derived from Jacopo.[253] Pacioli solves both problems by means of rules explained in the *Distinctio sexta* dealing with proportions [Pacioli 1494: 87ʳ⁻ᵛ]. The method for solving analogues of V.19.2 makes use of "the position of a thing" (for the salary of the second year), since "otherwise it would be very difficult". For the analogue of V.19.3, Pacioli gives essentially the same rule as Jacopo and Filippo Calandri, but he does not identify the quantity $\dfrac{(b+d)^3}{3(b+d)+(a+e)}$ as the product of the salaries of the first and fourth (respectively the second and the third) years. Strangely, he describes the rest of the

---

[251] A pure-number version of V.19.4 (with sums 26 and 39, which yields a solution in integer numbers) is found in another part of Benedetto's compilation, namely in his selection from Antonio de' Mazzinghi's *Fioretti* [ed. Arrighi 1967a: 22]. Even in this case the solution is algebraic, though different from Biagio's. Antonio was considered a disciple of Paolo dell'Abbaco.

[252] In a context where the *censo* was no longer remembered to represent an amount of money, a spontaneous replacement of the *cosa* by a *censo* is so unexpected that it is unbelievable (not least because the author of the *Trattato d'algibra* feels obliged to find the *cosa* from the *censo*, only to square it again). There seems to be no doubt that the *Trattato d'algibra* had its words not from those of *Biagio* (which would ask for such a spontaneous replacement) but directly from an environment closer to an Arabic source – from which follows that the problem type must probably be of Arabic origin, in spite of its apparent absence from extant Arabic sources.

[253] Even Pacioli's phrase "ala rate de li altri"/"at the rate of the others" in the analogue of V.19.2 *could* be a reminiscent of Jacopo's "a quella medesima ragione"/"at that same rate" in this problem, not least because Pacioli's phrase strictly speaking is not very meaningful when nothing is said about the salaries of the first three years increasing with a constant rate.

procedure (the finding of d and b from their sum and product) in terms which only make sense if the procedure is geometric – when determining d as

$$\frac{b+d}{2} + \sqrt{(\frac{b+d}{2})^2 - \frac{(b+d)^3}{3(b+d)+(a+e)}}$$

he speaks of the two occurrences of $^{b+d}\!/_2$ as the *different* halves of the $b+d$, in agreement with the use of this diagram:

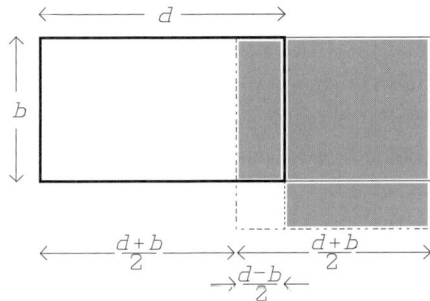

Subtracting the shaded areas, equal to the original rectangle and thus to the product of the sides, from the square on one half of their sum, leaves the square on the half of their difference; adding this to *the other* half of the sum produces the length of the rectangle, exactly as Pacioli claims. There is no reason to believe that Pacioli thought about this (given his style he would have explained), and therefore we must assume that he borrowed directly or indirectly from a source which still used this old geometric procedure.

Slightly later (fol. 88^r) Pacioli gives a rule corresponding to V.19.4, which is based on the observation that $\frac{a-d}{b-e} = \frac{a}{b}$. In his *Practica arithmetice, et mensurandi singularis*, Cardano [1539: fol. II ii^v] deals with the same situation, saying that Pacioli has also treated it; however, Cardano gives a "much better" alternative *per regulam*, proposing exactly the same rule as Jacopo – but insisting that it is his own invention and explaining why it works. Elsewhere in the same treatise [1539: fol. M iiii^v] he presents the two identities which Jacopo presupposes in V.19.3,

$$\frac{b+d}{2} + \sqrt{(\frac{b+d}{2})^2 - \frac{(b+d)^3}{3(b+d)+(a+e)}} = d \,,$$

$$\frac{b+d}{2} - \sqrt{(\frac{b+d}{2})^2 - \frac{(b+d)^3}{3(b+d)+(a+e)}} = b \,.$$

Just like Pacioli, he does not identify $\dfrac{(b+d)^3}{3(b+d)+(a+e)}$ as the product $a{\cdot}e = b{\cdot}d$; nor, however, does he distinguish two different halves of the sum.

The pure-number problem in *Alchune ragione* is of a new kind. It asks for five numbers in continuous proportion, whose sum is 110, while the third number is 10. This text gives a correct solution to this non-trivial biquadratic problem but offers no hint as to how it should be obtained.

These texts – the *Istratti*, Benedetto's extracts from Biagio and Antonio de' Mazzinghi, Filippo Calandri's *Raccolta di Ragioni*, Pacioli's *Summa* and Cardano's *Practica* – are too different to allow the reconstruction of a network (not to mention a stemma). All we can know with certainty is that the compiler of *Alchune ragione* did not himself invent the problem, and that he did not borrow it from any of the other texts which are considered here (see also below, note 354). However, they leave no doubt that problems of the kind which Jacopo dealt with in V.19 stayed in circulation during the fourteenth and fifteenth centuries; moreover, several of the authors differ from Jacopo in their approach in ways that cannot simply be explained as the result of algebraic reconstruction, suggesting that Jacopo was not the only channel through which they reached the Italian abbacus environment.

Referring to Jacopo as a "channel" presupposes that he was not the inventor. We may indeed be confident he was not. In V.19.1 he explains the basic facts about the continued proportion with three members. Afterwards, when things become more difficult, he explains nothing; given his general pedagogical orientation we may be sure that he omitted explanations because he had none. But even if Jacopo did not understand himself why his rules worked, we may still ask which mathematics was involved in them.

In V.19.1, all that is needed beyond the property of the continued proportion which Jacopo himself explains (namely $ac = b^2$) is the rule for finding two numbers from their sum and product; since Pacioli's solution makes indirect references to the geometric solution going back at least to the Old Babylonian scribe school and Near Eastern Bronze Age surveyors, we may assume that this was also Jacopo's underlying method, even though he only seems to know it as a numerical algorithm.[254]

In V.19. 2 and V.19.4, all that is required is the same kind of familiarity with basic polynomial algebra as is needed to see that the higher-order equations of Chapter 17 are reducible. This had been stock knowledge in Arabic algebra at least since al-Karajī and al-Samaw'al, and it was inherent in all writings that presented the sequence of algebraic powers as a geometric progression and also stated the rules for multiplying binomials –

---

[254] Pacioli's slip (or what it is) is not our only evidence that the geometric procedures were around. When finding the side of a square from the sum of the area and "all four sides", Raffaello Canacci [ed. Procissi 1954: 308] explains it in details. A similar explanation is found in Jean de Murs' *De arte mensurandi* (V.2 prop. 26, ed. [Busard 1998: 187*f*]), where Jean uses it to reduce a rectangle problem to the same square problem *which he does not treat* but which his text presupposes to be known. He must thus build on a source where it was present; since Abū Bakr as translated by Gherardo da Cremona [ed. Busard 1968: 94] offers a fallacious solution (based on the same reduction but not explaining it) and Fibonacci [ed. Boncompagni 1862: 66] when copying him discovers the fallacy but replaces it by the mere cover-up "et cetera", none of these works can be Jean's source.

thus in Ibn al-Yāsamīn's *Urjūza*, Ibn al-Bannā''s *Talkhīṣ* and al-Qalaṣādī's *Kašf*.

V.19.3 is more exacting and not to be solved by methods explained directly in works like the *Urjūza*, the *Talkhīṣ* and the *Kašf*. Nor does it however go beyond what can be found for instance in the *Fakhrī*, where al-Karajī makes use of the formula for the third power of a binomial [Woepcke 1853: 58] – at first exemplifying it on $(2+3)^3$, next applying it to show that $\sqrt[3]{2} + \sqrt[3]{54} = \sqrt[3]{128}$. All in all, it seems certain that the mathematical substance of the problems of V.19 is of (as yet unidentified) Arabic origin. According to the observations made in note 252, the same may hold for the dress of geometrically increasing wages.

Ch. 20. The Coin List

Coin lists were apparently common in the earliest abbacus treatises (unless the presence of a coin list gave particular value to an abbacus manuscript and augmented its survival chance). In [2003], Lucia Travaini published a global survey of known coin lists in *abbaco*- and *mercatura*-treatises (with complete editions of all of them). Three are found in abbacus works:

− in the Columbia Algorism, which can thereby be dated to c. 1285–90 (cf. note 70);
− in the *Liber habaci* from c. 1310 (a rather rudimentary list);
− in Jacopo's *Tractatus algorismi* (**V** as well as **M+F**).

One list appears in the manuscript Venice, Marciana Ital. IX. 18 (part I, from c. 1360), which combines abbacus matters not only with a coin list but also with a small treatise on the techniques of assaying and tables of composite interest, 5%–20% over 20 years [Van Egmond 1980: 204]; from its contents, the coin list appears to be from 1300–1305, and it is indeed dated 1305.[255] According to its contents, this manuscript is thus a money-dealers' manual with inserted abbacus material rather than a genuine abbacus treatise.[256] To this can now be added a short list found in *Alchune ragione*,[257] fol. 82^{r-v}, certainly descending, if not from Jacopo's lists of gold and silver coins then from a list he copied.[258]

Six lists are found in *libri di mercatura;* two of them are in the published specimens

---

[255] After the list, the compiler adds "don't rely on it" [Travaini 2003: 114], explaining both before the list and in the end that coins change: apparently he knew his list to be outdated when he copied it some 50 years after it was put together.

[256] A similar hybrid (though without a coin list) is the Castilian *Libro que enseña ensayar qualquier moneda* (ed. Córdoba de la Llave, in [Caunedo del Potro & Córdoba de le Llave 2000: 215–242].

[257] See note 250.

[258] Of Jacopo's gold coins, three are omitted, one (*perperi nuovi a carati 14*) is relocated, and the two *genovini* are inverted. Apart from that and two slightly different values, the lists are identical. Two of Jacopo's silver coins are omitted, there is one inversion and two slightly different values.

mentioned above, note 204 (Pegolotti, c. 1330, and *Datiniana*, c. 1385; the list in the latter manual is considerably earlier); a third from c. 1450 was included by Pacioli [1494: 210^v–224^v] in the end of the arithmetical part of his *Summa* (the coin list is on fol. 224^{r-v}). The dates of these merchants' manuals go from 1330 to c. 1450 – that is, they all postdate the bulk of abbacus treatises containing coin lists. One list is in letter form (though hardly a genuine private letter; the "letter" may be from 1444, the list itself one or two decades earlier). A final list is in a private notebook from 1315–16. It is an attractive hypothesis (but so far it can be no more) that it was the development of the *mercatura* genre around 1330 which eliminated the need to insert the same information in abbacus treatises.

Jacopo is not likely to have been a specialist in the matter – after all, he appears to have been a teacher and neither a merchant nor an accountant or a banker. However, he uses recent information (probably a complete recent list, perhaps similar to what is found in the notebook from 1315–16 and to the private material collected by Pegolotti over decades and eventually inserted in his *Pratica*): according to Travaini [2003: 102, 104], most of his coins antedate 1300, but he includes the Provençal *rinforzati* ("Rinfazzati [...] a denari 3 grani 15 de legha") introduced in 1302 (yet not about innovations taking place in 1303). As she observes, Provençal coins are very well represented, but Tuscan coins less well than in other lists.

For the identification of the various coins, I shall only refer to [Travaini 2003: 102–104, 235–313].

The inversion of the sections on silver coins and small coins was already discussed above (p. 8) as evidence for the character of the copying process. It is almost self-evident that the inversion is not repeated in **M+F**; further, three items have been forgotten in **V**, and there is evidence that a copyist has (mis-)repaired a numismatic technical term ("paglialocati") unknown to him. Apart from that, the lists in **V** and **F** are identical, whereas **M** contains supplementary lists of French, Genovese, Lombardian and other coins (M.20.3A, M.20.5A).

We may take note of the care with which the metrology is specified. As far as the silver coins are concerned, everything is standard, cf. note 172. The metrology for small coins also appears to be conventional, until we compare with the information given by other coin lists and discover that "*p denari, q grani* of alloy" means that the coin contains *p ounces, q denari* per pound, or (equivalently) *p denari, q grani* per half-ounce. The coin list in the Marciana manual follows the same system and explains its meaning.[259] The system was habitual in France, where the unit for measuring the fineness of silver was known as the *denier de fin, denier d'alloi* or *denier de loi* (pure silver being 12 *deniers*

---

[259] "Where you find that such and such a coin contains so many *denari* and so many grains, there you should understand that the pound contains as many ounces of pure silver as it says *denari* and as many *denari* as it says grains" [ed. Travaini 2003: 110].

de fin).[260] Both in the Marciana manual and in the French dictionaries this was meant as a purely relative measure, not as a specific weight. In Francesc Santcliment's *Summa de l'art d'aritmètica* [ed. Malet 1998: 330] we find a different interpretation, which may well correspond to the origin of the usage (since *denarius* had long functioned as a term for a twelfth of many different units)

> ès mester de saber que un marc de pes d'argent fi és a 12 diners de llei, enaixí 12 diners de llei és un marc de pes, i un diner de llei val 16 diners de pes. I 1 gra de llei val 16 grans de pes, i ⅛ gra de llei val 2 grans de pes,

that is,

> it is necessary to know that one mark of weight of pure silver is at 12 *diners* of alloy, as 12 *diners* of alloy are one mark of weight, and 1 *diner* of alloy is worth 16 *diners* of weight. And 1 grain of alloy is worth 16 grains of weight, and ⅛ of a grain of alloy is worth 2 grains of weight.

As we shall see (p. 126), this seems to have been Jacopo's understanding of the matter. The explanation in the Marciana manual suggests that its author was *not* familiar with the usage.

The gold metrology could also be considered odd. As mentioned both in the end of V.20.4 and above in note 172, $\frac{1}{24}$ of an ounce was a weight *denaro*, not a carat.[261] Actually, the established carat was 4 grains, that is, $\frac{1}{6}$ of a weight denaro, and 24 carats should therefore be 4 weight *denari*, that is, $\frac{1}{72}$ pound. This was indeed the weight of the gold *solidus* introduced in 312 CE by Constantin the Great[262] (which, though debased, stayed the standard coin of the Byzantine Empire under the names of νόμισμα and ὑπέρπυρος, Jacopos *perpero*). The fineness of gold could thus be measured by the contents of carats in $\frac{1}{72}$ pound; by generalization, the carat could be used in the sense of $\frac{1}{24}$ of a business partnership [Edler 1934: 63]. Since the French-influenced Marciana manual also states the *fiorino* to contain "*carati* 24 per ounce" [ed. Travaini 2003: 112] and the *Liber habaci* coin list [ed. Arrighi 1987: 156] apparently written in Provence says

---

[260] [*Dictionnaire universel françois et latin*: III, 223a; Diderot & d'Alembert 1751: IV, 828a], cf. also [Travaini 2003: 108]. The system is also used in *Trattato di tutta l'arte dell'abbaco* (**T**$_F$ fol. 139^v–140^r; **T**$_R$ fol. 61^{r-v}) and by Paolo Gherardi [ed. Arrighi 1987: 89].

[261] Since the *Vocabolario degli Accademici della Crusca* from 1612, many Italian general dictionaries explain the *carato* to be $\frac{1}{24}$ of an ounce of gold, but this is a secondary development which did not spread into genuine metrology (technical sources until the nineteenth century as reflected in [Zupko 1981: 81–83] contain no trace of this value). As a matter of fact, the *Crusca* quotes but two examples, both of them literary (Villani and Dante), none of which mentions the ounce, and both of which are thus irrelevant for its claim.

[262] Constantin's pound was of course the Roman *libra* of 12 ounces, which however corresponded fairly well in weight (and exactly in its subdivision) to the light pound which Jacopo makes use of.

that "the Lucca ounce of gold contains carats xxiii and a half" (etc.), we may assume that Jacopo took over a French-Provençal notion, just as he took over the terminology for the fineness of small coin – not least because his carat of $\frac{1}{24}$ of the ounce has a parallel in the later use in France of a *carat de prix* or *carat de poids* sometimes equal to $\frac{1}{24}$ of an ounce, sometimes $\frac{1}{24}$ of a mark (8 ounces)[263] [*Dictionnaire universel françois et latin*: II, 257a; Diderot & d'Alembert 1751: II, 672b], What we find in these dictionaries is clearly no literate's reconstruction but a rendition of actual technical customs; the former also distinguishes this *carat de prix/poids* (an analogue of Santcliment's *diner de llei*) from a *carat de fin*, a purely relative measure (the one we are familiar with nowadays, indeed an analogue of the Franco-Procençal *denier de fin*).

That the carat of $\frac{1}{24}$ of an ounce was also (at least in origin) meant as a relative measure is indicated in Pegolotti's *Pratica*. Its coin list [ed. Evans 1936: 287] starts by indicating that *fiorini d'oro* are "*a carati 24 d'oro fine per oncia*" ("at 24 carats of pure gold per ounce"). In the presentation of Byzantine coin [ed. Evans 1936: 40] we are also told that the *perpero* (that is, the ὑπέρπυρος) is "of alloy 11 carats of pure gold per ounce, and in the rest of the alloy until 24 carats there are 6 carats of pure silver and 7 carats of copper". Immediately afterwards it is explained, however, that the *perpero*, in weight, contains 24 carats; these are evidently absolute, and in fact identical with the traditional carat of 4 grains.

Specifications of the metrology were not customary in coin lists – indeed, nothing similar is found in any of the other lists mentioned above. Given Jacopo's predilection for pedagogical explanations we may guess that he inserted the explanations in a list he had borrowed from a specialist. On the other hand, the metrologies themselves must have been *used* in the original list. However, Jacopo adopts them in the next chapter ("carats per ounce" in V.21.2–3, also in V.14.23; "denari of alloy" in V.21.4,8); we thus have another piece of evidence that Jacopo had gone to Montpellier not in order to spread Tuscan knowledge but in order to learn, cf. p. 91.[264]

---

[263] The carat equal to $\frac{1}{24}$ mark (and hence, as it is explained, to $\frac{1}{3}$ of an ounce) is found in the Columbia Algorism [ed. Vogel 1977: 61], which also defines a corresponding grain. It is referred to as a *charato grosso*; the ordinary carat of $\frac{1}{144}$ ounce is referred to as a *charato picciolo*. The Umbrian *Livero del l'abbecho* [ed. Arrighi 1989: 133] speaks of it as a *charrata de marcha* and says that it is used when gold is sold in Venice.

    Even the *Libro di molte ragioni d'abacho* [ed. Arrighi 1973: 104*f*] from Lucca measures the fineness of gold in carats per ounce; as we shall see, at least the algebraic sections of this manuscript descend from Jacopo's *Tractatus*.

[264] Of course he also brought a mathematical upbringing which he had received at home. The fact that neither he nor any of the other Italians writing in Provence (Paolo Gherardi, the author of the *Trattato di tutta l'arte dell'abbaco*) took over the habit of introducing the rule of three by means of counterfactual number problems shows that the rule of three was part of that upbringing.

Ch. 21. Alloying Problems

In contrast to the money-dealers' manuals and Pegolotti's *Pratica* but in the likeness of abbacus books in general, Jacopo's treatise contains neither technical advice concerning the refining or alloying of bullion, nor guidance about assaying; all Jacopo takes up is the arithmetic of alloying.

We notice that Jacopo's alloying problems make use of the same metrologies as his coin list – V.21.2+2A+3 of that for gold, V.21.4+8 of that for small coin, V.21.5–6 of that for silver coin.

We also notice that all of Jacopo's alloying problems begin "I have ...", as do the non-Fibonacci problems about the same subject in the *Livero de l'abbecho* [ed. Arrighi 1989], cf. above, p. 43, in an idiom which Fibonacci had demonstrated to know but which he did not use himself. After Jacopo's times the same use of the first person characterizes alloying problems in many other abbacus treatises – for instance in Paolo Gherardi [ed. Arrighi 1987: 29–31, 89]:[265] in the Lucca *Libro de molte ragioni d'abacho* [ed. Arrighi 1973: 95–106];[266] in Giovanni de' Danti's *Tractato de l'algorismo* [ed. Arrighi 1985: 50–52]; in the *Libro di conti e mercatanzie* [ed. Gregori & Grugnetti 1998: 72–74];[267] in Francesco Bartoli's *Memoriale* [ed. Sesiano 1984b: 134*f*], a private notebook written in Avignon before 1425 and containing excerpts from earlier abbacus works[268]; in Piero della Francesca's *Trattato d'abaco* [ed. Arrighi 1970a: 56–59]; and (with the slight variation, also used at times by Piero, "Io mi trovo ...") in Pacioli's *Summa* [1494: 184^r–185^v]. It is also found in a Byzantine treatise of abbacus-type (Ψηφηφορικὰ ζητήματα καὶ προβλήματα, "Calculation Questions and Problems") from the early fourteenth century [ed. Vogel 1968a: 21–27], but here the first person singular is also used for other problem types (mostly but far from always dealing with possession or payments in gold coin).

Apart from those which initially declare an authorial intention ("I want to give you ..." etc.), the only other problems in **V** which start in the first person are V.14.5 and V.14.29, the first of which deals with exchange of money and the second with money in two purses. M.14.3=F.v.3, which also deals with exchange of money, starts "I have ...", whereas V.14.3 begins "Somebody has ...". V.16.17, an algebra problem about exchange of money (with no counterpart in **M+F**) starts "Somebody has ...". Paolo Gherardi has a similar alternation between the first and the third person in problems about

---

[265] In the first of these passages, the first person only initiates problems about gold, whereas a silver problem starts "There is somebody who has ...".

[266] Alternating with the formula "A man has ...".

[267] Gold problems only. Problems about silver are neutral or start "Somebody has ...".

[268] Ten instances of "I have" regarding gold as well as silver, and a single of "Somebody has" (about gold).

exchange (as he and certain other authors have in alloying problems). This might be a suggestion that problems beginning "I have .." share a common origin in a particular environment, distinct from that where abbacus problems in general were circulating (probably a money-dealers' environment). The similarity in this respect with the Byzantine treatise *could* suggest a Byzantine location. Since a Byzantine treatise from the next century [ed. Hunger & Vogel 1963] shows no trace of the style, this remains a conjecture – supported, however, by the observation that the oldest of these treatises is local Byzantine in its choice of coins referred to, whereas the younger one is under heavy Italian influence in this respect [Scholz 2001: 102].

As we observe, the two diagrams in this chapter are similar in structure to the one used in V.18.1 (and thereby to those of *Trattato di tutta l'arte dell'abbaco* – see above, p. 115), but they omit the circles.

The single sections call for the following remarks:

V.21.1. This transition between the coin list and the alloying problems is absent from **M+F**.

V.21.2. Like V.14.7+21+28, this is a problem of inverse proportionality. In such problems, we remember, Jacopo previously has made use of the inverse rule of three (without pointing out the inversion). Here, instead, he offers a concrete explanation and does not mention the rule of three.

M.21.2=F.vii.1 omits the final period, in which the general value of the procedure is pointed out. It is also a bit more terse, without differing from **V** in any conspicuous way.

M.21.2A=F.vii.2. This problem, absent from **V**, is so close to the previous one that there is no particular reason to believe it to be added by a later hand, as M.14.24A,27A,31A (=F.v.26,30–31,36) seemed to be (see above, p. 88). Neither is there, however, anything that speaks directly against this possibility. It may have been in Jacopo's original, but it may also be a later addition.

V.21.3. From this straightforward problem, M.21.3=F.vii.3 again omits the final statement that the method is of general value. After finding the fineness of the ingot to be $21\,^{581}/_{624}$ it adds that the fraction is "rather over $^3/_4$" – no impressing accuracy, since it is much closer to 1 ($^{581}/_{624} = 0.93...$). Otherwise, only details of the formulations differ from **V**.

V.21.4. In contrast to the preceding problems, this one offers no concrete interpretation of the initial calculation, only a mechanical application of the "alloying rule". One wonders whether Jacopo judged it too difficult to formulate the concrete explanation, or did not expect the reader to be able to grasp it. The proofs suggest that Jacopo followed Santcliment's understanding of the *denaro* of alloy as $^1/_{12}$ mark; a concrete explanation might therefore run more or less like this:

For each mark we take of the high-value coin there is an excess of 11–7 = 4 *denari*, and for each mark of the low-value coin there is a deficit of 7–4 = 3 *denari*. If we take 3 marks of the high-value coin and 4 of the low-value coin, the total excess should thus cancel the total deficit.

This is indeed slightly more complex than the concrete explanations offered in the preceding problems, but hardly more complex than the explanation of the alternative to the double false position given in V.14.5.

Once having determined how to make 7 marks of the requested alloy, Jacopo applies the partnership model in order to calculate how to make 100 marks. In the end there is another proof, a statement that the method is general, and a reference to the diagram.

M.21.4=F.vii.4 omits the diagram (and the corresponding reference), the first proof, the reference to a partnership, and the statement of general validity. Instead it observes after the final proof that "it would have been a pity if we had found more or less" (than the requested 700 *denari*). It also speaks of *biglione*/"bullion" all the time.

V.21.5. This time, the fineness is given in ounces per pound but the actual weights in marks. Since the number of *denari* of alloy per pound is the same as the number of ounces per pound, this should make no difference, but Jacopo manages to formulate things in a way which presupposes that 1 mark of bullion (that is, 8 ounces) may contain 9 respectively 10$\frac{1}{2}$ ounces of silver.

We observe that the determination of the amount of silver in the two marks of pure copper does not get any near to the idea that "nothing" can be treated as a number, irrespective of the use of 0 in this function in the tables of squares.

M.21.5=F.vii.5 speaks of the lingot as a *pane*, a "bread"; otherwise, it only differs from **V** in the details of the phrasing

V.21.6. Once more, the fineness is given in ounces per pound but the actual weights in marks. In the application of the alloying rule, this does not create problems, since the rule does not depend on the unit in which the fineness is defined; however, the proof has the same flaw as before – Jacopo manages to produce 160 ounces of an alloy which contains 190 ounces of pure silver. No alchemist ever did better!

M.21.6=F.vii.6 skips the cross-reference to the analogous second-last problem and the promise to be less verbose (for good reasons – M.21.4=F.vii.4 is *not* verbose). It also omits the diagram as well as the reference to it. Instead of the latter it has a commentary to the metrological problem, together with the statement of general validity omitted before:

> And you should know that in these alloyings it brings the same to say pounds as mark. And there where you have said 20 marks you might say 20 pounds, and the same comes to you in one way and the other, because one should take pounds where you take mark. But reckon with mark, because silver is weighed in marks. And in this way and by this rule you may alloy and prove whatever alloy is said to you.

The explanation would have been more to the point if it had said that instead of ounces per pound one might have said *denari* of alloy (per mark); but apparently the compiler of **M+F** was no more familiar with this usage than the author of the commentary in the Marciana manuscript.

V.21.7–8. Once more, the computations presuppose *denaro* of alloy to be $\frac{1}{12}$ of a mark, and once more Jacopo refers to a partnership. Nothing is said, we notice, about the problem being undetermined. It is hard to believe that Jacopo did not understand that other solutions were possible (inherent in his very calculation is that the required alloy can be made from the best and the poorest alloy alone, or from the two others alone); but even though Jacopo often attempts (with more or less luck) to explain why his procedures work, theoretical coherence and completeness were not unconditional aims of his.

M.21.7–8=F.vii.7 contains the preamble and the problem without any break. The initial reference to the diagram is conserved, but the diagram itself is eliminated together with the reference to it in the end. Omitted are also the intermediate proof, the partnership (**M+F** brings only the bare calculations) and the final indication that the reader can perform a proof.

In the end, **M** closes with the words *Explicit liber tractatus algorismi. Deo Gratias.* Indeed, **M+F** stops here.

Ch. 22. Supplementary Mixed Problems

V, instead, goes on with a mixed collection of 32 problems, 9 of which fall within the area of Chapter 14, the remaining 23 within that of Chapter 15. As pointed out above (p. 24), there are no genuine overlaps, the apparent repetitions turn out at closer inspection to consist of duly cross-referenced variations and supplements. If a later hand had glued a separate collection onto Jacopo's original treatise, this would hardly have been the case, the abbacus record as a whole is extremely rich in repetitions. To this comes the homogeneity of style and computational approach. We may be confident that Chapter 22 was really added by Jacopo himself. On the other hand there is at least one reason to believe that he did borrow a whole problem collection and adapted it to his own purpose. In Chapter 15, circles are made *a compasso*, if they are not simply said to be "round". In Chapter 22, instead, the current corresponding expression for the making of circles and circular shapes is *a sesto*. The *sesto* is nothing but a synonym for the compass (now obsolete in Italian, whence my translation "ambit", equally obsolete in English in this sense). Admittedly, it is possible that Jacopo had acquired a new linguistic habit before adding a supplementary chapter to his *Tractatus*, but it seems more plausible that he borrowed it from another treatise from which he culled material for this supplement – also because the term *diametro*, used regularly in Chapter 15, does not appear in Chapter 22.[269]

---

[269] It cannot be excluded that he culled from several sources. It may be significant that the problems

V.22.1. Partnership problems of this kind are quite common in abbacus treatises, and their solution follows the principle that the distribution of the gain is made proportionally to the product of investment and time. In order to put Jacopo's particular approach in perspective, a list of examples will be useful:

– Liber habaci [ed. Arrighi 1987: 146*f*];
– Paolo Gherardi [ed. Arrighi 1987: 39];
– The Lucca *Libro de molte ragione* [ed. Arrighi 1973: 75, 77];
– *Istratti di ragioni* [ed. Arrighi 1964: 98];
– *Libro di conti e mercatanzie* [ed. Gregori & Grugnetti 1998: 72–75];
– *Alchune ragioni* (see above, note 250), fol. 54ᵛ;
– Benedetto da Firenze's *Tractato d'abbacho* [ed. Arrighi 1974: 136*f*] (cf. note 63);
– Filippo Calandri's *Aritmetica* [ed. Arrighi 1969: 97*f*];
– Piero della Francesca's *Trattato d'abaco* [ed. Arrighi 1970a: 53*f*];
– The "Treviso arithmetic" (trans. D. E. Smith, in [Swetz 1987: 143–151]);
– Luca Pacioli's *Summa de arithmetica* [1494: 151ᵛ];
– The Pamiers algorism [partial ed. Sesiano 1984a: 47*f*];
– Barthélemy de Romans' Provençal *Compendy de la praticque des nombres* [ed. Spiesser 2003: 264];
– Francesc Santcliment's *Summa de l'art d'aritmètica* [ed. Malet 1998: 248, 257*f*];
– Francés Pellos's *Compendion de l'abaco* [ed. Lafont & Tournerie 1967: 132–134].

Most of these simply divide proportionally to the products of time and investment. The only exceptions are the *Liber habaci, Alchune ragione* and Benedetto da Firenze, where these products are used as the investments in a fictitious simple partnership; the *Libro de molte ragioni* and the *Libro di conti e mercatanzie*, both of which use the same trick as Jacopo and take the virtual interest of the investments as the investments in a fictitious partnership;[270] and Piero della Francesca, who calculates the virtual interests and divides according to the resulting proportions but does not refer to any new, simple partnership.[271] The *Libro de molte ragioni* and the *Libro di conti e mercatanzie* share another striking trait, namely that they specify the calendar that is used in exactly the same phrase, "a mutazione di milesimo a la natività e none alla incharnazione" ("the year changing at Christmas and not at the Incarnation"). Elsewhere, both Paolo Gherardi [ed. Arrighi

---

V.22.23–26, all about adjoining to or cuttings away from circles, do *not* use the *sesto* idiom.

[270] In a testament problem analogous to V.14.9, of which only the beginning is made, the *Libro de molte ragioni* [ed. Arrighi 1973: 200*f*] also hints at the use of the partnership model: "You should say that one has put in 1, and the other 2, and the other 4".

[271] Since composite interest is never involved, the use of the virtual interest simply implies a multiplication of all proportionals by a common factor corresponding to the interest rate, which of course changes nothing in the outcome.

1987: 37] and Barthélemy de Romans [ed. Spiesser 2003: 265] show familiarity with the use of the partnership as a functionally abstract model; but they are clearly less infatuated with it than Jacopo. Since the *Libro di conti e mercatanzie* obviously depends on the same sources as the *Libro de molte ragioni d'abacho* which, as we have seen (note 263) also agrees with Jacopo in the Franco-Provençal use of carats of $\frac{1}{24}$ ounce, only Benedetto da Firenze seems to be an exception to the rule that the reference to a fictitious partnership in this problem type was restricted to authors working within or manifestly inspired from Provence.[272]

V.22.2. This procedure is quite straightforward and not in need of any explanation. It is worth noticing, however, that the computation follows the rule of three without mentioning it. This is different from what we find in the Latin practical geometries – for instance in the *Geometria incerti auctoris* IV.11 [ed. Bubnov 1899: 324*f*], where the ratio between the "model height" and the "model shadow" is found; the same is done by Hugue de Saint Victor in the *Practica geometriae* [ed. Baron 1956] in the cases where he gives numerical examples – in others he just asks for comparison of the triangles in question (which still asks for the determination of a ratio).

Both the *Geometria incerti auctoris* and Hugue do give the method of a stick, but they prefer the use of instruments: an astrolabe, or a quadrant.

V.22.3. Even this procedure is quite elementary, once one has understood that the cross-section of the tower is meant to be square (or, less probable from the architectonic point of view, rectangular).

V.22.4. This is a more sophisticated variant of the problem about the goblet than the one we encountered in 14.25. The sophistication may be the reason that the method of a (false) position is mentioned explicitly here; until now, a "position" was only referred to in passing in 14.9 (once as *positione*, once as *propositione*).

For a modern reader, the argument may be easier to understand if expressed in symbols. The weight of the stem being S, that of the cup being C and that of the lid being L, we have that

$$C = (\tfrac{1}{3}+\tfrac{1}{4})\cdot(C+S) ,$$
$$L = (\tfrac{1}{4}+\tfrac{1}{5})\cdot(L+C) = 6 .$$

The first position corresponds to putting $L+C = 20x$, from which follows

---

[272] *Alchune ragione* is hardly to be counted as an exception. On fol. 56[v], a problem about two merchants loading sacks of wool on a ship shares so many phrases with Jacopo's V.14.14 that a close shared source is almost certain. Further, as we have seen, its coin list descends from Jacopo's, or from a list he copied (see p. 121). Also suggestive are the problem about five numbers in continued proportion (above, p. 120) and the two algebraic fragments contained in the compilation (below, p. 176 and note 253). So, much of this mixed compilation appears to point toward Jacopo's Provençal environment.

$$9x = 6 , \quad L+C = 20x = {}^{20 \cdot 6}\!/_9 = 13\tfrac{1}{3} ,$$

and since $L = 6$, the weight of the cup must be $C = 13\tfrac{1}{3} - 6 = 7\tfrac{1}{3}$.

Next we make the position $C+S = 12y$, from which follows

$$7y = 7\tfrac{1}{3} , \quad C+S = 12y = (12 \cdot 7\tfrac{1}{3})/7 = 12\tfrac{4}{7} ,$$

whence the weight of the stem is found to be $S = 12\tfrac{4}{7} - 7\tfrac{1}{3} = 5\tfrac{5}{21}$.

The same problem with the same numerical parameters is found in Giovanni de' Danti's *Tractato de l'algorismo* [ed. Arrighi 1985: 72]. Giovanni does not mention the position, but asks directly for a number whose $\tfrac{1}{4} + \tfrac{1}{5}$ is 6, and states that it is $13\tfrac{1}{3}$, etc.

V.22.5. The circular area was determined in V.15.4 from formula (7) (see p. 90), as one fourth of (diameter × perimeter). The rule and the example in V.22.5 is explicitly meant as an alternative, as is shown by the phrase "without espying the circulation around". It is noteworthy that this alternative is the Arabic rule (4), not the adaptation (4*) to Latin and European-vernacular ways to speak of fractions.

The "proof" is evidently not a proof but a numerical example. Only when read together with the following problem does it become a (kind of) proof. Probably a mere writing error (*prova* instead of *asempro*) is involved.

V.22.6. This determination of the circular area applies formula (7), and in this respect it is merely a numerical variant of V.15.4. In V.15.4, however, the given parameter is the perimeter, here it is the diameter. Independently of whether their determination of the circular area is based on formula (7) or (4*), it is the habit of abbacus geometries first to show how to find the diameter from the perimeter (if (7) is used) or the perimeter from the diameter (if (4) is used), and next the area. Jacopo's sequence here can therefore not have been copied from another treatise where it served to introduce the treatment of the circle; not only its didactical explanations but also its order are almost certainly due to Jacopo, and the order must reflect pedagogical intentions.

V.2.7–8. Problems asking for a pure number which, after subtraction or addition of specified aliquot part (occasionally, a sum of aliquot parts alone), fulfils certain arithmetical conditions, are quite common in abbacus treatises, even though Jacopo only brings them in this appendix. Some examples (translated into symbols[273]) are:

- *Liber abbaci* [ed. Boncompagni 1857a: 175, 422]
  - (a) $\tfrac{1}{3}x + \tfrac{1}{4}x + \tfrac{1}{5}x + \tfrac{1}{6}x = \sqrt{x}$; solved from the false position $x := 60$.
  - (b) $x - \tfrac{1}{3}x - \tfrac{1}{4}x - 4 = \sqrt{x}$, solved by the algebraic position $x := t$.
  - (c) $(x - \tfrac{1}{3}x - 6)^2 = 2x$, solved by a geometric argument leading to an analogue of *Elements* II.5.

- The Columbia Algorism [ed. Vogel 1977: 31]:
  - $(x - \tfrac{1}{3}x - \tfrac{1}{4}x)^2 = x$; solved from the false position $x := 12$.

---

[273] x stands for the number asked for, t for the algebraic *thing* and C for the algebraic *censo*.

- *Livero de l'abbecho* [ed. Arrighi 1989: 124, 125]:[274]
    - (a) $(x-\frac{1}{3}x-\frac{1}{4}x)^2 = x$; solved from the false position $x:= 12$.[275]
    - (b) $x-\frac{3}{5}x:= \frac{2}{7}$; solved from the false position $x:= 5$.
- Paolo Gherardi [ed. Arrighi 1987: 54]:
    $x-\frac{1}{3}x-\frac{1}{4}x-\frac{1}{6}x-\frac{1}{10}x = 11$. Solved from the false position $x:= 60$.
- *Trattato di tutta l'arte dell'abbacho* ($\mathbf{T_F}$ fols 157^r, 162^v):
    - (a) $x-(\frac{1}{3}x+\frac{1}{4}x+5) = (\frac{1}{3}x+\frac{1}{4}x+5)$; solved by the algebraic position $x:= t$.
    - (b) $x+\frac{1}{3}x+\frac{1}{4}x = \sqrt{x}$; solved by the algebraic position $x:= C$.
- *Istratti di ragioni* [ed. Arrighi 1964: 170] (problems probably from around 1340 and perhaps due to Paolo dell'abbaco, cf. note 167 and p. 117):
    $x-\frac{1}{4}x = \sqrt{10}$; gives the solution $x = \sqrt{10}+\frac{1}{3}\sqrt{10}$.[276]
- Biagio, *Chasi esenplari alla regola dell'algibra* in Benedetto da Firenze's selection [ed. Pieraccini 1983: 2, 5], probably from before 1340:
    - (a) $x-\frac{1}{3}x-\frac{1}{4}x = \sqrt{x}$, solved by the algebraic position $x:= C$.
    - (b) $x-\frac{1}{3}x-\frac{1}{4}x-4 = \sqrt{x}$, solved by the algebraic position $x:= C$.
    - (c) $(x+\frac{1}{3}x+\frac{1}{4}x)^2 = 20x$, solved by the algebraic position $x:= t$.
    - (d) $(x+\frac{1}{3}x+\frac{1}{4}x+4)^2 = 30x$, solved by the algebraic position $x:= t$.
- Benedetto da Firenze, a problem added before the previous one [ed. Pieraccini 1983: 1*f*] which Benedetto believes Biagio not to have considered:
    $\frac{1}{3}x+\frac{1}{4}x = \sqrt{x}$, solved by the algebraic position $x:= C$.

Five Arabic writings (two known in Latin translation only) are also worth mentioning:[277]

- The "other book" in which Gherardo da Cremona [ed. Hughes 1986: 257–261] found 21 problems which he attached to his translation of al-Khwārizmī's *Algebra*, and which have also crept into the Arabic fourteenth-century manuscript (cf. note 238):
    $(x-\frac{1}{3}x-\frac{1}{4}x-4)^2 = x+12$, solved by the algebraic position $x:= t$.
- The *Liber augmentis et diminutionis* [ed. Libri 1838: I, 305, 307, 314, 316, 318]:
    - (a) $x-\frac{1}{3}x-\frac{1}{4}x = 8$, solved by double false position and single false position.
    - (b) $x-\frac{1}{3}x-4-\frac{1}{4}(x-\frac{1}{3}x-4) = 20$, solved by double false position and by *regula recta* ($x:= t$).
    - (c) $x+\frac{1}{3}x+\frac{1}{4}(x+\frac{1}{3}x) = 30$, solved as (b).

---

[274] None of these problems are borrowed from the *Liber abbaci*.

[275] The text claims that this and similar problems are solved "by means of square roots", but no square root is needed in the solution, nor used.

[276] This trick – in symbols, to conclude directly from $\frac{p}{q}x = a$ that $x = a+\frac{q-p}{p}a$ – was known in Latin as *regula infusa*. It is rarely used in Christian-European sources but quite common in Arabic writings. Reinvention by Paolo/ps.-Paolo cannot be excluded.

[277] In all the Arabic writings, the "number" is a *māl/census*.

(d) $x+\frac{1}{3}x+4+\frac{1}{4}(x+\frac{1}{3}x+4) = 40$, solved as (b)

(d) $x+4+\frac{1}{2}(x+4)+5+\frac{1}{4}(x+4+\frac{1}{2}(x+4)+5) = 70$, solved as (b) and further by *regula infusa* (cf. note 276).

– Ibn Badr's *Ikhtiṣār al-jabr wa'l-muqābalah* [ed., trans. Sánchez Pérez 1916: 55, 63]

(a) $(x-\frac{1}{3}-\frac{1}{4})^2 = x+12$, solved by the algebraic position $x:=t$.

(b) $(x-\frac{1}{3}x-\frac{1}{4}x)^2 = x$, solved by the algebraic position $x:= C$.

– Abū Kāmil's *Algebra* [ed., trans. Sesiano 1993: 379; ed. trans. Levey 1966: 122; ed., trans. Chalhoub 2004: 90]:

$$(x-\frac{1}{3}x-2)^2 = x+24, \text{ solved from the algebraic position } x:= t.$$

– al-Karajī's *Fakhrī* – I render Woepcke's paraphrase [1853: 75–77, 79*f*]:

(a) $x+\frac{1}{2}x+4+\frac{1}{2}(x+1/2x+4)+4 = 20$; Woepcke does not describe the solution.

(b) $x-\frac{1}{2}x-3-\frac{1}{2}(x-\frac{1}{2}x-3)-3 = 10$; Woepcke does not describe the solution (but the *regula recta* with $x:= t$ is a reasonable guess).

(c) $x+\frac{1}{5}x+5-\frac{1}{3}(x+\frac{1}{5}x+5)-5 = 0$ (thus Woepcke); Woepcke does not describe the solution (we may guess at the *regula recta*).

(d) $x-\frac{1}{3}x-\frac{1}{4}x = 10$; Woepcke does not describe the solution (we may guess at the *regula recta*).

(e) $x-\frac{1}{3}x+\frac{1}{4}(x-\frac{1}{3}x) = 10$; Woepcke does not describe the solution (we may guess at the *regula recta*).

(f) $x+\frac{1}{11}x-\frac{1}{3}(x+\frac{1}{11}x)-\frac{1}{4}(x+\frac{1}{11}x) = 10$; Woepcke does not describe the solution (we may guess at the *regula recta*).

(g) $\frac{1}{2}x+\frac{1}{3}x+\frac{1}{4}x = 10$; apparently solved by the algebraic position $\frac{1}{2}x:= t$, which reduces the problem to the type $y+\frac{2}{3}y+\frac{1}{2}y = 10$. The next problem is very similar, for which reason I omit it here.

(h) $\frac{1}{4}x+\frac{1}{9}x+\frac{1}{10}x = 10$; apparently solved from the false position $x:= 180$.

Jacopo's V.22.7 coincides with what we find in the Columbia Algorism and in the *Livero de l'abbecho* (a); it is also solved in the same way, as is the close kin *Liber abbaci* (a). Ibn Badr (b) is the same problem but solved differently; the closely related Biagio (a) follows Ibn Badr's method; etc. Jacopo's V.22.8 has the same structure as *Livero de l'abbecho* (b) and is solved in the same way.

An assumption which imediately suggests itself is that the dress was originally a pure-number translation of problems about a tree with a certain part hidden underground and the visible part given absolutely (that is, problems of the same type as V.14.15). Fibonacci was evidently of the same opinion, since his problem (a) is preceded by three problems formulated about a tree but with a heading speaking about "a tree or a number".

It is also obvious, however, that the family of these number problems had grown far beyond the reach of the original tree problems. One outgrowth (the first?) consisted of increasingly sophisticated or adorned pure-number problems solved by arithmetical methods (like Jacopo's V.22 and its kin) or the *regula recta* (al-Karajī's (a)–(e), and the similar problems from the *Liber augmentis et diminutionis*); problems solved by *māl*-root algebra constituted another outgrowth. Especially the latter sub-family was not closed; one variation (found already in Abū Kāmil) replaced the aliquot fractions of the number that were to be subtracted by its square root. This type as well as al-Karajī's embedding resemble closely the variations of the tree problem we find in Mahāvīra's *Ganita-sāra-sangraha* (ed., trans. [Rangacārya 1912: 74–78, *passim*]), even though these variations deal with apparently concrete entities (camels and their square roots, etc.!). Still other problem types may have arisen as variants although we cannot be sure; in general, it is problematic to reconstruct from a Wittgensteinian "natural family" the historical process that engendered it – not least since a plurality of distinct processes may have been involved. With open-ended possibilities for variation it is next to impossible to draw the line between this type and other one-variable *al-jabr* problems; in particular the three fundamental mixed algebra cases (see p. 102) could have been picked out and put to new use as a particularly interesting subset. Jacopo's two small problems about a number and some aliquot parts may thus present us with a misted window to the prehistory of algebra.

Even if the similarity with Mahāvīra's problems and the link to the crystallization of *al-jabr* should be accidental, there is no doubt that the separation of the number problems from the original simple tree problems preceded Fibonacci and the adoption of abbacus-type mathematics in Christian Europe by quite a few centuries. Fibonacci's lumping of the two types is as good a guess as ours, but not built on better familiarity with relevant sources.

V.22.9. This problem is a numerical variant of V.14.18, see p. 79 (and cf. p. 76 on other instances of "combined works"). Beyond the numbers, the only difference is that the present solution multiplies before it adds, whereas 14.18 (as well as the similar 14.12–13) adds first. All four divide *in.*

V.22.10. Two problems in Chapter V.15 deal with the areas of rectangular figures: V.15.8 asking simply for the area of a terrain, and V.15.17 calculating the areas of a terrain and a house in order to know how many houses can find space in the terrain. Further, 14.26 contains a calculation similar to that of V.15.17 (V.14.30 being a three-dimensional analogue). The present question is of a new kind (but a very classical kind – we need only think of *Elements* II.14): to find a square with the same area as a given rectangle.

V.22.11. This is a numerical variant of V.15.9, with the difference that the present problem has an exact solution.

V.22.12. This problem is the complement of V.15.10 and shares its numerical values (which are also those of V.22.11). In both, the length of the rope is given; in V.15.10, the other given magnitude was the height of the tower, here it is the width of the moat.

V.22.13. Here as well as in V.22.14 and V.22.28, the manuscript has an indubitable *vinaio*, "wine-dealer", alternating with the equally impossible *vinaro*. As confirmed by numerous other abbacus books and by the fact that the object contains water and not wine, the meaning is obviously *vivaio*, a water basin (usually serving to keep living fish). Most likely a precursor manuscript has written *v* in a way that could be mistaken for *n*.

As regards the contents of the problem, we already encountered a stone falling "by chance" into water in the curiously mistaken V.15.18 (see above, p. 95). This time, the calculation is correct, apart from the subtraction of the thickness of the stone instead of the original depth of the water – also 3 *braccia* – when the increase is to be found.

V.22.14. This problem presents us with a new terminological peculiarity (returning in V.22.20, V.22.21 and V.22.28), namely the term *lo mino longho* for the diameter, almost certainly a miswriting for *miluogo* (now obsolete in Italian, but cf. French *milieu*); probably, an original "milogho" has been misread "milōgho", interpreted as "mi longho" and then expanded in order to transform "mi" into an apparent noun.[278]

As in V.15.4, the circular area is determined from formula (7) (see p. 90), as one fourth of (diameter × perimeter). The calculation is correct, though at the unmentioned condition that the water raises above the height of the stone.

V.22.15. In this problem, instead, the "Arabic" formula (4) is used for the determination of the circular area, in agreement with what is taught in V.22.5. One may thus suspect this problem (but perhaps not the previous one, in spite of the term *ascesto*?) to be drawn from the same source as V.22.5.

The computation is correct. The procedure has the advantage of making everything concretely meaningful (except where the rule of three is used), but the disadvantage of requiring a fair amount of mathematically superfluous computations. However, Jacopo might be right in not expecting his reader to understand why the result could be found simply as $\frac{(3\frac{1}{2}{}^2-3^2)\cdot 12}{3^2}$ (even if he should have been able to invent the shortcut himself, which is not certain). Unlike problems asking for the application of the rule of three, this one was not of a type that even a professional reckoner would encounter every day or week: mere rote learning of the method therefore had no purpose.

V.22.16. In most of the sources where it occurs, this dazzling and heavily overdetermined problem is usually told about a father leaving 1 monetary unit and $\frac{1}{d}$ of what remains to the oldest son (*d* being regularly but far from always either 7 or 10), 2 units and $\frac{1}{d}$ to the second, etc. As in V.14.8 (see p. 72) Jacopo thus uses a dress which usually goes together with a different mathematical structure, "nested-box problems" allowing stepwise

---

[278] This hypothesis is supported by the appearance of the form *mislongho* (along with the variants *miluogho*, *milungho*, and *diritto del miluogho*) in the *Libro de molte ragioni d'abacho* [ed. Arrighi 1973: 114*f*]. Paolo Gherardi [ed. Arrighi 1987: 68*f*] has *milugo* and *miluogho*, *Liber habaci* [ed. Arrighi 1987: 130*f*, 133, 141], "diritto del miluogho".

Cf. however ancient French *milöain* [Tobler-Lommatzsch VI: 53a], a similar alternative to *milieu*.

backward calculation (represented in Jacopo's *Tractatus* by V.14.22 and V.22.30).[279]

As we see, Jacopo does not seem to know why his rule works. A few writings do indicate a method: Either first-degree algebra or a double false position;[280] Jacopo, as we have seen, never uses the former of these outside the algebra chapter, and the latter never occurs in his treatise. In the first case, the total heritage is put equal to the *thing*, and the first two shares are then calculated and put equal; in the latter, two values for the total are posited, the two first shares are determined for each position, and the true value is found as usually by weighing the two false guesses in inverse proportion to the error (cf. p. 72). Neither calculation of course shows that the subsequent shares will have the same value as the former two, and the control which Jacopo makes would therefore have been adequate even if he had made use of one of these methods.

Curiously, a simple arithmetical solution (though with the same flaw) exists, which nobody (neither any medieval or Early Modern reckoner nor any modern historian) ever seems to have noticed. If we use Jacopo's numbers and dress, each visitor in the garden first picks as many oranges as his number in the sequence and then takes $\frac{1}{10}$ of the remainder (thereby leaving $\frac{9}{10}$ of the same remainder). The last picker (say, number N), leaving nothing, must therefore have had exactly N oranges at his disposal. The number of oranges each picker gets (Δ) must therefore equal that of the pickers – that is, N. The second-last visitor picks $N–1$ oranges before taking $\frac{1}{10}$ of the remainder, and this $\frac{1}{10}$ of the remainder must therefore be one. Therefore, the remainder must be 10; keeping one of these he leaves $10–1 = 9$ to the last picker; therefore $\Delta = 9$. With a similar argument, we get the solution $N = \Delta = d–1$ for any value $\frac{1}{d}$ of the fraction.[281] Obviously, this solution can only be constructed once the problem is already formulated; it could never give anybody the idea that the problem was possible.

That idea may originally have come from geometrical considerations based on pebble counters or the like, as shown in this diagram, reduced for convenience to $d = 6$ (arguments for the hypothesis follow below):

---

[279] For a full commentary to this problem and problem type, see [Høyrup 2007]. What follows is a brief abstract from this paper.

[280] Fibonacci, as we shall see, uses this kind of algebra albeit for the solution of a sophisticated variant. Later, Christoff Rudolff [ed. Stifel 1615: 416] does so for the basic type. The double false is used in the *Istratti di ragioni* [ed. Arrighi 1964: 140f].

[281] Jacopo, and many other abbacus authors with him, reveal in the statement of the problem that $\Delta = N$. None of them can therefore have known the present argument.

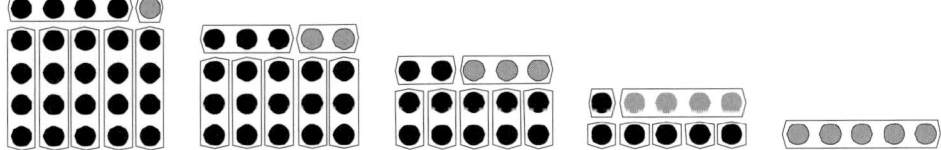

If we remove 1 small (grey) dot from a square pattern of $n \times n$ dots, what is left can be grouped as $n+1$ strips of $n-1$ (black) dots. Removal of one of these strips ($\frac{1}{n+1}$ of what is left) leaves a rectangular system of $n \times (n-1)$ dots. Removing 2 small (grey) dots from this rectangle leaves $n+1$ strips of $n-2$ (black) dots, and removing one of these strips (still $\frac{1}{n+1}$ of the remainder) leaves a rectangle $n \times (n-2)$ dots; etc. In symbols, and for $p = 0$, 1, 2, ...,

$$(n-p) \times n = (n-[p+1]) \times n + (p+1) + (n-[p+1]).$$

As we see, the absolutely defined contributions 1, 2, ..., n form the triangular number T_n, and the relatively defined contributions (the fractions $\frac{1}{n+1}$ of the successive remainders) the triangular number T_{n-1}. Splitting the square number n^2 into these two triangular numbers and looking at the configuration could well have inspired the invention.

Beyond variations of d, the abbacus books sometimes modify the absolutely defined contributions. The simplest modification multiplies all of them by the same factor (say, ε), whereby the sequence becomes ε, 2ε, 3ε, etc. This leaves N unchanged, whereas Δ (and the total $T = N\Delta$) is multiplied by ε (in the above diagram, it corresponds to ascribing the value ε instead of 1 to each dot). In another modulation, the sequence of absolute contributions starts at a later point $\alpha = n\varepsilon$, going on with $n\varepsilon+\varepsilon$, $n\varepsilon+2\varepsilon$, etc. This corresponds to leaving out the initial $n-1$ contributions (which no text explains, however), whence $N = (d-1)-(n-1) = d-n$, while $\Delta = d-1$. Finally, the fraction $\frac{1}{d}$ of what is at hand may be taken first, and the absolutely defined contributions (then always ε, 2ε, 3ε, ...) afterwards. In the above diagram, this corresponds to adding another row of $d-1$ dots on top, whence $N = d$, $\Delta = (d-1) \cdot \varepsilon$.

Three texts of a particular character consider more sophisticated generalizations of the problem type – namely the *Liber abbaci* [ed. Boncompagni 1857a: 279–281], Barthélémy de Romans' Provençal *Compendy* from c. 1467 [ed. Spiesser 2003: 391–423] and the collection of problems which Nicolas Chuquet added to his *Triparty* [ed. Marre 1881: 448–451].[282] In these sophisticated variants, the aliquot part $\frac{1}{d}$ may be replaced by an arbitrary fraction $\frac{p}{q}$, and ε is no longer presupposed to divide α; the fraction may still be taken either after or before the absolutely defined contributions. This has the consequence that the number of shares is not necessarily integer, and therefore the simple argument delineated above does not work. Instead, Fibonacci derives a solution to one

---

[282] Cardano's *Practica arithmeticae et mensurandi singularis* [1539: fol. FF ii^r] contains a faint echo of this sophisticated type but nothing more.

of the variants algebraically, taking the total T as the *thing* and equating the first two shares. Afterwards, he gives a formula, pretending that it comes from his algebra, which however it cannot do. He also gives formulae for other variants without algebra (since his formula makes use of the quantity α–ε and since he does not operate with negative numbers, he needs four different formulae). Barthélémy knows two sets of rules – obviously equivalent, but one expressed in terms of p and q, the other by means of the (not necessarily integer) "true denominator" $d = \text{}^q/_p$. The former set is close to but not fully identical with Fibonacci's formulae. Chuquet only states a rule valid for a simple case and solutions for the rest; Barthélémy (whom he cites) may well have been his only source for the sophisticated variants. In any case it is evident that the sophisticated problems and the appurtenant rules antedate Fibonacci's *Liber abbaci*, and also that neither Fibonacci nor Barthélémy knew how the rules they use had been (or could be) established.

For once, one may be certain – at least as certain as one can be in such matters – that the source for their knowledge was not located in the Arabic world. The sophisticated variants of the problem appear not to be found in any known Arabic text. More decisive, the simple variants themselves are also absent, but two texts contain a simplified version where N is given (thus allowing stepwise backward solution) but where the rule that is offered is none the less the one we know from Jacopo: Ibn al-Yāsamīn's *Talqīḥ al-afkār fī'l ʿamali bi rušūm al-ghubār* ("Fecundation of thoughts through use of *ghubār* numerals" – written in Marrakesh in c. 1190, and thus antedating the *Liber abbaci*), and *al-Maʿūna fī ʿilm al-ḥisāb al-hawāʾī* ("Assistance in the science of mental calculation") written by Ibn al-Hāʾim (1352–1412, active in Cairo, Mecca & Jerusalem, and familiar with Ibn al-Yāsamīn's work).[283] There can hardly be any doubt that this reflects a secondary import into the Arabic world from a place where the regular problem type was familiar, assimilated however to the more regular nested-box problem type – an assimilation which Jacopo effectuates solely by borrowing a dress normally going together with that type.

From the thirteenth century onward, if not before, the simple type was known in Byzantium. Two occurrences from the fourteenth and fifteenth centuries are similar to what we find in Jacopo's and other abbacus treatises and thus not very informative.[284] But the treatment in Maximos Planudes's late thirteenth-century *Calculus According to the Indians, Called the Great* [ed., trans. Allard 1981: 191–194] is interesting. It comes at the point where Planudes has finished his exposition of how to use the Indian numerals, and before his presentation of an indeterminate problem probably going back to late Greek

---

[283] I know about these texts thanks to the kind assistance of Mahdi Abdeljaouad (personal communication).

[284] One is in the above-mentioned problem collection from the early fourteenth century [ed., trans. Vogel 1968a: 102–105], another one in Elia Misrachi's *Sefer ha-mispar* from c. 1500 [ed., trans. Wertheim 1896: 59*f*].

Antiquity, namely to "find a rectangle equal in perimeter to another rectangle and a multiple of it in area" – that is, for a given n to find two rectangles $\square(a,b)$ and $\square(c,d)$ such that $a+b = c+d$, $n \cdot ab = cd$ (a, b, c and d being tacitly assumed to be integers).[285]

Before presenting an inheritance problem with $\alpha = \varepsilon = 1$, $\frac{1}{d} = \frac{1}{7}$, Planudes states a quasi-theorem:[286]

> When a unit is taken away from any square number, the left-over is measured by two numbers multiplied by each other, one smaller than the side of the square by a unit, the other larger than the same side by a unit. As for instance, if from 36 a unit is taken away, 35 is left. This is measured by 5 and 7, since the quintuple of 7 is 35. If again from 35 I take away the part of the larger number, that is the seventh, which is then 5 units, and yet 2 units, the left-over, which is then 28, is measured again by two numbers, one smaller than the said side by two units, the other larger by a unit, since the quadruple of 7 is 28. If again from the 28 I take away 3 units and its seventh, which is then 4, the left-over, which is then 21, is measured by the number which is three units less than the side and by the one which is larger by a unit, since the triple of 7 is 21. And always in this way.

No explicit reference to pebble counters is given, but the whole passage fits the above geometric explanation of Jacopo's problem to the slightest detail. Without support by either symbolic algebra or a geometric representation it is also difficult to see that the "theorem" holds for "any square number", and only the geometric diagram makes it evident that the procedure will continue in such a way that exactly nothing remains in the end.[287] Since this is the closest we get in any source to an argument revealing a possible basis for the idea which underlies the problem, it is a reasonable guess that Planudes presents us with the original underlying idea; given the location in his treatise where the problem turns up it is also reasonable to assume that it comes from a Byzantine tradition probably going back to (possibly late) Antiquity. Such a conjecture also agrees with the use of pebble counters in ancient Greek arithmetic (for instance in the treatment of the figurate numbers).[288]

---

[285] The problem and the first of two solutions offered by Planudes (the one he does not claim to be his own) are found in almost exactly the same words in the ps.-Heronian *Geometrica*, Ch. 24 [ed., trans. Heiberg 1912: 414–417].

[286] My translation is painfully literal, conserving all quasi-logical particles even when they offend the modern ear; a somewhat more free French translation can be found along with Allard's edition of the text.

[287] A calculation in symbols based on the corresponding sequence of identities $n \cdot (n-p+1)-p = (n+1) \cdot (n-p)$ *can* of course show it, but with much less ease. A purely verbal argument like that of Planudes and unsupported by a diagram would hardly give the idea.

[288] It is true that arithmetical argumentation based on counters is likely to have been widespread in the ancient Near East, for instance in Demotic and Seleucid mathematics, as revealed by the formulae by which triangular and various pyramidal numbers (sums of triangular, square and cubic numbers) are found. However, these sums spread to Indian and Arabic mathematics, whereas higher

Planudes's argument is of little help when we are confronted with the sophisticated variants.[289] Byzantine sources containing no traces of these, they were most likely invented in an environment which had learned about the simple variants from Byzantium. Since Fibonacci must have had acces to its results, and since it appears not to have been Arabic, the Ibero-Provençal area is the most likely candidate, even though Sicily is not quite to be excluded. Barthélémy's familiarity with the same rules corroborates the Ibero-Provençal conjecture.

V.22.17. The data of this problem repeat those of V.15.4 in different and slightly more abstract terms – but this time the diameter is asked for (together with the area), whereas V.15.4 only asks for the area (and then finds the diameter as an intermediate result).

V.22.18. This problem combines the types of V.15.7 (*leo in puteo*, p. 92) and 14.12–13 and 14.18 (the "meeting problem", see p. 76), and uses the same methods (apart from multiplying 480 by 800 before finding the sum, as also done in V.22.9, see p. 134); the calculation is thus correct – at least if we disregard the fallacy already pointed out in connection with V.15.7.

V.22.19–22. In V.15.20 and V.15.22 (see above, pp. 97 and 98), the height of the equilateral triangle was given, and its side/perimeter and area were asked for. Here, the side is given, and the height and area are asked for.

V.22.19 is formulated in abstract terms and prescribes that the height be measured from "the middle of one of the faces [...] to the opposite corner". The formulation is very close to what we find in the "geometric" treatment of the pentagon in *Trattato di tutta l'arte dell'abbacho* (**T**$_F$, fol. 137^v), cf. above, p. 96; the abstract formulation not supported by a numerical example corresponds to what we find in the late thirteenth-century Picardian *Pratike de geometrie* I.15 [ed. Victor 1979: 489]; the link suggested between these two treatises on p. 97 is thus confirmed, and seen to involve also this part of Jacopo's text.[290]

V.22.20, also formulated abstractly, explains how the height can be calculated from the side, or vice versa. The latter rule corresponds exactly to the one by which the side was

---

figurate numbers (pentagonal numbers etc.) did not. Since the "unknown inheritance" presents us with the same diffusion pattern as the latter, it is reasonable to assume that it arose in the same (namely, Greek) orbit as the higher figurate numbers. Cf. [Høyrup 2007: 42].

[289] If we accept fractional counters, any solution to a sophisticated variant *can* be justified *a posteriori* in a similar way, but the argument is tortuous, seems to depend miraculously on the numbers involved (although it does not), and it could never lead to *discovery* of the formula.

[290] Given the differences between the texts taken pairwise together, the link cannot be direct; what we see must have resulted from a common background – but a more specific background than what is shared by all abbacus-type geometries.

found from the height in V.15.20 and V.15.22 – also in the particular that $\frac{1}{3}$ of the squared side is found separately and then added "above".

V.22.21 applies the first of the abstract rules of V.22.20 to a numerical example. The approximation $\sqrt{300} \approx 17\frac{3}{10}$ may have been found by rounding from the usual approximation $17+\frac{300-289}{2\cdot17} = 17\frac{11}{34}$. The only approximate character of the value is pointed out in the same characteristic phrases as in Chapter V.15, leaving no reasonable doubt that the redaction of Chapter V.22 is due to the same hand as Chapter V.15.

V.22.22 applies the second rule from V.22.20. This time the approximation to $\sqrt{1200}$ is found according to the book. Once more, the only approximate character of the value is discussed in the familiar phrases.

V.22.23–26. Circles with inscribed and circumscribed squares (at times, equilateral triangles) and squares (at times equilateral triangles) with inscribed and circumscribed circles were to become a popular topic in abbacus geometries. They turn up, for instance, in Paolo Gherardi's *Libro di ragioni* [ed. Arrighi 1987: 76], in the *Trattato di tutta l'arte dell'abbacho* (T_F, fol. 135^r), in Tommaso della Gazzaia's *Praticha* [ed. Arrighi & Nanni 1982: 34], in the *Libro di conti e mercatanzie* [ed. Gregori & Grugnetti 1998: 121–123, 128], in the *Istratti di ragioni* [ed. Arrighi 1964: 55], in Orbetano de Montepulciano's fifteenth-century *Regole di geometria pratica* [ed. Simi 1991: 27*f*], in the fifteenth-century anonymous *Trattato di geometria pratica* [ed. Simi 1993: 75–79]. All of these, however, use the phrases "to put inside/outside the largest/smallest" circle, square or triangle, none of them uses Jacopo's idiom. One may suspect that the reformulation of inscription/ circumscription as adjoining/cutting away was his own invention, perhaps motivated by pedagogical considerations.

The configuration is also dealt with in the ps.-Heronian *Geometrica*, Chapter 24, ms. S [ed. Heiberg 1912: 428]. It was taken over by the Latin agrimensorial tradition – indeed a watered-down version of Greek practical geometry as known, e.g., from the treatises that were put together by the modern editor as *Geometrica*. It thus appears in the *Geometria incerti auctoris* IV.25 [ed. Bubnov 1899: 349]. Closer to Jacopo's times, it is still treated in *Artis cuiuslibet consummatio* I.28–30 [ed. Victor 1979: 196–202], from where it was borrowed into *Pratike de geometrie* I.26,28–30, with extension in V.1 [ed. Victor 1979: 498–500, 522–525]; the latter treatise uses phrases similar to those of the abbacus treatises, whereas its Latin model and its precursors speak of inscription and circumscription. Other differences exclude that the abbacus geometries were copied directly from the Picardian treatise, but the passage of this problem type from the Latin to the vernacular is obviously a single, though not simple, process (cf. also p. 97). The *Pratike*, though largely a simplifying translation of *Artis cuiuslibet consummatio*, can be seen to be guided at least on the present point by an already established pattern – a pattern which recurs in the later Italian abbacus geometries.

Arabic and Indian treatises regularly treat of regular polygons inscribed in or circumscribed about circles or other regular polygons, but squares with circles (or triangles with circles) do not stand out as particularly important; Savasorda and Fibonacci follow their lead. The abbacus way to approach the topic thus appears to reflect influence from the Latin tradition.

V.22.23. This piece of simple geometry calls for no supplementary commentaries.

V.22.24. Even this calculation is simple, but it is noteworthy that the circular area is found by the "Arabic" formula (4) which we already encountered in V.22.5 and V.22.15. The ultimately Latin origin of the interest in the configuration (which Jacopo had no reason to be aware of) did not prevent him (or his source) from applying methods of clearly non-Latin origin.

The merely approximate character of the value given for $\sqrt{200}$ is pointed out in the characteristic way we know from Chapter V.15 as well as V.22.21–22.

V.22.25. Once more, the "Arabic" formula (4) is applied in this simple computation. We may notice that the length of *the whole circumference* of the circle is written on each fourth of it in the diagram, whereas each single square side has its value indicated in the diagram for V.22.26.

V.22.26. Even here, the geometry is simple, and the approximate character of the square root is indicated in the usual way. However, the approximation is not obtained in the usual way.[291]

V.22.27. The areas of isosceles triangles were already calculated in several places – thus in V.15.22, and in V.22.19–22. But the present determination of that of a right triangle is the first approach to that subject. We notice that the method is presented in abstract terms before it is applied to the particular paradigmatic example, similarly to what happened in the group V.22.19–22.

V.22.28. In this determination of a volume from linear dimensions and of a linear dimension from a volume and other linear dimensions, Jacopo stays clear of the fallacies that sometimes affect his volume calculations – maybe because he avoids the shortcuts which the numerical parameters invite, and explains the concrete meaning of every intermediate result, maybe simply because he follows a decent model with these qualities. We notice that the circular area is determined by means of his favourite formula (7). Since

---

[291] Approximating from above, with 5 as the first guess, would yield $\sqrt{24\frac{1}{2}} \approx 4\frac{19}{20}$; if instead the first guess is 4, the standard approximation becomes $5\frac{1}{16}$. The only obvious way to obtain a denominator 17 seems to be the value $4\frac{1}{4}$ for the first guess, which however yields the first approximation $4\frac{16}{17} + (1\frac{1}{8})/17 = 5\frac{1}{136}$. Should we believe in a simple calculational error committed by Jacopo or his source?

other authors writing in Provence shared this preference, this choice does not prove that he constructed the problem himself – but it is a supplementary argument that Chapter V.22 is no late addition (though not as certain a proof that Jacopo himself was responsible for its inclusion in the treatise as the phraseology surrounding the approximate square roots).

V.22.29. Problems about successive profits on an unknown capital were well known, and might be constructed in ways that called for the application of varying tools. A simple type – given initial capital and final capital+profits with constant profit rate after two years – was presented in the algebraic problem V.16.9. In the particular dress of "two travels", we encountered it in the equally algebraic V.16.13; as mentioned in note 103, the *Liber abbaci* contains two other instances of the same dress – dealt with before Fibonacci approaches algebra, and solved by means of proportion theory and *Elements* II.6.

The present very simple problem uses a superfluous partnership dress. The mathematical type, however, is that of "nested boxes", and the problem *could* have been solved by stepwise backward calculation. Instead Jacopo chooses a single false position, which indeed is often more convenient.

Jacopo's solution is correct, but his argument for choosing 60 as his position is fallacious, borrowed from cases where all fractions are to be taken of *the same* number. In the present case, a position 3 would have been just as adequate.

V.22.30. This problem dress probably arose in the wake of Christianity and Islam, religions which make charity a duty; it seemingly disappeared in Europe when the Reformation condemned the idea of buying salvation through good deeds (in Catholic areas, the Counter-Reformation may have made jokes of this kind somewhat dangerous). Max Weber in memoriam, we might baptize it "pre-Modern merchant's nightmare".

The earliest appearance of the type I know of is in Ananias of Širak's seventh-century problem collection [ed., trans. Kokian 1919: 116]. It is also, for instance, in al-Karajī's *Fakhrī* [Woepcke 1853: 75] and (with arithmetically increasing gifts) in Ibn Badr's *Ikhtiṣar* [ed., trans. Sánchez Pérez 1916: 78] and in the *Liber augmenti et diminutionis* [ed. Libri 1838: I, 326]. Fibonacci may have found the dress too far removed from what real merchants from his times and town would do; in any case, he speaks of generic "expenditures" instead of charity [ed. Boncompagni 1857a: 329].

Even this problem is of the type "nested boxes". Its structure does not allow solution by means of a single false position. Since Jacopo avoids the double false position (which Fibonacci applies here) and never uses algebra outside the algebra chapter (not even in the shape of *regula recta*, see p. 105), he has to resort to stepwise backward calculation.

V.22.31–32. These two problems are comparable to V.22.7–8, discussed above, pp. 131*ff.* We notice, however, that the present problems only include the aliquot fractions of the number, not the number itself. Among all the problems considered in connection with the previous group, only one of Fibonacci's problems and the one which Benedetto adds to Biagio's list are similar on this account – but their two problems are not linear. Though not the least innovative in their method, the two problems that close Jacopo's treatise are thus unusual in the details of their statement.

General Observations

The preceding detailed presentation of the contents of Jacopo's *Tractatus* may seem even more perplexing than late medieval metrology. In consequence, a birds' eye view of a few topics may be useful.

First, there is the question of the relation of **M+F** to **V**. I shall not repeat the arguments that show **V** (including Chapters 16–19 and 22) to be quite close to the shared archetype of the two manuscripts and **M+F** to be the outcome of a thorough rewriting, touching also on the style; this was already done in pp. 14*ff.* However, some of the differences between the two manuscripts must antedate the rewriting. As we have seen (for instance, in note 32), some passages in **F** which do not agree with what we find in the corresponding place in **V** nonetheless exhibit orthographic features which otherwise characterize **V** but have mostly been eliminated from **F** (having disappeared totally in **M**). They *might* go back to a different version produced by Jacopo; it is much more plausible, however, that the compiler drew on supplementary material produced in the same Provençal environment as Jacopo's treatise, like the left-right ordered schemes of higher products (cf. note 195 and pp. 23 and 54), or that such material had already been inserted in the copy of Jacopo's treatise which reached the compiler. The attempted faithfulness of **V** (as evident, for instance, in the conservation of the empty spaces in V.16.10, cf. p. 8) makes it less plausible that genuine reformulations had taken place in this branch of the stemma.

In any case, several problems that are present in **M+F** but absent from **V** are almost certainly not due to Jacopo – thus at least M.14.24A=F.v.26, M.14.27A=F.v.30+31 and M.14.31A=F.v.36. Some erroneous rewritings are also unlikely to be his – thus the erroneous recalculation in M.11.4=F.iv.3, see p. 63, and the miscorrection in M.15.21A(20)=F.vi.18 of the erroneous second approximation to the square root (see p. 22). However, under the assumption that **V** should descend from an expanded version of a precursor of **M+F** it cannot be totally excluded that the former of these errors was in the earliest precursor of **M+F** and not introduced by the later hand that undertook the stylistic rewriting and the addition of M.14.24A=F.v.26 and M.14.27A–B=F.v.30–31 (and possibly of other problems present in **M+F** but absent from **V**). Since the recalculation is in the same spirit as the *welsche Praktik* it is much more likely to have been introduced at the same moment as the cluster M.14.24A=F.v.26, M.14.27A–B=F.v.30–31 and M.14.31A=F.v.36.

All in all, the following stemma seems to correspond to the process producing the three extant manuscripts **V**, **M** and **F** – **V″** being the manuscript from which **V** is a faithful copy, and **V′** a precursor for **V″** where the coin lists were still in order (cf. p. 8) and already containing Chapters 16–19 and 22:

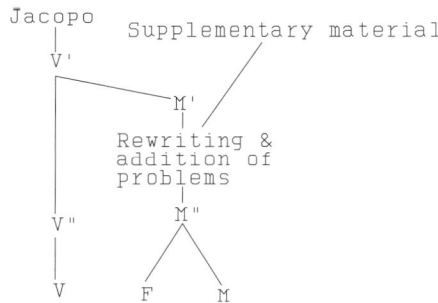

As we shall see on p. 166, **V′** must antedate 1328 by quite a few years, for which reason it seems reasonable to believe it identical with or very close to Jacopo's autograph.

Next, we may try to characterize the contents of **V** in relation to other abbacus treatises. Which topics are dealt with, which are left out?

Many abbacus books start directly with the rule of three or with some other problem type. Jacopo instead, after his introduction, presents us with the Hindu-Arabic numerals and the place value notation, in agreement with the title of his treatise: *a treatise of algorism*. On the other hand, he skips what follows next to this introduction in the Latin algorisms and some of those abbacus writings which also start with the number system, namely algorithms for numerical computation – while **M+F** illustrates the division *a danda* without explaining anything.

When discussing the *Livero de l'abbecho* (pp. 32*ff*) we identified in its Chapters 1–10, 12–14 the basic abbacus subjects, those which correspond to the curriculum of an abbacus school and to the future real needs of the students of this school. All of these are covered by Jacopo, with two exceptions: Chapter 14, dealing with the final discharge of a composite debt – according to Benedetto da Firenze often the last topic dealt with in school, see note 64 and preceding text; and Chapter 6, barter of goods with known prices without the intervention of money.[292]

The *welsche Praktik* was of certain practical use, but many abbacus books did not

---

[292] Actually, in the *Livero de l'abbecho*, only two problems deal with goods, the others with bullion or coin. However, if the values of two types of coin in terms of a third type is known and the two first are to be exchanged, the principle is the same (composition of a direct and an inverse rule of three) – a particular case of the rule of five. The *Livero de l'abbecho* also deals with cases where the mutual relations between three or more coins are known in sequence, which are mathematically similar.

present it; probably abbacus teachers – whose genuine practice was the teaching of mathematics, not trade – deemed it mathematically unsatisfactory, preferring the rule of three and the single false position. Jacopo follows the same pattern, even though some of his tricks approach it in spirit – for instance the informal iteration in V.14.2. In contrast, as pointed out, it is used in several of the extra problems contained in **M+F**.

Many abbacus books also present the double false position. Jacopo, as we have seen, does not (nor does **M+F**). Instead, V.14.5 introduces a concretely meaningful alternative.

Certain other difficult matters that turn up regularly in abbacus books are also absent: the "purchase of a horse", the "finding of a purse", and the related linear problem of men asking money from each other (the topics of Chapters 16–19 of the *Livero de l'abbecho*, cf. above, p. 34), the "100 fowls" (see note 79), and the two doves equally distant from a cup (see p. 99). Absent are also problems of genuine riddle character. Jacopo prefers to teach problems for which *a rule* can be given, and mostly only such where he believes to understand why the rule works.[293] His text does not give the impression of empty showing-off which we receive from the *Livero de l'abbecho*.

At times, it is true, his understanding is illusory: his spatial intuition is poor, and his suggested second-order approximation to square roots builds on a conceptual short-circuit. He certainly did well when not venturing into more sophisticated matters. But what he does he usually does quite well, not only mathematically but also when judged in a pedagogical perspective. It was for good reasons that his treatise was still copied in the mid-fifteenth century and his name (mis)used more than 200 years after the completion of his treatise (see note 116).

---

[293] V.22.16 about the unknown heritage is almost certainly an exception: in that case Jacopo is satisfied with the verification of the result. In Chapter V.19 he probably understood nothing but the first problem (and perhaps only the initial step, explained in his usual pedagogical ways).

Algebra

The introduction referred (on p. 3) to the challenge which Jacopo's algebra, as described by Karpinski, presents to conventional thinking about the history of pre-Renaissance algebra. Since the detailed presentation of the contents of the algebraic chapters of his treatise (pp. 100*ff*) did not touch on this topic, it is now time to take it up. As it turns out, the challenge is even stronger than could be guessed from Karpinski's description. In order to see that, we shall first have to look closely at the general character of Jacopo's own algebra compared to the Latin presentations of the topic – the translations of al-Khwārizmī prepared by Robert of Chester and Gherardo da Cremona, the anonymous fourteenth-century translation of Abū Kāmil's algebra, and the *Liber abbaci*.[294] Afterwards, we shall need to compare it both to a wide range of Arabic treatises and to a selection of Italian algebraic fourteenth-century algebraic writings.

Jacopo's Algebra

As we have seen, Jacopo's Chapters 16–17 – the algebra section proper – gives rules for the following cases – *C* still stands for *censo*, *t* for *thing* (*cosa*), *n* for *number* (*numero*), *K* for *cube* (*cubo*), *CC* for *censo di censo*, i.e., the fourth power of *t*:[295]

---

[294] Here and elsewhere I disregard the brief excerpts "de libro qui dicitur gleba mutabilia" in *Liber Alchorizmi de pratica arismetice* [ed. Boncompagni 1857b: 112*f*]. They are not in Allard's partial edition of the *Liber Alchorizmi* [1992] but they are present in manuscripts that are as distant from each other in the stemma as possible – see [Høyrup 1998b: 16 n.7]; there is thus no doubt that they were present in the original and have not been interpolated. But the few paragraphs in question can hardly count as a presentation of the field and appear to have had no impact whatsoever.

[295] The Latin translations of al-Khwārizmī (but not that of Abū Kāmil, nor the *Liber abbaci*) refer to the numbers as *dragmas*, but this idiom is absent from Jacopo's formulation of the rules. Similarly, Jacopo refers to the first power of the unknown as *thing* (*cosa*), never as *root* (*radix*) as do the Latin translations as well as the *Liber abbaci* when the rules are stated. Cf. the above "aside on Arabic algebra", pp. 101*ff*.

(1) $\alpha t = n$	(3) $\alpha C = \beta t$	(5) $\beta t = \alpha C + n$
(2) $\alpha C = n$	(4) $\alpha C + \beta t = n$	(6) $\alpha C = \beta t + n$

(7) $\alpha K = n$	(12) $\alpha K = \beta C + \gamma t$	(17) $\alpha CC + \beta K = \gamma C$
(8) $\alpha K = \beta t$	(13) $\alpha CC = n$	(18) $\beta K = \alpha CC + \gamma C$
(9) $\alpha K = \beta C$	(14) $\alpha CC = \beta t$	(19) $\alpha CC = \beta K + \gamma C$
(10) $\alpha K + \beta C = \gamma t$	(15) $\alpha CC = \beta C$	(20) $\alpha CC + \beta C = n$
(11) $\beta C = \alpha K + \gamma t$	(16) $\alpha CC = \beta K$	

The first six cases are the traditional first- and second-degree cases, familiar since al-Khwārizmī's *Kitāb al-mukhtaṣar fī ḥisāb al-jabr wa'l-muqābalah*. The remaining ones are all reducible to homogeneous problems or to second-degree problems, and thus nothing new compared to what had been done in the Arabic world for at least three centuries. As mentioned on p. 3, the order of the six fundamental cases differs, both from that of al-Khwārizmī (the extant Arabic text as well as the Latin translations)[296] and Abū Kāmil (both have 3-2-1-4-5-6) and from that of Fibonacci (who has 3-2-1-4-6-5). Jacopo's higher cases, as we see, are ordered group-wise according to the same principles as the groups (2)–(3) and (4)–(5)–(6).[297]

    Another noteworthy characteristic is that all cases are defined as non-normalized problems (that is, the coefficient of the highest power is not supposed to be 1), the first step of each rule being thus a normalization. In the Latin treatises, all cases except "roots equal number" (where the normalized equation *is* the solution[298]) are defined as normalized problems, and the rules are formulated correspondingly. All also teach how to proceed when a non-normalized problem is encountered, but this is done outside the regime of rules – cf. al-Khwārizmī's treatment of case 4 as quoted on p. 102.

---

[296] In an appendix to Robert of Chester's translation of al-Khwārizmī [ed. Hughes 1989:67], the six rules are given in Jacopo's order; but the shared archetype for all three extant manuscripts appears to be from mid-fifteenth–century Germany, and the appendix makes use of the abbreviations for number, *coss* and *zensus* that were current in German *coss* algebra at that moment. It clearly descends from Italian abbacus algebra.

[297] That is, $\alpha x^n = \beta x^p$, n fixed (either 3 or 4), p increasing from 0 to $n-1$; and equation groups obtained from the group (4)–(6) by multiplication by x or x^2. According to this principle, (20), the biquadratic obtained from (4), should obviously be followed by two other biquadratic equations, (21*) $\beta C = \alpha CC + n$ and (22*) $\alpha CC = \beta C + n$. Jacopo must have intended to include one of them, since the text announces "15 rules which [...] lead back to the six rules from before". Whether he forgot himself or the omission was due to an early copyist cannot be decided (a late copyist can be excluded, see note 317).

[298] Fibonacci actually defines even this case in normalized form – but gives no example and thus escapes the absurdity.

The Examples

For each of the first six cases, as we have seen, Jacopo gives at least one, sometimes two or three examples after setting forth the rule in abstract form. For the remaining cases, only the rules and no examples are given. For ease of reference, translations of the statements of cases (1)–(6) may be listed together:

1a.   (= V.16.2) Make two parts of 10 for me, so that when the larger is divided in the smaller, 100 results from it.

1b.   (= V.16.3) There are three partners, who have gained 30 *libre*. The first partner put in 10 *libre*. The second put in 20 *libre*. The third put in so much that 15 *libre* of this gain was due to him. I want to know how much the third partner put in, and how much gain is due to (each) one of those two other partners.

2.   (= V.16.5) Find me two numbers that are in proportion as is 2 of 3: and when each (of them) is multiplied by itself, and one multiplication is detracted from the other, 20 remains. I want to know which are these numbers.

3.   (= V.16.7) Find me 2 numbers that are in proportion as is 4 of 9. And when one is multiplied against the other, it makes as much as when they are joined together. I want to know which are these numbers.

4a.   (- V.16.9) Someone lent to another 100 *libre* at the term of 2 years, to make (up at) the end of year. And when it came to the end of the two years, then that one gave back to him *libre* 150. I want to know at which rate the *libra* was lent a month.

4b.   (= V.16.10) There are two men that have *denari*. The first says to the second, if you gave me 14 of your *denari*, and I threw them together with mine, I should have 4 times as much as you. The second says to the first: if you gave me the root of your *denari*, I should have 30 *denari*. I want to know how much each man had.

5a.   (= V.16.12) Make two parts of 10 for me, so that when the larger is multiplied against the smaller, it shall make 20. I ask how much each part will be.

5b.   (= V.16.13–14) Somebody makes two voyages, and in the first voyage he gains 12. And in the second voyage he gains at that same rate as he did in the first. And when his voyages were completed, he found himself with 54, gains and capital together. I want to know with how much he set out.[299]

5c.   (= V.16.15) Make two parts of 10 for me, so that when one is multiplied against the other and above the said multiplication is joined the difference which there is from one part to the other, it makes 22.

6.   (= V.16.17) Somebody has 40 gold *fiorini* and changed them to *venetiani*. And then from those *venetiani* he grasped 60 and changed them back into *fiorini* at one *venetiano* more per *fiorino* than he changed them at first for me. And when he has changed thus, that one found that the *venetiani* which remained with him when he detracted 60, and the *fiorini* he got for the 60 *venetiani*, joined together made 100. I want to know how much was worth the *fiorino* in *venetiani*.

The first observation to make concerning Jacopo's examples is that none of them are stated in terms of number, *things* and *censi* (afterwards, of course, a "position" is made identifying some magnitude with the *thing*; without this position, no reduction to the

---

[299] Both solutions are shown to be valid.

corresponding case could result). In the Latin treatises, in contrast, the basic examples are always stated directly in the same number-roots-*census* terms as the rules.

Second, we notice that three of Jacopo's pure-number examples (*viz* 1a, 5a and 5c) follow the pattern of the "divided ten", familiar since al-Khwārizmī's treatise and abundantly represented in the *Liber abbaci* (cf. p. 107). (2) and (3), those in which the ratio between two unknown numbers is given, on the other hand, are of a type with no such precedent.[300] For any given polynomial equation with a single unknown it is of course easy to create an example of this kind, thereby adding cheap apparent complexity.

Further, we should be struck by the abundant presence of problems (5 out of 10 problems) that pretend to deal with commercial questions – *mu'āmalāt*-problems ("problems dealing with social life"), in the classification of Arabic mathematics. The only problem type belonging to this category which we find in the Latin algebra translations is the one where a given sum of money is distributed evenly first among an unknown number x of people, next among $x+1$ [ed. Hughes 1986: 255], with given difference between the shares in the two situations. Among the problems treated in the algebra section of the *Liber abbaci* at most some 8 percent belong to the *mu'āmalāt* category: 4 variants of the problem type just mentioned, one problem treating of the purchase of unspecified goods, and one referring to interest and commercial profit.

Finally, we should take notice of the presence of the square root of an amount of real money in (4b); this is without parallel even in the non-algebraic chapters of the *Liber abbaci*, where *mu'āmalāt* problems abound.[301]

Peculiar Methods

In the main, the methods used by Jacopo of course coincide with what we know from the Latin works. But some differences can be observed here and there. We may look at the solution to (1b) – a wonderful example of superfluous application of algebra – in which several idiosyncrasies are represented (fols 36ᵛ–37ʳ):

---

[300] There is an analogue of Jacopo's superficially similar problem (1a = V.16.2) in al-Khwārizmī's treatise [ed. Hughes 1986: 248], repeated by Abū Kāmil [ed. Sesiano 1993: 360]; but like Jacopo's (1a) these problems speak of division, not of "proportion", and like Jacopo's they are primarily divided-ten problems.

[301] The problems in *Liber abbaci*, Chapter 12, Part 3 ("Questions of trees and similar things") that involve square roots all treat of *numbers*: "On finding a certain number, of which $\frac{1}{6}$ $\frac{1}{5}$ $\frac{1}{4}$ $\frac{1}{3}$ of the same is the root of the same number" [ed. Boncompagni 1857a: 175], etc.

Elsewhere in the medieval world, problems involving the square root of real entities may go together with problems that consider their product – thus in Mahāvīra's *Gaṇita-sāra-saṅgraha* [ed. Raṅgācārya 1912: 75–85]. Of this latter type, several specimens are present in the *Liber abbaci*, namely a number of problems about three or five men finding *bizanti* respectively having *denari*, where relations between the *products* of their possessions taken pairwise are given [ed. Boncompagni 1857a: 204–206, 281]. As mentioned above (p. 35), several of these were borrowed into Chapter 19 of the *Livero de l'abbacho*.

Do thus, if we want to know how much the third partner put in, posit that the third put in a thing. Next one shall aggregate that which the first and the second put in, that is, *libre* 10 and *libre* 20, which are 30. And you will get that there are three partners, and that the first puts in the partnership 10 *libre*. The second puts in 20 *libre*. The third puts in a thing. So that the principal of the partnership is 30 *libre* and a thing. And they have gained 30 *libre*. Now if we want to know how much of this gain is due to the third partner, when we have posited that he put in a thing, then you ought to multiply a thing times that which they have gained, and divide in the total principal of the partnership. And therefore we have to multiply 30 times a thing. It makes 30 things, which you ought to divide in the principal of the partnership, that is, by 30 and a thing, and that which results from it, as much is due to the third partner. And this we do not need to divide, because we know that 15 *libre* of it is due to him. And therefore multiply 15 times 30 and a thing. It makes 450 and 15 things. Hence 450 numbers and 15 things equal 30 things. Restore each part, that is, you shall remove from each part 15 things. And you will get that 15 things equal 450 numbers. And therefore you shall divide the numbers in the things, that is, 450 in 15, from which results 30. And as much is the thing. And we posited that the third partner put in a thing, so that he comes to have put in 30 *libre*. The second 20 *libre*. The first 10 *libre*. And if you should want to know how much of it is due to the first and to the second, then remove from 30 *libre* 15 of them which are due to the third. 15 *libre* are left. And you will say that there are 2 partners who have gained 15 *libre*. And the first put in 10 *libre*. And the second put in 20 *libre*. How much of it is due to (each) one. Do thus, and say, 20 *libre* and 10 *libre* are 30 *libre*, and this is the principal of the partnership. Now multiply for the first, who put in 10 *libre*, 10 times 15 which they have gained. It makes 150. Divide in 30, from which results 5 *libre*. And as much is due to the first. And then for the second, multiply 20 times 15, which makes 300 *libre*. Divide in 30, from which results 10 *libre*, and as much is due to the second partner. And it is done, and it goes well. And thus the similar computations are done.

Let us first look at the start of the procedure, the one that leads to the determination of what the third partner put in. It makes use of the "partnership rule": the share of each partner in the profit is found by multiplying first his share of the capital by the total profit, and dividing next the outcome by the total capital of the partnership,

$$p_i = \frac{c_i \cdot P}{C} .$$

The second part of the procedure, the one determining the shares of the first two partners by means of a fictitious new partnership, shows that Jacopo's use of the commercial partnership as a general model or functionally abstract representation for proportional distribution spreads even to his algebra chapter.

Other Idiosyncrasies

Al-Khwārizmī's "algebra" was entitled "Concise Book about *al-jabr* and *al-muqābalah* Calculation", *al-jabr* being derived from the verb *jabara* (mostly translated "to restore") and *al-muqābalah* from the verb *qabila* ("to accept" etc., the non-technical meaning of *muqābalah* being "encounter", "comparison", etc.) – cf. above, p. 105. *Al-jabr* and *al-muqābalah* must hence be central operations for the discipline – and it must be significant that Jacopo does not use the terms in the same way as the Latin algebra writings.

In these, *restaurare* (the translation of *jabara*) designates the cancellation of a subtractive term by addition. As we have seen in the detailed commentary to Chapter 16, Jacopo uses the corresponding term *ristorare* both in this function and for the cancellation of an additive term. In Abū Bakr's *Liber mensurationum* as translated by Gherardo da Cremona [ed. Busard 1968] *restaurare* is further used a couple of times (#7, #55) in the function of multiplicative completion (changing $\frac{2}{5}$ and $\frac{1}{4}$ into 1 through multiplication by $2\frac{1}{2}$ and 4, respectively). Cancellation of an additive term, on the other hand, is nowhere spoken of in this way in any of the Latin treatises but instead as *opponere*, the Latin equivalent of *qabila* (from which *muqābalah* is derived).

Opporre, the vernacular match of *opponere*, is absent from Jacopo's text, but that probably does not mean that it contains no equivalent of *qabila/muqābalah*. Indeed, in Raffaello Canacci's *Ragionamenti d'algebra* [ed. Procissi 1954: 302] we read, in a passage ascribed to Guglielmo de Lunis, that *elmelchel* (the neighbour of *geber*, i.e., *jabr*, in Canacci/Guglielmo's text and thus certainly a transcription of *al-muqābalah*[302]) means "exempio hovvero aghuaglamento", "example or equation", whereas *elchel* (another noun derived from *qabila*) is *oppositione*, explained as the opposition or counterposition of the two members of the simplified equation. This term (in the form *raoguaglamento*) is indeed used in the end of Jacopo's example (5b), precisely in the sense of "equation".[303]

A final characteristic by which Jacopo's treatise differs from all Latin algebra writings is the complete absence of geometric proofs for the correctness of the rules by means of which the cases 4–6 are solved.

The *Fondaco* Problems

The four problems of Chapter 19 would be considered algebraic by us, but they do not make use of the technique of *thing* and *censo* (fols 43ᵛ–45ᵛ); in consequence, Jacopo and his contemporaries would not see them as belonging under the heading *regola della cosa*. Since they were discussed in depth above, pp. 115–121, close examination at this point is superfluous.

---

[302] As pointed out to me by Ulrich Rebstock (personal communication), the transcription (with assimilated *b*) appears to render a Mozarabic pronunciation.

Canacci's explanation is similar to but not directly copied from a passage in Benedetto da Firenze's exposition of "La reghola de algebra amuchabale" [ed. Salomone 1982: 1*f*], which lends credibility to their common reference to Guglielmo de Lunis. Nothing similar is found in the Latin version of al-Khwārizmī's algebra contained in MS Lyell 52 (Bodleian Library Oxford) [ed. Kaunzner 1986]; this version is therefore *not* likely to represent Guglielmo's translation, as sometimes claimed – cf. also [Kaunzner 1985: 11*f*].

[303] In Biagio's problem collection as rendered by Benedetto, *aguagliamento* is also an "equation", and the corresponding verb *ragugliare* is used repeatedly about the process of producing the simplified equation [ed. Pieraccini 1983: 4, 21, 22].

Abbreviations and Notation

It is a general and noteworthy characteristic of Jacopo's algebra (or at least of manuscript **V**, but there are good reasons to believe the manuscript to be true to the original in this regard) that it avoids all abbreviations in the technical algebraic terminology, as if the author was conscious of introducing a new field of knowledge where readers would be unfamiliar with the terminology and therefore unable to expand abbreviations correctly – cf. above, p. 9. *A fortiori*, nothing in his algebra reminds us even vaguely of algebraic symbolism or syncopation. Early in the treatise, however, we find an unusual variant of the Roman numerals – for instance in the explanation of 400,000 as $_{cccc}^{m}$ (see p. 51). This way to put the "denominator" above the number being denominated coincides exactly with the algebraic notation found in Maghreb writings from the twelfth century onward [Abdeljaouad 2002: 11*f*; Souissi 1969: 92 n. 2]. Moreover, the usage turns up in tables that are oriented from right to left, as in Arabic writing. However, the same multiplicative system is used by Diophantos and other ancient Greeks when they write multiples of aliquot parts, and by Middle Kingdom Egyptian scribes for the writing of large numbers [Sethe 1916: 9]; the notation could therefore be considered near at hand, and independent invention cannot be excluded.

Jacopo's Possible Sources: Arabic Writings on Algebra

Jacopo's algebra does not depend on, *a fortiori* is not derived, neither from Fibonacci nor from the Latin translations of al-Khwārizmī (or Abū Kāmil[304]) – that much should already be clear. That it is ultimately derived from Arabic *al-jabr* is no less certain. In consequence, Jacopo's algebra confronts us with a hitherto unknown channel from the Arabic world and its mathematics.

This conclusion raises two difficult questions. Firstly, Jacopo's algebra, being fundamentally different from the Latin translations of al-Khwārizmī and Abū Kāmil, is also fundamentally different from their Arabic originals, and his Arabic inspiration must therefore be of a different kind; secondly, his treatise contains no single Arabism, and direct use of Arabic sources on his part can thus be safely excluded. We must therefore ask, firstly, which kind of Arabic material provided his ultimate inspiration? And secondly, where in the Romance-speaking world did he find an environment using this material actively?

The two questions must be addressed one by one. At first we shall therefore look at a larger range of Arabic algebraic writings in relation to the parameters where Jacopo's

---

[304] This is in fact already excluded for chronological reasons – the fourteenth-century manuscript being the translator's own original, see [Sesiano 1993: 315–317].

algebra differs from the Latin treatises. Afterwards we shall examine Italian writings. The following Arabic writings are taken into consideration:

– Abū Bakr, *Liber mensurationum* (*Kitāb al-misāḥah*?). Terminological considerations suggest an early date (c. 800?), either for the treatise itself or for an original which it follows fairly closely. [Ed. Busard 1968, trans. Gherardo da Cremona].

– Al-Khwārizmī, *Kitāb al-mukhtasar fī ḥisāb al-jabr wa'l-muqābalah*. Written in Baghdad, earlier ninth century. [Ed. Hughes 1986, trans. Gherardo da Cremona].

– Ibn Turk, *Kitāb al-jabr wa'l-muqābalah* (extant fragment containing geometrical proofs). Roughly contemporary with al-Khwārizmī. [Ed., trans. Sayılı 1962].

– Thābit Ibn Qurra, *Qawl fī tasḥīh masāʾil al-jabr bi'l-barāhīn al-handasīyah*. Written in Baghdad, later ninth century. [Ed., trans. Luckey 1941].

– Abū Kāmil, *Risālah fi'l-jabr wa'l-muqābalah*. Late ninth or early tenth century. The author's surname al-Miṣrī means "the Egyptian", but that does not prove that Abū Kāmil actually lived there. [Ed. Sesiano 1993, trans. anon.].

– Al-Karajī, *Kāfī fi'l-hisāb*. Written in Baghdad, c. 1011. [Ed., trans. Hochheim 1878].

– Al-Karajī, *Fakhrī fi'l-jabr wa'l-muqābalah*. Written in Baghdad, c. 1011. Paraphrase [Woepcke 1853].

– Al-Khayyāmī, *Risālah fi'l-barāhīn ʿalā masāʾil al-jabr wa'l-muqābalah*. Written in Samarkand, c. 1070. [Ed., trans. Rashed & Djebbar 1981].

– Ibn al-Yāsamīn, *Urjūzah fi'l-jabr wa'l-muqābalah*. Written in Morocco (or possibly Seville?) before 1190. [Ed., trans. Abdeljaouad 2005].

– Ibn al-Bannāʾ, *Talkhīs aʿmāl al-ḥisāb*. Written in Morocco in the later thirteenth century. [Ed., trans. Souissi 1969].

– Ibn Badr, *Ikhtiṣār al-jabr wa'l-muqābalah*. Written before 1343 (and after Abū Kāmil's times), perhaps in Muslim Spain. [Ed., trans. Sánchez Pérez 1916].

– Al-Qalasādī, *Kašf al-asrār ʿan ʿilm ḥurūf al-ghubār*. Written in Cairo in 1448, but the author had studied and taught in al-Andalus and the Maghreb. [Ed., trans. Souissi 1988].

– Bahaʾ al-Dīn al-ʿĀmilī, *Khulāsah al-ḥisāb*. Written in the late sixteenth or the early seventeenth century; the author was born in Syria and died in Iran. [Ed., trans. Nesselmann 1843].

Occasionally, information concerning other works is borrowed from the secondary literature.

The Order of the Six Cases

As already mentioned, al-Khwārizmī as well as Abū Kāmil present the six fundamental cases in the order 3-2-1-4-5-6 (Jacopo's order being 1-2-3-4-5-6). This "classical order" recurs in Ibn al-Bannāʾ's presentation of the cases in the *Talkhīs*, in al-Qalaṣādī's *Kašf*, in Ibn Badr's *Ikhtisār*, and in Ibn al-Yāsamīn's *Urjūzah*.

Al-Karajī arranges things differently. In the *Kāfī* as well as the *Fakhrī*, his order is

1-3-2-4-5-6. The same pattern is found with al-Samaw'al and al-Kašī [Djebbar 1981: 60*f*] and in Bahā' al-Dīn al-'Amilī's *Khulāṣah* from c. 1600. In his *solution* of the equations, Ibn al-Bannā' follows the pattern 3-2-1-4-6-5 (that of the *Liber abbaci*).

Jacopo's order is referred to around 1500 by al-Māridīnī in his commentary to Ibn al-Yāsamīn's *Urjūzah* as the one that is used "in the Orient", and it is indeed that of al-Misṣīsī, al-Bīrūnī, al-Khayyāmī and Šaraf al-Dīn al-Ṭūsī [Djebbar 1981: 60]. Not only there, however; Abū'l-Qāsim al-Qurašī, born in Seville in the twelfth century and active in Bejaïa, has the same order [Djebbar 1988: 107].

Normalization

Al-Khwārizmī's original text, like the Latin translations, defines all cases except "things made equal to number" in normalized form and gives corresponding rules.[305] This also applies to Ibn Turk's and Thābit's demonstrations of the correctness of the rules, and to al-Khayyāmī's algebra [ed., trans. Rashed & Djebbar 1981]. Al-Karajī's *Kāfī* confronts us with a mixed situation: the three simple cases (1–3) are non-normalized (definitions as well as rules); case (4) is defined as non-normalized, but its rule presupposes normalization; the two remaining composite cases are defined only through normalized paradigmatic examples, and the formulation of the rules presupposes this normalization. The *Talkhīṣ* and the *Kašf* treat the simple cases as the *Kāfī;* they give no explicit definitions of the composite cases, but give rules that presuppose normalization. Ibn Badr gives non-normalized definitions for all cases, and corresponding rules for the simple cases; his rules for the composite cases apply to the normalized equation; as far as can be judged from the very concise versified text, this is also Ibn al-Yāsamīn's intention (only the cases 1–3 and 6 are quite explicit).[306] Only Bahā' al-Dīn states all definitions as well as all rules in non-normalized form, as does Jacopo.

Examples

Basic examples formulated in the same terms as the rules, i.e., dealing with a *māl* ("possession", the equivalent of Jacopo's *censo*) and its *jidhr* ("[square] root"), are found

---

[305] The Arabic manuscript published first by Rosen [1831] and later by Mušarrafa & Ahmad [1939] defines the cases in non-normalized form, even though its rules presuppose normalized equations. However, Gherardo's extreme grammatical faithfulness in other respects should guarantee his reliability even on this account. The different pattern of the published Arabic text is thus an innovation – an adaptation of the original to changing customs within the field (a partial adaptation only, the rules being unchanged and the resulting totality thus incoherent). Indeed, comparison of the published Arabic version with Gherardo's and Robert of Chester's Latin translations shows that it must have been submitted to at least three successive revisions, two of which have also affected Robert's Arabic text – see [Høyrup 1998a: 172*f*].

[306] However, Ibn al-Yāsamīn has a very explicit discussion of how to treat non-normalized mixed problems, either through division by the coefficient of the possession or by multiplication (the Babylonian-Diophantine method).

in almost all the Arabic works I have looked at – in al-Khwārizmī's, Abū Kāmil's and al-Khayyāmī's treatises, in al-Karajī's *Kāfī* and *Fakhrī*, in al-Qalaṣādī's *Kašf* and in Ibn Badr's *Ikhtiṣār*. Only Ibn al-Yāsamīn's *Urjūza*, Ibn al-Bannā''s *Talkhīṣ* and Bahā' al-Dīn's *Khulāṣah* contain no examples of this kind – but the *Urjūzah* and the *Talkhīṣ* because they give no examples at all.[307]

The divided ten turns up everywhere (except where no examples are given), from al-Khwārizmī and Abū Kāmil to Bahā' al-Dīn. Problems where two unknown numbers are given in proportion are as absent from the Arabic treatises I have inspected as from the Latin ones.

Abū Kāmil, like al-Khwārizmī, deals with the division of a given amount of money between first x, then $x+p$ men, but apart from that none of the two treat of *mu'āmalāt*-problems in the properly algebraic parts of their treatises. Most other treatises keep *mu'āmalāt* matters wholly apart from their algebra. The only exceptions among the works I have inspected are the *Fakhrī* and Ibn Badr's and Bahā' al-Dīn's treatises. Ibn Badr, after a large number of divided-ten and *māl-jidhr* problems, has others dealing with the remuneration of a capital, dowries,[308] the mixing of grain, the distribution of booty among soldiers, travels of couriers, and reciprocal gifts (three or four of each type). Of Bahā' al-Dīn's illustrations of the six fundamental cases, two deal with pure numbers and four with feigned *mu'āmalāt* (that is, with "recreational") problems. In a later chapter listing nine problems that can be resolved by more than one method, the share of recreational problems is the same.

Square Roots of Real Money

One of Jacopo's problems – (4b), the only one of his *mu'āmalāt* problems that belongs to a familiar recreational type – refers to the square root of an amount of real money. From a purely formal point of view this is highly traditional, the basic *al-jabr* cases being defined as problems dealing with a *māl* or "possession" and its square root, and treating the known number as a number of dirhams. But already in al-Khwārizmī's time this had become a formality. It is true that he gives the value not only of the root when it has been found but also of the *māl*, remembering thus that once this had been the real unknown quantity of the problem. But stating the case "*māl* made equal to number" in normalized form (and defining first the *root* as one of the number types and next the *māl* as the product of this number by itself [ed. Hughes 1986: 233*f*]) he clearly considers the *root* as the unknown proper – in perfect agreement indeed with his later identification of the root with the *šay'* or *thing*. From al-Khwārizmī onward we may thus claim that the *root*

---

[307] The *Khulāṣah* does contain a first-degree problem about a *māl*, but apparently meant to stand for real money.

[308] Capital as well as dowry are designated *māl*, but the problem texts show that real invested money and real dowries are meant.

was a square root of formal, not real money.

Roots of real money are absent from almost all of the Arabic algebra writings I have examined – the only exceptions being al-Karajī, who in the *Fakhrī* once takes the root of an unknown price and twice of unknown wages, and Ibn Badr, who twice takes the root of a dowry. However, the *Liber mahumaleth*, a Latin composition made in Spain during the twelfth century, contains at least two algebraic problems of the kind: in one, the square roots of a capital and a profit are taken, in another the square root of a wage [Sesiano 1988: 80, 83].

In order to find copious square roots of real entities (not only money but also, for instance, a swarm of bees, the arrows fired by Arjuna, or a horde of elephants) we have to go to India.

Commercial Calculation within Algebra

Jacopo employs the rule of three (expanded into the partnership rule) as a tool for algebraic computation; further, he uses the commercial partnership as a functionally abstract representation for proportional distributions. I have never noticed anything similar in an Arabic treatise – al-Khwārizmī *presents* the rule of three in a separate chapter (said to deal with *muʿāmalāt*) after the algebra proper and before the geometry, but this is a different matter.

Jabr and *Muqabalah*

Jacopo's use of the equivalent of *jabr* (*ristorare*) and of the likely equivalent of *muqābalah* (*raoguaglamento*) differs from al-Khwārizmī's use of the original terms (which is also the main usage of Abū Kāmil, and that of Ibn al-Bannāʾ, al-Qalaṣādī and Bahāʾ al-Dīn). However, the Arabic usage is far from uniform.

Firstly, Abū Bakr's *Liber mensurationum*, whose multiplicative use of *restaurare* was mentioned above, uses the phrase *restaura et oppone* repeatedly in situations where no subtraction is to be made. The meaning of "opposition" is clearly in concordance with Canacci's explanation, namely *to form a (simplified) equation* – and thus with Jacopo's usage. Similar ambiguities are found in Ibn Badr [Sánchez Pérez 1916: 24 n. 1].

In the *Fakhrī* [Woepcke 1853: 64], *jabr* is initially explained as the elimination of additive as well as subtractive terms, just as in Jacopo's treatise.[309] *Muqābalah*, on its part, is explained as the formation of a simplified equation where two terms are equal to one (or vice versa) – that is, the formation of one of the equations that define the basic cases. In the *Kāfī* [ed. trans. Hochheim 1878: III, 10], *jabr* is said to include also multiplicative completion (as it does in the *Liber mensurationum*). For the rest, this text seems to be ambiguous (as far as can be judged from the translation). Perhaps it means

---

[309] According to Jeffrey Oaks (personal communication), the term is only *used* about the elimination of subtractive terms in the ensuing problem solutions.

to leave the elimination of an additive term unnamed and uses *muqābalah* as the *Fakhrī*; perhaps this latter term is meant instead to designate the removal which leads to the formation of the simplified equation.[310]

Geometric Proof

Geometric proofs for the correctness of the rules for the three composite cases are given by al-Khwārizmī and Ibn Turk, and (with new ones added) by Abū Kāmil and in the *Fakhrī*. They are absent from the *Kāfī*, from the treatises belonging to the Maghreb school (Ibn al-Yāsamīn, Ibn al-Bannā', al-Qalaṣādī) and from those of Ibn Badr and Baha' al-Dīn.

Polynomial Algebra and Geometric Progressions

As mentioned on p. 116, I have seen nothing similar to Jacopo's four *fondaco* problems in Arabic works; nor have I ever received a positive answer when asking others who might know better, in spite of the evidence for an Arabic origin of the problem type discussed in note 252. In any case, the basic underlying theory – that which also allows one to see that Jacopo's cases (7) through (20) can be solved – was known at least since al-Karajī and al-Samaw'al (see p. 121), and part of it was inherent in all writings that presented the sequence of algebraic powers as a geometric progression and also stated the rules for multiplying binomials – thus in the *Urjūzah*, the *Talkhīṣ* and the *Kašf*.[311]

Summing Up

Almost every seeming idiosyncrasy we find in Jacopo's algebra can be found in Arabic writings (the exceptions being the use of the rule of three and the partnership structure as tools for algebra, the examples asking for numbers in given proportion, and the idea that wages increase by default in geometric progression). But they never occur together in treatises I have inspected. Those which are furthest removed from Jacopo are al-Khwārizmī and Abū Kāmil. The exponents of the Maghreb school are somewhat closer (in their omission of geometric proofs and, hypothetically, in the similarity between their algebraic notation and Jacopo's multiplicative writing of Roman numerals). But Jacopo's

---

[310] As Georges Saliba [1972] has argued, the *Fakhrī* usage appears to be the original one; the ambiguity in the *Kāfī* illustrates the way in which the new interpretation as the subtractive counterpart of *jabr* can have come about.

Raffaello Canacci, in the passage where he explains *elmelchel* to stand for "exemple or equation" [ed. Procissi 1954: 302], states that *elchel* (*al-qābila*, according to the parallel) stands for "opposition", explained to be the simplified equation.

[311] With hindsight, not only "part" but all that is required to resolve all of Jacopo's *fondaco* problems was implied. But hindsight may amount to historiographical blindness: Cardano's solution to the third-degree equation is "implied" in Old Babylonian "algebra", in the sense that he combines tricks that were in use in that discipline; but it took more than three millennia to discover that it could be done.

order of cases, his use of the *jabr-* and *muqābalah*-equivalents, his square roots of real money and his ample use of *mu'āmalāt*-problems within the algebra links him to (some middle ground between) al-Karajī's writings, Ibn Badr's possibly Iberian *Compendium of Algebra*, the certainly Iberian *Liber mahamaleth*, and Bahā' al-Dīn's *Essence of the Art of Calculation*; his consistent presentation of non-normalized cases is only shared with the latter much younger work. In other, more explicit words: We do not know the kind of Arabic algebra that provided him with his ultimate inspiration, but it was certainly different from those (scholarly or "high") currents that have so far been investigated by historians of mathematics; we may also conclude with fair certainty that it was linked to an institution that taught algebra as integrated in *mu'āmalāt*-mathematics.

Jacopo's Possible Sources: a Look at the Next Italian Generation

The next question about the sources for Jacopo's algebra is: where in the Romance-speaking world did Jacopo find an environment actively engaged in algebra?

However, an answer to this question (indirect and partially negative as it will be) can only be given if we look closely at the still extant Italian expositions of algebra written during the decades that followed immediately after Jacopo's treatise.

One of these is contained in Paolo Gherardi's *Libro di ragioni* [ed. Arrighi 1987], written in Montpellier in 1328, often referred to above (henceforth **G**). Two others are contained in the *Libro di molte ragioni d'abaco* [ed. Arrighi 1973], a conglomerate written by several hands in Lucca in c. 1330 [ed. Arrighi 1973], also referred to regularly already. Its fols 80ᵛ–81ᵛ (pp. 194–197) contain a section on "le reghole dell'aligibra amichabile" (henceforth **L**); another section on "le reghole della chosa con asenpri" is found on fols 50ʳ–52ʳ (pp. 108–114; henceforth **C**).

Somewhat later but so closely related to one or more members of the first generation that they can inform us about it are two other items: **A**, the *Trattato dell'alcibra amuchabile* from c. 1365 [ed. Simi 1994]; and **P**, the anonymous *Libro di conti e mercatanzie* [ed. Gregori & Grugnetti 1998] kept today in the Biblioteca Palatina of Parma and probably compiled in the Tuscan-Emilian area – according to problems dealing with partnership and interest in the years immediately after 1389–95.

All of these depend to some extent on what we know from **V**, that is, on Jacopo. The earliest extant vernacular algebra that does not depend on him – and the earliest vernacular work dedicated exclusively to algebra – is the *Aliabraa argibra*, which according to one manuscript was written by an otherwise unidentified Master Dardi from Pisa in 1344 (henceforth **D**; on a *possible* identification of its author, see note 333). Dardi's work is analyzed in some depth in the next section. Slightly earlier and also independent of Jacopo is a treatise written by Giovanni di Davizzo, from which however nothing but a fragment (**Z**) survives, whose importance only becomes clear when we compare it with **V** as well as **D**.

Case	V	G	L	C	A	P	D	Z
$\alpha t = n$	1.R,E_1,n	1.R,E_1,n	1.R,E_1,n	1.R,E_1,p	1.R,E_1,n	1.R,E_1,p	1.R,E_{1-1},p	1.R.p
$\alpha C = n$	2.R,E_1,p	2.R,E_2,n	2.R,E_2,n	2.R,E_1,n	2.R,E_1,p	2.R,E_1,[a],p	2.R,E_1,p	2.R.p
$\alpha C = \beta t$	3.R,E_1,p	3.R,E_1,n	3.R,E_1,p	3.R,E_2,p	3.R,E_1,p	3.R,E_1,p	3.R,E_1,p	3.R.p
$\alpha C+\beta t = n$	4.R,E_1,n	4.R,E_1,n	4.R,E_1,n	4.R,E_1,n	4.R,E_1,n	4.R,E_1,p	4.R,E_1p	4.R.n
$\beta t = \alpha C+n$	5.R,E_1,n	5.R,E_1,n	5.R,E_1,p	5.R,E_1,[b],n	5.R,E_1,n	5.R,E_1,p	5.R,E_1,p	5.R.n
$\alpha C = \beta t+n$	6.R,E_1,n	6.R,E_2,n	6.Omitted[c]	6.R,E_3,n	6.R,E_1,n	6.R,E_2,p	6.R,E_4,[d],p	6.R.n
$\alpha K = n$	7.R,p	7.R,E_1,p	7.R,n	7.R,p	7.R,E_1,p	7.R,E_1,p	7.R,E_1,p	7.R.n
$\alpha K = \beta t$	8.R,p	9.R,E_1,p	8.R,n	8.R,p	8.R,E_1,p	9.R,E_1,[e],p	8.R,E_1,p	8.R.p
$\alpha K = \beta C$	9.R,p	10.R,E_1,p	9.R,p	9.R,p	9.R,E_1,p	10.R,E_1,p	9.R,E_1,p	9.R.p
$\alpha K+\beta C = \gamma t$	10.R,n	15.R,E_1,n	10.R,[f],p	14.R,n	15.R,n	15.R,E_1,p	14.R,E_1,p	
$\beta C = \alpha K+\gamma t$	11.R,n		11.R,n	15.R,n	16.R,n			
$\alpha K+\gamma t = \beta C$					14.R,E_1,n	16.R,E_1,n	15.R,E_{3-4},p	10.R.n
$\alpha K = \beta C+\gamma t$	12.R,n	11.R,E_1,n	12.R,[g],n	16.R,p	10.R,E_1,n	11.R,E_1,p	16.[h]R,E_2,p	11.R.n
$\alpha K = \sqrt{n}$		8.R,E_1,p			11.R,E_1,p	8.R,E_1,n	21.R,E_2,p	
$\alpha K = \beta t+n$		12.X,E_1,n			12.X,E_1,[i],n	12.X,E_1,p		
$\alpha K = \beta C+n$		13.X,E_1,n			13.X,E_1,n	13.X,E_1,p		
$\alpha K = \gamma t+\beta C+n$		14.X,E_1,n				14.X,E_1,n		
$\alpha CC = n$	13.R,n		13.R,p	11.R,p	17.R,n	17.R,E_1,n	11.R,E_1,p	12.R.p
$\alpha CC = \beta t$	14.R,p		12.R,p		18.R,p	18.R,E_1,p	12.R,E_2,p	13.R.p
$\alpha CC = \beta C$	15.R,p		13.R,p		19.R,p	19.R,E_1,p	13.R,E_3,p	14.R.p
$\alpha CC = \beta K$	16.R,p		10.R,p		20.R,p	22.E_1,p	10.[j]R,E_1,p	
$\alpha CC+\beta K = \gamma C$	17.R,n				21.R,n			15.R.n
$\beta K = \alpha CC+\gamma C$	18.R,n				22.R,n			16.R.n
$\alpha CC = \beta K+\gamma C$	19.R,n				23.R,n			17.R.n
$\alpha CC+\beta C = n$	20.R,n				24.R,n			18.R.n
$\alpha CC+n = \gamma C$						20.R[k],E_1,n		
$\alpha C = \sqrt{n}$						21.R,E_1,n		
$\alpha C = n+\sqrt{v}$						23.X,E_1,p		
$\alpha K+\beta C+\gamma t = n$						24.S,,E_1	A1.S,E_1,p	
$\alpha CC+\beta K+\gamma C+\delta t = n$						25.S,E_1,n	A2.E_1,p	
$\gamma t+\alpha CC = \beta K$?								19.X.?

a. With the difference that $\frac{1}{3}+\frac{1}{4}$ has been replaced by $\frac{7}{12}$.

b. In the end of the solution, the compiler of **C** tinkers with the double solution which was present in his original. In the short collection of further illustrative examples, **C** also has the problem E_1 of **V**.

c. Absent; but since the ensuing text refers to "6 reghole", this is clearly by involuntary omission.

d. E_4 in this line is closely related to E_3.

e. With a copying error in the statement which might look like being inspired by E_2.

f. The rule should read "Quando li chubi ⟨e li censi⟩ sono egualj alle cose [...]".

g. The rule should read "Quando li chubi sono egualj ⟨a' censi⟩ e alle chose [...]".

h. Formulated $\beta C+\gamma t = \alpha K$.

i. Correcting a lacuna in the statement, which should read "Trouami 2 numeri che tale parte sia l'uno dell'altro come 2 di 3 e, multiprichato il primo per se medesimo et poi ⟨per⟩ quello numero faccia tanto quanto e più 12".

j. Formulated $\beta K = \alpha CC$.

k. With a copying error, "traendone" instead of "più".

The scheme on p. 160 summarizes some important features of these presentations of algebra. If a work has a rule for a particular case, it is marked R when the rule is true; X when it is false and constructed merely as an illegitimate imitation of the solution to a similar-looking second-degree problem; and S when the rule is valid only in a special case modelled for instance after Jacopo's example (4a), from which the rule has been guessed (S$_n$ if stated for the normalized case). The presence of examples is indicated by E, marked by subscript digits (E$_{12}$ thus indicates that two examples are given; E$_1$ and E$_2$ in the same row but different columns indicate that the examples are different, E$_1$ and E$_{1*}$ that they are identical apart from the choice of numerical parameters). The letters "p" and "n" indicate whether the division by which the equation is normalized is expressed as "partire per" or "partire in"; we shall see that this "neutral mutation" is an interesting parameter.[312] As above, K stands for *cubo*, C for *censo*, CC for *censo di censo*, t for *cosa*, and n for *numero* (in whatever spellings the manuscripts may use); Greek letters for coefficients (implied by the plurals *cubi*, *censi*, and *cose*). We notice immediately that all works have the six fundamental cases in the same characteristic "non-Latin" (including non-FIbonacci) order as Jacopo.

Paolo Gherardi

Let us first concentrate on the column for **G**, Gherardi's algebra from 1328, composed in that very town where Jacopo had written 21 years before him. Gherardi, as we see, follows Jacopo fairly closely in the six fundamental cases. The differences are the following:

- Gherardi never gives more than one example;
- he replaces Jacopo's pure-number example for case (2) with a different pure-number example;
- in example (4), he divides the amount borrowed in Jacopo's (4a) by 5;
- in Jacopo's example (5b), he changes the given numbers in such a way that the result becomes irrational, and omits the second solution even though his rule mentions it;
- he replaces Jacopo's example (6) by a pure-number version of the problem of dividing a given quantity (here 100), first among x, then among x+p (here x+5) persons and adding the two results: $\frac{100}{t} + \frac{100}{t-5} = 20$. The description of the procedure refers to a number diagram[313]

---

[312] Etymologically, as mentioned in note 35, "partire *a* in *b*" refers to the division of the quantity *a* into *b* equal parts, and "partire *a* per *b*" to the numerical computation; but I have never remarked any reference to the "parts" in question in any Italian *abbaco* writing which divides "in" – the etymology must already have been forgotten. Any systematic choice of one or the other formulation (for instance, Jacopo dividing always the product of circular diameter and perimeter *in* 4 in order to find the area, and the perimeter invariably *per* $3\frac{1}{7}$ in order to find the diameter) therefore points to a source in time or space where the distinction was still semantically alive.

[313] The diagram is actually missing from the manuscript, but it can be reconstructed from the verbal

$$100 \diagdown 1 \text{ cosa}$$
$$100 \diagup \diagdown 1 \text{ cosa piu } 5$$

in a way (with "cross-multiplication" and all the other operations needed to add fractions) that implies underlying operations with the "formal" fractions $\frac{100}{1\ cosa}$ and $\frac{100}{1\ cosa\ piu\ 5}$.

Further on, major differences turn up:

– Gherardi leaves out all fourth-degree cases;
– he introduces $\alpha K = \sqrt{n}$ as a case on its own;
– he introduces three irreducible third-degree cases, giving false rules fashioned after those for the second degree – solving for instance the case $\alpha K = \beta t + n$ as if it had been $\alpha C = \beta t + n$;
– the higher-degree rules are illustrated by examples, all of which are pure-number problems of the kind that could easily be constructed *ad hoc* ("to find two or three numbers in given proportion so that ...").

The illustrations corresponding to the false rules all lead to solutions containing irrational roots. That allowed the fraud to go undetected, since no approximate value of these solutions was computed – approximation was not the custom in abbacus algebra, even Jacopo when finding correctly a monthly interest of $\sqrt{600}–20$ *denari* in his example (4a) leaves it there. In contrast, approximations to irrational results were habitual in abbacus geometry; geometrical computation was indeed supposed to lead to practically relevant results, whereas algebra (at least beyond the *regula recta*) had no such purpose. For most of those abbacus writers who took it up it served as a way to display mathematical dexterity and skill, a small minority seems also to have taken it up because of genuine theoretical interest.[314]

---

description and coincides with what is known from later manuscripts – see [Van Egmond 1978: 169 n. 11].

[314] The pertinence of this harsh judgement is illustrated by the protracted survival of the false rules (the number of which was to expand over the next 150 years). One may wonder that mathematicians would find pleasure in inventing and propagating such falsities – but such wonder would be misguided: in general, the abbacus masters (and even the abbacus writers) were not "mathematicians", they were teachers advertising and selling their abilities in a free market, where cheating the customers (parents of potential students or communal councils) successfully was just as efficient as convincing them honestly. Fraud was no obstacle to success, nor was true mathematical insight a necessity; what was all-important was the inability of competitors to unmask whatever deceit was made. Tartaglia's fortunes and misfortunes illustrate the point well.

It would be equally misguided to see the interest in the falsely solved higher-degree cases as an expression of laudable though for the time being unfulfillable mathematical theoretical ambition. Theoretical ambition only reveals itself with those abbacus algebraists who inquired into the validity of the solutions that were floating around and either discarded the invalid ones tacitly or discussed the conditions for solvability (as did Luca Pacioli [1494:150ʳ]). Conscious fraud has never had

The Lucca Manuscript

The two algebraic components of this conglomerate (**L** and **C**) are closer to **V**, and largely to be described as somewhat free abridgments of Jacopo's algebra.[315] The changes they introduce in the numerical parameters of certain examples do not change the character of these. Two of the examples where Gherardi differs from Jacopo are shared with **L**, but both are too simple to prove particular affinity.

Trattato dell'alcibra amuchabile

While sharing the title with **L**, this *Trattato* (**A**) is much closer to **V** in those cases and problems that have a counterpart in that treatise than are **L** and **C**; it has all of Jacopo's examples with identical parameters, deviating mainly at the level of orthography; however, where Jacopo left spaces open in example (4b) in order to insert later the result of $4 \cdot \sqrt{54}$ (cf. above, p. 8), **A** has the correct result "radicie di 864". As we see, it even agrees strictly with **V** in the decision whether to divide *in* or *per*; both must hence descend by careful copying from a common archetype (which can hardly be anything but Jacopo's original manuscript or an early copy).

With two exceptions, however – *viz* Gherardi's only four-term case[316] and Jacopo's case 10, $\alpha K + \beta C = \gamma t$ – **A** has all those examples for the higher-degree cases which we find with Gherardi, including his false rules for irreducible cases; and it has no other new examples. However, the agreement is not verbatim as with Jacopo, which speaks against use of **G** in the preparation of **A** – a compiler who copies one treatise verbatim is not likely to treat another one freely. **A** also contains a rule and an example for the reducible case $\alpha K + \gamma t = \beta C$, which **A** distinguishes from its mirror image $\beta C = \alpha K + \gamma t$; only the latter and not the former shape is present in **V**. Those higher-degree rules that are found in **V** but not in **G** (including the just-mentioned $\beta C = \alpha K + \gamma t$) follow **V** and are equally devoid of examples. The two biquadratic cases that are missing in **V** are also absent from **A**.[317]

---

anything to do with theoretical ambition (in any historical context where this concept itself makes sense), nor has the mindless copying of what one does not understand. But both have often been linked to social ambition.

Evidently, theoretical ambition does not exclude social ambition – even Fields medal winners may only take the trouble to publish because of that, however great their theoretical interest. But one should not be mistaken for the other.

[315] Evidently, it cannot be excluded that they descend from a model to which Jacopo was very close and not from Jacopo's own manuscript. In that case, however, Jacopo himself should have copied the statement in V.17.16 that "fifteen cases" precede without noticing that his original contained only fourteen.

[316] $\alpha K = \gamma t + \beta C + n$, solved as if it had been $\alpha K = (n+\gamma)t + \beta C$, $t = \sqrt{\frac{\gamma - n}{a} + \left(\frac{b}{2a}\right)^2} + \frac{b}{2a}$.

[317] Since **A** has no reference to "15 cases which [...] lead back" to the basic rules, this observation

So far, only the middle part of the tripartite *Trattato dell'alcibra amuchabile* was spoken of. The first part starts by presenting the sign rules ("più via più fa più e meno via meno fa più ...", "plus times plus makes plus, and less times less makes plus ...") and then goes on to teach operations with roots – number times root, root times root, products of binomials containing roots and the division of a number or one such binomial by another binomial. For the product of binomial by binomial, a diagram is introduced to illustrate the procedure – for instance, for $(5+\sqrt{20})\cdot(5-\sqrt{20})$:

$$
\begin{array}{llll}
5 & e & piu & \text{R} \; di \; 20 \\
\text{via} & & & \\
5 & e & meno & \text{R} \; di \; 20
\end{array} \; .
$$

As was usual in algebraic manuscripts from the Maghreb [Abdeljaouad 2002], the diagram stands outside the running text and recapitulates what is done by rhetorical means in the text. For the division of a number by a binomial, for instance 100 by $10+\sqrt{20}$, we find the similar diagram

$$
\frac{10 \quad piu \quad \text{R} \; di \; 20}{10 \quad meno \quad \text{R} \; di \; 20} \; ,
$$

which serves to illustrate that both dividend and divisor are to be multiplied by $1-\sqrt{20}$. Whether the writer thinks in terms of formal fractions is not clear at this point.

However, in the third part [ed. Simi 1994: 41*f*] we find Gherardi's example for the sixth case; the addition $\frac{100}{t} + \frac{100}{t-5}$ – thus the words here – is to be performed "in the mode of a fraction", explained with the parallel $\frac{24}{4} + \frac{24}{6}$.

The Parma Manuscript

The algebra section of the Parma *Libro di conti e mercatanzie* (**P**) is closer to **G** than is **A**, also in its treatment of such cases as had been dealt with by Jacopo. But in the illustration of the case $\alpha C = \beta t + n$ (still the problem $\frac{100}{t} + \frac{100}{t-5} = 20$) it has the explicit formal fractions of **A** (distorted in the beginning in a way that suggests that the writer did not understand) and not Gherardi's diagram. It also has the case $\alpha K + \gamma t = \beta C$ that was absent from **G** but present in **A**, with the same example as **A** – but the mirror case $\beta C = \alpha K + \gamma t$ is absent from **P** though present in **A**. Gherardi's only four-term problem $(\alpha K = \gamma t + \beta C + n)$, absent from *A*, is present in **P**.

P also provides examples to four of those fourth-degree rules which had none in **A**; three of these are of the usual facile pure-number type, but one $(\alpha CC = n)$ is illustrated

---

excludes that **V** could descend from a model also inspiring **A**: the error committed by Jacopo or an early copyist of his is repeated in **A**. Given **G**'s dependence on an intermediary between Jacopo's original and **A**, Gherardi must therefore also depend on Jacopo and not (or not exclusively) on a common archetype. Cf. also the below section on Giovanni di Davizzo, p. 176, according to which the distribution *in*/*per* found in **V** errs in two cases from the canon prevailing in the environment where Jacopo found his inspiration; if **A** was inspired directly from here, it is very unlikely that exactly the same errors should be committed.

by a (still facile) geometric question – to find the side of an equilateral triangle with given area. Further we find a biquadratic that was omitted in **V** (and **A**), and more examples involving roots of numbers ($\alpha C = n+\sqrt{}v$ being solved by taking the root of the right-hand terms separately!). The four-term problem and the three problems involving roots of numbers are all normalized by division *in*, where all other normalizations are *per*.

The two cases $\alpha K+\beta C+\gamma t = n$ and $\alpha CC+\beta K+\gamma C+\delta t = n$ are of a new kind. The rules are still false, but they are not copied from rules for second-degree cases – and they work for the examples that are given. The former example coincides with Jacopo's example (4a), with the difference that the 100 *libre* are lent for three, not two years – but the capital still grows to 150 *libre*, which speaks in favour of inspiration from Jacopo's or some related text (starting with 100 *libre*, on the other hand, seems to have been the standard, and thus does not tell much). In the latter example, 100 *libre* are lent for four years and grow to 160 *libre*. The rules, (complicated as they look because the *thing* is put equal to the interest in *denari* per month of one *libra*) appear to be constructed from the solutions that may be found from $\sqrt[3]{150/100}$ and $\sqrt[4]{160/100}$.[318] The fraud is certainly more

---

[318] We may look at the first example – a capital grows in three years with composite interest from 100 £ to 150 £. If the value of the capital after 1 year – or, even simpler, the value of 1 £ after one year – had been taken as the *thing*, we would have been led to a homogeneous equation,
$$t^3 = 1500000 \quad \text{respectively} \quad t^3 = 1\tfrac{1}{2} \;.$$
Instead, **P** takes the monthly interest of 1 ß expressed in δ as the *thing* (as Jacopo had also done). The yearly interest of 1 £ is therefore $\tfrac{1}{20}$ £ *thing*. This leads to the equation
$$100+15t+\tfrac{3}{4}C+\tfrac{1}{80}K = 150 \;.$$
The rule used to solve it is
$$t = \sqrt[3]{\left(\tfrac{\gamma/\alpha}{\beta/\alpha}\right)^3 + \tfrac{n}{\alpha}} - \tfrac{\gamma/\alpha}{\beta/\alpha} \;,$$
– or rather, since the rule first tells to divide by [the coefficient of] the cubes and afterwards speaks only of the resulting coefficients,
$$t = \sqrt[3]{\left(\tfrac{\gamma'}{\beta'}\right)^3 + n'} - \tfrac{\gamma'}{\beta'} \;,$$
where $\beta' = \beta/\alpha$, etc.

In order to see that this guess requires nothing beyond some training in the arithmetic of polynomials and awareness that a different position for the *thing* leads to a homogeneous equation, we may for simplicity consider the homogeneous equation
$$(t+\phi)^3 = \mu$$
(in the actual problem, $\phi = 20$, $\mu = 12000$). Performing the multiplication we get
$$\phi^3+3\phi^2 t+3\phi C+K = \mu \quad \text{or} \quad 3\phi^2 t+3\phi C+K = \mu-\phi^3 \;,$$
which should correspond to
$$\gamma' t+\beta' C+K = n' \;.$$
Therefore, $\phi = \gamma'/\beta'$, $n' = \mu-\phi^3$, $\mu = \phi^3+n' = (\gamma'/\beta')^3+n'$. Now, the solution obtained from the homogeneous equation is
$$t = \sqrt[3]{\mu} - \phi \;,$$
that is,

intelligent than that behind Gherardi's formulae – but it remains a fraud, and was almost certainly recognized as such by its inventor (who was certainly not the compiler of **P**).[319]

Lines of Ancestry and Descent

We have now come to the point where it is possible to construct a first approximate stemma showing the connections between the various Italian treatises discussed so far (the vertical axis corresponds to time, Jacopo writing in 1307, **G** being from 1328 and **V** from c. 1450). On top, we have Jacopo's original writing. **V′** is the shared archetype for **V**, **M**, **F** and **A** – possibly identical with Jacopo's original work.[320] **V″** is the faithful copy from which **V** is made (cf. p. 8). **A″** is the shared archetype for **A**, **L** and **C**, which must still have been very faithful to **V′** and can have contained none of the false rules, nor examples for the higher-degree rules. **C′** is the common ancestor of **L** and **C** (since everything that is in **C** is also in **L** they are likely to have a common ancestor not very different from **C** but already free with respect to **A″**). **A′** is a common ancestor of **A** and **G**, faithful to **V′** in the parts coming from Jacopo but already provided with examples for some of the higher-degree cases and false rules for some irreducible cases.

$$t = \sqrt[3]{(\tfrac{\gamma'}{\beta'})^3 + n'} - \tfrac{\gamma'}{\beta'} \,,$$

exactly the requested rule. Whoever invented the rule must have done so from a numerical example, but following the numerical steps precisely and seeing from which operations the coefficients arise it would not be too difficult to see that the 20 of our example results, in the words of the rule, "when the *things* are divided by the *censi*"; similarly for the rest of the rule.

This reconstruction has good chances to correspond to historical events. In Biagio's collection, we find both the three- and the four-year problem, though with numbers that lead to a rational solution. In both cases, the value of the capital after one year is posited as the *thing* [ed. Pieraccini 1983: 69–72, 84*f*]. Elsewhere in the collection he shows his actual interest in new rules which can be shown to hold on the example produced, by producing one for the case $\alpha CC + \beta t = n$ which actually holds for the example he gives ($CC + t = 18$) but not in general – as Benedetto comments, "in questa si perde il nostro maestro", "here our master gets lost" [ed. Pieraccini 1983: 25*f*]. According to Benedetto [ed. Pieraccini 1983: 1], Biagio died before c. 1340, which would fit chronologically (irrespective of whether Biagio himself or somebody else following his lead produced the rules we know **P** and also, as we shall see imminently, from a treatise from 1344).

[319] Compilers of texts like **P** were probably quite unaware of the fraud; they merely repeated what they believed to be good algebra.

[320] There is no reason to assume that the line leading toward **A** and the one leading toward **M′** and further toward **M+F** split off at the same point, whence the black ellipse after **V′** – since that part of **V** which is in **A** is absent from **M+F** we have no possibility to track such details. What we *can* know with fair certainty (cf. note 230 and surrounding text) is that the line leading toward **A** only split off when the beginning of the algebra chapter had already been lost in copying: **A** indeed starts the chapter in question exactly as does **V**. Possibly, Jacopo's autograph could have become **V′** by losing a sheet.

G′ is an ancestor to **G** from which **P** descends (the agreement of **P** and **A** in the case $\alpha K + \gamma t = \beta C$ appears to exclude direct descent of **P** from **G**). The extra cases in **P**, for instance involving square roots of numbers (and the prevalence of division *in* in the cases not shared with **A**, whereas division *per* is its standard choice in the shared cases) suggests that these have been borrowed from an unidentified source or area (labelled "**?**") and not created between **G′** and **P** as generalizations of the case $\alpha K = \sqrt{n}$.[321] It is likely that the latter problem (shared by **G** and **A**) has been adopted into **A′** from the same area.

Crosswise contamination is not to be totally excluded, but the distribution of shared versus particular features in the various treatises makes substantial importance of such influences unlikely. The stemma suggested here should be fairly close to the actual process.

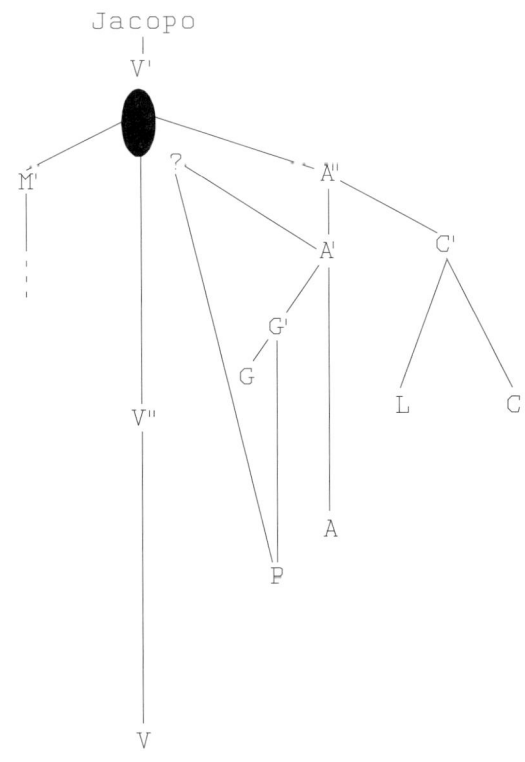

This implies, firstly, that everything written on algebra in Italian vernacular in the first generation after Jacopo depended on his work (or at least on **V′**), with further but limited influence from the "area **?**". This excludes the existence of an Italian environment practising algebra before Jacopo's times. As also argued above for other reasons, Jacopo must have gone abroad *in order to learn* – and his whole treatise indeed suggests that he was very conscious of presenting knowledge that was *new* to his public. Secondly, since **A**, **L** and **C** are all written in Tuscan with no traces of non-Tuscan orthography, even **A″** and **A′** are likely to have been written in Tuscan area; if this is so, then Gherardi must have sought his inspiration in Italian writings[322] and found little of algebraic

---

[321] Perhaps with the exception of #23, the one which finds the root of $n + \sqrt{v}$ as $\sqrt{n} + \sqrt{\sqrt{v}}$, and which divides *per*. This could be an independent misshaped addition.

[322] In the introductory passage [ed. Arrighi 1987: 15] he also presents himself as being from Florence.

interest in Montpellier.[323] But if there was no strong environment practising algebra in Montpellier in 1328, there can hardly have been any in 1307. If what he encountered was similar to what we can find in the *Trattato di tutta l'arte dell'abacho*, then he may have become aware of the importance of the subject in Montpellier, but he may have needed to borrow from somewhere else in order to make a coherent exposition.

This gives us no direct answer to the question concerning the localization of that *somewhere else*, that Romance-speaking area from which Jacopo drew his detailed knowledge of algebra. Indirectly, however, things begin to narrow down: if Italy and Provence are excluded, little beyond Catalonia remains – easily reached from Montpellier, and at the time involved in intense trading relations with the Arabic world as far as Egypt, and also an obvious channel for Ibero-Islamic influences.[324] Alternatively, the Iberian peninsula at large may be thought of[325] – the Castilian *Libro de arismética que es dicho alguarismo* from 1393 [ed. Caunedo del Potro & Córdoba de la Llave 2000],

---

[323] Pure veneration for Jacopo can be excluded, since his name does not appear in Gherardi's treatise. Since Jacopo knows none of the false rules (according to the style of his work he would have mentioned it if he knew about them and understood them to be false), even they are not likely to have been known in Montpellier in 1307.

Though Gherardi seems to have found little in Montpellier, he is still likely to have found *something*. In the draft manuscript of the *Trattato di tutta l'arte dell'abacho*, a sheet written in a different hand ($\mathbf{T}_F$ fol. 171^v) contains the beginning of a (not very) systematic presentation of *le regole della cosa* – with the rules for Gherardi's (and Jacopo's) cases no. 1, 2, 7 and 4 (in this order), with examples for the first three. Those for no. 1 and 7 are unlike those of both Gherardi and Jacopo and even simpler ($(x·3)÷4 = 20$ and $(x·x)·x = 12$, respectively), but that for no. 2 coincides precisely with that of Gherardi.

The main text of the *Trattato* contains a number of problems whose solutions refer to *cosa* or to *censo* as well as *cosa*; a convenient survey can be found in [Cassinet 2001: 124–128], to which should be added a problem with the structure $x+y = 16$, $(\frac{3}{4}x)^2 = (\frac{3}{4}y)^2–20$ on fol. 158^v. One of these problems ($\mathbf{T}_F$ fol. 162^v) has the structure

$$(x+\tfrac{1}{3}x+\tfrac{1}{4}x)^2 = \sqrt{x}$$

and another one ($\mathbf{T}_F$ fol. 157^r) involves $(x–\frac{1}{3}x–\frac{1}{4}x)$. Even though both $(x+\frac{1}{3}x+\frac{1}{4}x)$ and $(x–\frac{1}{3}x–\frac{1}{4}x)$ occur often in the record (see. pp. 131*f*) their particularly strong presence in this Provençal treatise bolsters the assumption that the foreign hand that added this rudimentary presentation worked in the same area, and that Gherardi also borrowed his correction to Jacopo from there.

[324] It is worth observing in this connection that the semantic distinction between "partire in" and "partire per" (see note 312) is still fairly present in Francesc Santcliment's Catalan *Summa de l'art d'aritmètica* from 1482 [ed. Malet 1998]. Thus, fol. 27^v, "digues: que partisses 589 en 6 parts", "say, you divide 589 in 6 parts", versus fol. 32^r, "no es nenguna altra cosa partir per 25, ho per 35 ho 57 ho 77 [...] sino partir per 12 ho per 19", "it is no different to divide by 25, or by 35 or by 57 or by 77 [...] than to divide 12 by 19".

[325] Sicily seems less likely but is perhaps not to be totally excluded – Fibonacci [ed. Boncompagni 1857a: 1] lists it along with Egypt, Syria, Greece (i.e., Byzantium) and Provence as one of the places where he had pursued the study of the "nine Indian figures" and what belonged together with them after having been introduced to the topic in Bejaïa.

astonishingly close to Jacopo in many formulations, is even closer to the various extant fifteenth-century Catalan-Provençal algorisms (and closer to these than to the Italian counterparts); it contains no algebra, which prevents us from drawing too definite conclusions.

Maestro Dardi da Pisa

Dardi's *Aliabraa argibra*, apparently from 1344, is the first full-scale vernacular algebra that does *not* depend on Jacopo (as will be argued below); it thus represents a different strand in the beginning of Italian vernacular algebra. It is also the earliest extant vernacular work devoted solely to algebra – and it is more than four times as long as the *Trattato dell'alcibra amuchabile* from c. 1365, also solely algebraic.[326] Like Jacopo's treatise, it contains no single Arabism (unless we count the two versions of the word "algebra" in the title as such). As it turns out, its independence of Jacopo does not preclude its being informative about some of the questions left open in the preceding section. The following concentrates on the aspects of the work that are relevant in this respect, but can only do so on the condition of presenting the treatise in more general terms.[327]

Its basic structure is fairly similar to the first two parts of **A**. However, at first comes an introduction and an index listing all 194+4 cases to be dealt with.[328] The sign rules of **A** are missing – but Dardi proves[329] when arriving to the point where it is first needed that "meno via meno fa più" (using the example $(10–2)·(10–2)$). Instead, the index is followed directly by a "Treatise on the rules which belong to the multiplications, the divisions, the summations and the subtractions of roots".[330] Then comes a

---

[326] I used the Vatican manuscript Chigi M.VIII.170 from c. 1395 (**D₁**); Raffaella Franci's edition [2001] of the Siena manuscript I.VII.17 from c. 1470 (**D₂**); and Warren Van Egmond's personal transcription of the Arizona manuscript, written in Mantova in 1429 (**D₃**); the datings of **D₁** and **D₂** are based on watermarks and according to [Van Egmond 1980]; that of **D₃** is stated in the manuscript. **D₁** is in Venetian, **D₂** in Tuscan, **D₃** as far as I can judge in a northern dialect not too different from Venetian. For further information, see [Van Egmond 1983] and [Hughes 1987]. I thank Raffaella Franci for supplementary information on **D₂** and Van Egmond for giving me access to his transcription of **D₃**.

Of a fourth manuscript from c. 1495 (Florence, Biblioteca Mediceo-Laurenziana, Ash. 1199) I have only seen an extract containing the irregular cases in [Libri 1838: III, 349–356]; to judge from this it appears to be very close to the Siena manuscript.

[327] For other aspects of the treatise, see [Van Egmond 1983], [Hughes 1987] and [Franci 2001: 1–33].

[328] The index is absent from **D₂**, but the introduction promises to bring it and leaves three empty pages – the obvious intention being to insert it once the equally promised corresponding folio numbers were known. In **D₁**, the introduction and the first page of the index is missing, and the first folio number is 2.

[329] **D₂** p. 44; **D₁** fol. 5ᵛ; **D₃** fol. 11ᵛ.

[330] **D₁** fol. 3ᵛ; **D2** p. 38. **D₃** does not have this general caption but separate captions for the single

presentation of the six fundamental cases, with geometric demonstrations (**A** has nothing similar), and finally a presentation of 194 "regular" and 4 "irregular" cases, all with rules and one or more examples. The distinction regular/irregular is made in the introduction; a note to the index distinguishes in different terms, namely between cases governed by general respectively non-general rules.

In **D₁**, the following abbreviations are made use of consistently: ç for *censo*, c for *cosa*, *nũo* for *numero*, ℞ for *radice*, m̃ for *meno*; the notation for multiples of ç and c emulates that for fractions, writing the "denominator" below the "numerator" with a stroke in between – for instance, $\frac{10}{c}$ for "10 things". ç, c and ℞ are also used in the later manuscripts **D₂** and **D₃**; the fraction-like notation does not occur in **D₂** but often (not always) in **D₃**; it therefore seems plausible that it was used in Dardi's original.[331] It is indeed also used in the slightly earlier *Trattato di tutta l'arte dell'abacho* (**T_R**, fol. 159ʳ), though as $\frac{18}{cose}$, without abbreviation.

Chapter 1: Calculating with Roots

In the chapter on roots, we find diagrams illustrating the multiplication of binomials similar to those appearing in **A** – for instance, for $(3–\sqrt5)\cdot(3–\sqrt5)$,[332]

We notice that Dardi's diagram is fuller than that of **A**, which makes it implausible that **A** should have simply borrowed from him.

When looking at the explanation of how to divide a number by a binomial we find greater differences. In order to divide 8 by $3+\sqrt4$, Dardi first makes the calculation $(3+\sqrt4)\cdot(3–\sqrt4) = 5$ and infers from this that *5 divided by* $3+\sqrt4$ *gives* $3–\sqrt4$. *What*, he

---

sections.

[331] In general, **D₁** is not only earlier in time than **D₂** and **D₃** but also indubitably closer to the text of their common archetype in various respects. One example is the reference to the rule of three in the passage of **D₁** quoted below (note 333) and the absence of the reference in **D₂**; since **D₂** cites it when referring backwards to the passage (p. 62, corresponding to **D₁** fol. 14ʳ), it must have been present in the common archetype (it is indeed also found in **D₃**). Another example is the use of the term *adequation* in **D₁**, corresponding to *dequazione* in **D₂**; they are indistinguishable in the definite form *ladequation/ladequazione*, which explains that one of the manuscripts has misunderstood the intended term of the original; but in one place (p. 77) **D₂** has an unexpected and indubitable *adequazione*, which can therefore be assumed to be the original form (indeed, **D₃** also uses *adequation*).

In single readings, **D₃** often seems better than **D₁**, but at the level of overall structure (captions etc.) **D₁** is apparently to be preferred. Since Dardi *could* be identical with one Ziio Dardi present in Venice in 1346 [Hughes 1987: 170], even the language of **D₁** might be closest to the original.

Globally, the differences between the three manuscripts are fairly modest.

[332] **D₂** p. 45; **D₁** fol. 6ʳ; **D₃** fol. 12ʳ.

next asks, will result if *8* is divided similarly, finding the answer by means of the rule of three (5, 3–√4 and 8 being the three numbers involved).[333]

Chapter 2: the Six Fundamental Cases

The chapter proving the correctness of the second-degree rules has no counterpart in **A**, nor in any of the other Italian treatises discussed so far; they are also very rare in later abbacus writings. The demonstrations descend from those found in al-Khwārizmī's algebra, but their style is as different as it would be if somebody not familiar with the received conventions governing the use of letters in geometric diagrams were to relate al-Khwāriz-mī's proofs from memory to somebody not too well versed in geometry. As an illustration (which should speak for itself as soon as it is confronted

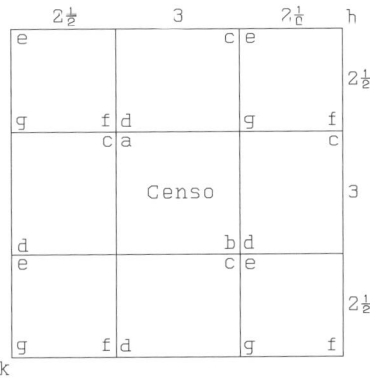

with any version of al-Khwārizmī's text – cf. p. 102) I translate the beginning of the first proof verbatim (repeating the grammatical inconsistencies of the text)[334] and reproducing also the first diagram :

> How 1 *ç* and 10 *c* are proved to be equal to 39. Since the *c*, which is said to be ℞ of the *ç*, the *ç* now comes to be a quadrangular and equilateral surface, that is, with 4 corners and four equal and straight sides. Now we shall make a square with equal sides and right corners, and we shall say that the *ç* is its surface, which is *ab*, and since the *c* is the ℞ of the *ç*, it comes to be the sides of the said square, and since to the *ç* $\frac{10}{c}$ are added, we divide this $\frac{10}{c}$ in 4

---

[333] I render the text of **D₁** (fol. 12ᵛ; similarly **D₃** fol. 19ᵛ); punctuation and diacritics have been adjusted/added:

> Se tu volessi partir nũo in ℞ e nũo, serave a partir 8 in 3 e ℞ de 4, tu die moltiplicar 3 e ℞ de 4 per 3 m̃ ℞ de 4, che monterà 5. Adonqua a partir 5 in 3 e ℞ de 4 te ne vien 3 m̃ ℞ de 4 perché ogne nũo moltiplicado per un'altro nũo, la moltiplication che ne vien partida per quel nũo sì ne vien l'altro nũo moltiplicado per quello. Adunqua partando 5 in 3 e ℞ de 4 sì ne vien 3 m̃ ℞ de 4, e partando 5 in 3 m̃ ℞ de 4 sì ne vien l'altra parte, zoè 3 e ℞ de 4, e inperzò diremo che questo 5 sia partidor, e metteremo questo partimento alla regla del 3, e diremo, se 5, a partir in 3 e ℞ de 4, ne ven 3 m̃ ℞ de 4, che ne vegnirà de 8, e moltiplica 3 m̃ ℞ de 4 via 8, che monta 24 m̃ ℞ de 256, la qual moltiplication parti in 5, che ne vien 4⅘ per lo nũo. Ora resta a partir ℞ de 256 ^meno^ in 5, che ne vien ℞ de 10%₂₅, che a partir ℞ in nũo el se die redur lo nũo a ℞ , zoè lo 5 redutto in ℞ monta ℞ de 25. E così avemo che a partir 8 in 3 e ℞ de 4 sì ne vien 4⅘ men ℞ de 10%₂₅.

D₂ omits the explicit reference to the rule of three, but as observed in note 331 it must have been present in the common archetype.

All three manuscripts also make use of the rule of three at a later point, see note 341.

[334] **D₂**, pp. 68*f*; **D₁** fols 16ᵛ–17ʳ; **D₃** fols 24ᵛ–25ᵣ.

parts, which comes to be $\frac{2\ \ 1}{c\ \ 2}$ each,[335] and since the c comes to be the sides of the ς, we shall place each of these four parts along ς, each along its own side of ς, the surface of each being cd, and outside each of the corners of ς falls an equilateral quadrangle with right corners, which as side will have the breadth of the c, that is, 2½, which breadth, or length, multiplied by itself amounts to 6¼, that is, ef, [...].

A closer look at some textual details reveals that the chapter has been adopted from the same environment as Jacopo's algebra (which was not a priori to be expected, given that Jacopo brings no geometric proofs). Dardi's rule for the fifth case runs as follows in $\mathbf{D_1}$:[336]

> Quando li ς e'l numero è equali ale c, el se die partir tutta l'adequation per la quantità dei ς, e po partir le c in 2, e una de queste mità, zoè la quantità de una de queste parte, moltiplica in si medesima, e de quella moltiplication trazi lo numero e la $\mathbf{R}$ de quello che roman zonzi all'altra mità dela quantità dele c, e tanto vegnirà a valer la c, e *sappi che in algune raxon te convegnirà responder esser la c per lo primo modo, zoè la mità dela quantità dele c più* $\mathbf{R}$ *de quello che roman, e algun fiade per lo secondo modo, zoè la mità dela quantità dele c m̃ la* $\mathbf{R}$ *de quello che roman, e algune se pò responder per tutte e 2 li modi, com'io te mostrerò.*

Jacopo's corresponding rule (V.16.11) is not very similar (except, by necessity, in mathematical substance):

> Quando le cose sonno oguali ali censi et al numero, se vole partire nelli censi, et poi dimezzare le cose et multiprichare per se medesimo et cavare el numero, et la radice de quello che romane, et poi el dimezzamento dele cose vale la cosa. Overo el dimezzamento dele chose meno la radice de quello che remane.

However, when Jacopo comes to present the double solution of example (5b), we find the following passage (V.16.14, emphasis added):

> Siché tu vedi che all'uno modo et all'altro sta bene. Et però quella così facta regola è molto da lodare, che ce dà doi responsioni et così sta bene all'una come all'altro. Ma *abbi a mente che tucte le ragioni che reduchono a questa regola non si possono respondere per doi responsioni se non ad certe. Et tali sonno che te conviene pigliare l'una responsione, et tale l'altra. Cioè a dire che a tali ragioni te converà rispondere che vaglia la cosa el dimezzamento dele cose meno la radice de rimanente. Et a tale te converrà dire la radice de remanente e più el dimezzamento dele cose.* Onde ogni volta che te venisse questo co'tale raoguaglamento, trova in prima l'una responsione. Et se non te venisse vera, de certo si piglia l'altra senza dubio. Et averai la vera responsione.

The similarities between the two italicized passages are too particular to allow explanation merely from shared general vocabulary and style. However, several reasons speak against Dardi making direct use of Jacopo's text: not least the total absence of shared examples

---

[335] We notice that Dardi extends his fraction-like notation into an "ascending continued fraction"; indeed, $\frac{2\ \ 1}{c\ \ 2}$ means $\frac{2}{c}$ plus $\frac{1}{2}$ of $\frac{1}{c}$.

[336] Fol. 16^r, emphasis added; similarly $\mathbf{D_2}$ p. 66 and $\mathbf{D_3}$ fol. 24^r.

and of anything similar to Jacopo's *fondaco* problems from the *Aliabraa argibra*. Moreover, *if* Dardi had found the italicized passage in Jacopo interesting and moved it to the rule (because the examples he promises only come in the following chapter), he would not have changed its finer texture as seen in the excerpt;[337] nor would he have had any reason to invent the term *adequation* in replacement of *raoguaglamento* if using Jacopo's treatise. In consequence, Dardi must have drawn his inspiration for this chapter from the very environment which Jacopo had once drawn on. And he must have kept fairly close to his direct source: only thoughtlessly faithful copying explains the sudden appearance of "78 dramme, zoè numeri" in the example illustrating the fourth case (**D₁** fol. 16ʳ, similarly **D₂** p. 65) – up to this point, all numbers have been nothing but *numeri*.

Chapter 3: 194+4 Regular and Irregular Cases

As mentioned, the final chapter presents 194 "regular" cases with rules, only a small selection of which are listed in the scheme on p. 160. A very large part of them involve radicals, not only roots of numbers but also of *things, censi, cubi* and *censi di censo* – thus, for instance, no. 59, $\alpha t = \sqrt[3]{\beta C}$, and no. 123, $\sqrt{\beta t} + \sqrt[3]{n} = \alpha t$ (notation as in the scheme). Apart from two slips, convincingly explained in [Van Egmond 1983: 417], all are solved correctly, and all are provided with an illustrative example (at times two or, with rules allowing a double solution, three examples[338]). All are pure-number problems, almost half of them of the fraudulently complicated type asking for two or three numbers in given proportion; a good fourth asks for a single number fulfilling conditions fashioned in agreement with the equation type, some 15 percent deal with a divided ten. The order of the six fundamental cases is the same as in the other treatises we have looked at, which corroborates the conclusion that Jacopo and Dardi were inspired from the same area. Even the order of the next three cases coincides with that of Jacopo – but since these are just the simplest higher-degree cases (cubes equal to number/*things*/*censo*), this agreement is hardly significant. After that, Dardi's order is wholly his own, as are most of his problem types.

In **D₁** and **D₂**, the four "irregular" cases are inserted between the regular cases no. 182 and 183, after the observation that all equations up to this point contain no more than three terms.[339] In contrast, the regular cases from no. 183 onwards all correspond to

---

[337] When we are able to compare Dardi's text with another one deriving from the same source, as Dardi's first irregular case with the corresponding case in **P** (see presently), Dardi can be seen to change at most the wording of the single phrases while conserving their order and mutual relation (but since **P** is later and hence more likely than **D** to have changed with regard to the original source, Dardi may well be even more faithful).

[338] Even this, we notice, corresponds to Jacopo's treatment of the six fundamental cases, three examples showing that case (5) is sometimes solved by one solution, sometimes by the other, sometimes by both.

[339] In **D₃**, the irregular cases come after the last regular case, but the observation that all preceding

four-term equations. The rules for the irregular cases are presented at this point as "adapted solely to their problems, and with the properties these possess";[340] they are included all the same because they may turn up in certain problems. This, together with the separate numbering, suggests that Dardi has adopted the group wholesale and inserted it into the main body of his treatise. The character of the examples supports this inference. Two of them (no. 1 and no. 2) are identical with examples (24) and (25) from **P**, which means that they are the only problems in Dardi's treatise that do not treat of pure numbers (but of lending with interest, as we remember), and that they are directly inspired by Jacopo's example (4a).[341] The other two, $\alpha t+\beta C+\gamma CC = n+\delta K$ and $\alpha t+\gamma CC = n+\beta C+\delta K$, are based on the divided ten; had it not been for their constituting a closed group together with the former two, they could have been Dardi's invention; as things actually stand, this is very unlikely.[342]

Dependency or Independence

Dardi's fraction-like notation for monomials was known in Avignon in 1334. His many rules involving radicals and roots of numbers show that he shares in the inspiration coming from "area ?". They do not tell whether he only received general inspiration and used that as a starting point for something going far beyond what his source tradition had done, or he borrowed in large scale. Some details in the chapter on roots suggest dependency on a model,[343] and the importance of a model for several features of the presentation

---

cases involve at most three terms is found in the appropriate place, namely on fol. 113^r – together with the observation that so far everything has been done in agreement with rules of general validity, which makes little sense if cases governed by rules of a different kind do not follow immediately. The order of **D₁** and **D2** must therefore be the original order.

[340] "[...] reghulati solamente alle loro ragione, e di quelle proprietà delle quale elle sono ordinate" (**D₂** p. 269; similarly **D₁** fol. 102^r). The wording in **D₃** (fol. 121^r) is slightly different but equivalent.

[341] On one point, Dardi's presentation of these cases differs from that of **P**. All three manuscripts of the *Aliabraa argibra* make use of the rule of three as an alternative way to find the value of the capital after two, three and four years from its value after one year (**D₁** fol. 100^{r,v}, **D₂** fol. 98^{r,v}, **D₃** fol. 121^v, 122^r). This alternative is not presented in **P** and could be Dardi's own contribution (but it might also have been lost in the transmission toward **P**).

[342] Raffaella Franci [2002b: 96–98] supposes that **P** and a very similar treatment of algebra in MS 2Qq E13 (1398, d), Biblioteca Comunale di Palermo, which I have not seen, represent a synthesis combining material borrowed from Jacopo, Gherardi and Dardi. If this is meant to imply that the four irregular cases were Dardi's own invention and the borrowing made from his treatise, it is implausible. Quite apart from the above considerations speaking against Dardi's authorship, it would be strange that only these four wrong rules were borrowed and nothing else.

[343] Thus, a number of procedures are illustrated by polynomials containing rational roots (e.g., $36/(\sqrt{4}+\sqrt{9}+\sqrt{16})$, treating them *as if* they were surds ("intendando de queste R discrete como s'elle fosse indiscrete" – **D₁** fol. 3^v, similarly **D₂** p. 62), the obvious point being that this allows control of the correctness of the result; however, such control is never made, nor is any other advantage

of the six fundamental cases was already discussed. But the main body of the last chapter, the regular cases 1–194, may still have been structured by Dardi. Of the single cases, quite a few had been dealt with before, as we have seen, and Dardi may plausibly have known about that, just as he knew about the way to construct pseudo-complex examples by asking for numbers in given proportion (while copying no examples directly from predecessors known to us, neither from Jacopo nor from Gherardi, nor from the *Trattato di tutta l'arte dell'abacho* including the inserted fragment mentioned in note 323); yet no evidence contradicts a conjecture that most were devised by Dardi.

As argued, the principle of creating new algebraic cases involving roots will have been inspired from the unidentified "area ?". For the use of diagrams for the multiplication of binomials, Dardi seems to have shared his inspiration with **A**; **A** and **G** (and hence their shared archetype **A′**) make use of the related calculations with formal fractions. Finally, the order of the fundamental cases, the discussion of the double solution to the fifth case and the use of the rule of three as an algebraic tool shows affinity with Jacopo, while, as we have seen, the details of Dardi's text speak against direct borrowing; even Jacopo and Dardi hence share a source of inspiration.

Occam's razor is a dangerous weapon – wielding it without caution was what led to the assumption that *abbaco* algebra had to come from Fibonacci. But ad hoc multiplication of explanatory entities beyond what is needed remains gratuitous, and a reasonable working hypothesis is that all these unidentifiable sources of shared inspiration belong to the same area – that is, our "area ?" (in which case this area can hardly be Montpellier itself). The only extra entity we may be forced to accept could be the one which, in the wake of the success of Jacopo's higher-degree cases, invented **P**'s and

---

taken of the choice of rational roots, except an unproven statement that the result coming from a particular calculation ($\sqrt{40\frac{24}{25}}+\sqrt{92\frac{4}{25}}+\sqrt{5\frac{19}{25}}-\sqrt{163\frac{21}{25}}-\sqrt{10\frac{9}{25}}$) *can* be reduced.

Omitting a proof when copying or missing the opportunity to make it when borrowing a style that prepares for it (or when using a model where such a thing has already happened at an earlier stage of transmission) may easily happen; but that the author prepares it repeatedly on his own initiative and then himself omits it each time is not very likely. Cf. below, note 353 and surrounding text, on the occurrence of the same phenomenon with Giovanni di Davizzo.

The trick is not only used but also explained in the anonymous *Trattato d'algibra* [ed. Franci & Pancanti 1988: 6] from the 1390s, whose aberrant use of the term *censo* was mentioned above (note 252) as evidence for its non-Italian affinities. The same treatise makes use of diagrams and schemes for the multiplication of binomials and of formal fractions containing algebraic binomials in the numerator as well as the denominator. It also explains the reduction of higher cases. It treats all 22 basic and reducible cases (all provided with examples; the first 20 are in the same order as Jacopo), and expresses all normalizations by way of division *in*. Its does not bring the false rules, instead it introduces [ed. Franci & Pancanti 1988: 98] a concept "cube root with addition *a*", the "cube root of 44 with addition 5" being the solution to the equation $n^3 = 44+5n$ (i.e., 4). This allows the author to solve not only this equation type (which actually is not done) but also (by means of transformations that are taught) equations containing cubes, *censi* and number.

Dardi's irregular cases – which we may designate **I**. These various observations cause the
addition of new elements and links to our
stemma, without changing anything (except the
age ascribed to "area **?**"[344]) in what was
already drawn up (in the interest of simplification, the branch leading toward **M+F** is omitted
this time).

An Instructive Fragment: Giovanni di Davizzo

Alchune ragione, written in 1424,[345] contains six pages with the heading "Algebra"
(possibly more, see presently), said to be copied
from a book written by Giovanni di Davizzo
de l'abacho da Firenze on September 15th,
1339. Giovanni must have given this information in his incipit in the same way as Jacopo –
this is indeed the place where such information,
when at all given, appears in the *abbaco* treatises. We can therefore safely assume it to be
reliable.[346]

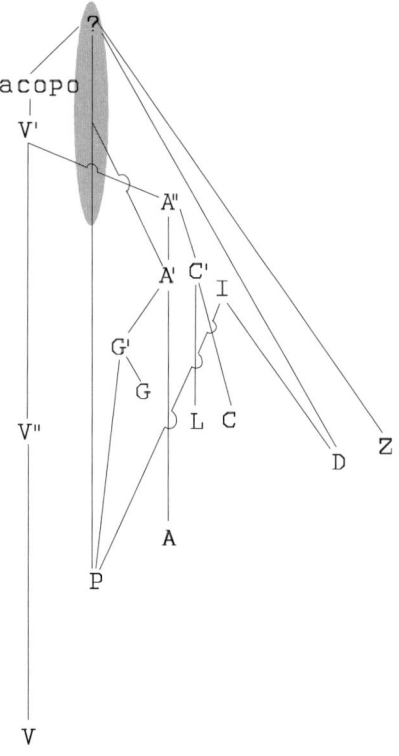

The first three pages (fols 28^v–29^v, original
foliation) contain sign rules and rules for operations with monomials and binomials. Next
follow rules for nineteen algebraic cases (fols 29^v–31^r). After that comes a sequence of
examples (fols 31^r–32^r) which are not likely to be from Giovanni's hand; it is implausible

---

[344] As suggested in the diagram, however, an "area" or environment from which inspiration is drawn
may well function for decades or even longer. It need not, as must a particular treatise, appear as
a single point in the stemma. There is no reason to believe that everybody who received inspiration
from what was done in "area **?**" was inspired at the same moment, nor to assume that the algebraic
practice *within* this area underwent no change. **I**, the place where the new false rules of **D** and **P**
originated, *could* thus be located within the "area **?**"; on p. 178, we shall encounter evidence
suggesting that it was.

[345] See above, note 250.

[346] Giovanni (fl. 1339–1344) belonged to a Florentine abbacist family, whose activity spanned almost
the whole fourteenth century – his father, his brother and two nephews were also abbacus masters,
see [Ulivi 2002b: 39, 197, 200].

but not impossible that the heading "Algebra" was intended to cover even these.[347]

We shall first concentrate on the rules for the algebraic cases. According to [Franci 2002b: 87], the list contains all of Jacopo's twenty rules plus the two that are missing. This is mistaken, as can be seen in the scheme on p. 160, column **Z**. What we find is Jacopo's list of twenty, with two omissions (no. 10, no. 16) and with no. 11 being replaced by the mirror case $\alpha K + \gamma t = \beta C$.[348] These eighteen cases are numbered. The last, nineteenth case is unnumbered, and only partly legible – it equates $\gamma t + \alpha CC$ with a right-hand side that contains at least βK but perhaps more terms, and is thus neither to be found in **V** nor in any of the other treatises we have examined. The reason it is in part illegible is that a piece of paper has been glued over this rule, probably by someone who discovered that it was wrong; the paper has been removed, but the humid glue has made the paper almost as dark as the ink.[349]

The wording of the rules is mostly identical with that of Jacopo, but there are a fair number of deviations; sometimes different expressions are used, sometimes, as mentioned, Jacopo's cases appear in mirror form. However, the decision whether to divide *per* or *in* is the same in all cases except three. If Giovanni had copied from Jacopo (whether directly or indirectly), there is no reason that agreement should be higher concerning this choice than in the rest of the wording. Instead, he must like Dardi be independent of Jacopo, but also (like Dardi, and with less independent initiative) draw on the same environment as Jacopo.

Confrontation of Jacopo's and Giovanni's lists of rules allows us to decipher the canon that governs the choice *in/per*. It is quite simple: two-term equations (those that can be reduced to homogeneous problems) divide *per*, while three-term equations (those which reduce to mixed second-degree equations) divide *in*. Jacopo, or a copyist between him and **V′**, errs twice (no. 1 and no. 13); Giovanni, or his fifteenth-century copyist, errs once

---

[347] The presentation of algebra is located within a long sequence of problems about finding numbers but just before the ones that make use of *cosa* and *censo*. It is therefore likely that the compiler discovered the need to present the tool for solving problems of this kind (and the conceptual framework within which they belong), and found an appropriate exposition in Giovanni's treatise.

[348] Even Jacopo's no. 18 is actually reversed, Giovanni's case no. 16 being $\alpha CC + \gamma C = \beta K$. Like no. 11 of both, this case reduces to case no. 5. Since the mirror image of Jacopo's no. 18 does not appear separately in other treatises, I have not given it a separate line in the scheme.

[349] The headline "Algebra" stands outside the normal text frame, and must hence be a later addition. It is written in the same bright red ink as the numbering of the cases and the indication of paragraphs in the introduction – an ink type which is found nowhere else in the manuscript, although paragraphs are also indicated in other places in what perhaps was once red). Even the numbering of cases must therefore be a secondary addition, almost certainly made after the discovery that rule no. 19 was wrong. It is thus by mere accident that **Z** has a distinction between numbered and unnumbered cases which looks like Dardi's certainly genuine distinction between numbered rules of general validity and unnumbered rules that only hold in special cases.

(no. 7). **G** and **A** both obey the canon in their new, falsely solved cases. Since they do not repair Jacopo's two errors, we may conclude that the canon was not known to the intermediate copyists (**A″**, **A′**, **G′**), which means that the false rules are drawn from some other source where it was known and respected – that is, probably, the very area or environment where Jacopo and Giovanni (or somebody from whom he borrows) found their inspiration. Since Jacopo appears not to have known about these false rules, they could represent an invention made by somebody moving within that area, perhaps after Jacopo's time, perhaps simply without influencing the material used by Jacopo.

The formulation of the sign rules coincides verbatim with that of **A**, which need not tell very much – the order "++, —, +–, –+" could be considered "natural", and the phrases themselves leave little room for variation.[350] The rules for multiplying monomials are no more informative, beginning with the products $n \cdot K$, $n \cdot C$ and $n \cdot t$, then (after the insertion of the sign rules) going on rather disorderly with $t \cdot t$, $C \cdot C$, $t \cdot C$, etc. Divisions are more interesting. Giovanni starts by stating that number divided by thing becomes number, number divided by *censo* becomes root, thing divided by *censo* becomes number, number divided by cube becomes cube root ... and ends, after another thirteen calculations of the kind,[351] by asserting that number divided by cube of cube of cube of cube becomes cube root of cube root of cube root of cube root. Close scrutiny reveals that the mathematical mistakes constitute a system – a rather ingenious but unfortunately incoherent experiment aiming, in modern terms, at extending the semi-group of non-negative powers of the algebraic *thing* into a complete group. Not possessing negative exponents, Giovanni expresses t^p as the pth "root", composing such "roots" additively in the way the positive powers are composed ("cube of cube" meaning $t^3 \cdot t^3$, not $(t^3)^3$); the "first root" is identified with number. The invention is likely to be Giovanni's own – it is difficult to see how it could be adopted for any algebraic purpose; but it may none the less reflect inspiration from an environment very interested in "roots" and experimenting with the power series of algebraic unknowns.[352]

---

[350] An edition and translation of the part of Giovanni's text which precedes the presentation of the cases will be found in [Høyrup 2006].

[351] Only *censo* of cube (meaning $t^2 \cdot t^3$, not $(t^3)^2$) divided by cube, which leads to no negative exponent, is given correctly as *censo*. However, when repeating the list in his *Trattato d'abaco*, Piero della Francesca [ed. Arrighi 1970a: 84*f*] has "partire censo de cubo per cubo di chubo ne v[e]ne numero". This appears to be the original version; somewhere in the copying process *cubo di cubo* has been reduced to *cubo* simply, giving "partire zenso di chubo per chubo viene numero". Noticing the error, somebody – presumably the writer of the fifteenth-century manuscript – discovered that this was wrong, and stated the correct result of the new division.

[352] Without being adopted for proper algebraic *use* (which would immediately have exposed the inconsistencies), the idea was borrowed by several other authors. Apart from being copied in 1424 by the source where Giovanni is directly quoted and by Piero della Francesca, it appears in almost the same words as Piero's in Mariotto di Giovanni Guiducci's *Libro d'arismetricha* from c. 1465 –

The rules for adding, subtracting, multiplying and dividing square roots and for multiplying or dividing binomials are correct, and in so far uninformative. It is noteworthy, however, that five out of nine examples[353] operate with the roots of square numbers, *without taking advantage of this particular choice*, exactly as Dardi. As argued in the case of the latter, this must mean that the idea is borrowed from elsewhere (and once again borrowed badly). The idea is sufficiently unexpected to allow the conclusion that the two must have borrowed the inspiration from the same source tradition – though certainly not precisely the same source, given how different they are on almost all other accounts.

Giovanni is definitelytmuch more similar to Jacopo than to Dardi, and the two appear to have used very similar sources though hardly precisely the same source. It is quite possible, indeed, that Giovanni's treatise contained examples which were omitted by the fifteenth-century compiler as not necessary for his purpose.[354] The part of the Giovanni-excerpt which precedes the rules may therefore be similar to the introduction that can be presumed to have been lost from Jacopo's algebra (cf. notes 230 and 317). Indeed, Giovanni's introduction contains exactly the 66 lines normally found on two pages of V. Irrespective of the conscientious copying process leading to V, Jacopo's original need evidently not have had exactly the same number of lines to a page, nor is his introduction

---

see [Giusti 1993: 205]; the Piero-Guiducci-version is the outcome of tinkering with Giovanni's system by somebody who understood the idea well enough for tinkering without destroying its consistency (but not for seeing that it could not work). The system also turns up in Bento Fernandes' *Tratado da arte de arismetica* from 1555 – see [Silva 2006: 14].

It may be noticed that both Piero and Bento Fernandes also bring the false solutions to higher-degree algebraic cases. Though seemingly unrelated in origin, the two phenomena served the same purposes and survived under the same conditions.

[353] Namely $\sqrt{9}\cdot\sqrt{9}$; $\sqrt{25}/\sqrt{9}$; $(5+\sqrt{4})\cdot(5-\sqrt{9})$; $(7+\sqrt{9})\cdot(7+\sqrt{9})$; $35/(\sqrt{4}+\sqrt{9})$.

[354] We may also remember that the *Alchune ragione* contains a problem belonging to the same family as Jacopo's *fondaco* problems but with a structure which is neither known from Jacopo nor from the other treatises I have discussed – see above, p. 120. We cannot know whether the compiler (or rather, the compilers, the manuscript is in several hands) borrowed the problem in question from Giovanni, but the reference to the date (likely to come from the colophon) makes it almost certain that he had access to Giovanni's treatise as a whole; it is a fair guess (and can be nothing more) that Giovanni is the source, and that Giovanni had taken the problem from that same environment where Jacopo found his *fondaco* problems.

It should be observed, however, that fols 92ʳ–93ᵛ contain a second introduction to algebra which almost certainly is *not* borrowed from Giovanni, and which contains rules for equations with radicals ($\alpha t = \sqrt{n}$, $n = \sqrt{(\alpha t)}$, $\alpha C = \sqrt{n}$, $n = \sqrt{(\alpha C)}$, $\gamma C+\beta K+\alpha C = \sqrt{n}$). Here we also find Dardi's second irregular case (with no explanation that it is not generally valid), and the same symbolic notation as known from the Maghreb: *co* (meaning *cosa*) or □ (standing for *censo*) written *above* the coefficient – thus the inverse of the notation known from *Trattato di tutta l'arte* and from Dardi, and without their fraction line, and similar to Jacopo's multiplicative writing of Roman numerals – see pp. 51 and 153.

likely to have coincided precisely with Giovanni's. Even with this proviso, however, the size of Giovanni's introduction is compatible with the hypothesis that **V′** is either a copy of Jacopo's original having lost a sheet or identical with this mutilated original.

Summing Up

It should be firmly established by now that the algebra section of **V** belongs to the early fourteenth century, and thus that it is quite reasonable to trust both the ascription to Jacopo da Firenze and the date 1307; it should also be obvious that it does not draw even minimally on the preceding Latin treatises on algebra, neither on the translations from the Arabic nor on the *Liber abbaci*. In spite of his indubitable ultimate inspiration from the Arabic world it should also be incontrovertible that Jacopo has not drawn his material from the levels or types of Arabic algebra which have so far been examined by historians of mathematics; further, there can be no doubt that his access to the Arabic inspiration is indirect, mediated by a Romance-speaking, non-Italian environment already engaging in algebra.

Finally, it should be clear that for the next three decades, all known Italian expositions of algebra drew on Jacopo's treatise, adopting at most supplementary inspiration from other sources.[355] It was argued that the source for such additional matters (labelled "area ?") was also the area where Jacopo had found his inspiration, and that even Giovanni and Dardi, writing respectively 32 and 37 years after Jacopo, appear to have learned from this environment or area.

The existence of "area ?" followed from indirect arguments and, as far as its being a single area is concerned, from plying Occam's razor. However, several of the lines connecting "?" with known Italian writings in the revised stemma on p. 176 represent multiple inspirations: For instance, **V′** and **D** share the order of the basic cases, the way the double solution to the fifth case is spoken of, and the use of the rule of three as an algebraic method. More decisive perhaps, Giovanni follows Jacopo's statement of the rules as precisely as can be done if no direct manuscript copying is involved while sharing with Dardi the futile predilection for taking roots of square instead of non-square numbers. Rejection of the assumption of one unitary area of inspiration would therefore force us to accept that each author belonging to the first generation of Italian vernacular algebra was inspired by several or all of a multiplicity of direct sources – a multiplicity of

---

[355] This statement disregards the scattered appearance of algebraic methods in the *Trattato di tutta l'arte* and the fragment of an ordered exposition of the topic contained in the manuscript **T$_F$**, cf. note 323. It also disregards Biagio's problem collection, which is likely to antedate 1340 and is also independent of Jacopo, but which has also learned from the area where examples were constructed around numbers in given proportion and where problems of types like $\alpha CC = n$, $\alpha K = \beta t$ and $\alpha C = \sqrt{n}$ were dealt with. We do not know, indeed, whether Biagio had originally produced these problems as part of a coherent exposition of the algebraic art (cf. note 358).

Romance-speaking sources, moreover, given the absence of Arabisms in the texts.

Since the only Romance-speaking area outside Italy where the next 150 years offer any evidence of algebraic interest is the Catalan-Provençal region (or perhaps the larger Iberian area[356]), and since Montpellier itself appears not to have been a rich source, it seems reasonable to assume that "area ?" was indeed one area, to be identified with, located in or encompassing the Catalan region (see also note 324 with preceding text).[357] On the other hand the appearance of algebraic problem solutions in the *Trattato di tutta l'arte* suggests that Avignon, and Provence in general, were peripheral to rather than outside the area in question.

Within this area, most of that by which the first generation of Italian algebra goes beyond al-Khwārizmī will already have been known either fully unfolded or in germ: polynomial algebra, the use of computational diagrams, the beginnings of formal computation. The easy way to create problems looking more complex than they are may have originated here together with the interest in equations involving roots of numbers and perhaps other radicals. The carrying environment is likely to have been close to the teaching of commercial mathematics, given the generalized use of the rule of three and of the partnership structure and the preponderance of *mu'āmalāt* problems in **V**.

Quite independently of this we may notice that several of the points where the first generation of Italian vernacular algebra writers goes beyond al-Khwārizmī were to become centrally important when, in Karpinski's words (above, p. 4), two centuries of *abbaco* algebra "bore fruit in the sixteenth century in the achievements of Scipione del Ferro, Ferrari, Tartaglia, Cardan and Bombelli": *viz* polynomial algebra making use of schemes and diagrams, the use of standard abbreviations in formal operations preparing the genuine symbolic operations of Descartes – and even the ambition to solve irreducible higher-degree problems notwithstanding the fraud it had led to. The mathematical competence of a Jacopo and a Paolo Gherardi and even a Dardi is likely to have been well below that of Fibonacci, and many of the abbacus teachers may hardly deserve a characterization as "mathematicians"; but collectively *they* were the ones who prepared the algebraic take-off of the sixteenth century and that total transformation of the mathematical enterprise which it brought about in the seventeenth and eighteenth centuries.

Jacopo's own role for all this was apparently soon forgotten, and probably for acceptable reasons. He was apparently the first to introduce the solution to the six fundamental cases and to (most of) those cases of the third and fourth degree that can

---

[356] Indeed, sixteenth-century printed books on practical arithmetic from Spain and Portugal offer evidence of a much richer abbacus-type background than known surviving manuscript sources would make us suspect. See, for instance, [Silva 2006] and [Høyrup 2002b].

[357] Cf. also [Freudenthal 1993: 70–72] on the dwindling importance of Montpellier for cultural transmission from the Arabic world and on the increasing prominence of Barcelona around 1305.

be reduced by simple means, and his order for the basic cases was to remain the standard until Pacioli [1494: 144^v–145^r] returned to Fibonacci's ordering. He also appears to have introduced the habit of applying algebra to *mu'āmalāt*-problems. But Dardi's and Giovanni di Davizzo's examples show us that others could get access to the same material. At first, none the less, Jacopo's material spread – perhaps just because it was easily at hand – but soon more ambitious writers started looking for more. To that second phase belong polynomial algebra based on diagrams and schemes and formal calculations using standard abbreviations. Jacopo may thus have been a pioneer, but he had little to do with the decisive contributions of abbacus algebra to the shaping of modern algebra.

However, a less respectable reason may have contributed to Jacopo's suppression from memory. In [1985: 32], Franci and Toti Rigatelli quoted Benedetto da Firenze for claiming that Maestro Biagio († c. 1340, a contemporary and friend of Paolo dell'Abbaco) had been the first "who adapted this treatise [on algebra] to a good practice" after Fibonacci.[358] Since abbacus schools were competitors, and since Biagio and Paolo (also mentioned with great veneration by Benedetto) may have been the founding fathers of the Lungarno school were Benedetto was taught [Ulivi 2002b: 38, 203], he or his teachers could have found it adequate to eliminate such masters from memory as did not belong to their own professional family. In this respect, as in the proliferation of false rules, the direct market orientation of knowledge worked against the "Mertonian" norms for decent scientific behaviour. Then as now.

---

[358] Ed. [Pieraccini 1983: 1]. Franci and Toti Rigatelli interpret this as a statement that Biagio was the first to apply algebra regularly as a tool for business problems. This reading is less than certain, Biagio is said in the passage in question to be the first for a very long time (since Fibonacci, it is then specified) to treat of this *praticha* (here seemingly meaning the algebraic field or discipline) and reduce this *trattato di praticha* (now either a complete abbacus treatise or a treatise specifically on algebra) into a good *praticha* (here perhaps meaning a good treatise in abbacus style?); but the ambiguity is not relevant for the present discussion.

The problem collection from Biagio's hand [ed. Pieraccini 1983] which Benedetto includes in his *Trattato di praticha d'arismetricha* contains no introduction to the field, and Benedetto's presentation does not suggest that Biagio had written any; his problems are similar in kind to those which we know from Jacopo or which are found in the *Trattato di tutta l'arte*.

Jacopo's Material and Influence

So much for Jacopo's algebra, its probable roots and its influence. However, since it seems likely that Jacopo has made use of an existing treatise on that subject, and since those treatises which appear to follow in his algebraic footsteps during the next decades are not particularly close to him on other accounts (with the partial exception of the *Libro di molti ragioni*), the algebra constitutes a case *per se* from which we cannot generalize.

We have seen from the way the circle was dealt with and from metrology (pp. 91 and 124) that Jacopo (and other Italians with him) had gone to Provence, not in order to spread Italian knowledge but in order to supplement the knowledge they had acquired at home. So, to the extent Jacopo inserted *new* knowledge in his treatise we may assume he found it in Montpellier and its surroundings.

However, neither Provence nor that Italy where he grew up were carriers of an autochthonous and perennial mathematical tradition. We may therefore legitimately ask for the more distant roots of Jacopo's mathematics (and, with that, of abbacus mathematics in general).

Ultimately, of course, the Hindu-Arabic numerals came from the Arabic world (and, behind that, from India). As we have seen, however, the shapes of Jacopo's numerals point more specifically to the Iberian and Maghreb areas, whereas his explanation of the place-value system borrows from Sacrobosco's *Algorismus vulgaris*.

Agreement with what we find in *Trattato di tutta l'arte dell'abacho* and other treatises written in Provence allows us to conclude that Jacopo follows specific Provençal models for much of what he presents in Chapters 5–10; the presence in **M+F** and the *Trattato de tutta l'arte* of schemes oriented from right to left (and a similar scheme in Chapter V.3) allows us to conclude that this Provençal style, on its part, built on Arabic inspiration. Most of what follows in Chapters 11–15 and 20–22 belongs to an "abbacus culture" which already Fibonacci had drawn upon, and which in Jacopo's times is likely to have encompassed the whole arc from Tuscany over Provence to Catalonia, and perhaps also the Arabic parts of the Mediterranean and Byzantium – with further roots in a commercial culture ranging at least to India. Certain traces in the texts (not least the fact that the etymological reason for the distinction between divisions *in* and divisions *per* was still remembered in Barcelona in the late fifteenth century but neither understood in Provence not in Italy in the early fourteenth) suggest that the original core of the Romance branch of this culture was Catalan rather than Provençal or Italian, but the problems and formulae coming from Latin agrimensorial and post-agrimensorial geometry in Jacopo's Chapter 15 show with high certainty that *he* learned from this broad abbacus culture in the shape it had taken on in Montpellier, or in any case in Provence. On the other hand, this eclecticism appears mainly (perhaps only) to affect geometry – at least, there are no certain traces in **V** of the particular methods used occasionally in the mercantile chapter of the late thirteenth-century Picardian *Pratike de geometrie* [ed., trans. Victor 1979: 550–601].[359]

It is sometimes supposed that abbacus culture was also influenced by Hebrew traditions – see, for instance, [Caunedo del Potro & Córdoba della Llave 2000: 35–38]. That some Hebrew writers influenced "Christian" mathematics is beyond doubt: Savasorda's geometrical *Collection* was translated into Latin by Plato of Tivoli [ed. Curtze 1902: 10–182], and its author counted among "the Ancients" by Alberti in his *Ludi matematici* [ed. Grayson 1973: 151]; Abraham Ibn Ezra's and Levi Ben Geršon's influence is also well established. That some Hebrew writings contained material of the kind that also went into the abbacus books is also clear, although this material is shared by Arabic and other sources – thus Ibn Ezra's twelfth-century *Sefer ha-mispar* [ed., trans. Silberberg 1895], written in Lucca or Rome in c. 1146 [Sela 2001: 96]; the problem collection accompanying Ben Geršon's *Sefer maaseh hoshev* [ed. Simonson 2000]; and Elia Misrachi's *Sefer ha-mispar* [ed., trans. Wertheim 1896] (early sixteenth-century). Exactly these three works also show that it is justified to speak of a Hebrew *tradition* – not so much because Misrachi borrows directly from Ibn Ezra and cites him as because of a stylistic particularity. All three, instead of speaking of anonymous persons, regularly give names to them – namely Reuben, Simeon, Levi and Judah (when four are needed), that is, those of Jacob's first four sons; their father, when he occurs, is Jacob.[360]

---

[359] On the other hand, the *welsche Praktik*, which turns up a couple of times in **M+F**, could have such affinities. It may be no accident that the first appearance is in M.14.27A–B=F.v.30–31, dealing with the Champagne fairs.

In later abbacus books, variants of the river-passage riddle turn up often, see [Franci 2002a]; they need not necessarily descend from the two versions in the Carolingian *Propositiones ad acuendos iuvenes* [ed. Folkerts 1978: 54*f*], but if not they almost certainly come from the environment on which the Carolinian compiler had drawn.

[360] For instance, [ed., trans. Silberberg 1895: 60], [ed. trans. Wertheim 1896: 48] – this is one of the problems where Misrachi builds directly on Ibn Ezra, but the style is also found in other problems which do not come from that source.

Most of Ben Geršon's problems are pure-number problems, and among those which are not, some are in the normal *muʿāmalāt-* and abbacus style, dealing with anonymous "travellers", "merchants" etc. But a whole sequence of problems about "Reuben" and "Simeon" is found in [Simonson (ed.) 2000: 269–275].

The habit could be borrowed from the Babylonian Talmud, where Jacob and in particular his oldest sons occasionally serve in examples which have no background in the events of Genesis – thus Reuben and Simeon having married two sisters, and Levi and Judah two strangers (Yebamoth, [ed., trans. Slotki 1964: 28b]), or Reuben selling all his lands to Simeon, who then sells one of the fields to Levi (Baba Ḳamma [ed., trans. Kirzner 1964: 8b]).

In problems about alloying, Jacopo and many other abbacus writers took over the characteristic opening "I have ..." (see p. 126); if they had borrowed from Hebrew writings, one should therefore also expect them to copy *their* favourite style in the borrowed material. However, Jacopo never uses it, and it is in general very rare in the abbacus record; I have only noticed it in a single treatise, namely Muscharello's *Algorismus* [ed. Chiarini et al 1972: 154–158, 193] – first in three problems dealing with the settling of accounts, where the protagonists are, respectively, Piero+Martino, Rinaldi+Simoni and Roberto+Martino, next in one dealing with four gamblers named Piero, Martino, Antonio and Francischo. Apart from this treatise written in Nola (close to Naples) in 1478, abbacus masters seem not to have taken inspiration from Hebrew writers.[361]

The conclusions that can be drawn about Jacopo's long-term influence are almost as tentative as those concerning his immediate sources. We know that somebody still invested in having a nice copy of his original treatise made as late as c. 1450; that somebody took the care to rewrite the treatise while conserving the author's name, thus producing a treatise **M+F** of which surviving copies were made in c. 1410 and at an undetermined date. Even in 1513, Jacopo had to lend his name to a short treatise which however borrowed nothing from him but the introduction (see note 116). As a person or at least as a name, he was thus remembered for a couple of centuries. His "philosophical" introduction spread further, and actually quite widely (see p. 46). However, if we ask for the influence of the mathematics Jacopo undertook to teach, things are much more blurred. Even when an unmistakeable error like Jacopo's formula for the area of the regular pentagon (V.15.9) turns up in later writings, we cannot be sure that Jacopo was the source – the (mis-)repair of the same formula in the *Trattato di tutta l'arte* (not suspect of drawing on Jacopo) suggests it to have been widely used. Various passages and techniques in the *Libro di molti ragioni*, Tommaso della Gazzaia's *Praticha* and a number of other abbacus books could look as if inspired from Jacopo, but the similarities are rarely more striking than, for instance, certain parallels between the Columbia Algorism and the Castilian *Libro de arismética que es dicho alguarismo* (and certainly less striking than the agreement between the algebra in **V** and the corresponding part of **A**); all we can known is thus that Jacopo's *Tractatus* and, for instance, the *Libro di molti ragioni* belong together within a narrower sub-culture, embedded within the wider Italian branch of abbacus culture. Given the habit of abbacus writers to borrow and mix freely (and their likely tendency to feature those predecessors who belonged to their own school and neglect others, see page 182), it is unlikely that we shall ever come to know more. The network of interacting actors was too dense and the percentage of surviving manuscripts too small even in Italy – not to speak of the Iberian area and Provence.

---

[361] The opposite, as we know, is not true. Mordechai Finzi not only copied and revised a Hebrew translation of Abū Kāmil's algebra [Lévy 2007: 91] but also translated Dardi's *Aliabraa argibra* in 1474 from **D₃** [Van Egmond 1983: 419; Lévy 2007: 101].

Jacopo is interesting as an exponent of the process which got the Italian abbacus tradition started in earnest, and through the evidence he offers for the importance of Provençal and Iberian influence. It is difficult to know whether his general personal influence in this process was as large as his influence on the earliest beginning of Italian abbacus *algebra* appears to have been – but after all this question may also be rather unimportant.

THE VATICAN MANUSCRIPT
EDITION AND TRANSLATION

Edition and Translation Principles

Below follows a complete edition of Jacopo's text as contained in manuscript **V**, that is, Vat. Lat. 4826, accompanied by an English translation. A corresponding edition of manuscripts **M** and **F** is given without translation in the Appendix (p. 377). For those problems from Chapters 14, 15 and 21 of manuscript **M** (respectively Chapters v, vi and vii of manuscript **F**) which have no counterpart in **V**, a translation (following **M** except when **F** is clearly better) is inserted in the corresponding place in the translation of **V**.

The division of the text into chapters is mine (although some of the divisions are made explicit by the text); the division of the chapters into paragraphs follows the initials of the manuscript, whereas the numbering of the single problems (and other paragraphs) is added. The initials of the manuscript are rendered as boldface.

In the interest of readability, I have inserted punctuation, keeping it however at the strict minimum of comma and period. Standard abbreviations for monetary or metrological units are rendered as abbreviations; the primary purpose of this is to convey an impression of the total lack of systematicity in the use of these symbols. As explained in note 142, £ renders the abbreviation for *libra*/*libre*, ß that for *soldo*/*soldi* and δ that for *denaro*/*denari* (another abbreviation for *denaro*/*-i* which is also used in **V** and which is closer to the full writing is rendered deñ). Further, *fiorino*/*fiorini* is rendered *f*, ŝ^t stands for *staio*/*staia* (a capacity measure, see p. 115), b̃^{r} for *braccio*/*braccia* and g̃^{r} for *grano*/*grani* (abbreviated only in the coin list). The sole abbreviated *radice* appears as ℞.

I have dissolved the remaining abbreviations of the manuscript. When spellings vary I have tried to agree each time with the latest occurrence of the full writing – but I may have erred occasionally on that account.[362] Word separations are normalized in agreement with modern Italian habits; in cases where gemination indicates an intended pronunciation as one word, the separation is made by an apostrophe instead of full separation – thus "a'cciò" instead of "a cciò". Accents and apostrophes have been added in agreement with contemporary Italian orthography, and likewise in conjugated forms of *avere* where modern Italian has an *h* that is absent in the manuscript (*ò*, *à*, *ànno* where modern Italian would have *ho*, *ha*, and *hanno*). All accents and apostrophes are evidently absent from the manuscript.

Like other medieval manuscripts, this one uses letter shapes that look like later *u*,

---

[362] In cases where full spellings are rare (thus *sopra*/*sopre*, *multiplichare*/*multiprichare*/...), one spelling may thus be used for long, and then suddenly be replaced by another orthography; nothing can be concluded from this phenomenon, which only reflects my editorial principles.

v, *i* and *j*. However, these are pairwise mere graphical variants of the same letter, used both as a consonant and a vowel.[363] I consistently rendered representatives of the former pair as *v* when used as a consonant and with *u* when used as a vowel – a system invented only in the late Renaissance. With two exceptions, the pair *i*/*j* is always rendered as *i*. The first exception is in the name Jacobus in the incipit, the only time the letter serves as a consonant. The second is when it represents the Roman numeral 1 and stands as the last in a sequence (thus j, vij, xiij, etc.); this rule is borrowed from the manuscript (it was the norm of the times).

A particular problem is presented by the *si* of the manuscript. It may be both a personal pronoun (modern *si*) and an adverb (modern *sì*, often to be translated "then" or "such that"). In most Italian dialects the pronoun is only the third person (singular and plural); in older Tuscan, however, it may also be the first person plural, for which reason *si dobiamo* could in principle mean "we shall". However, in cases where the pronominal function is indubitable, the manuscript almost always has the spelling *se*.[364] Since there is no single instance of *se* which *could* be a first person plural personal pronoun, *si dobiamo* must certainly be interpreted *sì dobiamo*, "then we shall". Similarly, there are 18 instances of the construction *si multiplica* (with varying spellings of the imperative) and none of *se multiplica*, for which reason the interpretation must certainly be *sì multiplica*, "then multiply";[365] accordingly for other imperatives.

Passages in ⟨ ⟩ repair copyist's omissions, in the translation also copyist's errors; the occasional superscript letters (⟨ ⟩^{M+F}, ⟨ ⟩^{M}, ⟨ ⟩^{A}) refer to a manuscript or manuscript group on which the restitution is based. Letters, words and passages in { } are present in the manuscript by error; those that are deleted by the copyist are ~~struck out~~ in the text edition and omitted from the translation; words or passages that were at first omitted by the copyist and afterwards inserted above the line are marked ^ ^, whereas insertion in the margin is marked * *. Editorial comments are in [ ], added words in the translation in ( ). Passages in italics in the edition correspond to the use of red ink in the manuscript.

The English translation attempts to keep very close to the text, and to render always in the same way the same phrase or term when used in the same function, even in cases where this implies some awkwardness. The purpose is in part to reflect the medium which Jacopo had at his disposal – a vernacular which was not yet fashioned as an adequate tool for technical discourse; in part it is to render that imprecision of the conceptual structure which was a consequence of the character of the language (but which probably

---

[363] The shape a modern reader tends to read as *v* only stands in initial position, in all other positions the letter *u*+*v* is written *u*. The letter *i*+*j* actually exists in three distinct graphical variants, *j* standing only in final position, *J* only in initial position, and *i* in initial, intermediate or final position.

[364] Thus, there are 108 cases of *se sa*, *se fanno* and *se faccia*, but only 6 corresponding with *si*.

[365] This interpretation also agrees with a spelling "sie multiplica" in F.vi.8.

had other roots as well). However, the principle is not followed to the pedantic extremes which I felt it necessary to accept when rendering Babylonian mathematical texts in [Høyrup 2002a] – lax as it is and still *in statu nascendi*, the conceptual structure of Jacopo's mathematics is after all not very different from ours. Thus, *libra* is left untranslated when standing for a monetary unit, and translated "pound" when standing for a weight. In the translation, abbreviations for monetary and metrological units are dissolved. Numbers are rendered as they stand in the text, whether as number words, as Roman or as Hindu-Arabic numerals.

Notes to the edition stand as end notes to the single problems or subsections, marked by superscript letters.

All diagrams and illustrations are redrawn by me.

The Text

(fol. 1ʳ)

[1. Incipit and General Introduction]

¹·¹ *Incipit tractatus algorismi, huius autem artis novem sunt speties, silicet, numeratio, addictio, subtractio, ⟨mediatio,⟩ᵃ duplatio, multiplicatio, divixio, progrexio, et radicum extractio. Conpilatus a magistro Jacobo de Florentia apud Montem Phesulanum, anno domini m°ccc° vij° in mense septenbris.*

¹·¹ The treatise of algorism begins, which art consist of 9 species, namely, numeration, addition, subtraction, ⟨mediation⟩,ᵃ duplation, multiplication, division, progression, and root extraction. Composed by master Jacopo of Florence in Montpellier, in the year 1307 in the month of september.

ᵃ Inserted in agreement with **M+F**, and needed in order to fill out the number of nine species.

¹·² Conciossia cosa che tucte quelle cose che la humana generatione de questo secolo sanno et possono sapere, si fanno per duy principale vie, le quale vie sonno queste. La prima si è senno, et la seconda si è la scienza. E ciascheuna di queste due vie si à secho duy gentile et nobile conpagne. L'una si è gratia di Dio. Et l'altra si è cognoscenza per ragione. Et le conpagne dela scienza, si è l'una l'amaestramento dele scripture. Et l'altra si è intendimento con bono ingegnio. Et secondamente che dice la santa scriptura, el senno è el più nobile thesoro che sia al mondo. Et dovete sapere che Salamone, che fo quasi el più savio homo di tucto el mondo, sì adomando al nostro signore Idio in sua gioventudine che gli desse senno. Et el nostro signore gli disse che el suo domando fo el più alto domando che egli avesse possuto avere domandato. Onde gle dede el terzo ⟨del⟩ᴹ⁺ᶠ

¹·² Admittedly, all those things which the human race of this world know and are able to know, are obtained in two main ways, which ways are these. The first is discernment, the second is science. And each of these two ways is accompanied by two gentle and noble partners. One is the grace of God. And the other is knowledge by reason. And of the partners of science, one is mastery of what has been written. And the other is understanding with good intelligence. And according to what the Holy Scripture says, discernment is the noblest treasure that there is in the world. And you shall know that Solomon, who was close to being the wisest man of all the world, asked the Lord in his youth to give him discernment. And our Lord said to him that his request was the highest request that he could have asked. Wherefore he gave him one third of the discernment of Adam,

senno de Adam, et questo senno fo per grazia de Dio. Ancho dice la santa scriptura che tucti li homini che ancho fuorono non dimandarono a Dio nigiuno più bello né più alto domandamento de quello perciò che tucti li boni et perfecti doni de Dio descendono da quello domandamento. È vera cosa che l'omo po nominare lo senno et la scienza, l'uno senno naturale, et l'altro scienza accidentale. Et dovete sapere che tucto ciò che li homini fanno naturalmente et accidentalmente, sì è che el nostro padre à conceduto a sapere per la sua santissima virtù et grazia et misericordia. Et però noi siamo tucti tenuti di rendare grazia a lui che è si dolce padre e signore che ci à dato a conosciare tanta soctilità. Et però noi nel suo santissimo nome et al suo sanctissimo honore sì incomenzaremo el nostro tractato, lo quale è dicto algorismus. Et sappiate che si chiama algorismus perché questa scienza fo principalmente facta in Arabia, et quelli che la troverono forono simigliantemente arabi. Et l'arte è dicta in lingua arabia algho, e'l numero è dicto rismus, et perciò è dicto alghorismus. Lo quale destingue in cinque capitoli, li quali vi mostrarò manifestamente nel nostro tractato ordinatamente secondo la dicta materia, sicchomo *(fol. 1ʳ)* domanda la decta scienza. Et incominciamo a honore e venerentia del nostro signore Gesu Christo[a] et dela sua sanctissima matre Vergine Maria et de tucta la corte celestiale, et con lo adiuto de nostri predecessori, et a honore di tucti magistri et scolari de questa scienza. Et de qualunqua altra bona persona vedesse et legesse questo tractato

and this discernment was by grace of God. The Holy Scripture also says that no man until now asked God for any request more beautiful or higher than that, since all God's good and pure gifts descend from this request. It is true that one may call discernment and science, one natural discernment, the other accidental science. And you shall know that everything men do naturally and by accident, our Father has granted (them) to know in his most holy virtue and grace and compassion. And therefore we are all obliged to thank Him who is such a sweet Father and Lord, who has given us to know so much subtlety. And therefore in His most holy name and His most holy honour we begin our treatise, which is called algorism. And know that it is called algorism because this science was first made in Arabia, and those who found it were similarly Arabs. And art in Arabic is called algo, and the number is called rismus, and so it is called algorism. Which falls into five chapters, which I shall show you manifestly in our treatise ordered according to the said matter, as *(fol. 1ʳ)* the said science asks for. And we begin in the honour and reverence of our Lord Jesus Christ and his most holy mother Virgin Mary and the whole celestial court, and with the assistance of our predecessors, and in honour of all masters and scholars of this science. And of every other honest person who might see and read this treatise with dedication and sense.[a]Abbreviated writing making use of Greek letters (but no more than could be seen in church paintings – Χρ).

devoto e ragionevolmente.

^{1.3} **O**ra mostraremo la proprietade de sopre-dicti cinque capitoli secondo che dice Boetio nell'Arismetica sua. Lo primo capi-tolo si è moltiplicare. Lo secondo capitolo si è dividere. Lo terzo capitolo si è li nu-meri rocti. Lo quarto capitolo si sonno le regole. Lo quinto capitolo si è el generale intendimento che se tra de dicti quattro capitoli. Et devete sapere che li dicti cinque capitoli ànno in loro molti divisioni et menbri, sicomo di moltiplicare de doy o de tre o de quattro o de cinque o de più figure. El dividere si è in rocti sani e rocti in rocti. Sonno moltiplicare, dividere, giungere, sobtrare, e dire quale è più l'uno rocto che l'altro, overo quanto meno. Et quale è di conoscerli vedendoli scripti per figure. Le regole ànno in loro molte manere et inten-dimenti et sottilitadi, le quali udirete ordina-tamente secondo loro natura che è dicta.

^{1.3} Now we shall show the properties of the five chapters spoken of above according to what Boethius says in his *Arithmetic*. The first chapter is to multiply. The second chapter is to divide. The third chapter is broken numbers. The fourth chapter are the rules. The fifth chapter is the general under-standing which is drawn from the said four chapters. And you shall know that the said five chapters have many subdivisions and sections, such as multiplying by two or three or four or five or more figures (i.e., digits). Division falls in integer fractions and fractions in fractions. They are to multiply, to divide, to join, to subtract, and to say which fraction is greater that the other, or how much smaller. And to recog-nize what they are, seeing them written by figures. The rules comprise many routines and insights and subtleties, which you will hear in orderly manner according to their nature which is explained.

^{1.4} **S**icomo in questo tractato lu intellecto e'l bono ingegno si ce dona a sapere la grande soctilitade dele profetie et dele philosifie et dele celestiali scripture et deli temporali, et che ci dona a sapere anco più innanzi, che per intellecto et per bono ingegno et sottile si fanno li homini molte sperientie et congielationi [*sic*, read "con-pilationi"]^{M+F} de tractati, li quali non forono ancora facti per altri homini, et sanno fare molti artifitii et argomenti de scripture che ce affinino le cose che forono facte per li

^{1.4} As in this treatise the mind and good intelligence grants us to know the great subtlety of the prophecies and the philo-sophies and the celestial and temporal writings, it will grant us to know even more henceforth, since by mind and good and subtle intelligence men make many inves-tigations and compose many treatises which were not made by other people, and know to make many artifices and written argu-ments which for us bring to greater perfec-tion things that were made by the first men.

primi homini. Dunqua sicomo dicto avemo de sopre, lu nostro tractato si chiama in lingua arabia alghorismio, perciò che debiamo scrivere le dece figure del dicto alghorismio secondo la costumanza delli arabi, perciò che forono trovatori de questa scienza. Cioè che noi debiamo scrivere a retroso et legere a dericto secondo noi, cioè a dire che ce dobiamo comenzare a scrivere dal minore numero et legere del magiore numero.

Hence as we have said above, our treatise is called in Arabic algorism, and so we should write the ten figures of the said algorism according to the custom of the Arabs, since they were those who found this science. That is, we shall write backwards and read to the right according to (what is customary with) us, that is to say, we shall begin by writing from the smallest number and read from the greatest number.

(fol. 2ʳ)

[2. Introduction of the Numerals and the Role of Zero]

[2.1] **Queste** sonno le nostre figure del'abocho, co le quali tu poi scrivere qualunqua numero tu voli, o de quantunqua quantità se fusse. Et queste sonno le figure dell'arte vecchia et dela nova.

[2.1] These are our abbaco figures, by means of which you may write whatever number you wish, or of whatever quantity it were. And these are the figures of the old art and the new.

[2.2] **Ancora** scrivaremo qui disocto como ellevano le dicte figure. Et perché se intendano meglio et più apertamente sì le scrivaremo per figure, et simegliantemente per lectere, perché senza alchuno maesterio l'omo per se medesmo le possa intendare. Et dovete sapere, et chosì è che el zero per se solo non significa nulla, ma bene à potentia di fare significare quando è accompagnato, ma non ogni volta ma secondo dove ello è posto, o denanzi o dercto. Cioè

[2.2] Further we shall write here below how the said figures denote. And so that they may be understood better and more clearly we shall write them by figures, and similarly by letters, so that one may understand by himself without any master teaching him. And you shall know, and it is thus that the zero by itself does not signify anything, but it surely has the power to make signify when it is accompanied, but not always but according to where it is put, either before

se el zero è posto denanzi a un'altra figura non à potentia di dare significare niente,[a] ma quando è posto dercto ala figura sì à potentia de fare significare sceconda quella figura che è. Cioè fusse al lato a 1, significa 10, et se fusse al lato a 2, significa 20. Et se fusse al lato a tre, significa 30. Et chosì secondo la figura che fa significare.

or behind. That is, if the zero is put before[a] another figure it does not have the power to make signify anything, but if placed behind the figure then it has the power to give to signify according to which figure it is. That is, if it were beside 1, it signifies 10, and if it were beside 2, it signifies 20. And if it were beside three, it signifies 30. And thus according to the figure which it makes signify.

[a] "Denanzi"/"Before", we notice, is to the left.

(fol. 2ʳ)

[3. Tabulated Writing of Numbers, with Corresponding Roman or Semi-Roman Writings]

9	8	7	6	5	4	3	2	1
viiij	viij	vij	vj	v	iv	iij	ij	j
17	16	15	14	13	12	11	10	
xvij	xvj	xv	xiiij	xiij	xij	xj	x	
24	23	22	21	20	19	18		
xxiiij	xxiij	xxij	xxj	xx	xviiij	xviij		
30	29	28	27	26	25			
xxx	xxviiij	xxviij	xxvij	xxvj	xxv			
35	34	33	32	31				
xxv	xxxiiij	xxxiij	xxxij	xxxj				
40	39	38	37	36				
xl	xxxviiij	xxxviij	xxxvij	xxxvj				
46	45	44	43	42	41			
xlvj	xlv	xliiij	xliij	xlij	xlj			
90	80	70	60	50	49			
lxxxx	lxxx	lxx	lx	l	xlviiij			

600	500	400	300	200	100
c	c	c	c	c	c
vj	v	iiij	iij	ij	j

(fol. 3ʳ)

1000	900	800	700	600	500
	c	c	c	c	c
m	viiij	viij	vij	vj	v
10000	9000	8000	7000	6000	
m	m	m	m	m	
x	viiij	viij	vij	vj	
60000	50000	40000	30000	20000	
m	m	m	m	m	
lx	l	xl	xxx	xx	
100000	90000	80000	70000		
m	m	m	m		
c	lxxxx	lxxx	lxx		
500000	400000	300000	200000		
c	m	m	m		
v	cccc	ccc	cc		
1000000	900000	800000	700000		
m	m	m	m		
m	vcccc	vccc	vcc		

[4. Explanation and Exemplification of the Place-Value Principle]

4.1 Dovete sapere che una sola figura, cioè figura de uno posta nel primo grado, significa uno. Et quando ella è posta nel secondo grado significa dece, et in terzo grado cento. Et in quarto grado significa mille, et in quinto grado significa decemilia. Et in sexto grado significha ~~significa~~ centomilia, et in septimo grado significa mille migliara. Et così advene de qualunqua figura tu poni nel suo grado. Et similmente qualunqua

4.1 You shall know that one single figure, that is the figure of one put in the first rank, signifies one. And when it is put in the second rank it signifies ten, and in third rank hundred. And in fourth rank it signifies thousand, and in fifth rank it signifies ten thousand. And in sixth rank it signifies a hundred thousand, and in seventh rank it signifies a thousand thousand. And thus happens with every figure which you put

figura tu poni in sequente loco significa
dece cotanti che le figure che sonno possate
denanzi[a] secondo le figure che fosse.

in its rank. And similarly whatever figure
you put in the following place signifies ten
times as much as the figures that are put
before[a] according to the figures it may be.

[a] Now "denanzi"/"before" is to the right. The plural "sonno possati" excludes a reading "before which it is placed".

(fol. 3ᵛ)

4.2 Ancora el mostraremo per altro modo, et diremo, una sola figura qualunqua figura si fosse leva unitade, cioè da diece insuso. Et quando fosseno due figure, la prima figura lieva dece, la seconda unitade, legendo per dericto modo. Et quando sonno tre figure, la prima leva centonara, la seconda dicine, la terza unità. Et quando sonno quattro figure, la prima leva migliara, la seconda centinara, la terza dicine, la quarta unitade. Et quando sonno cinque figure, la prima leva dicine di migliara, la seconda unitade di migliara, la terza centonara, la quarta dicine, la quinta unitade. Et quando sonno sey figure, la prima leva centonara di migliara, la seconda dicine di migliara, la terza unitade di migliara, la quarta centonara, la quinta dicine, la sexta unitade. Et quando sonno sette figure, la prima leva migliara de migliara, la seconda centonara de migliara, la terza dicine di migliara, la quarta unitade de migliara, la quinta centonara, la sexta dicine, la septima unitade.

4.2 Further we shall show it in a different way, and we shall say, a single figure, whatever figure it might be, denotes unities, that is, from ten upwards [sic]. And when there were two figures, the first figure denotes tens, the second unities, reading in the forward way. And when there are three figures, the first denotes hundreds, the second tens, the third unities. And when there are four figures, the first denotes thousands, the second hundreds, the third tens, the fourth unities. And when there are five figures, the first denotes tens of thousands, the second unities of thousands, the third hundreds, the fourth tens, the fifth unities. And when there are six figures, the first denotes hundreds of thousands, the second tens of thousands, the third unities of thousands, the fourth hundreds, the fifth tens, the sixth unities. And when there are seven figures, the first denotes thousands of thousands, the second hundreds of thousands, the third tens of thousands, the fourth unities of thousands, the fifth hundreds, the sixth tens, the seventh unities.

4.3 Et dovete sapere che la figura nel primo grado rappresenta tante unità quante la

4.3 And you shall know that the figure in the first rank represents as many unities as

figura medesma. Et nel secondo grado tante dicine quante la figura medesema. Et nel terzo grado tante centonara quante la figura medesma. Et nel quarto grado tante migliara quante la figura medesma. Et nel quinto grado tante dicine di migliara quante la figura medesma. Et nel sexto grado tante centonara di migliara quante la figura medesma. Et nel septimo grado tante migliare de migliara quante la figura medesma. Et così, per ogni figura che tu ve accressi, rappresenta dece contanti che quelle che sonno passate inanzi per la regola.

the figure itself. And in the second rank as many tens as the figure itself. And in the third rank as many hundreds as the figure itself. And in the fourth rank as many thousands as the figure itself. And in the fifth rank as many tens of thousands as the figure itself. And in the sixth rank as many hundreds of thousands as the figure itself. And in the seventh rank as many thousands of thousands as the figure itself. And thus for every figure that you augment there, by the rule it represents ten times as much as those that came before.

[4.4] **O**ra abbiamo dicto delo representamento dele figure da una infine in vij. Ora diremo da sette infino in infinito. Et diremo così, dovete sapere che le figure da una infino in sette ⟨representano ciascheuna per sua regola, ma da sette infino in infinito⟩[M+F] representano et levano tucte per una regola.[a] Et in questo modo scrivi le figure che tu voli levare per ordine, et poi te comenza da ꝑ la parte dela unitade et conta sette figure, et poi fa uno punto. Et queste figure rappresentano sì **(fol. 4ʳ)** como dicto abiamo di sopre. Et poi fa uno punto a ogni quattro figure. Et tanti punti quanti trovarai, tante migliara di migliara representaranno, et questo serà secondamente che figura et che in qual loco fosse posta.

[4.4] Now we have spoken about what is represented by the figures from one until vij. Now we shall speak ⟨of what is represented by figures in rank⟩ from seven until the infinite. And we shall say thus, you shall know that the figures from one until seven ⟨represent each after its own rule, but from seven until the infinite⟩[M+F] they represent and denote all by one and the same rule.[a] And in this way you write the figures that you want to denote in order, and then you begin by the unities and count seven figures, and then you make a point. And these figures represent in the **(fol. 4ʳ)** way we have said above. And then make a point for every four figures. And as many points as you find, so many thousands of thousands will they represent, and this will be according to the figure and to the place where it was put.

[a] The phrase "per una regola"/"by one and the same rule", recurs repeatedly in **V**. In **M+F**, this is its only appearance.

^{4.5} **Et** in ciò daremo uno exemplo. Ora me di' quanto relevano queste nove figure 987.644321. Sappi che relevano nove cento ottantasette migliara de migliara, et secento quarantaquattro migliara, et trecento vintiuno. Et per questa regola relevano le più.

^{4.5} And to this we shall give an example. Now say me how much these nine figures 987.644321 denote. Know that they denote nine hundred eighty-seven thousands of thousands, and six hundred forty-four thousands, and three hundred twenty-one. And by this rule (follows what) most denote.

^{4.6} **Dime** quanto relevano queste quindice figure che io te mostro et pongo qui, 23456789.8765432. Sappi che relevano dumilia trecento quarantacinque migliara de migliara ^de migliara^, et sei milia settecento ottantanove migliara de migliara. Et ottomilia settecento sexantacinque migliara, et quattrocento trentadoi. Et per questo modo rilevarebbono quante figure fosseno per infino in ogni quantitade.

^{4.6} Say my how much these fifteen figures denote which I show you and put here, 23456789.8765432. Know that they denote two thousand three hundred forty-five thousands of thousands of thousands, and six thousand seven hundred eighty-nine thousands of thousands. And eight thousand seven hundred sixty-five thousands, and four hundred thirty-two. And in this way as many figures denote as they might be until any quantity.

^{4.7} **Et** dovete sapere che la figura che si ritrova nel primo grado representa se medesma. Et questo se intende, se la figura che se retrovarà nel primo grado sera uno, relevarà uno. Et se serà doi, releverà doi, et così relevarà secondo la figura che serà per infino in nove. Et la figura che se trovarà nel secondo grado o loco^a farà levare la prima tante dicine quante unitade era prima, et la seconda serà unità. Et pongoti qui lo exemplo. Cioè questa figura releva doi, 2, che è nel primo loco, et poi pognendone un'altra nel secondo grado, cioè così, 24, leva vintiquattro. Et vedi apertamente che fa levare dece tanti che prima. Prima diciva doi, ora dice vinti, cioè

^{4.7} And you shall know that the figure that stands in the first rank represents itself. And this is to be understood so, that if the figure which stands in the first rank is one, it will denote one. And if it is two, it will denote two, and thus denote according to the figure which it is until nine. And the figure that stands in the second rank or place^a will make (those in) the first denote as many tens as they were unities at first, and the second will be unities. And here I propose to you the example. That is, this figure denotes two, 2, which is in the first place, and then if we put another one in the second rank, that is thus, 24, it denotes twenty-four. And you see clearly that it

direbbe vinti se vi fosse al lato al doi uno zero, che el zero per se medesemo non leva nulla, ma bene à potentia de fare levare como io t'ò dicto di sopre, secondo nel grado che è posto. Ora te dico che queste figure de sopre levano vintiquattro perché el dui sta per vinti, et el quattro, perché è nel secondo grado, releva quattro, siché però fanno vintiquattro. Et **(fol. 4ᵛ)** così ponendogle al lato un'altra figura farà levare quello dui denanti dece cotanti che prima, et la terza leva quello che è. Et diciamo che stesse in questo modo 248. Queste relevano secondo che io t'ѳò[b] dicto, et provale et multiplica quelle doi de prima che diciva vintiquattro per doi [*sic,* read dece] faranno ducento quaranta, et la terza per se sola releva quello che è, che è otto. Et però quelle tre figure relevano ducento quarantaotto. Et così se le figure fosseno quattro, cioè 2486, relevano dece tanta che prima, che dece via 248 fa dumilia quactrocento ottanta. Et la quarta che tu hai posta è uno sey, et però fa ottantasey. Et così ogni figura che tu vi poni relevano dece tanti che prima. Et così va secondo le figure che sonno nel modo dicto.

denotes ten times as much as at first. At first it said two, now it says twenty, that is, it would say twenty if beside two there had been a zero, as the zero by itself denotes nothing, but surely has the power to make denote, as I have said to you above, according to the rank it is put in. Now I say that these figures from above denote twenty-four because the two denotes twenty, and the four, because it is in the second rank, denotes four, so that they therefore make twenty-four. And **(fol. 4ᵛ)** thus putting another figure beside it will make this two before denote ten times as much as at first, and the third denotes that which it is. And let us say that there stood in this way 248. These denote according to what I have said to you. And verify it, and multiply these two from before that said twenty-four by ⟨ten⟩, they make two hundred forty, and the third by itself alone denotes what it is, which is eight. And therefore these three figures denote two hundred forty-eight. And thus, if the figures had been four, that is 2486, they denote ten times as much as before, as ten times 248 makes two thousand four hundred eighty. And the fourth that you have put (there) is a six, and therefore it makes eighty-six. And thus every figure that you put there denote ten times as much as before. And thus it goes according to the figures that are there in the said way.

---

[a] Now, as we see, ranks are counted from the left.
[b] This is one of the examples of orthographic fidelity. The scribe begins a full writing "te ò" and then discovers that his manuscript has the contraction "t'ò", and corrects.

[5. Introduction to the Multiplication Tables]

5.1 Ora qui appresso insegnaremo molti-plicare l'una figura contra all'altra, et queste se chiamano librectine minori. Et simile insegniaremo moltiplicare una figura contra a doi, et doi contra altro du, et ancora contra a tre, et queste se chiamano librettine magiori. Et così a poco a pocho verrai a inprendere de questa arte et de questa scienza. Senza le quali librectine non poi mai venire a perfectione, però che le dicte librectine te insegnano fare tre cose, cio⟨è⟩ moltiplicare et partire et raccoglere, le qual cose sonno el fondamento de questa scien-za. Ora cominciamo al nome di Dio et prima porremo le librettine minori.

5.1 Now next to here we shall teach to multiply one figure against the other, and this is called minor booklets. And similarly we shall teach to multiply one figure against two, and two against two other, and further against three, and these are called major booklets. And thus you will gradually come to learn this art and this science. Without which booklets you will never arrive at perfection, because the said book-lets teach you to do three things, that is, to multiply and to divide and to aggregate, which things are the fundament of this science. Now we begin in the name of God and first we propose the minor booklets.

[6. Multiplication Tables, Including Multiples of *soldi* Expressed in *libre* and *soldi*]

1	via	1	fa	1		2	via	3	fa	6
2		2		4		2		4		8
3		3		9		2		5		10
4		4		16		2		6		12
5		5		25		2		7		14
6		6		36		2		8		16
7		7		49		2		9		18
8		8		64		2		10		20
9		9		81						
10		10		100						

(fol. 5ʳ)

via				via		
3	4	12		5	6	30
3	5	15		5	7	35
3	6	18		5	8	40
3	7	21		5	9	45
3	8	24		5	10	50
3	9	27				
3	10	30				
				6	7	42
	via			6	8	48
				6	9	54
4	5	20		6	10	60
4	6	24				
4	7	28				
4	8	32		7	8	56
4	9	36		7	9	63
4	10	40		7	10	70
				8	9	72
				8	10	80
				9	10	90
				10	10	100
				10	100	1000
				10	1000	10000

5.2 **Q**ueste che sonno qui di socto {di} che incomenza doi via dece fa vinti et tre via dece 30, et de renpecto a queste recomenza per lo retroso de esse e dice così, dece via cento, mille, et 10 via 90 fa 900 etcetera sì se chiamono librettine magiori, senza le quali non si po venire a perfectione de questa scienza e arte.

2	10	20		10	100	1000
3	10	30		10	90	900
4	10	40		10	80	800
5	10	50		10	70	700
6	10	60		10	60	600
7	10	70		10	50	500
8	10	80		10	40	400
9	10	90		10	30	300
10	10	100		10	20	200
					{100}	

[2] These that are here below, which begin two times ten makes twenty and three times ten 30, and opposite of these begins anew backwards (with respect) to these and says thus, ten times hundred, thousand, and 10 times 90 makes 900 etc., are called major booklets, without which it is impossible to come to perfection in this science and art.

(fol. 5ᵛ)

	via				via	
2	11	22		11	100	1100
3	11	33		11	90	990
4	11	44		11	80	880
5	11	55		11	70	770
6	11	66		11	60	660
7	11	77		11	50	550
8	11	88		11	40	440
9	11	99		11	30	330
10	11	110		11	20	220
2	12	24		12	100	1200
3	12	36		12	90	1080
4	12	48		12	80	960
5	12	60		12	70	840
6	12	72		12	60	720
7	12	84		12	50	600
8	12	96		12	40	480
9	12	108		12	30	360
10	12	120		12	20	240
	via				via	
2	13	26		13	100	1300
3	13	39		13	90	1170
4	13	52		13	80	1040
5	13	65		13	70	910
6	13	78		13	60	780
7	13	91		13	50	650
8	13	104		13	40	520
9	13	117		13	30	390
10	13	130		13	20	260

(fol. 6ʳ)

	via				via	
2	14	28		14	100	1400
3	14	42		14	90	1260
4	14	56		14	80	1120
5	14	70		14	70	980
6	14	84		14	60	840
7	14	98		14	50	700
8	14	112		14	40	560
9	14	126		14	30	420
10	14	140		14	20	280
	via					
2	15	30		15	100	1500
3	15	45		15	90	1350
4	15	60		15	80	1200
5	15	75		15	70	1050
6	15	90		15	60	900
7	15	105		15	50	750
8	15	120		15	40	600
9	15	135		15	30	450
10	15	150		15	20	300
2	16	32		16	100	1600
3	16	48		16	90	1440
4	16	64		16	80	1280
5	16	80		16	70	1120
6	16	96		16	60	960
7	16	112		16	50	800
8	16	128		16	40	640
9	16	144		16	30	480
10	16	160		16	20	320

(fol. 6ᵛ)

	via				via	
2	17	34		17	100	1700
3	17	51		17	90	1530
4	17	68		17	80	1360
5	17	89		17	70	1190
6	17	102		17	60	1020
7	17	119		17	50	850
8	17	136		17	40	680
9	17	153		17	30	510
10	17	170		17	20	340
2	18	36		18	100	1800
3	18	54		18	90	1620
4	18	72		18	80	1440
5	18	90		18	70	1260
6	18	108		18	60	1080
7	18	126		18	50	900
8	18	144		18	40	720
9	18	162		18	30	540
10	18	180		18	20	360
2	19	38		19	100	1900
3	19	57		19	90	1710
4	19	76		19	80	1520
5	19	95		19	70	1330
6	19	114		19	60	1140
7	19	133		19	50	950
8	19	152		19	40	760
9	19	171		19	30	570
10	19	190		19	20	380

(fol. 7ʳ)

	via				via	
2	23	46		23	100	2300
3	23	69		23	90	2070
4	23	92		23	80	1840
5	23	115		23	70	1610
6	23	138		23	60	1380
7	23	161		23	50	1150
8	23	184		23	40	920
9	23	207		23	30	690
10	23	230		23	20	460
2	29	58		29	100	2900
3	29	87		29	90	2610
4	29	116		29	80	2320
5	29	145		29	70	2030
6	29	174		29	60	1740
7	29	203		29	50	1450
8	29	232		29	40	1160
9	29	261[a]		29	30	870
10	29	290		29	20	580
2	31	62		31	100	3100
3	31	93		31	90	2790
4	31	124		31	80	2480
5	31	155		31	70	2170
6	31	186		31	60	1860
7	31	217		31	50	1550
8	31	248		31	40	1240
9	31	279		31	30	930
10	31	310		31	20	620

[a] Corrected in the manuscript from "2?1"; correspondingly, the product of 29 and 90 has been corrected from "2510" into "2710". Obviously, one column was used for the construction of the other.

(fol. 7ᵛ)

	via				via	
2	37	74		37	100	3700
3	37	111		37	90	3330
4	37	148		37	80	2960
5	37	185		37	70	2590
6	37	222		37	60	2220
7	37	259		37	50	1850
8	37	296		37	40	1480
9	37	333		37	30	1110
10	37	370		37	20	740
2	41	82		41	100	4100
3	41	123		41	90	3690
4	41	164		41	80	3280
5	41	205		41	70	2870
6	41	246		41	60	2460
7	41	287		41	50	2050
8	41	328		41	40	1640
9	41	369[a]		41	30	1230
10	41	410		41	20	820
2	43	86		43	100	4300
3	43	129		43	90	3870
4	43	172		43	80	3440
5	43	219		43	70	3010
6	43	258		43	60	2580
7	43	301		43	50	2150
8	43	344		43	40	1720
9	43	387		43	30	1290
10	43	430		43	20	860

[a] This number, as well as 41×90, has been corrected – apparently from 379 and 3790, respectively.

(fol. 8ʳ)

	via				via	
2	47	94		47	100	4700
3	47	141		47	90	4230
4	47	188		47	80	3760
5	47	235		47	70	3290
6	47	282		47	60	2820
7	47	329		47	50	2350
8	47	376		47	40	1880
9	47	423		47	30	1410
10	47	470		47	20	940
11	12	132		11	1 £	11 £ s
11	13	143		11	19 s	10 £ 9 s
11	14	154		11	18 s	9 £ 18 s
11	15	165		11	17 s	9 £ 7 s
11	16	176		11	16 s	8 £ 16 s
11	17	187		11	15 s	8 £ 5 s
11	18	198		11	14 s	7 £ 14 s
11	19	209		11	13 s	7 £ 3 s
11	20	220		11	12 s	6 £ 12 s
12	13	156		12	1 £	12 £ ß
12	14	168		12[a]	19 s	10 £ 8 ß
12	15	180		12[b]	18 s	10 £ 18 ß
12	16	192		12	17 s	10 £ 4 ß
12	17	204		12	16 s	9 £ 12 ß
12	18	216		12	15 s	9 £ — ß
12	19	228		12	14 s	8 £ 8 ß
12	20	240		12	13 s	7 £ 16 ß

[a] The result of this should be 11 £ 8 ß.
[b] The result of this should be 10 £ 16 ß.

(fol. 8ᵛ)

via				via		
13	14	182		13	1 £	13 £ ß
13	15	195		13	19 ß	12 £ 7 ß
13	16	208		13	18 ß	11 £ 14 ß
13	17	221		13	17 ß	11 £ 1 ß
13	18	234		13	16 ß	10 £ 8 ß
13	19	247		13	15 ß	9 £ 15 ß
13	20	260		13	14 ß	9 £ 2 ß
via						
14	15	210		14	1 £	14 £ – ß
14	16	224		14	19 ß	13 £ 6 ß
14	17	238		14	18 ß	12 £ 12 ß
14	18	252		14	17 ß	11 £ 18 ß
14	19	266		14	16 ß	11 £ 4 ß
14	20	280		14	15 ß	10 £ 10ß
15	16	240		15	1 £	15 £ ß
15	17	255		15	19 ß	14 £ 5 ß
15	18	270		15	18 ß	13 £ 10 ß
15	19	285		15	17 ß	12 £ 15 ß
15	20	300		15	16 ß	12 £ 0 ß
16	17	272		16	1 £	16 £ – ß
16	18	288		16	19 ß	15 £ 4 ß
16	19	304		16	18 ß	14 £ 8 ß
16	20	320		16	17 ß	13 £ 12 ß
17	18	306		17	1 £	17 £ 0 ß
17	19	323		17	19 ß	16 £ 3 ß
17	20	340		17	18 ß	15 £ 6 ß

(fol. 9ʳ)

via				via		
18	19	342		18	1 £	18 £ 0 ß
18	20	360		18	19 ß	17 £ 2 ß
19	20	380		19	1 £	19 £ 0 ß
20	20	400		20	1 £	20 £ 0 ß
20	21	420ª		20	21 ß	21 £ 0 ß
20	22	440		20	22 ß	22 £ 0 ß
20	23	460		20	23 ß	23 £ 0 ß
20	24	480		20	24 ß	24 £ 0 ß
20	25	500		20	25 ß	25 £ 0 ß
20	26	520		20	26 ß	26 £ 0 ß
2	20	40		2000	2000	4000000
3	30	90		3000	3000	9000000
4	40	160		4000	4000	16000000
5	50	250		5000	5000	25000000
6	60	360		6000	6000	36000000
7	70	490		7000	7000	49000000
8	80	640		8000	8000	64000000
9	90	810		9000	9000	81000000
10	100	1000		10000	10000	100000000
20	20	400		20000	20000	400000000
30	30	900		30000	30000	900000000
40	40	1600		40000	40000	1600000000
50	50	2500		50000	50000	2500000000
60	60	3600		60000	60000	3600000000
70	70	4900		70000	70000	4900000000
80	80	6400		80000	80000	6400000000
90	90	8100		90000	90000	8100000000
100	100	10000		10000⟨0⟩	10000⟨0⟩	10000000000

ª "420" and "440" are illegible because of a water damage to the manuscript.

(fol. 9ᵛ)

[7. Tables of Higher Squares]

	via			2 ¼	12 ¼	20 ¼
200	200	40000		1 ½	3 ½	4 ½
300	300	90000		×	×	×
400	400	160000		1 ½	3 ½	4 ½
500	500	250000				
600	600	360000				
700	700	490000				
800	800	640000		30 ¼	42 ¼	56 ¼
900	900	810000		5 ½	6 ½	7 ½
1000	1000	1000000{0}		×	×	×
				5 ½	6 ½	7 ½

121	144	169	196	225	256
1 1	1 2	1 3	1 4	1 5	1 6
×	×	×	×		
1 1	1 2	1 3	1 4	1 5	1 6
4	0	7	7	0	4
289	**324**	**361**	**400**	**441**	**489**
1 7	1 8	1 9	2 0	2 1	2 2
1 7	1 8	1 9	2 0	2 1	2 2
10 [sic]	0	1	4	0	7
529	**576**	**625**	**676**	**729**	**784**
2 3	2 4	2 5	2 6	2 7	2 8
2 3	2 4	2 5	2 6	2 7	2 8
7	0	4	1	0	1
841	**900**	**961**	**1024**	**1089**	**1156**
2 9	3 0	3 1	3 2	3 3	3 4
2 9	3 0	3 1	3 2	3 3	3 4
4	0	7	7	0	4

(fol. 10ʳ)

1225	1296	1369	1444	1521	1600
3 5 3 5 1	3 6 3 6 0	3 7 3 7 1	3 8 3 8 4	3 9 3 9 0	4 0 4 0 7
1681	1764	1849	1936	2025	2116
4 1 4 1 7	4 2 4 2 0	4 3 4 3 4	4 4 4 4 1	4 5 4 5 0	4 6 4 6 1
2209	2304	2401	2500	2601	2704
4 7 4 7 4	4 8 4 8 0	4 9 4 9 7	5 0 5 0 7	5 1 5 1 0	5 2 5 2 4
2809	2916	3025	3136	3249	3364
5 3 5 3 1	5 4 5 4 0	5 5 5 5 1	5 6 5 6 4	5 7 5 7 0	5 8 5 8 7
3481	3600	3721	3844	3969	4096
5 9 5 9 7	6 0 6 0 0	6 1 6 1 4	6 2 6 2 1	6 3 6 3 0	6 4 6 4 1

(fol. 10ᵛ)

4225	4396	4489	4624	4761	4900
6 5 6 5 4	6 6 6 6 0	6 7 6 7 7	6 8 6 8 7	6 9 6 9 0	7 0 7 0 4
5041	5184	5329	5476	5625	5776
7 1 7 1 1	7 2 7 2 0	7 3 7 3 1	7 4 7 4 4	7 5 7 5 0	7 6 7 6 7
5929	6084	6241	6400	6561	6724
7 7 7 7 7	7 8 7 8 0	7 9 7 9 4	8 0 8 0 1	8 1 8 1 0	8 2 8 2 1
6889	7056	7225	7396	7569	7744
8 3 8 3 4	8 4 8 4 0	8 5 8 5 7	8 6 8 6 7	8 7 8 7 0	8 8 8 8 4
7921	8100	8281	8464	8649	8836
8 9 8 9 1	9 0 9 0 0	9 1 9 1 1	9 2 9 2 4	9 3 9 3 0	9 4 9 4 7

(fol. 11ʳ)

9025	9216	9409	9604	9801	10000
9 5 9 5 7	9 6 9 6 0	9 7 9 7 4	9 8 9 8 1	9 9 9 9 0	100 100 1
12100	14400	16900	19600	22500	25600
110 110	120 120	130 130	140 140	150 150	160 160
28900	32400	36100	40000	44100	48400
170 170	180 180	190 190	200 200	210 210	220 220
52900	57600	62500	67600	72900	78400
230 230	240 240	250 250	260 260	270 270	280 280
84100	90000	96100	102400	108900	115600
290 290	300 300	310 310	320 320	330 330	340 340

(fol. 11ᵛ)

122500	129600	136900	144400	152100	160000
350 350	360 360	370 370	380 380	390 390	400 400
168100	176400	184900	193600	202500	211600
410 410	420 420	430 430	440 440	450 450	460 460
220900	230400	240100	250000	260100	270400
470 470	480 480	490 490	500 500	510 510	520 520
280900	291600	302500	313600	324900	336400
530 530	540 540	550 550	560 560	570 570	580 580
348100	360000	372100	384400	396900	409600
590 590	600 600	610 610	620 630	630 630	640 640

(fol. 12ʳ)

2	10	20		10	100
3	10	30		10	90
4	10	40		10	80
5	10	50		10	70
6	10	60		10	60
7	10	70		10	50
8	10	80		10	40
9	10	90		10	30
10	10	100		10	20
					{100}

422500	435600	448900	462400	476100	490000
650	660	670	680	690	700
650	660	670	680	690	700
504100	518900	532900	547600	562500	577600
710	720	730	740	750	760
710	720	730	740	750	760
592900	608400	624100	640000	656100	672400
770	780	790	800	810	820
770	780	790	800	810	820
688900	705600	722500	739600	756900	774400
830	840	850	860	870	880
830	840	850	860	870	880
792100	810000	828100	846400	864900	883600
890	900	910	920	930	940
890	900	910	920	930	940
902500	921600	940900	960400	980100	1000000
950	960	970	980	990	1000
950	960	970	980	990	1000

(fol. 12ᵛ)

72¼	90¼	110¼	132¼	156¼	182¼
8½ 8½	9½ 9½	10½ 10½	11½ 11½	12½ 12½	13½ 13½
210¼ 14½ 14½	240¼ 15½ 15½	272¼ 16½ 16½	306¼ 17½ 17½	342¼ 18½ 18½	380¼ 19½ 19½

[8. Divisions *a regolo*]

[8.1] Qui appresso insegneremo partire inne numeri che sono più necessarii secondo che porremo qui appresso. Et prima cominciamo a partire in dui, et poi in tre, et poi in quactro, et poi in cinque et ~~secondo~~ sequentemente, secondo vedete soctoscripto.

[8.1] Next to here we shall teach to divide in the numbers that are most necessary according to what we put next to here. And first we begin by dividing in two, and then in three, and then in four, and then in five, and so on, according to what you see written below.

2	473212345678910		643912345678910	5
0	236606172839455	0	128782469135782	
1	118303086419727	2	025756493827156	
1	559151543209863	1	405151298765431	
1	779575771604931	1	281030259753086	
1	889787885802465	1	256206051950617	
1	944893942901232	2	251241210390123	
0	972446971450616	3	450248242078024	
0	486223485725308	4	690049648415604	
0	243111742862654	4	938009929683120	
3	346712345678910		793612345678910	6
2	115537448559636[a]	2	132268724279818	
2	705179149419878[b]	0	355378120713303	
1	901729716506629[c]	3	059229686785550	
2	633909905502209	2	508371614464258[j]	
1	877969968500736	0	418061935746043	
1	625989989500245	1	069676989291007	
1	541996663166748	5	178279498215167	
1	513998887722249	1	863046583035861	
2	504666299240749[d]	1	310507763839310	
4	586512345678910		846512345678910	7
2	146628086419727	1	120930335096987	
3	536657021604231[e]	2	160132905013855	
3	884164255401232	2	308590415001979	
0	971041063850308	3	329798630714568	
0	242760265962577	5	475685518678509[k]	
1	060690066490644	6	782240788381929[l]	
0	265172516622671[f]	3	968891541197418	
3	066293149155667[g]	4	566984505885342[m]	
3	766574282288916[h]	2	652426356983680[n]	

[a] This results from 346612345678910÷3.

[b] Error for 705179149519878

[c] This results from 2705189149519878÷3 – whence follows that the error in the previous line is a copying error, while those committed here are computational.

[d] This results from 1513998897722249÷3

[e] Error for 536657021604931, which is used for the following division and hence is no copying error.

[f] Error for 265172516622661.

[g] Copying error for 066293129155667.

[h] This results from 3066297129155667÷4.

[j] Error for 509871614464258.

[k] This results from 3329798630749568÷7.

[l] This results from 5475685518673509÷7.

[m] This results from 3968891541197396÷7.

[n] Copying error for 652426357983680, the result of 4566984595885362÷7.

(fol. 13^r)

8	675712345678910		876512345678910	11
6	084464043209863	0^k	079682940516264	
7	760558005401232	2	552698449137842	
0	970068500675154^a	5	232063495376167	
2	121258562584394	10	475642135943289^l	
2	265157320323049	6	952331103267571	
1	283144665040381	10	632030100297051	
5	160293083130047^b	6	966548190936095	
7	645036635391255	3	633322562812372	
7	956629579423906^c	1	330302051164761	
9	598412345678910		765412345678910	12
0	066490260630990	10	063784362139075^m	
6	007387806736776	11	838648696844922	
0	667387534081864^d	2	986554058070410	
4	074154170453540	10	248879504839200	
8	452672685605948^e	0	854073292069100^n	
2	916963631733994^f	8	071172774339091	
0	324107070192666	3	652597731144929^p	
8	036011896676962^g	0	306048977599577^q	
6	892890210741884	5	025504081466631	
10	987612345678910		654312345678910	13
~~0~~	~~066490260630990~~^{h}			
0	098761234567891	0^r	050331718898377	
1	109876123456789^j	2	680787055299875^s	
9	100987612345678	2	206214388869221	
8	910098761234567	7	169708791451478^t	
7	891009876123456	12	551516060880882	
6	789100987612345	7	965501235452375	
5	678910098761234	4	612730864265567	
4	567891009876123	4	354825451097351	
3	456789100987612	9	334986573161334	

^a   This results from 760548005401232÷8.

^b   Error for 1603930831300473.

^c   Error for 955629579423906

^d   Error for 667487534081864.

^e   This results from 4074054170453540 ÷ 9.

^f   This results from 84252672685605948÷9.

^g   This it the result of 324107070092666÷9.

^h   Repeats the first line from the divisions by 9.

^j   Copying error for 009987612345678.

^k   Copying error for 6.

^l   Copying error for 475642135943287.

^m   This results from 765412345668910÷12.

ⁿ   This results from 10248879504829200÷12.

^p   Copying error for 672597731144929.

^q   This results from 3672587731144929÷12.

^r   Copying error for 9.

^s   This results from 8850231718898377÷13.

^t   This results from 2206214288869221÷13.

(fol. 13ᵛ)

14	543212345678910		456712345678910	17
12	038800881834207	4	026159559745818[e]	
5	859914348702443	1	236832855967401[f]	
1	418565310621603	0	072754873874553	
11	101326093615828	0	004279698463209	
0	792951863829702	8	000251805791943[g]	
12	056639418844985[a]	11[h]	470603047399526	
13	861188529917498	15	674741355729381[j]	
0	990084894994107	16	922043609160551	
3	070720349642436	11	995414329950620	
15	432212345678910		567812345678910	18
5	028807489711927	0	031545130315495	
12	335253832647461	15	001752618350860[k]	
11	822350255509830	4	833430701019492	
10	788156683633988[b]	12	268523927833860[l]	
3	719210445575599	2	681584662101880[m]	
4	247947363038373	13	148976925672326	
13	283196490869224	10	730498773648462[n]	
14	885546432724614	16	506138820758247[p]	
14	992369762181640	1	922007712264347	
16	345612345678910		678912345678910	19
14	021600771604931[c]	12	035733228719942[q]	
3	876350048225308	2	633459590985260	
12	242271878014081	5	138693136367645[r]	
0[d]	765141992375880	17	270452796124612[s]	
8	047821374523492	16	908918568217084[t]	
4	502988835907718	5	889943082537741	
6	281436802244232	12	309997004344091	
8	392589800140264	15	647894579176004	
8	524536862508766	4	823573398904000	

^a This results from 792951863829802÷14.
^b This results from 11822350254509830÷15.
^c At first written 021600771649, then the scribe discovers he has omitted a digit and corrects and completes as 021600771604931.
^d Error for 1.
^e Copying error for 026158549745818, which results from 444695345678910÷17.
^f Copying error for 236832855867401.
^g This results from 4280698463209÷17.
^h Error for 1.
^j Copying error for 674741355729383.
^k This results from 031547130315495÷18.
^l This results from 4833430701009492÷18.
^m Copying error for 681584662101881, which results from 12268523917833860÷18.
ⁿ This results from 13148977925672326÷18.
^p Copying error for 596138820758247.
^q Copying error for 035732228719942.
^r Copying error for 138603136367645.
^s This results from 5138693126367645÷19.
^t This results from 17269452796124612÷19.

(fol. 14ʳ)

23	789112345678910		888812345678910	41
4	034309232420822	38	216312496231192[u]	
21	175404749235689[a]	4	932104938932468[w]	
12	920669771905899[b]	26	120295242412962[x]	
4	561724772682865[c]	39	637080396156453[y]	
16	198335859689863[d]	0	966758058442833[z]	
0	904275472160081[e]	4	023559464840009[A]	
6	039316324876525	7	098136133288782[B]	
29	892212345678910		777712345678910	43
5	031765942954445[f]	31	018086356876253[C]	
9	173474687688084	3	721350845508750	
23	316326334058209[g]	38	086554670825784[D]	
11	804011252891662[h]	2	885733829554279[E]	
19	407034870789367	31	067110089059401	
23	669208898923736[j]	4	722490932303079[F]	
9	471352003411163[k]	21	109360254239606[G]	
31	934512345678910		666612345678910	47
24	030145559537706[l]	29	014283241397423[H]	
28	775165985791538	41	617323047689306	
6	928231154380372[m]	0	885474745695518[J]	
20	223491295302592[n]	29	018839888206287	
0	652370654687180[p]	39	617422125280984	
12	021044214667328	18	842923875005978	
16	387775619824752[q]	25	400913273936299[K]	
37	999912345678910		555512345678910	48
22	027024657991324[r]	30	011573173868310	
10	595324990756522	6	625241107788923	
15	286360104885311[s]	43	138025854328935[L]	
30	413144868510413	7	898708871965186	
36	821976881040821[t]	2	164556434832608	
23	995188564352454	16	045094925725679	
23	648518609847363	15	334272810952618	

a   Error for 175404749235687.

b   Copying error for 920669771705899.

c   This results from 12919669771705899÷23.

d   Copying error for 198335859681863.

e   Error for 7042754721600081.

f   Copying error for 030765942954445

g   This results from 9173463687688084.

h   This results from 23316326333858209÷29.

j   Copying error for 669208098923736, itself an error for 669208098992736

k   This results from 13669208098923736÷29.

l   This results from 934512345668910÷31.

m   This results from 28775165785791538÷31.

n   This results from 6928230154380372÷31.

p   This results from 20223490295302592÷31.

q   This results from 12021044214567328÷31.

r   This results from 999912345679010÷37.

s   Copying error for 286360134885311.

t   This results from 30413144598496513÷37.

"   This results from 8868812345478910 : 41.

w   This results from 382163024962311192÷41.

x   This results from 4932104938931468÷41.

y   This results from 26120296242414612÷41.

z   This results from 39637080396156153÷41.

A   This results from 965938058440373÷41.

B   This results from 4023581464840069÷41.

C   This results from 777713345678910÷43.

D   This results from 3721850845508750÷43.

E   Copying error for 885733829554274, which results from 38086554670833784÷43.

F   This results from 31067110089032401÷43.

G   This results from 4702490932303079÷43.

H   Copying error for 014183241397423.

J   This results from 41617313047689346÷47.

K   This results from 18842923875006078÷47.

L   This results from 6625241007788923÷48.

(fol. 14ᵛ)

[9. Graphic Schemes Illustrating the Arithmetic of Fractions]

Multiplicare numeri rotti l'uno
contra l'altro et vedere quello che
fa. E prima diciamo chossì

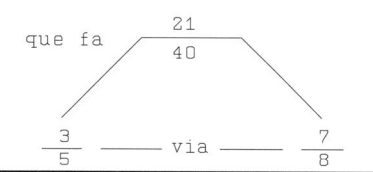

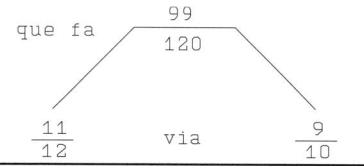

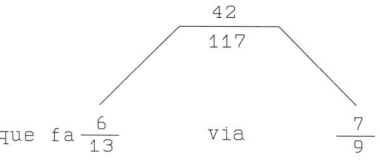

Giongi inseme numero rocto a uno
numero rotto

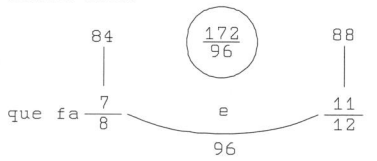

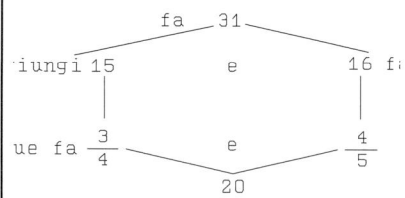

Parti 31 in 20 che ne viene uno e
$^{11}/_{20}$. Et cotanto fa gionti inseme $^3/_4$
e $^4/_5$.

Quanto è più l'uno numero rocto che
l'altro numero rocto, cioè quale è
più

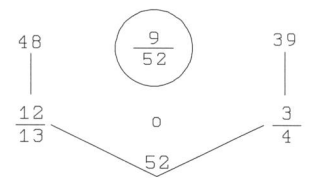

Trai l'uno numero dell'altra, cioè,
el minore del magiore, 39 de 48,
resta {resta} 9, parti 9 in 52, che
ne vene como tu vedi $^9/_{52}$. Et cotanto è più $^{12}/_{13}$ che $^3/_4$.

Parti numero rotto in uno altro
numero rotto, cioè

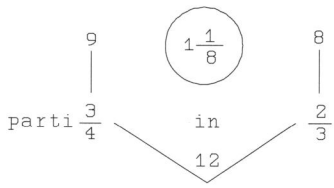

Trai l'uno numero rotto dell'altro
numero rotto e dire quello che remane

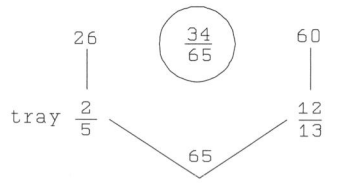

To multiply broken numbers one against the other and to see what it makes. And first we say thus

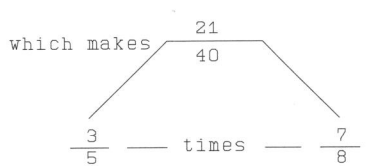

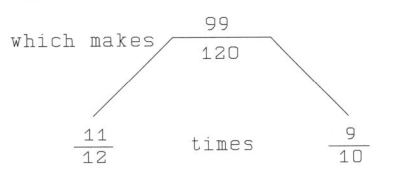

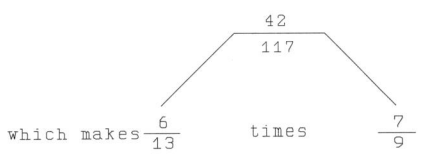

Join together a broken number to a broken number

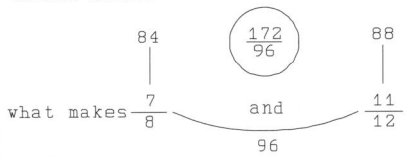

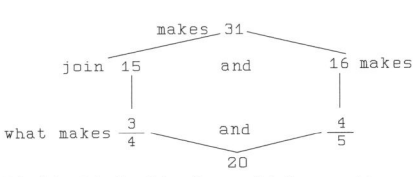

Divide 31 in 20, from which results one and $^{11}/_{20}$. And as much makes joined together $^3/_4$ and $^4/_5$

How much one broken number is more than another broken number, that is, which is more

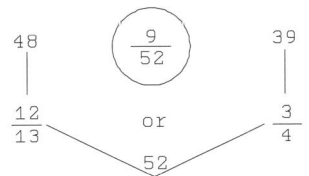

Detract one number from the other, that is, the smaller from the larger, 39 from 48, 9 is left, divide 9 in 52, from which results as you see $^9/_{52}$. And as much is $^{12}/_{13}$ more than $^3/_4$

Divide a broken number in another broken number, that is

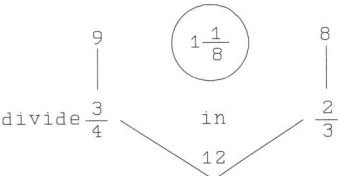

Detract one broken number from another broken number and say what remains

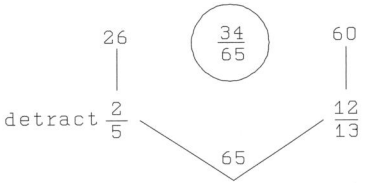

(fol. 15ʳ)

[10. Examples Explaining the Arithmetic of Fractions]

[10.1] Abiamo dicto dele multiplicationi et dele divisioni et de tucto quello che intorno a ciò è di necessità. Ora lasciamo questo, et dirremo per propria et legitima forma et regola sopre tucti manere de numeri rocti, sicomo proponemmo denanti nel prolagho, perciò che danno argomento ale altre ragioni et ⟨senza⟩$^{M+F}$ esse non si po fare soctilmente ne inprendere questa arte.

[10.1] We have spoken about the multiplications and the divisions and of all that is necessary concerning this. Now we leave this, and we shall speak in proper and legitimate diagram and rule about all routines about broken numbers, such as we proposed before in the prologue, since they give tools for the other computations, and ⟨without⟩$^{M+F}$ them this art cannot be subtly exercised nor learnt.

[10.2] Primamente comenzaremo nel nome di Dio. Et diremo così, dimme quando è gionto inseme $\frac{1}{2}$ e $\frac{1}{3}$. Fa così, et di' così, $\frac{1}{2}$ e $\frac{1}{3}$ si trova in sei, perciò che 2 via 3 fa 6. Et piglia el mezzo e lo terzo di 6, che fanno gionti inseme 5, et parti 5 in 6, che ne vene $\frac{5}{6}$, et cotanto fa gionti inseme $\frac{1}{2}$ e $\frac{1}{3}$. Et in questo modo se fanno tucte le simile ragioni de qualunqua rocto se fosse.

[10.2] Firstly we shall begin in the name of God. And we shall say thus, say me, how much is joined together $\frac{1}{2}$ and $\frac{1}{3}$. Do thus, and say thus, $\frac{1}{2}$ and $\frac{1}{3}$ one finds in six, since 2 times 3 makes 6. And seize the half and the third of 6, which joined together make 5, and divide 5 in 6, from which results $\frac{5}{6}$, and as much makes $\frac{1}{2}$ and $\frac{1}{3}$ joined together. And in this way all the similar computations are made with whatever fraction it were.

[10.3] Dime quando è gionto inseme $\frac{1}{3}$ e $\frac{1}{4}$. Fa così como di sopre t'ò dicto, et di' così, $\frac{1}{3}$ e $\frac{1}{4}$ si trova in 12. Et poi piglia $\frac{1}{3}$ di 12 che è 4, et el $\frac{1}{4}$ di 12 si è 3. Agiongni inseme 4 e 3, che sonno 7, et poi parti 7 in 12, che ne vene $\frac{7}{12}$. Et cotanto fa gionto inseme $\frac{1}{3}$ e $\frac{1}{4}$, cioè $\frac{7}{12}$. Et è facta.

[10.3] Say me, how much is joined together $\frac{1}{3}$ and $\frac{1}{4}$. Do thus as I have said to you above, and say thus, $\frac{1}{3}$ and $\frac{1}{4}$ one finds in 12. And then seize $\frac{1}{3}$ of 12 which is 4, and the $\frac{1}{4}$ of 12 is 3. Join together 4 and 3, which are 7, and then divide 7 in 12, from which results $\frac{7}{12}$. And as much makes joined together $\frac{1}{3}$ and $\frac{1}{4}$, that is, $\frac{7}{12}$. And it is done.

10.4 **Dime** quando è gionto inseme $\frac{1}{2}$ $\frac{1}{3}$ $\frac{1}{4}$. Fa così como te dico de sopre, et di', $\frac{1}{2}$ e $\frac{1}{3}$ e $\frac{1}{4}$ se trova in 24. Et poi piglia el $\frac{1}{2}$ di 24, che sonno 12, et poi piglia $\frac{1}{3}$ de 24 che sonno 8. Et poi piglia el $\frac{1}{4}$ de 24, che sono 6. Ora agiogni inseme questi tre numeri, cioè 12, 8 e 6, che fanno 26. Ora parti 26 in 24, che ne vene uno et $\frac{2}{24}$. Et cotanto fa gionto inseme $\frac{1}{2}$ $\frac{1}{3}$ $\frac{1}{4}$, cioè $1\frac{2}{24}$, et così ⟨se⟩ fanno le simili ragioni.

10.4 Say me, how much is joined together $\frac{1}{2}$ $\frac{1}{3}$ $\frac{1}{4}$. Do thus I say to you above, and say, $\frac{1}{2}$ and $\frac{1}{3}$ and $\frac{1}{4}$ one finds in 24. And then seize the $\frac{1}{2}$ of 24, which are 12, and then seize the $\frac{1}{3}$ of 24, which are 8. And then seize $\frac{1}{4}$ of 24, which are 6. Now join together these three numbers, that is, 12, 8 and 6, which make 26. Now divide 26 in 24, from which results one and $\frac{2}{24}$. And as much makes joined together $\frac{1}{2}$ $\frac{1}{3}$ $\frac{1}{4}$, that is, $1\frac{2}{24}$, and thus the similar computations are done.

10.5 **Dime** quando è gionto inseme $\frac{1}{4}$, $\frac{1}{5}$ e $\frac{1}{6}$ e $\frac{1}{10}$. Fa così, trova uno numero che abia $\frac{1}{4}$ e $\frac{1}{5}$ e $\frac{1}{6}$ e $\frac{1}{10}$, et questo numero è 60. Ora piglia el $\frac{1}{4}$ de 60, che sonno 15. Et poi piglia el $\frac{1}{5}$, che sonno 12, et poi piglia el $\frac{1}{6}$, che sonno 10. Et poi piglia el $\frac{1}{10}$, che sonno 6. Ora agiogni inseme questi numeri, che sonno 43 in tucto. Ora parti 43 in 60, che ne vene $\frac{43}{60}$. Et cotanto fanno gionti inseme $\frac{1}{4}$ e $\frac{1}{5}$ e $\frac{1}{6}$ e $\frac{1}{10}$. Et così vedi apertamente che si fanno tucte queste ragioni a uno modo.

10.5 Say me, how much is joined together $\frac{1}{4}$, $\frac{1}{5}$ and $\frac{1}{6}$ and $\frac{1}{10}$. Do thus, find a number that possesses $\frac{1}{4}$ and $\frac{1}{5}$ and $\frac{1}{6}$ and $\frac{1}{10}$, and this number is 60. Now seize the $\frac{1}{4}$ of 60, which are 15. And then seize the $\frac{1}{5}$, which are 12, and then seize the $\frac{1}{6}$, which are 10. And then seize the $\frac{1}{10}$, which are 6. Now join together these numbers, which are 43 in all. Now divide 43 in 60, from which results $\frac{43}{60}$. And as much make $\frac{1}{4}$ and $\frac{1}{5}$ and $\frac{1}{6}$ and $\frac{1}{10}$ joined together. And thus you see clearly that all these computations are done in one and the same way.

10.6 **Et** sappi che quando tu voli scrivere uno rocto, et diciamo che fosse uno quarto, scrivi sempre uno de sopre et el quarto de **(fol. 15')** sotto, et poi fa una verga in mezzo, et cossì fa ~~depoi~~^di tutti^ li rotti che tu voli scrivere como tu vedi che per lo adercto abiamo posto et per lo innanzi porremo.

10.6 And know that when you want to write a fraction, and let us say it were one fourth, always write one above and the fourth **(fol. 15')** below, and then make a dash in between, and do thus with all the fractions that you want to write as you see that we have written behind and shall do henceforth.

$^{10.7}$ **D**ime quando sonno gionti asseme $\frac{2}{3}$ e $\frac{3}{4}$. Fa così, di', $\frac{2}{3}$ e $\frac{3}{4}$ si trovano in 12, et poi prendi $\frac{2}{3}$ de 12, che sonno 8, et prendi $\frac{3}{4}$ de 12, che sonno 9. Poi agiongi inseme 8 e 9, che sonno 17, et parti 17 in 12, che ne vene 1 sano et più $\frac{5}{12}$. Et così diremo che $\frac{2}{3}$ e $\frac{3}{4}$ sonno gionti inseme j et $\frac{5}{12}$, et sta bene.

$^{10.7}$ Say me, how much are joined together $\frac{2}{3}$ and $\frac{3}{4}$. Do thus, say, $\frac{2}{3}$ and $\frac{3}{4}$ are found in 12, and then take $\frac{2}{3}$ of 12, which are 8, and take $\frac{3}{4}$ of 12, which are 9. Then join together 8 and 9, which are 17, and divide 17 in 12, from which results 1 integer and added $\frac{5}{12}$. And thus we shall say that $\frac{2}{3}$ and $\frac{3}{4}$ joined together are j and added $\frac{5}{12}$, and it goes well.

$^{10.8}$ **D**ime quando sonno gionti inseme $\frac{4}{5}$ et $\frac{5}{6}$. Di' così como di sopre abiamo dicto, $\frac{4}{5}$ e $\frac{5}{6}$ si trovano in 30. Ora prendi $\frac{4}{5}$ di 30, che sonno 24, et prendi $\frac{5}{6}$ de 30, che sonno 25. Ora agiogni inseme 24 et 25, che fanno 49, et parti 49 in 30, che ne vene uno sano et $\frac{19}{30}$, et cotanto fanno agionti inseme $\frac{4}{5}$ e $\frac{5}{6}$, cioè j et $\frac{19}{30}$, et sta bene.

$^{10.8}$ Say me, how much are joined together $\frac{4}{5}$ and $\frac{5}{6}$. Say thus as we have said above, $\frac{4}{5}$ and $\frac{5}{6}$ are found in 30. Now take $\frac{4}{5}$ of 30, which are 24, and take $\frac{5}{6}$ of 30, which are 25. Now join together 24 and 25, which make 49, and divide 49 in 30, from which results one integer and $\frac{19}{30}$, and as much make $\frac{4}{5}$ and $\frac{5}{6}$ joined together, that is, j and $\frac{19}{30}$, and it goes well.

$^{10.9}$ **D**ime quando sonno agionti inseme $\frac{1}{2}$ $\frac{1}{3}$ $\frac{1}{4}$. Fa così et di', $\frac{1}{2}$ et $\frac{1}{3}$ et $\frac{1}{4}$ se trova in 12. Ora prendi el mezzo de 12, che è 6, et prendi el terzo de 12, che è 4, et prendi el quarto de 12, che è 3. Agiogni inseme, che fanno 13, et parti 13 in 12, che ne vene uno sano et $\frac{1}{12}$. Et così ⟨sonno⟩ gionti inseme $\frac{1}{2}$ $\frac{1}{3}$ $\frac{1}{4}$.

$^{10.9}$ Say me, how much are $\frac{1}{2}$ $\frac{1}{3}$ $\frac{1}{4}$ joined together. Do thus and say, $\frac{1}{2}$ and $\frac{1}{3}$ and $\frac{1}{4}$ one finds in 12. Now take the half of 12, which is 6, and take the third of 12, which is 4, and take the fourth of 12, which is 3. Join together, which make 13, and divide 13 in 12, from which results one whole and $\frac{1}{12}$. And thus $\frac{1}{2}$ $\frac{1}{3}$ $\frac{1}{4}$ ⟨are⟩ joined together.

$^{10.10}$ Ancora diremo, giongi inseme $\frac{2}{3}$ $\frac{3}{4}$ $\frac{7}{8}$ $\frac{5}{12}$. Trova uno numero che abia $\frac{1}{3}$ $\frac{1}{4}$ $\frac{7}{8}$ $\frac{5}{12}$, et questo numero si è ancora 24. Ora prendi $\frac{2}{3}$ de 24, che sonno 16. Et prendi $\frac{3}{4}$ de 24, che sonno 18, et prendi $\frac{7}{8}$ de 24, che sonno

$^{10.10}$ Further we shall say, join together $\frac{2}{3}$ $\frac{3}{4}$ $\frac{7}{8}$ $\frac{5}{12}$. Find a number that possesses $\frac{1}{3}$ $\frac{1}{4}$ $\frac{7}{8}$ $\frac{5}{12}$, and this number is again 24. Now take $\frac{2}{3}$ of 24, which are 16. And take $\frac{3}{4}$ of 24, which are 18, and take $\frac{7}{8}$ of 24, which are

21, et prendi $\frac{5}{12}$ de 24, che sonno 10. Agiongi ogni cosa inseme, che fanno 65. Ora parti 65 in 24, che ne vene doi interi et $\frac{17}{24}$, et sta bene. Et così diciamo che gionti inseme $\frac{2}{3}$ $\frac{3}{4}$ $\frac{7}{8}$ $\frac{5}{12}$ fanno 2 saldi et $\frac{17}{24}$. Et così fa de tucti rotti de quanti fosseno, che tucti se fanno per una regola.

21, and take $\frac{5}{12}$ of 24, which are 10. Join everything together, which make 65. Now divide 65 in 24, from which results two integer and $\frac{17}{24}$, and it goes well. And thus we say that $\frac{2}{3}$ $\frac{3}{4}$ $\frac{7}{8}$ $\frac{5}{12}$ joined together make 2 unbroken and $\frac{17}{24}$. And do thus with all fractions as many as they might be, as all are done by one and the same rule.

10.11 Abiamo dicto del giongimento de numeri rotti. Ora dirremo **(fol. 16ʳ)** de trare l'uno numero rotto dell'altro, et sapere quanto è lo rimanente. Et primamente dirremo così, tray $\frac{2}{3}$ de $\frac{11}{12}$ et dimme quanto è rimanente. Fa così, al modo che tu ài facto li rotti passati, cioè che tu trovi uno numero che abia $\frac{1}{3}$ e $\frac{1}{12}$, et questo numero è 12. Ora piglia $\frac{2}{3}$ de 12, che è 8, et poi piglia $\frac{11}{12}$ de 12, che sonno 11, et poi tray 8 de 11, et reman'te 3, et questi 3 sonno duodecimi, et cotanto remane tracto $\frac{2}{3}$ de $\frac{11}{12}$. Remane $\frac{3}{12}$, et questi $\frac{3}{12}$ schifa, cioè, dà el terzo a tri si remane uno. Et dà el terzo a 12 remane 4. Et però questi $\frac{3}{12}$ sonno $\frac{1}{4}$. Cotanto rimane, et è facta.

10.11 We have spoken of the joining of broken numbers. Now we shall speak **(fol. 16ʳ)** of detracting one broken number from the other, and knowing how much is the remainder. And firstly we shall say thus, detract $\frac{2}{3}$ from $\frac{11}{12}$ and say me how much is the remainder. Do thus, in the way you have made the previous fractions, that is that you find a number which possesses $\frac{1}{3}$ and $\frac{11}{12}$, and this number is 12. Now seize $\frac{2}{3}$ of 12, which is 8, and then seize $\frac{11}{12}$ of 12, which are 11, and then detract 8 from 11, and 3 remains for you, and these 3 are twelfths, and as much remains when $\frac{2}{3}$ is detracted from $\frac{11}{12}$. $\frac{3}{12}$ remains, and simplify these $\frac{3}{12}$, that is, give the third to three, one remains. And give the third to 12, 4 remains. And therefore these $\frac{3}{12}$ are $\frac{1}{4}$. As much remains, and it is done.

10.12 Ancora diremo così, tray $\frac{2}{7}$ de $\frac{9}{11}$. Ancora fa così, trova uno numero che abia settimo e undecimo, et questo si è 77, però che 7 via 11 fa 77. Ora piglia $\frac{2}{7}$ de 77, che sonno 22. Et poi piglia $\frac{9}{11}$ de 77, che sonno 63, et tray 22 de 63. Reman'ti 41, et questi sonno $\frac{41}{77}$. Et però diciamo così, che tracti $\frac{2}{7}$

10.12 Further we shall say thus, detract $\frac{2}{7}$ from $\frac{9}{11}$. Again do thus, find a number that possesses seventh and eleventh, and this is 77, because 7 times 11 makes 77. Now seize $\frac{2}{7}$ of 77, which are 22. And then seize $\frac{9}{11}$ of 77, which are 63, and detract 22 from 63. 41 remains for you, and these

de $\frac{9}{11}$ remane $\frac{41}{77}$. Et è facta, et così se fa le simile ragioni.

are $\frac{41}{77}$. And therefore we say thus, that when $\frac{2}{7}$ are detracted from $\frac{9}{11}$, $\frac{41}{77}$ remains. And it is done, and thus one makes the similar computations.

---

^{10.13} Ancora diremo, trai $\frac{1}{13}$ de $\frac{1}{7}$. Fa como te dico de sopra, trova uno numero che abia tredecimo e settimo, et questo numero si è 91, però che 7 via 13 fa 91. Ora arecha a nonantuneximi $\frac{1}{13}$ e $\frac{1}{7}$, et di' così, quanto è el tridecimo de 91, et trovi che è 7. Et questi sono $\frac{7}{91}$. Et poi sappi quanto è il settimo de 91, et trovi che è 13. Ancora $\frac{13}{91}$. Ora trai $\frac{7}{91}$ de $\frac{13}{91}$, et reman'ti $\frac{6}{91}$. Et cotanto te remane, tracto che tu hai $\frac{1}{13}$ de $\frac{1}{7}$, cioè $\frac{6}{91}$ appunto. E sta bene.

^{10.13} Further we shall say, detract $\frac{1}{13}$ from $\frac{1}{7}$. Do as I say to you above, find a number that possesses thirteenth and seventh, and this number is 91, because 7 times 13 makes 91. Now bring $\frac{1}{13}$ and $\frac{1}{7}$ to ninety-oneths, and say thus, how much is the thirteenth of 91, and you find that it is 7. And these are $\frac{7}{91}$. And then know how much is the seventh of 91, and you find that it is 13. Again, $\frac{13}{91}$. Now detract $\frac{7}{91}$ from $\frac{13}{91}$, and $\frac{6}{91}$ remains for you. And as much remains for you, when you have detracted $\frac{1}{13}$ from $\frac{1}{7}$, that is, precisely $\frac{6}{91}$. And it goes well.

---

^{10.14} **A**biamo dicto del giongimento et del sobtraemento de numeri rotti. Ora diremo che parte è l'uno numero dell'altro. Et primamente diremo così, quanto è $\frac{1}{3}$ de $\frac{5}{7}$. Fa così, como tu vidi, $\frac{1}{3}$ se scrive, et ponse uno de sopre et tre de sotto, et una verga **(fol. 16ᵛ)** ⟨in⟩^{M+F} mezzo, et $\frac{5}{7}$ se scrive 5 de sopre et el 7 de sotto, et una vergha in mezzo. Ora multiplica le parti de sopre, cioè l'uno et el cinque, l'uno contra all'altro, che fa 1 via 5, fa 5. Et similmente multiplica le parti de sotto, cioè el 3 contra el 7, che fa 21. Et poi poni 5 de sopre a 21, et fa una vergha in mezzo, et serà $\frac{5}{21}$. Et $\frac{5}{21}$ dirremo che serà el terzo de $\frac{5}{7}$, et sta bene. Et così fa l'altre, cioè che ⟨multipli-

^{10.14} We have spoken of the joining and the detraction of broken numbers. Now we shall say which part one number is of the other. And firstly we shall say thus, how much is $\frac{1}{3}$ of $\frac{5}{7}$. Do thus, as you see, $\frac{1}{3}$ is written and one is put above and three beneath, and a dash **(fol. 16ᵛ)** ⟨in⟩^{M+F} between, and $\frac{5}{7}$ is written 5 above and the 7 beneath, and a dash in between. Now multiply the parts above, that is, the one and the five, one against the other, which makes 1 times 5, it makes 5. And similarly multiply the parts beneath, that is, the 3 against the 7, which makes 21. And then put 5 above 21, and make a dash in between, and it will be $\frac{5}{21}$. And we shall say that the third of $\frac{5}{7}$

chi⟩ sempre le parti de sopre l'una contra l'altra, et similmente quelle de sotto, como tu vedi che abiamo fatto testé.

will be $\frac{5}{21}$, and it goes well. And do thus the other (computations), that is that you always ⟨multiply⟩ the parts above one against the other, and similarly those beneath, as you see that we have done just now.

$^{10.15}$ **Dime** quanto ^sonno^ $\frac{3}{4}$ de $\frac{9}{10}$. Multiplica le parti de sopre, cioè 3 via 9, che fa 27, et poi multiplica le parti de sotto, cioè 4 via 10, che fa 40, et ài $\frac{27}{40}$. Et diremo che $\frac{3}{4}$ de $\frac{9}{10}$ sonno $\frac{27}{40}$, et sta bene. Et così se fa de qualunqua rotto se fosse.

$^{10.15}$ Say me, how much are $\frac{3}{4}$ of $\frac{9}{10}$. Multiply the parts above, that is, 3 times 9, which makes 27, and then multiply the parts beneath, that is, 4 times 10, which makes 40, and you get $\frac{27}{40}$. And we shall say that $\frac{3}{4}$ of $\frac{9}{10}$ are $\frac{27}{40}$, and it goes well. And thus is done with whatever fraction it might be.

$^{10.16}$ **Ora** mostraremo quale è più l'uno numero rotto che l'altro, et quanto è più. Et prima comenciamo così, dime quale è più, et quanto, o $\frac{2}{3}$ o $\frac{7}{8}$. Fa così, di', terzo et ottavo se trova in 24, et questi sonno vintiquattressimi. Ora piglia $\frac{2}{3}$ de 24, che sonno 16. Et piglia $\frac{7}{8}$ de 24, che sonno 21. Et vedi apertamente che $\frac{7}{8}$ sonno più che $\frac{2}{3}$ 5, et questi 5 sonno vintiquactreximi. Et però diremo che $\frac{7}{8}$ sonno più che $\frac{2}{3}$ $\frac{5}{24}$. Et sta bene, et così se fanno tucte le altre simile, de qualunqua rotto se fosse.

$^{10.16}$ Now we shall show which is larger of one broken number and another, and how much larger it is. And first we begin thus, say me, which is larger, and how much, either $\frac{2}{3}$ or $\frac{7}{8}$. Do thus, say, third and eighth one finds in 24, and these are twenty-fourths. Now seize $\frac{2}{3}$ of 24, which are 16. And seize $\frac{7}{8}$ of 24, which are 21. And you see clearly that $\frac{7}{8}$ are 5 larger than $\frac{2}{3}$, and these 5 are twenty-fourths And therefore we shall say that $\frac{7}{8}$ are $\frac{5}{24}$ larger than $\frac{2}{3}$. And it goes well, and thus are done all the other similar (computations), with whatever fraction it might be.

$^{10.17}$ **Dime** quale è più e quanto, o $\frac{5}{6}$ o $\frac{4}{5}$. Fa pure al modo dicto, trova uno numero che abia sexto et quinto, et questo numero è 30. Ora piglia $\frac{5}{6}$ de 30, che sonno 25, et

$^{10.17}$ Say me, which is larger and how much, either $\frac{5}{6}$ or $\frac{4}{5}$. Do once again in the said way. Find a number that possesses sixth and fifth, and this number is 30. Now

piglia $\frac{4}{5}$ de 30, che sonno 24, et vedi che $\frac{5}{6}$ sonno più che $\frac{4}{5}$ uno. Et questo uno si è trenteximo, cioè $\frac{1}{30}$. Et diremo che $\frac{5}{6}$ sonno più che $\frac{4}{5}$ $\frac{1}{30}$, et sta bene.

seize $\frac{5}{6}$ of 30, which are 25, and seize $\frac{4}{5}$ of 30, which are 24, and you see that $\frac{5}{6}$ are one larger than $\frac{4}{5}$. And this one is a thirtieth, that is, $\frac{1}{30}$. And we shall say that $\frac{5}{6}$ are $\frac{1}{30}$ larger than $\frac{4}{5}$, and it goes well.

10.18 **D**ime quale è meno, e quanto, o $\frac{7}{8}$ o $\frac{8}{9}$. Fa così, trova uno numero che abia ottavo et nono, et questo è 72, però che **(fol. 17ʳ)** otto via nove fa 72. Ora piglia $\frac{7}{8}$ de 72, che sonno 63, che ogni ottavo è 9, et 7 via 9 fa 63. Poi piglia $\frac{8}{9}$ de 72, che sonno 64, che ogni nono sonno otto, et 8 via 8 fa 64. Et vedi che $\frac{8}{9}$ sonno più che ~~uno~~ $\frac{7}{8}$ 1. Et questo uno è settantadueximo. Et perciò diremo che $\frac{8}{9}$ sonno più che $\frac{7}{8}$ $\frac{1}{72}$. Ed è facta, e sta bene, et così se fanno tucte se [*sic*, read "le"] simile ragioni.

10.18 Say me, which is smaller, and how much, either $\frac{7}{8}$ or $\frac{8}{9}$. Do thus, find a number that possesses eighth and ninth, and this is 72, because **(fol. 17ʳ)** eight times nine makes 72. Now seize $\frac{7}{8}$ of 72, which are 63, as every eighth is 9, and 7 times 9 makes 63. Then seize $\frac{8}{9}$ of 72, which are 64, as every ninth are eight, and 8 times 8 makes 64. And you see that $\frac{8}{9}$ are 1 larger than $\frac{7}{8}$. And this one is a seventy-second. And so we shall say that $\frac{7}{8}$ are $\frac{1}{72}$ more than $\frac{7}{8}$. And it is done, and it goes well, and thus all ⟨the⟩ similar computations are done.

[11. The Rule of Three, with Examples]

11.1 **A**biamo dicto de rotti abastanza, però che dele simili ragioni de rotti tucte se fanno a uno modo et per una regola. E però non ne diremo più al punte. Et incominciaremo ad fare et ad mostrare alcune ragioni secondo che appresso diremo.

11.1 We have said enough about fractions, because of the similar computations with fractions all are done in one and the same way and by one and the same rule. And therefore we shall say no more about them here. And we shall begin by doing and showing some computations according to what we shall say soon.

11.2 ⟨**S**⟩e ci fosse data alcuna ragione nela quale se proponesse tre cose, sì debiamo

11.2 If some computation should be given to us in which three things were proposed,

multiplicare sempre la cosa che noi vogliamo sapere contra a quella che non è simegliante, et parti nel'altra cosa, cioè, nell'altra che remane.

then we should always multiply the thing that we want to know against that which is not similar, and divide in the other thing, that is, in the other that remains.

11.3 Vogliote dare l'exemplo ala dicta regola, et vo' dire chosì, vij tornisi vagliono viiij parigini.[a] Dimmi quanto varranno 20 tornisi. Fa così, la cosa che tu voli sapere si è quello che varranno 20 tornisi. Et la non simegliante si è quello che vale vij tornisi, cioè, vagliono 9 parigini. Et però dobiamo multiplicare 9 parigini via 20, fanno 180 parigini, et parti in 7, che è la terza chosa. Parti 180, che ne viene 25 et $\frac{5}{7}$. Et 25 parigini et $\frac{5}{7}$ varrano 20 tornesi. Et così se fanno le simili ragioni.

11.3 I want to give you the example to the said rule, and I want to say thus, vij tornesi are worth viiij parigini.[a] Say me, how much will 20 tornesi be worth. Do thus, the thing that you want to know is that which 20 tornesi will be worth. And the not similar (thing) is that which vij tornesi are worth, that is, they are worth 9 parigini. And therefore we should multiply 9 parigini times 20, they make 180 parigini, and divide in 7, which is the third thing. Divide 180, from which results 25 and $\frac{5}{7}$. And 25 parigini and $\frac{5}{7}$ will 20 tornesi be worth. And thus the similar computations are done.

[a] "Parigini" are minted in Paris, "tornesi" in Tours.

11.4 Ancora diremo chosì, 7 £ di tornesi vagliono 9 £ de parigini. Che varrano 120 £ de tornesi. Fa così como de sopre abiamo dicto, 120 £ via 9 £ de parigini fanno 1080 £ de parigini. Et parti per 7 £ de tornesi, cioè, parti 1080 in 7, che ne viene 154 £ et 5 ß et 8 δ e $\frac{4}{7}$. Et cotanto diremo che vagliono le 120 £ de (**fol. 17ᵛ**) tornesi, cioè £ 154, ß 5, δ 8 $\frac{4}{7}$ de parigini.

11.4 Further we shall say thus, 7 libre of tornesi are worth 9 libre of parigini. What will 120 libre of tornesi be worth. Do thus as we have said above. 120 libre times 9 libre of parigini make 1080 libre of parigini. And divide by 7 libre of tornesi, that is, divide 1080 in 7, from which results 154 libre and 5 soldi and 8 denari and $\frac{4}{7}$. And as much shall we say that the 120 libre of (**fol. 17ᵛ**) tornesi are worth, that is, libre 154, soldi 5, denari 8$\frac{4}{7}$ of parigini.

11.5 Ancora diremo uno altro exemplo ala dicta ragione overo regola. Et diremo così, 7 £ de tornesi vagliono 9 de parigini. Dim-

11.5 Further we shall say another example to the said computation or rule. And we shall say thus, 7 libre of tornesi are worth

me, per £ 150 ß 13 δ 4 de tornesi, quanti parigini aremo. Abiamo a fare chosì, la chosa che noi vogliamo sapere si è, quanti parigini aremo per le £ 150 ß 13 δ 4 de tornesi, et quella che non è simegliante a quelle si è le 9 £ de parigini. Et però debiamo multiplicare 9 via 150 £ ß 13 δ 4, che fanno 1356 £. Et parti 1356 per la terza cosa, che è 7, che ne vene 193 £ et ß 14 et deñ 3 et $\frac{3}{7}$. Et tanti parigini aremo per le £ 150 ß 13 δ 4 de tornesi.

9 of *parigini*. Say me, for *libre* 150, *soldi* 13, *denari* 4 of *tornesi*, how many *parigini* shall we get. We have to do thus, the thing that we want to know is, how many *parigini* shall we get for the *libre* 150, *soldi* 13, *denari* 4 of *tornesi*, and that which not is similar to these is the 9 *libre* of *parigini*. And therefore we shall multiply 9 times 150 *libre*, *soldi* 13, *denari* 4, which make 1356 *libre*. And divide 1356 by the third thing, which is 7, from which results 193 *libre* and *soldi* 14 and *denari* 3 and $\frac{3}{7}$. And so many *parigini* shall we get for the *libre* 150, *soldi* 13, *denari* 4 of *tornesi*.

11.6 Ancora diremo così, se 5 via 5 facesse 26, dime quanto farebbe 7 via 7 a quella medesema ragione. Fa così, et di', 5 via 5 fa 25. Et io dico che fa 26. Et 7 via 7 fa 49. Et però diremo, se 25 vale 26, che varrà 49. Dobiamo multiplicare per la ragione io t'ò dicta adietro. 26 via 49, che ⟨fanno⟩^{M+F} 1274, et parti in 25, che ne viene 50 e $\frac{24}{25}$, et cotanto faremo 7 via 7 a quella medesema ragione. Et sta bene. Et così se fanno tucte le simile ragioni.

11.6 Further we shall say thus, if 5 times 5 would make 26, say me how much would 7 times 7 make at this same rate. Do thus, and say, 5 times 5 makes 25. And I say that it makes 26. And 7 times 7 makes 49. And therefore we shall say, if 25 is worth 26, what will 49 be worth. We shall multiply after the computation I have said to you earlier. 26 times 49, which ⟨make⟩^{M+F} 1274, and divide in 25, from which results 50 and $\frac{24}{25}$, and as much will we make 7 times 7 at this same rate. And it goes well. And thus are done all the similar computations.

11.7 Ancora diremo, se 3 via 4 facesse 13, quanto farebbe 7 via 9 a quella medesema ragione. Fa così et di', 3 via 4 fa 12, et io dico che fa 13. Et 7 via 9 fa 63. Et però dobiame multiplicare 13 via 63, che fa 819, et parti in 12 che è la terza cosa, che ne vene 68 et $\frac{1}{4}$. Et tanto farebbe 7 via 9 se

11.7 Further we shall say, if 3 times 4 would make 13, how much would 7 times 9 make at this same rate. Do thus and say, 3 times 4 makes 12, and I say that it makes 13. And 7 times 9 makes 63. And therefore we shall multiply 13 times 63, which makes 819, and divide in 12, which is the third

3 via 4 facesse 13. Et sta bene. Et così tucte fa [*sic*, read "se"] fanno per una reghola.

thing, from which results 68 and $\frac{1}{4}$. And so much would 7 times 9 make if 3 times 4 made 13. And it goes well. And thus all ⟨are⟩ found by one and the same rule.

11.8 Se ce fosse data alcuna ragione, la quale se proponesse in tre cose, et dall'una dele doi parti denanzi avesse rotto, sì dobiamo multiplicare ambo le parti denanzi per tale numero quanto è quello rotto.

11.8 If some computation was given to us, which was proposed in three things, and if in one of the two parts before there was a fraction, then we shall multiply both parts before by such a number as this fraction is.

11.9 Ora daremo uno exemplo ala dicta regola, et vo' dire così, tornesi $3\frac{1}{3}$ vagliono 4 parigini. Voglio sapere che varranno 25 tornesi. Fa così come di sopre io t'ò dicto, multiplica le parti **(fol. 18ʳ)** denanti per lo rotto, cioè per tre, et di' così, 3 via $3\frac{1}{3}$ fa 10. Et poi multiplica 3 via 4, che fa 12. Et poi di' così, 10 tornesi vagliono 12 parigini, che varranno 25 tornesi. Multiplica 12 via 25, che fanno 300, et parti in 10, che ne vene 30 parigini. Et cotanto varranno 25 tornesi. Et sta bene. Et così fa tucte le simili ragioni, che tucte se fanno per una regola, de che apertamente tu poi vedere l'exemplo dele sopredicte ragioni.

11.9 Now we shall give an example to the said rule, and I want to say thus, *tornesi* $3\frac{1}{3}$ are worth 4 *parigini*. I want to know what 25 *tornesi* will be worth. Do thus as I have said to you above. Multiply the parts **(fol. 18ʳ)** before by the fraction, that is, by three, and say thus, 3 times $3\frac{1}{3}$ makes 10. And then multiply 3 times 4, which makes 12. And then say thus, 10 *tornesi* are worth 12 *parigini*, what will 25 *tornesi* be worth. Multiply 12 times 25, which make 300, and divide in 10, from which results 30 *parigini*. And as much will 25 *tornesi* be worth. And it goes well. And thus do all the similar computations, as all are done by one and the same rule, to which you can see clearly the example from the computations told above.

11.10 Ancora daremo un'altro exemplo ala dicta regola. Et diremo così, tornesi 4 et uno quarto vagliono 6 parigini. Dime per 100 £ de parigini quanti tornesi aremo. Fa così como de sopre io t'ò dicto, multiplica

11.10 Further we shall give another example to the said rule. And we shall say thus, *tornesi* 4 and one fourth are worth 6 *parigini*. Say me, for 100 *libre* of *parigini* how many *tornesi* shall we get. Do thus as I

amedoi le parti denanti per 4, perché lo rotto è quarto. Et di' così, 4 via $4\frac{1}{4}$ fa 17, et poi 4 via 6 fa 24. Ora diremo che 17 tornesi vagliono 24 parigini. Et noi vogliamo sapere che varranno le £ 100 de parigini. Et però dobiamo multiplicare 17 via 100 £ de parigini, che fanno 1700 £, et parti in 24, che ne vene 70 £ et 16 ß et 8 deñ. Et diremo che per le £ 100 de parigini aremo £ 70 ß 16 deñ 8 de tornesi a ragione che tornesi 4 e $\frac{1}{4}$ vaglino 6 parigini. Et sta bene.

have said to you above, multiply both parts before by 4, because the fraction is fourth. And say thus, 4 times $4\frac{1}{4}$ makes 17, and then 4 times 6 makes 24. Now we shall say that 17 *tornesi* are worth 24 *parigini*. And we want to know what the *libre* 100 of *parigini* will be worth. And therefore we shall multiply 17 times 100 *libre* of *parigini*, which make 1700 *libre*, and divide in 24, from which results 70 *libre* and 16 *soldi* and 8 *denari*. And we shall say that for the *libre* 100 of *parigini* we shall get *libre* 70 *soldi* 16 *denari* 8 of *tornesi* at the rate that *tornesi* 4 and $\frac{1}{4}$ are worth 6 *parigini*. And it goes well.

[11.11] Se ce fosse data alcuna ragione nela quale se proponesse tre cose, et d'ambo le parti denanzi si proponesse rotto, sì dobiamo sapere in que numero sonno [*sic*, read "sano"] si trovano quelli rotti, et sì dobiamo multiplicare ammedoi le parti denanzi per quello numero nel quale quelli doi rotti si trovano. Et daremoti lo exemplo qui appresso.

[11.11] If some computation should be given to us, in which three things were proposed, and in both parts before a fraction was proposed, then we shall know in which ⟨integer⟩ number these fractions are found, and then we shall multiply both the parts before by that number in which these two fractions are found. And we shall give you the example next to here.

[11.12] Vogliote dare una ragione ala dicta regola. Et vo' dire così, tornisi $2\frac{2}{3}$ vagliono parigini 3 et $\frac{3}{4}$. Voglio sapere che varranno le £ 200 de tornesi. Fa così come dice la nostra regola, et di' così, li rotti sonno $\frac{2}{3}$ et $\frac{3}{4}$. Et $\frac{2}{3}$ et $\frac{3}{4}$ si trova in 12. Et però dobiamo multiplicare ambo li rotti denanzi per 12. Et di' così, 12 via tornesi $2\frac{2}{3}$ fanno 32 tornesi. Et poi multiplica 12 via parigini $3\frac{3}{4}$, fanno 45 parigini. Et però

[11.12] I want to give you a computation to the said rule. And I want to say thus, *tornesi* $2\frac{2}{3}$ are worth *parigini* 3 and $\frac{3}{4}$. I want to know what *libre* 200 of *tornesi* will be worth. Do thus as our rule says, and say thus, the fractions are $\frac{2}{3}$ and $\frac{3}{4}$. And $\frac{2}{3}$ and $\frac{3}{4}$ one finds in 12. And therefore we shall multiply both fractions before by 12. And say thus, 12 times *tornesi* $2\frac{2}{3}$ make 32 *tornesi*. And then multiply 12 times *pari-*

diremo così, 32 tornesi vagliono **(fol. 18ᵛ)** 45 parigini. Et altretanto è a dire che 32 tornesi ^vagliono^ 45 parigini quando tornesi 3 $\frac{2}{3}$ vagliono parigini 3 et $\frac{3}{4}$. Et vogliamo sapere quanto varranno le £ 200 de tornesi. Et però multiplica 45 via 200 £ de parigini, che fanno £ 900 de parigini, et parti in 32, che ne vene £ 281 ß 5 de parigini. Et è fatta. Et così se fanno le simili.

gini 3 $\frac{3}{4}$, they make 45 *parigini*. And therefore we shall say thus, 32 *tornesi* are worth **(fol. 18ᵛ)** 45 *parigini*. And it is the same as saying that 32 *tornesi* are worth 45 *parigini* when *tornesi* 3 $\frac{2}{3}$ are worth *parigini* 3 and $\frac{3}{4}$. And we want to know how much the *libre* 200 of *tornesi* will be worth. And therefore multiply 45 times 200 *libre* of *parigini*, which make *libre* 900 of *parigini*, and divide in 32, from which results *libre* 281, *soldi* 5 of *parigini*. And it is done. And thus are done the similar (computations).

11.13 Ancora diremo così, tornesi 3 $\frac{1}{2}$ vagliono parigini 4 et $\frac{1}{3}$. Quanto varranno £ 20 parigini. Fa così como de sopre abiamo dicto, et sappi $\frac{1}{2}$ et $\frac{1}{3}$ in que numero se trova, et trovi che questo numero è 6. Et però multiplica 6 via tornesi 3 et $\frac{1}{2}$, fanno tornesi 21. Et di', 6 via parigini 4 $\frac{1}{3}$ fanno 26 parigini. Et noi vogliamo sapere che varrano 20 parigini. Et però dirai così, 21 tornesi vagliono 26 parigini, que varranno 20 parigini. Multiplica 21 via 20, fa 420, et parti in 26, che ne vene 16 $\frac{2}{\langle 1 \rangle 3}$. Et diremo che 20 parigini varranno tornesi 16 et $\frac{2}{\langle 1 \rangle 3}$. Et è facta.

11.13 Further we shall say thus, *tornesi* 3 $\frac{1}{2}$ are worth *parigini* 4 and $\frac{1}{3}$. How much will *libre* 20 of *parigini* be worth. Do thus as we have said above, and know in which number one finds $\frac{1}{2}$ and $\frac{1}{3}$, and you find that this number is 6. And therefore multiply 6 times *tornesi* 3 and $\frac{1}{2}$, they make *tornesi* 21. And say, 6 times *parigini* 4 $\frac{1}{3}$ make 26 *parigini*. And we want to know what 20 *parigini* will be worth. And therefore you will say thus, 21 *tornesi* are worth 26 *parigini*, what will 20 *parigini* be worth. Multiply 21 times 20, it makes 420, and divide in 26, from which results 16 $\frac{2}{\langle 1 \rangle 3}$. And we shall say that 20 *parigini* will be worth *tornesi* 16 and $\frac{2}{\langle 1 \rangle 3}$. And it is done.

11.14 Se noi avessemo a multiplicare numero sano et rocto contra a numero sano et rotto, sì dobiamo arecare a rocti ambo le parti, cioè a quelli rotti che la ragione parla. Et poi multiplicare quelli numeri che ti vengo-

11.14 If we had to multiply an integer and broken number against an integer and broken number, then we shall bring to fractions both parts, that is, to those fractions which the computation speaks about.

no l'uno contra l'altro, et partire per le figure de quelli rotti, multiplicata ancho l'una contra l'altra. Et quello che ne vene, cotanto fa.

And then multiply these numbers that you get, one against the other, and divide by the figures of these fractions, also multiplied one against the other. And that which results from it, as much does it make.

[11.15] Ora diremo lo exemplo ala dicta regola, et vo' dire così, moltiplica 3 et $\frac{1}{3}$ via 3$\frac{1}{3}$. Fa così, arecha a rotti como de sopre io t'ò dicto amendoi le parti, cioè a terzi, et di' così, 3 via 3 et $\frac{1}{3}$ fa 10. Et 3 via 3 et $\frac{1}{3}$ fa 10. Ora moltiplica 10 via 10, fa 100. Ora moltiplica la figura delo rotto l'una contra all'altra. Cioè 3 via 3, fa 9. Parti 100 in 9, che ne vene 11 et $\frac{1}{9}$. Et cotanto fa moltiplicato 3 et $\frac{1}{3}$ via 3 et $\frac{1}{3}$. Cioè 11 et $\frac{1}{9}$. Et così fa tucte le simili ragioni.

[11.15] Now we shall say the example to the said rule, and I want to say thus, multiply 3 and $\frac{1}{3}$ times 3$\frac{1}{3}$. Do thus, bring to fractions both parts, as I have said to you above, that is, to thirds, and say thus, 3 times 3 and $\frac{1}{3}$ makes 10. And 3 times 3 and $\frac{1}{3}$ makes 10. Now multiply 10 times 10, it makes 100. Now multiply the figure of the fraction, one against the other. That is, 3 times 3, it makes 9. Divide 100 in 9, from which results 11 and $\frac{1}{9}$. And as much makes multiplied 3 and $\frac{1}{3}$ times 3 and $\frac{1}{3}$. That is, 11 and $\frac{1}{9}$. And do thus all the similar computations.

[12. Computations of Non-Compound Interest]

[12.1] Se ce fosse data alcuna ragione de merito che dicesse così, la £ è prestata a cotanti denari el mese. Et noi volessemo sapere le cotante £ che guadagneranno in cotanto tempo. Sì dobiamo sapere quanto vale la libra in tucto el termine. Et moltiplicare poi contra ala somma dele libre.

[12.1] If some computation should be given to us about interest which said thus, the *libra* is lent at so and so many *denari* a month. And we wanted to know what so and so many *libre* earn in so and so much time. Then we shall know how much the *libra* is worth in all the term. And multiply then against the total of the *libre*.

(fol. 19ʳ)

[12.2] Asempro ala dicta regola. Et vo' dire

[12.2] Example to the said rule. And I want

così, la £ guadagna el mese 3 deñ. Dimme quanta guadagnaranno le 100 £ in sei mesi. Fa così como di sopre abiamo dicto, sappi quanto guadagna la libra in questo tempo, cioè in questi sey mesi. Moltiplica 6 via 3, fa 18 deñ. Et cotanto guadagna la £ in sey mesi. Et se voi sapere quanto guadagnaranno le 100 {le 100} £, moltiplica 18 via 100 denari, fa 1800, che sonno £ 7 ß 10. Et cotanto guadagnaranno le 100 libre in sey mesi. Et sta bene. Et così se fanno tucte le simegliante ragioni.

to say thus, the *libra* earns 3 *denari* a month. Say me, how much will the 100 *libre* earn in six months. Do thus as we have said above. Know how much earns the *libra* in this time, that is, in these six months. Multiply 6 times 3, it makes 18 *denari*. And as much earns the *libra* in six months. And if you want to know how much the 100 {the 100} *libre* earn, multiply 18 times 100 *denari*, it makes 1800, which are *libra* 7, *soldi* 10. And as much will the 100 *libre* earn in six months. And it goes well. And thus are done all the similar computations.

12.3 **Se** ce fosse data alcuna ragione de merito, cioè la £ è prestata al mese overo guadagna cotanti deñ el mese, e noi volessimo sapere le quante £ guadagnano el dì uno deñ, dobiamo partire 30 per tante parte quanti deñ guadagna la £ el mese, et questo se fa perché el mese è 30 dì. Et quello che ne vene, le cotante £ guadagna el dì uno deñ.

12.3 If some computation should be given to us about interest, that is, the *libra* is lent at or earns so and so many *denari* a month, and we wanted to know how many *libre* earn one *denaro* a day, we shall divide 30 by so many parts as the *denari* which the *libra* earns a month, and this is done because a month is 30 days. And that which results from it, so many *libre* earn one *denaro* a day.

12.4 **A**sempro ala dicta regola, et vo' dire così, la £ è prestata al mese a 3 deñ. Vo' sapere le quante £ guadagnaranno el dì uno deñ. Fa como la regola dice, parti 30 per 3, che ne vene 10. Et 10 £ guadagnaranno el dì uno deñ.

12.4 Example to the said rule, and I want to say thus, the *libra* is lent at 3 *denari* a month. I want to know how many *libre* will earn one *denaro* a day. Do as the rule says, divide 30 by 3, from which results 10. And 10 *libre* will earn one *denaro* a day.

12.5 **Se** ce fosse dicta alcuna ragione de merito. Cioè el cetinaro[a] guadagna l'anno

12.5 If some computation should be said to us about interest. That is, the centinaio[a]

cotante £. Et noi volessimo sapere le quante libre guadagnano el dì uno deñ, sì dobiamo partire 150 per le tante parti quante guadagna l'anno el centinaro.

ª Here, 1 *centinaro* = 100 *libre*.

^{12.6} **Et** vo' dire così, el centinaro guadagna l'anno 12 £. Voglio sapere le quante £ guadagnaranno el dì uno deñ. Fa così come dice la regola. Parti 150 £ in 12, che ne vene £ 12 ß 10. Et le contante £ guadagnaranno el dì uno deñ. Et è fatta. Et chosì se fanno le simile ragioni.

^{12.7} **Se** ce fosse data alcuna ragione de merito. Cioè la £ è prestata a cotanti deñ el mese. Et noi volessimo sapere le cotante **(fol. 19ᵛ)** libre in quanto tempo seranno doppie senza fare capo d'anno.ª Sì dobiamo partire 20 anni per tante parti quanti guadagnia la £ el mese. Et appresso dirrò l'assemplo.

ª *Fare capo d'anno*/"to make (up accounts at) the end of year" was the standard term for computation with composite interest.

^{12.8} **Asemplo** ala dicta regola. Et diremo così, la £ è prestata al mese a 3 deñ. Vo' sapere in quanto tempo seranno doppie le 100 libre. Fa così come dice la regola nostra, parti 20 per 3, che ne viene $6\frac{2}{3}$. Et in sei anni et otto mesi seranno doppie le 100 libre a non fare capo d'anno. Et sta bene. Et chosì se fanno le simili ragioni.

earns so and so many *libre* a year. And we wanted to know how many *libre* earn one *denaro* in one day, then we shall divide 150 by as many parts as the *centinaio* earns a year.

^{12.6} And I want to say thus, the *centinaio* earns 12 *libre* a year. I want to know how many *libre* earn one *denaro* a day. Do thus as the rule says. Divide 150 *libre* in 12, from which results *libre* 12 *soldi* 10. And so many *libre* earn one *denaro* a day. And it is done. And thus the similar computations are done.

^{12.7} If some computation should be given to us about interest. That is, the *libra* is lent at so and so many *denari* a month. And we wanted to know how many **(fol. 19ᵛ)** *libre* will be doubled in how much time without making (up accounts at the) end of yearª. Then we shall divide 20 years by as many parts as a *libra* earns a month. And next I shall say the example.

^{12.8} Example to the said rule. And we shall say thus, the *libra* is lent at 3 *denari* a month. I want to know in how much time the 100 *libre* will be doubled. Do thus as our rule says, divide 20 by 3, from which results $6\frac{2}{3}$. And in six years and eight months will the 100 *libre* be doubled without making (up accounts at the) end of year. And it goes well. And thus the similar

computations are done.

^{12.9} **Se** ce fosse data alcuna ragione de merito. Cioè el centinaio guadagna l'anno cotante £. Et noi volessimo sapere le 100 £ in quanto tempo seranno doppie. Sì dobiamo partire 100 per tante parti quanti guadagna l'anno el centinaro.

^{12.9} If some computation should be given to us about interest. That is, the *centinaio* earns so and so many *libre* a year. And we wanted to know in how much time the 100 *libre* will be doubled. Then we shall divide 100 by as many parts as the *centinaio* earns in a year.

^{12.10} **Asemplo** ala dicta regola. Et diremo così, el centinaro guadagna l'anno 6 £. Vo' sapere in quanto tempo seranno doppie le 100 libre. Fa così come dice la regola. Parti 100 in 6, che ne vene $16\frac{2}{3}$. Et in 16 anni et $\frac{2}{3}$ seranno doppie le 100 £. Et sta bene. Et chosì se fanno tucte le simiglianti ragioni de ogni quantità che fosse.

^{12.10} Example to the said rule. And we shall say thus, the *centinaio* earns 6 *libre* a year. I want to know in how much time the 100 *libre* will be doubled. Do thus as the rule says. Divide 100 in 6, from which results $16\frac{2}{3}$. And in 16 years and $\frac{2}{3}$ will the 100 *libre* be doubled. And it goes well. And thus are done all the similar computations about any quantity it might be.

^{12.11} **Se** ce fusse data alcuna ragione de merito. Cioè el centinaro guadagna l'anno cotante £. Et noi volessimo sapere quanto guadagnarà el centinaro in uno dì. Sì dobiamo pigliare li $\frac{2}{3}$ di quella quantità quanto guadagna l'anno il centinaro. Et quello che ne vene, cotanti deñ guadagna el centinaro el dì.

^{12.11} If some computation should be given to us about interest. That is, the *centinaio* earns so and so many *libre* a year. And we wanted to know how much the *centinaio* earns in one day. Then we shall seize $\frac{2}{3}$ of that quantity which the *centinaio* earns a year. And that which results from it, so many *denari* earns the *centinaio* a day.

^{12.12} **Asemplo** ala dicta regola. Et vo' dire così, el centinaro guadagna l'anno 12 £. Vo' sapere quanto guadagnarà el dì el centinaro. Fa como te dico di sopre, piglia $\frac{2}{3}$ di 12, che sonno 8, et otto diremmo che

^{12.12} Example to the said rule. And I want to say thus, the *centinaio* earns 12 *libre* a year. I want to know how much the *centinaio* earns a day. Do as I say to you above, seize $\frac{2}{3}$ of 12, which are 8, and eight we

guadagna el dì el centinaro. Et se fusse
prestato overo guadagnasse l'anno el centi-
naro 16 £, prendi $\frac{2}{3}$ di 16, che è 10 et $\frac{2}{3}$.
Et 10 deñ et $\frac{2}{3}$ guadagna ogni dì il cen-
ti<na>ro.

shall say that the *centinaio* earns a day.
And if the *centinaio* were lent at or earned
16 *libre* a year, take $\frac{2}{3}$ of 16, which is 10
and $\frac{2}{3}$. And 10 *denari* and $\frac{2}{3}$ does the
centinaio earn each day.

[13. Problems involving metrological shortcuts]

[13.1] **Se** ce fusse data alcuna ragione in
questo modo, cioè, lo carcho del pepe, o
de qualunqua altra cosa fusse, la quale è
300 £, vale cotante £ o cotanti soldi o
cotanti deñ. Et noi volessimo sapere **(fol.
20ʳ)** quanto varrà la £. Sì dovete sapere che
per ogni £ che vale lo carcho, vale la £ $\frac{4}{5}$
de denaro. Et per ogni ß che vale lo carco,
la £ vale $\frac{1}{25}$ de denaro, cioè che vale $\frac{12}{300}$
de denaro, che a schifare viene appunto ~~ai~~ $\frac{1}{25}$
de denaro come te dico. ⟨Et per ogni de-
naro che vale lo carcho vale la £ $\frac{1}{300}$ de
denaro⟩ᴹ⁺ᶠ

[13.1] **If** some computation should be given
to us in this way, that is, the load of pepper, or of whatever thing it else might be,
which is 300 pounds, is worth so and so
many *libre* or so and so many *soldi* or so
and so many *denari*. And we wanted to
know **(fol. 20ʳ)** how much is worth the
pound. Then you shall know that for each
libra that the load is worth, the pound is
worth $\frac{4}{5}$ of *denaro*. And for each *soldo* that
the load is worth, the pound is worth $\frac{1}{25}$
of *denaro*, that is that it is worth $\frac{12}{300}$ of
denaro, which by simplifying becomes
precisely $\frac{1}{25}$ of *denaro* as I say to you.
⟨And for each *denaro* that the load is
worth, is the *libra* worth $\frac{1}{300}$ of *denaro*⟩ᴹ⁺ᶠ

[13.2] **Ora** diremo l'assemplo ala dicta regola,
et diremo così, lo carcho del pepe val 18
£. Vo' sapere quanto varrà la £. Fa chosì
como già t'ò dicto, che per ogni £ che vale
lo carcho, va⟨le⟩ᴹ⁺ᶠ la £ $\frac{4}{5}$ de denaro. Et
però fa così, moltiplica 18 via $\frac{4}{5}$ de denaro,
fa $\frac{72}{5}$ de denaro, che sonno a partire in
cinque, 14 denari et $\frac{2}{5}$. Et cotanti denari
vale la £. Et è fatta. Et così se fanno le
simili. Et se io dicesse, que varrà la £,
valendo lo carcho 17 ß, sappi che varrà $\frac{17}{25}$

[13.2] **Now** we shall say the example to the
said rule, and we shall say thus, the load
of pepper is worth 18 *libre*. I want to know
how much will be worth the pound. Do
thus as I have already said to you, that for
each *libra* that the load is worth, the pound
is worth $\frac{4}{5}$ of *denaro*. And therefore do
thus, multiply 18 times $\frac{4}{5}$ of *denaro*, it
makes $\frac{72}{5}$ of *denaro*, which are, when
dividing in five, 14 *denari* and $\frac{2}{5}$. And so
many *denari* is worth the pound. And it is

de denaro. Et simegliante se io dicesse, lo carcho vale vinti denari, quanto varrà la £. Et io t'ò dicto che {varrà}, a uno denaro lo carcho, vale la £ $\frac{1}{300}$ de denaro. Et però ⟨a⟩ 20 deñ lo carcho vale la £ $\frac{20}{300}$ de denaro, che sonno a schifare, viene $\frac{1}{15}$ de denaro. Et sta bene.

done. And thus are done the similar. And if I said, what will be worth the pound, the load being worth 17 *soldi*, know that it will be worth $\frac{17}{25}$ of *denaro*. And similarly if I said, the load is worth twenty *denari*, how much will be worth the pound. And I have said to you that {...}, at one *denaro* the load, the pound is worth $\frac{1}{300}$ of *denaro*. And therefore ⟨at⟩ 20 *denari* the load, the pound is worth $\frac{20}{300}$ of *denaro*, which are to simplify, $\frac{1}{15}$ of *denaro* results. And it goes well.

13.3 Ancora mostraremo questa regola per altro modo. Et diremo così, sappi che per quanti deñ vale la £, poni suso el quarto de quello che sonno, et tanti deñ quanti te vene, tante £ vale lo carcho. Et questo se intende essendo lo carcho 300 £ como noi abiamo dicto. Et diremo chosì, la £ vale 20 deñ. Vo' sapere quanto varrà lo carcho. Fa così como dice la regola, poni el $\frac{1}{4}$ sopre a 20, che è 5, et fa 25. Et 25 libre vale lo carcho de deñ 20 la £.

13.3 Further we shall show this rule in another way. And we shall say thus, know that for as many *denari* as is worth the pound, put on top the fourth of that which they are, and so many *denari* as result for you, so many *libre* is worth the load. And this is to be understood, the load being 300 pounds, as we have said. And we shall say thus, the pound is worth 20 *denari*. I want to know how much will be worth the load. Do thus as the rule says, put $\frac{1}{4}$ above 20, which is 5, and it makes 25. And 25 *libre* is worth the load, at 20 *denari* the pound.

13.4 Ancora diremo, la £ del pepe vale ß x deñ 8, vo' sapere quanto varrà lo carcho. Fa così, arrecha a deñ li soldi che vale la £, che sonno x ß et 8 deñ, 128 deñ. Porrai suso el quarto, che è 32. Et sonno in tucto 160, et 160 £ vale lo carcho. Et in questo modo fa tucte le simili, arrecha sempre li soldi a deñ et poni suso el quarto, et quelle sonno libre. Et cotante libre vale lo carcho.

13.4 Further we shall say, the pound of pepper is worth *soldi* x, *denari* 8, I want to know how much will be worth the load. Do thus, bring to *denari* the *soldi* that is worth the pound, which are x *soldi* and 8 *denari*, 128 *denari*. Put on top the fourth, which is 32. And in all they are 160, and 160 *libre* is worth the load. And in this way do all the similar, always bring the *soldi* to *denari* and put on top the fourth, and those are *libre*. And so many *libre* is worth the load.

[13.5] **Ancora** diremo, la libra del pepe vale 17 deñ, vo' sapere quanto varrà lo carcho. Fa como io t'ò dicto, poni el quarto sopra a 17, che è 4 et $\frac{1}{4}$, fa 21 et $\frac{1}{4}$. Et 21 £ et $\frac{1}{4}$ varrà lo carcho. **(fol. 20ᵛ)** Et sta bene, cioè £ 21 ß 5.

[13.5] Further we shall say, the pound of pepper is worth 17 *denari*, I want to know how much will be worth the load. Do as I have said to you, put the fourth above 17, which is 4 and $\frac{1}{4}$, it makes 21 and $\frac{1}{4}$. And 21 *libre* and $\frac{1}{4}$ will be worth the load. **(fol. 20ᵛ)** And it goes well, that is, *libre* 21 *soldi* 5.

[13.6] **Ora** diremo sopra a questa regola el contrario, et diremo chosì, sappi che quando lo carcho vale una quantità de £, et tu volexi sapere quanto vale la £ con breve modo, fa così, trai el quinto de quelle £ quanto vale lo carcho. Et quello che ti remane, tanto vale lo carcho [*sic*, read "tanti denari varrà la libra"]$^{M+F}$. Et in ciò daremo l'assemplo. Lo carcho vale 40 £, dimme quanto varrà la £. Fa cosa [*sic*, read "così"] como di sopra t'ò dicto, el quinto de 40 sonno 8 £, trai de 40 8, resta in 32 . Et 32 deñ varrà la £. Et sta bene, et per questo modo fa tucte le altre de quantunqua quantità se fosse.

[13.6] Now we shall say about this rule the opposite, and we shall say thus, know that when the load is worth a quantity of *libre*, and you wanted to know how much is worth the pound in a short way, do thus, detract the fifth from those *libre* which is worth the load. And that which remains for you, so ⟨many *denari* will be worth the pound⟩$^{M+F}$. And in this we shall give the example. The load is worth 40 *libre*, say me, how much will be worth the pound. Do ⟨thus⟩ as I have said to you above. The fifth of 40 are 8 *libre*, detract 8 from 40, 32 is left. And 32 *denari* will be worth the pound. And it goes well, and in this way do all the others, about whatever quantity it might be.

[13.7] **Ancora** diremo, lo carcho vale 57 £, dimme quanto varrà la £. Fa como de sopra, piglia el quinto de 57 £, che sonno £ 11 et $\frac{2}{5}$, ⟨trai⟩$^{M+F}$ de 57 £, che resta £ 45

[13.7] Further we shall say, the load is worth 57 *libre*, say me, how much will be worth the pound. Do as above, seize the fifth of 57 *libre*, which are *libre* 11 and $\frac{2}{5}$, ⟨de-

et $\frac{3}{5}$. Et 45 deñ et $\frac{3}{5}$ vale la £.

tract)$^{M+F}$ from 57 *libre*, *libre* 45 and $\frac{3}{5}$ is left. And 45 *denari* and $\frac{3}{5}$ is worth the pound.

^{13.8} **A**biamo dicto del carcho lo quale è 300 £. Ora perciò che in alcuna parte lo carcho se conta più o meno di 300 £, et simeglian-temente el quintale se conta meno o più de 100 £, sì diremo una generale regola delo carcho, et del quintale, et quantunqua £ se fusse l'uno et l'altro.

^{13.8} We have spoken of the load which is 300 pounds. Now since in some regions the load is counted more or less than 300 pounds, and similarly the quintal is counted less or more than 100 pounds, then we shall say a general rule about the load, and about the quintal, and however many pounds one and the other might be.

^{13.9} **E**t diremo in questo modo. Uno quin-tale pogniamo che se faccia in alcuna parte £ 104. Et lo carcho è 3 quintali. Et però verrà a essere lo carcho £ 312. Et pogniamo che sappi quanto vale lo dicto carcho. Et voli sapere quanto vale la libra. Sì ne da-remo questa regola.

^{13.9} And we shall say in this way. Let us posit that one quintal is made in some region pounds 104. And the load is 3 quin-tals. And therefore the load will come to be 312 pounds. And let us posit that you know how much is worth the said load. And you want to know how much is worth the pound. Then we shall give this rule for it.

^{13.10} **D**ovete sapere che per ogni ß che vale lo carcho vale la £ uno grano. Et per ogni ß che vale el quintale de libre 104, sì vale la £ 3 grani. Et sappiate che secondo questa regola, el denaro si è 26 grani, et perché el quarto di 104 si è 26. Et sappi che de quantunqua £ è el quintale, prendi el $\frac{1}{4}$, et tanti grani vale el denaro. Et se noi dices-semo che lo carcho fusse 324 £, serebe el quintale 108, et varrebbe el denaro grani 27, perché 27 è el quarto di 108. Et così faresti de quantunqua £ **(fol. 21ʳ)** fosse el

^{13.10} You should know that for each *soldo* the load is worth, the pound is worth one grain. And for each *soldo* that the quintal of pounds 104 is worth, the pound is worth 3 grains. And know that according to this rule, the *denaro* is 26 grani, because the fourth of 104 is 26. And know that of however many pounds the quintal is, take $\frac{1}{4}$, and so many grains is worth the *denaro*. And if we said that the load were 324 pounds, the quintal would be 108, and the *denaro* would be worth grains 27, because

quintale. Et poi sai secondo el proponimento che noi abbiamo fatto che 26 grani
sono uno denaro. Sì poi arecare li grani a
denari, secondo che mostraremo innanzi per
più assempli per meglio intendare.

27 is the fourth of 108. And thus you
should do however many pounds **(fol. 21ʳ)**
the quintal was. And then you know according to the statement we have made that
26 grains are one *denaro*. Then you can
bring the grains to *denari*, according to
what we shall show further on by more
examples in order to understand better.

[13.11] **Ora** diremo uno assempro ala dicta
regola, et vo' dire chosì, lo carcho vale £
13 ß 8, vo' sapere quanto varrà la £. Di'
chosì, 13 £ e 8 ß sonno 268 ß. Et secondo
che noi abbiamo dicto, ogni ß vale uno
grano.[a] Et però vale la £ 268 grani. Ora se
voli arecarli a denari, parti 268 per 26, che
ne viene 10 et $\frac{4}{13}$ de denaro. Et cotanto
varrà la libra. Et se noi avessemo dicto
che'l quintale fusse de £ 108, sì averesti
cont⟨at⟩^{M+F}o ogni denaro 27 grani, però che
è el $\frac{1}{4}$ de 108. Et se io avesse dicto che'l
quintale fusse 102, sì contaremmo ogni
denaro grani 25 $\frac{1}{2}$, però che sempre prendi
el quarto de quello che pesa el quintale. Et
tanti grani quanti te vengono, tanti grani
vale el denaro. Et sempre prendi el terzo
de quanto pesa lo carcho. Et tanto pesa el
quintale. Et questa regola de grani è facta
per le libre spezzate, che sonno più o meno
de 100. Et per li ß spezzati, più de libra.[b]

[13.11] Now we shall say an example to the
said rule, and I want to say thus, the load
is worth *libre* 13, *soldi* 8, I want to know
how much will be worth the pound. Say
thus, 13 *libre* and 8 *soldi* are 268 *soldi*.
And according to what we have said, (for)
each *soldo* (which the load is worth, the
libra)[a] is worth one grain. And therefore the
pound is worth 268 grains. Now if you
want to bring them to *denari*, divide 268
by 26, from which results 10 and $\frac{4}{13}$ of
denaro. And as much will be worth the
pound. And if we had said that the quintal
were of pounds 108, then you would have
counted each *denaro* 27 grains, because it
is $\frac{1}{4}$ of 108. And if I had said that the
quintal were 102, then we would count
each *denaro* grains 25$\frac{1}{2}$, because you
always take the fourth of that which the
quintal weighs. And so many grains as
result for you, so many is worth the *denaro*. And always take the third of what the
load weighs. And so much weighs the
quintal. And this rule of grains is made for
the broken pounds, which are more or less
than 100. And for the broken *soldi*, in
addition to the pounds.[b]

^a That is, "⟨per⟩ ogni soldo ⟨che vale lo carcho⟩ vale ⟨la libra⟩ uno grano.". Since exactly the same ellipsis is found in **M+F**, it is likely to be genuine.
^b The last observation makes little sense – the previous calculations regard the weight unit *libra*, that is, the pound, which is *not* subdivided into *soldi*. Since the observation is also found in **M+F**, Jacopo himself seems to be responsible.

^{13.12} Ancora diremo chosì, el quintale lo quale è libre 104, vale £ 4 ß 12. Dimme quanto varrà la libra. Ora sappi, como dicto abbiamo, che per ogni ß che vale el quintale vale ⟨la⟩ libra 3 grani. Et però di' chosì, le libre 4 et ß 12 sonno ß 92. E però moltiplica 3 via 92 grani, fa 276 grani. Et tanto vale la libra. Ora parti 276 grani per 26, che ne vene 10 denari et $\frac{8}{13}$. Et tanti deñ vale la £. Et è fatta, et sta bene.

^{13.12} Further we shall say thus, the quintal which is 104 pounds, is worth *libre* 4, *soldi* 12. Say me, how much will be worth the pound. Now know, as we have said, that for each *soldo* which is worth the quintal, is the pound worth 3 grains. And therefore say thus, the *libre* 4 and *soldi* 12 are *soldi* 92. And therefore multiply 3 times 92 grains, it makes 276 grains. And so much is worth the pound. Now divide 276 grains by 26, from which results 10 *denari* and $\frac{8}{13}$. And so many *denari* is worth the pound. And it is done, and it goes well.

^{13.13} Lo carcho vale libre 3 et ß 5, dimme quanto varrà la libra. Fa como io t'ò dicto, libre 3 ß 5 sonno 65 ß, dunqua varrà la libra 65 grani. Et parti in 26, che ne vene **(fol. 21ᵛ)** ij et $\frac{1}{2}$. Et se lo carcho fosse 108 libre, sì parteresti 65 in 27, che ne verrebbe dui denari et $\frac{11}{27}$. Et tanto vale la £. Et cossì fa tucte le simili ragioni.

^{13.13} The load is worth *libre* 3 and *soldi* 5, say me how much will be worth the pound. Do as I have said to you. *Libre* 3, *soldi* 5 are 65 *soldi*, hence the pound will be worth 65 grains. And divide in 26, from which results **(fol. 21ᵛ)** ij and $\frac{1}{2}$. And if the load were 108 pounds, then you would divide 65 in 27, from which would result two *denari* and $\frac{11}{27}$. And so much is worth the pound. And do thus all the similar computations.

[14. Mixed Problems, Including Partnership, Exchange and Genuine "Recreational" Problems]

^{14.1} El sonno tre conpagni che fanno conpagnia inseme. E l'uno conpagnio mette in

^{14.1} There are three partners who make partnership together. And one partner puts

corpo dela conpagnia libre 150. El secondo
conpagnio mecte libre 230. El terzo conpagnio mecte libre 420. Ora vene in capo
de uno tempo, ànno guadagniato 100 libre,
et vogliono partire. Vo' sapere quanto
toccha per uno. Fa così, agiongni inseme
tucto quello che ànno messo in corpo dela
conpagnia. Cioè le libre 150, et libre 230,
et libre 420, che fanno in tucto libre 800.
Ora parti 100 libre che ànno guadagniato
in 800 parti, che ne vene ß ij deñ 6 per £.
Ora moltiplica 150 via ß ij deñ 6, che fanno
£ 18 ß 15. Et tanto dè avere el primo conpagnio, che mise in conpagnia £ 150. Ora
multiplica 230 via ß ij deñ 6, che fanno
libre 28 ß 15. Et tanto dè avere el secondo
conpagnio, che misse in conpagnia libre
230. Ora multiplica 420 via ß ij deñ 6, che
fanno libre 52 ß 10. Et cotanto dè avere el
terzo conpagnio, che misse in conpagnia
420 libre. Et se la voi provare, agiongi
inseme tucte queste parti, cioè libre 18 ß
15, et libre 28 ß 15, et libre 52 ß 10, che
fanno in tucto £ 100. Et vedi che sta bene.
Et così fa tucte le simili ragioni.

into the principal of the partnership *libre*
150. The second partner puts in *libre* 230.
The third partner puts in *libre* 420. Now it
occurs after a certain time, that they have
earned 100 *libre*, and want to divide. I want
to know how much is due to (each) one.
Do thus, join together all that which they
have put into the principal of the partnership. That is, the *libre* 150, and *libre* 230,
and *libre* 420, which make in all *libre* 800.
Now divide 100 *libre* that they have earned
in 800 parts, from which results *soldi* ij,
denari 6 per *libra*. Now multiply 150 times
soldi ij, *denari* 6, which make *libre* 18,
soldi 15. And so much shall the first partner have, who put into partnership *libre*
150. Now multiply 230 times *soldi* ij,
denari 6, which make *libre* 28, *soldi* 15.
And so much shall the second partner have,
who put into partnership *libre* 230. Now
multiply 420 times *soldi* ij, *denari* 6, which
make *libre* 52, *soldi* 10. And as much shall
third partner have, who put into partnership
420 *libre*. And if you want to verify it, join
together all these parts, that is, *libre* 18,
soldi 15, and *libre* 28, *soldi* 15, and *libre*
52, *soldi* 10, which make in all *libre* 100.
And you see that it goes well. And do thus
all the similar computations.

[14.2] Uno mercatante dè avere da uno altro
£ 200 de ^qui^ a duo mesi e mezzo. Dice
questo merchatante, damme ogi questi deñ,
et scontoti li denari toi ad ragione de deñ
ij per £ el mese. Dimme quanto glie dè
dare innanzi per le dicte £ 200. Fa così, in

[14.2] A merchant shall have from another
libre 200 within two months and a half
from now. This merchant says, give me this
money[a] today, and I discount your money
at the rate of *denari* ij per *libra* a month.
Say me, how much shall he give him in

doi mese et mezzo a doi deñ per £ vale la £ 5 deñ. Fa così, apponti ale 195 £, et sappi quanto vagliono de merito a deñ 5 per £, che vaglıono £ 4 ß 1 deñ 3. **(fol. 22ʳ)** A-giongi sopra a 195 £, che fanno $\langle£\rangle^{M+F}$ 199 ß 1 deñ 3. Mancate infino in 200 £ ß 18 deñ 9, che vagliono de merito deñ 5. Resta ß 18 deñ 4, et è fatta. Cioè che dè avere per le £ 200, libre 195 ß 18 deñ 4. Et così se fa le simegliante ragioni.

advance for the said *libre* 200. Do thus, in two and a half month at two *denari* per *libra* the *libra* is worth 5 *denari*. Do thus, put yourself at the 195 *libre*, and know how much they are worth in interest at *denari* 5 per *libra*, and they are worth *libre* 4, *soldi* 1, *denari* 3. **(fol. 22ʳ)** Join above 195 *libre*, which make $\langle libre\rangle^{M+F}$ 199, *soldo* 1, *denari* 3. Until 200 *libre*, *soldi* 18, *denari* 9 are lacking for you, that in interest are worth *denari* 5. *Soldi* 18, *denari* 4 is left, and it is done. That is, he shall have for the *libre* 200, *libre* 195, *soldi* 18, *denari* 4. And thus one does the similar computations.

ᵃ Here and in the following, I shall translate *denari* as "money" whenever it is clear that it has this generic meaning.

14.3 Uno à a'ffare uno pagamento in Bolo-gnia de £ 100 de bolognini piccioli. Et a Bolognia vale el bolognino grosso deñ 13 et $\frac{1}{3}$ de bolognino $\langle picciolo\rangle^{M+F}$. Et in Firenze vale el dicto bolognino deñ 15 et $\frac{1}{4}$. Et a Bologna vale el fiorino ß 31 deñ 6 de bolognini piccioli. Et in Firenze vale el dicto fiorino ß 39 deñ 6 dela moneta di Firenze. Vo' sapere quale me mette meglio aportare a Bolognia, partendomi di Firenze, per fare el dicto pagamento, o *f* d'oro, o bolognini grossi, et quanto me mettarà meglio ale dicte £ 100. Fa così, sappi pri-mamente quanti bolognini grossi gli con-vene portare per fare el dicto pagamento. Et multiplica 100 via 15 et $\frac{1}{4}$, che fa 1525, et parti per 13 et $\frac{1}{3}$, che ne vene £ 114 ß 7 deñ 6 de bolognini. Et tanto glie convene portare de bolognini grossi. Ora sappiamo quanto glie convene portare in *f* d'oro. Et multiplica 100 via 39 et $\frac{1}{2}$, che fa 3950. Et

14.3 Somebody has to make in Bologna a payment of *libre* 100 of *bolognini piccioli*. And in Bologna the *bolognino grosso* is worth *denari* 13 and $\frac{1}{3}$ of *bolognino* $\langle picciolo\rangle^{M+F}$. And in Florence the said *bolognino* is worth *denari* 15 and $\frac{1}{4}$. And in Bologna the *fiorino* is worth *soldi* 31, *denari* 6 of *bolognini piccioli*. And in Florence the said *fiorino* is worth *soldi* 39, *denari* 6 of the coin of Florence. I want to know what is better for me to carry to Bologna, starting from Florence, in order to make the said payment, either gold *fiorini*, or *bolognini grossi*, and how much it will be better for me at the said *libre* 100. Do thus, know firstly how many *bo-lognini grossi* it suits him to carry in order to make the said payment. And multiply 100 times 15 and $\frac{1}{4}$, which makes 1525, and divide by 13 and $\frac{1}{3}$, from which results *libre* 114, *soldi* 7, *denari* 6 of *bolognini*.

parti per 31 et $\frac{1}{2}$, che ne vene £ 125 ß 7 deñ 11 $\frac{15}{36}$.[a] Et tanto gle convene portare in *f* d'oro. Et però serà meglio aportare bolognini grossi che fiorini d'oro. Et vense a vantagiare como tu vedi £ 11 ß — deñ 5 et $\frac{15}{36}$ appunto. Et così se fanno le simili ragioni.

And so much will it suit him to carry in *bolognini grossi*. Now let us know how much it suits him to carry in gold *fiorini*. And multiply 100 times 39 and $\frac{1}{2}$, which makes 3950. And divide by 31 and $\frac{1}{2}$, from which results *libre* 125 *soldi* 7 *denari* 11 $\frac{15}{36}$ [a]. And so much will it suit him to carry in gold *fiorini*. And therefore it will be better to carry *bolognini grossi* than gold *fiorini*. And the advantage as you see comes to be precisely *libre* 11, *soldi* —, *denari* 5 and $\frac{15}{36}$. And thus the similar computations are done.

[a] Error for $11\,{}^{15}\!/_{63}$.

14.4 Un soldo de provenzini vale deñ 40 de pisani. Et soldo deli imperiali vale 32 δ pisani. Vo' sapere, per 200 de pisani, quante arò de queste due monete mischiate inseme. Fa così, agiongi inseme 40 et 32 che fanno 72 de⟨nari⟩, che sonno ß vj. Ora parti 200 libre che tu ài in 6, che ne vene £ 33 ß 6 δ 8. Et cotante arai de queste doi monete mischiate inseme. Et se la voli provare, sappi quello che vale le £ 33 ß 6 δ 8 de provenzini per deñ 40 el soldo, che vagliono £ 111 ß 2 δ 2 et $\frac{2}{3}$. Et poi sappi quello che vagliono £ 33 ß 6 δ 8 de imperiali, che vagliono £ 88 ß 17 deñ 9 et $\frac{1}{3}$. Et agiongi inseme, che fanno libre 200. Et sta bene, et così se fanno le simeglianti ragioni de qualunqua moneta se fosse et d'ogni quantità.

14.4 One *soldo* of *provenzini* is worth *denari* 40 of *pisani*. And (the) *soldo* of the *imperiali* is worth 32 *denari pisani*. I want to know, for 200 of *pisani*, how many I shall get of these two coins mixed together. Do thus, join together 40 and 32 which make 72, which are *soldi* vj. Now divide 200 *libre* that you have in 6, from which results *libre* 33, *soldi* 6, *denari* 8. And so many shall you get of these two coins mixed together. And if you want to verify it, know what is worth the *libre* 33, *soldi* 6, *denari* 8 of *provenzini* at *denari* 40 the *soldo*, which are worth *libre* 111, *soldi* 2, *denari* 2 and $\frac{2}{3}$. And then know what are worth *libre* 33, *soldi* 6, *denari* 8 of *imperiali*, which are worth *libre* 88, *soldi* 17, *denari* 9 and $\frac{1}{3}$. And join together, which make *libre* 200. And it goes well, and thus are done the similar computations for whatever coin it might be and for every quantity.

(fol. 22ᵛ)

14.5 Io ò *f* novi e fiorini vecchi. Et el *f* vecchio vale ß 35, et el *f* novo vale ß 37. Et io ò cambiati *f* 100 tra novi et vecchi, et ò ne aute £ 178. Vo' sapere quanti *f* novi et quanti fiorini vecchi io avia. Fo così, poni caso che fusseno tucti de una de queste ragioni, cioè tucte et 100 de qualunqua ragione tu voli. Et diciamo che siano tucti et 100 fiorini vecchi. Et sappi quanto vagliono per ß 35 l'uno, che vaglono 175. Ora di' così, da 175 infino in 178 si à £ 3, che sono ß 60. Ora parti ß 60 nella differenza del pregio che è dall'uno *f* all'altro, cioè da 35 ß infino in 37, che è 2. Parti 60 in 2, ne vene 30. Et 30 *f* dirremo che siano stati el contrario de quelli che noi ponemo che fusseno, che noi dicemo che fusseno tucti vecchi. Et però diremo che questi 30 siano stati novi, e'l resto infino in 100, che è 70, siano stati vecchi. Et così dico che forono. Et se la voi provare, sappi quello che vale 30 *f* novi, che vagliono per ß 37 l'uno, £ 55 ß 10. Et sappi quello che vagliono *f* 70 per ß 35 l'uno, che vagliono £ 122 ß 10. Agiongi inseme, et fanno £ 178, et sta bene. Et simigliantemente te serebbe venuto se tu avesse facto che fusseno stati tucti novi. Et provala. Sappi quello che vale 100 *f* novi per ß 37 l'uno, che vagliono £ 185. Et tu dici che n'avesti de tucti et cento £ 178, siché te verrebe più £ 7, che sonno ß 140. Et parti anchora questi soldi nela diferenzia che è dall'uno all'altro, como facesti in prima, cioè in 2, che ne vene 70. Et così diremo che 70

14.5 I have new *fiorini* and old *fiorini*. And the *old fiorino* is worth *soldi* 35, and the new *fiorino* is worth *soldi* 37. And I have changed 100 *fiorini* new and old together, and I have got for them *libre* 178. I want to know how many new *fiorini* and how many old *fiorini* I had. Do thus, posit the case that all were of one of these rates, that is, all 100 of whatever rate you want. And let us say that they are all 100 old *fiorini*. And know how much they are worth for *soldi* 35 each, they are worth 175. Now say thus, from 175 until 178 there is *libre* 3, which are *soldi* 60. Now divide *soldi* 60 in the price difference which there is from one *fiorino* to the other, that is, from 35 *soldi* until 37, which is 2. Divide 60 in 2, 30 results. And 30 *fiorini* shall we say have been of the opposite (sort) of those which we posited that they were, and we said they were all old. And therefore we shall say that these 30 have been new, and the rest until 100, which is 70, have been old. And thus I say that they were. And if you want to verify it, know that which is worth 30 new *fiorini*, which for *soldi* 37 each are worth *libre* 55, *soldi* 10. And know that which *fiorini* 70 are worth at *soldi* 35 each, which are worth *libre* 122, *soldi* 10. Join together, and they make *libre* 178, and it goes well. And it would have resulted similarly for you if you had done (so) that they had all been new. And verify it. Know that which 100 new *fiorini* is worth at *soldi* 37 each, which are worth *libre* 185. And

fiorini fusseno li vecchi, et ⟨el⟩ resto fino in 100 fosseno i novi. Sicho in ogni modo vidi che sta bene, et ài la provata.

you say that you had for all hundred *libre* 178, so that you would get *libre* 7 more, which are *soldi* 140. And divide again these *soldi* in the difference which there is from one to the other as you did at first, that is, in 2, from which results 70. And thus we shall say that 70 *fiorini* were the old and ⟨the⟩ rest until 100 were the new. So that in each way you see that it goes well, and you have verified it.

[14.6] **El** fiorino dell'oro vale a Genova ß 14 de genovino, et la ghuglino vale deñ 12, et in Firenze deñ 33. Vo' sapere quanto varrà el *f* di Firenze dela moneta di Firenze a quella medesima ragione. Fa così, multiplica 14 via 33 ß, che fanno 462 ß, et parti in 12, che ne vene ß 38 deñ 6. Et cotanto varrà el fiorino de Firenze de quella moneta a quella medesma ragione. Et ⟨è⟩ fatta et sta bene. Et così se fanno tucte le simile ragioni.

[14.6] In Genoa, the gold *fiorino* is worth *soldi* 14 of *genovino*, and the *guglino* is worth (in Genoa) *denari* 12, and in Florence *denari* 33. I want to know how much is worth the *fiorino* from Florence in the coin of Florence at this same rate. Do thus, multiply 14 times 33 *soldi*, which make 462 *soldi*, and divide in 12, from which results *soldi* 38 *denari* 6. And as much will the *fiorino* from Florence be worth in that coin at that same rate. And ⟨it is⟩ done and it goes well. And thus all the similar computations are done.

(fol. 23ʳ)

[14.7] **Uno** mercatante prestò a uno suo amico una libra d'oro, la quale teniva once 2 de ramo. Quando venne in capo de uno tempo, el mercatante glela radomando la dicta £ d'oro. Et el bono homo dice, io non ho de quello oro così fino como tu me prestasti. Ma io ho oro che tene once 3 de ramo per £. Vo' sapere quanto oro costui glie debia rendare de suo per questa £ che gli avia

[14.7] A merchant lends to a friend of his one pound of gold, which contained ounces 2 of copper. When a certain time had passed, the merchant asked him to get the said pound of gold back. And the good man says, I do not have as fine gold as that which you lent me. But I have gold which contains ounces 3 of copper per pound. I want to know how much gold this one shall

prestata. Fa così, abacti del primo oro, che teniva oncie 2 de ramo per libra, once 2, resta necto once 10. Et abacti de questo che tene oncie 3 de ramo per £, once 3, resta once 9 netto. Ora di' così, se 9 once vagliono 10 once, que varranno le 12 oncie. Multiplica 10 via 12, che fa 120, et parti in 9, che ne vene oncie 13 et $\frac{1}{3}$, et oncie 13 et $\frac{1}{3}$ d'oro de quello che tene once 3 de ramo glie debbia rendare per 12 once che glie ne prestò d'oro che teneva once 2 de rame per £. Et è fatta. Et così se fanno le simili ragioni.

give him back of his for this pound which he had lent him. Do thus, strike off from the first gold, which contained ounces 2 of copper per pound, ounces 2, ounces 10 is left net. And strike off from that which contains ounces 3 of copper per pound ounces 3, ounces 9 is left net. Now say thus, if 9 ounces are worth 10 ounces, what will be worth the 12 ounces. Multiply 10 times 12, which makes 120, and divide in 9, from which results ounces 13 and $\frac{1}{3}$, and ounces 13 and $\frac{1}{3}$ of gold of that which contains ounces 3 of copper shall he give him back for 12 ounces that he lent him of gold that contained ounces 2 of copper per pound. And it is done. And thus the similar computations are done.

14.8 Una coppa pesa 14 once per questo modo, che el nappo è d'oro et pesa oncie 7. E'l gambo è d'argento et pesa oncie 4. E'l pede è de ramo et pesa oncie 3. Ora vene ch'io fo fondare questa coppa inseme, ogni cosa mescolato. Et quando è chosì fonduta, et io ne spiccho uno pezzo, el quale pesa once 6. Vo' sapere quanto va de ciascheuno de questi metalli, cioè quanto oro, quanto argento et quanto ramo. Fo così, agiongi inseme primamente l'oro, l'argento e'l rame de questa coppa, ch'è in tucto once 14, tucto mescolato inseme. Et el pezzo che tu ài spicchato si è oncie 6. Et però multiplica 6 via 7 oncie d'oro, fa oncie 42 d'oro. Et partilo in 14, che ne

14.8 A goblet weighs 14 ounces in this way that the cup is of gold and weighs ounces 7. And the stem is of silver and weighs ounces 4. And the foot is of copper and weighs ounces 3. Now it occurs that I have this goblet fused, everything mixed together. And when it is thus fused, I detach a piece from it, which weighs ounces 6. I want to know how much goes (into it) of each of these metals, that is, how much gold, how much silver and how much copper. Do thus, join together firstly the gold, the silver and the copper of this goblet, which in all is ounces 14, all mixed together. And the piece which you have detached is ounces 6. And therefore multi-

vene oncie 3 d'oro. Et tanto oro {ch}è in questo pezzo dele oncie 6. Et multiplica 6 via 4 oncie de argento, fa oncie 24 d'argento, et partilo in 14, che ne vene oncia j et $\frac{5}{7}$. Et però dirai che ve abia oncia j et $\frac{5}{7}$ de argento in questo pezzo. Et poi multiplica 6 via 3 oncie de rame, che fa once 18 de rame, et parti in 14, che ne vene oncia j et $\frac{2}{7}$ de rame. Et cotanto n'ebbe in quello pezzo dele once 6. **(fol. 23ᵛ)** Et se la voi provare, agiongi inseme oncie 3 d'oro et oncia j et $\frac{5}{7}$ de argento et oncia j et $\frac{2}{7}$ de rame. Et fa in tucto oncie 6, como tu di' che pesa el pezzo che tu spicchasti. Et sta bene. Et così se ⟨fanno⟩ tucte le simili ragioni.

ply 6 times 7 ounces of gold, it makes ounces 42 of gold. And divide it in 14, from which results ounces 3 of gold. And so much gold {...} is there is in this piece of ounces 6. And multiply 6 times 4 ounces of silver, it makes ounces 24 of silver, and divide it in 14, from which results ounce j and $\frac{5}{7}$. And therefore you will say that there are ounce j and $\frac{5}{7}$ of silver in this piece. And then multiply 6 times 3 ounces of copper, which makes ounces 18 of copper, and divide in 14, from which results ounce j and $\frac{2}{7}$ of copper. And as much was there in this piece of ounces 6. **(fol. 23ᵛ)** And if you want to verify it, join together ounces 3 of gold and ounce j and $\frac{5}{7}$ of silver and ounce j and $\frac{2}{7}$ of copper. And it makes in all ounces 6, as you say that the piece which you detached weighs. And it goes well. And thus are ⟨made⟩ all the similar computations.

14.8A **[F.v.9]**

14.8A The mark of silver, which is 8 ounces, costs me 66 *soldi* of *tornesi*. Now it happens that I have the said silver melted and refined. And when I take it from the fire I weigh it and find that each mark decreases by $\frac{3}{4}$ ounce, that is, that each mark becomes ounces 7 and $\frac{1}{4}$. Say me how much it suits me to sell the mark in order to reconstitute my capital. Do thus, say, 8 ounces of silver are worth *soldi* 66; what will ounces 7 and $\frac{1}{4}$ be worth. Multiply 8 times 66 *soldi*, they make *libre* 26 and *soldi* 8. And divide by 7 and $\frac{1}{4}$ in this way. Say, 4 times 7 and $\frac{1}{4}$ make 29. And say, 4 times

26 *libre* and *soldi* 8 make *libre* 105 and *soldi* 12. And divide *libre* 105 and *soldi* 12 by 29, from which results *soldi* 72 and *denari* 9 and $\frac{27}{29}$ of *denaro*, And at so much will it suit him to sell the mark of silver in order to reconstitute his capital, that is, *soldi* 72 and *denari* 9 and $\frac{27}{29}$ of *denaro*. And it is done. Thus do all the similar.

14.9   Uno homo sta gravemente et vole fare testamento. Et à una sua donna, la quale è grossa. Et costui lascia che se la donna sua fa fanciullo maschio, lascia a'llui li $\frac{2}{3}$ de tucta la robba sua et ala donna lascia el $\frac{1}{3}$. Et se la donna fa fanciulla femina, lascia ala fanciulla el $\frac{1}{3}$. Et ala donna li $\frac{2}{3}$ de tucto el suo avere. Ora advene che'l bono homo passò de questa vita, et quanno venne in capo del tempo, la donna partorì et fece uno fanciullo maschio, et una fanciulla femina. Vo' sapere in que modo se debba partire questo avere, che costui lasciò, cho ogni uno abia sua ragione, cioè el fanciullo, la donna, et la fanciulla femina. Che tu vedi che non si po partire nel modo che lasciò el testatore. Abiamo a fare chosì, et questo

14.9   A man is ill and wants to make testament. And he has a wife, who is pregnant. And he devises that if his wife makes a male child, he leaves to him $\frac{2}{3}$ of everything of his, and to the wife he leaves $\frac{1}{3}$. And if the wife makes a female child, he leaves to the girl $\frac{1}{3}$. And to the wife $\frac{2}{3}$ of all his possession. Now it happened that the good man departed from this life, and in due time the wife gave birth and made a male child, and a female child. I want to know in which way one shall divide this possession which this man left, everyone having his or her share, that is, the boy, the wife, and the girl, as you see that one cannot divide in the way the testator devised. We have to do thus, and this is its

è la sua regola. Fa primamente positione d'uno et di' chosì, quando la fanciulla femina dovesse avere uno, la donna arebbe ad avere due. Et quando la donna avesse ad avere due, el fanciullo maschio arebbe ad avere quattro. Però che tu vedi che'l padre de' fanciulli lascia al fanciullo dui tanti che ala madre. Et ala madre lascia dui tanti che ala fanciulla femina. Et però è bona propositione quella che noi abiamo facta de sopra, cioè, che la fanciulla avesse uno, la matre dui, e'l fanciullo quactro. Et però di' così, tra tucti e tre costoro ànno 7, cioè, el fanciullo 4, la matre 2, et la fanciulla j. Et d'ogni quantità che costoro avessero a partire, de ogni 7 el fanciullo arebbe ad avere 4, la matre 2, ày 6, et la fanciulla j, ày 7. Et però possiamo dire avere arrechata questa ragione a una conpagnia, et possiamo dire così, el sonno 3 conpagni che fanno conpagnia. Et l'uno conpagnio mette 4, l'altro conpagnio mette 2, et l'altro conpagnio mette j. Et ànno guadagnato tanto quanto cho [*sic*, read ciò] che vale tucta la robba che costui à lasciata, vo' sapere quanto toccha per uno. Io te dico che noi abiamo arechata **(fol. 24ʳ)** questa ragione a una conpagnia. Et però se vole fare per quello modo dela conpagnia che mostrata abiamo adietro in questo.

rule. Firstly make a position of one and say thus, if the girl should have one, the wife were to have two. And if the wife were to have two, the boy were to have four. Because you see that the father of the children leaves to the boy twice as much as to the mother. And to the mother twice as much as to the girl. And therefore the proposition that we have made above is valid, that is that the girl should have one, the mother two, and the boy four. And therefore say thus, all three together these have 7, that is, the boy 4, the mother 2, and the girl j. And of every quantity that they would have to divide, of every 7 the boy were to have 4, the mother 2, you have 6, and the girl j, you have 7. And therefore we can say to have brought this computation to a partnership, and we can say thus, there are 3 partners who make partnership. And one partner puts in 4, the other partner puts in 2, and the other partner puts in j. And they have earned as much as that which everything is worth which this man has left, I want to know how much is due to (each) one. I say to you that we have brought **(fol. 24ʳ)** this computation to a partnership. And therefore it shall be done after that way of the partnership which we have shown earlier in this (treatise).

14.10 **Ora** poniamo che questo iudicamento fusse *f* 1400. Dimme quanto ne dè avere la matre, quanto el figlolo maschio, et quanto la fanciulla femina. Fa chosì, agiongi inseme 4 et 2 et j, che fa 7, et questo è

14.10 Now let us posit that this bequest were 1400 *fiorini*. Say me, how much shall the mother have of it, how much the son, and how much the girl. Do thus, join together 4 and 2 and j, which makes 7, and this is

el partitore et è el corpo dela conpagnia. Ora multiplica 4 via 1400 *f*, che fa 5600 *f*, et parti in 7, che ne vene *f* 800 d'oro. Et tanto deba avere el fanciullo maschio. Et poi multiplica 2 via 1400 *f*, fa 2800 *f*. Parti in 7, che ne vene *f* 400, et tanto dè avere la matre. Et poi multiplica j via 1400 *f*, fa 1400 *f*, parti in 7, che ne vene *f* 200. Et tanto dè avere la fanciulla femina. Et è facta, et sta bene. Et in questo modo se fanno tucte le simigliante ragioni.

the divisor and is the principal of the partnership. Now multiply 4 times 1400 *fiorini*, which makes 5600 *fiorini*, and divide in 7, from which results gold *fiorini* 800. And so much shall the boy have. And then multiply 2 times 1400 fiorini, it makes 2800 *fiorini*. Divide in 7, from which results *fiorini* 400, and so much shall the mother have. And then multiply j times 1400 *fiorini*, it makes 1400 *fiorini*, divide in 7, from which results *fiorini* 200. And so much shall the girl have. And it is done, and it goes well. And in this way are done all the similar computations.

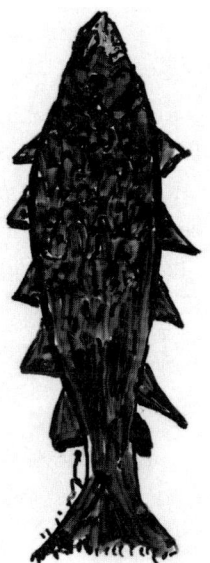

14.11 Uno pesce pesa, la testa el $\frac{1}{3}$ de tucto el pesce, et la coda pesa el $\frac{1}{4}$ de tucto el pescie. Et el corpo de mezzo pesa oncie 8. Dime quanto pesa la testa, quanto la coda, et quanto pesa tucto el pescie. Fa così et di', $\frac{1}{3}$ et $\frac{1}{4}$ se trova in 12. Et ^piglia^ el $\frac{1}{3}$ e'l $\frac{1}{4}$ de 12, che fanno giunti inseme 7. Et di', da 7 infino in 12 sonno 5, et questo è el partitore. Ora perché el corpo pesa oncie 8, multiplica 8 via 12, fa 96, parti in 5, che ne vene 19 et $\frac{1}{5}$. Et cotanto pesa tucto el pesce. Cioè oncie 19 et $\frac{1}{5}$. Et se la voli provare, et sapere quanto pesa ogni uno di per se, togli el $\frac{1}{3}$ di 19 et $\frac{1}{5}$, che è 6 et $\frac{2}{5}$. Et cotanto pesa la testa. Et poi togli el $\frac{1}{4}$ de 19 et $\frac{1}{5}$, che è 4 et $\frac{4}{5}$. Et cotanto pesa la coda. Ora agiongi inseme

14.11 A fish weighs, the head the $\frac{1}{3}$ of the whole fish, and the tail weighs the $\frac{1}{4}$ of the whole fish. And the body in middle weighs ounces 8. Say me, how much weighs the head, how much the tail, and how much weighs the whole fish. Do thus and say, $\frac{1}{3}$ and $\frac{1}{4}$ one finds in 12. And seize the $\frac{1}{3}$ and the $\frac{1}{4}$ of 12, which joined together make 7. And say, from 7 until 12 are 5, and this is the divisor. Now because the body weighs ounces 8, multiply 8 times 12, it makes 96, divide in 5, from which results 19 and $\frac{1}{5}$. And as much weighs the whole fish. That is, ounces 19 and $\frac{1}{5}$. And if you want to verify it, and know how much weighs each one by itself, grasp $\frac{1}{3}$ of 19 and $\frac{1}{5}$, which is 6 and $\frac{2}{5}$.

6 et $\frac{2}{5}$ et 4 et $\frac{4}{5}$, che fanno 11 et $\frac{1}{5}$. Trailo de 19 et $\frac{1}{5}$, resta 8 oncie, et octo oncie pesa quello de mezzo. Et sta bene, et così se fanno tucte.

And as much weighs the head. And then grasp the $\frac{1}{4}$ of 19 and $\frac{1}{5}$, which is 4 and $\frac{4}{5}$. And as much weighs the tail. Now join together 6 and $\frac{2}{5}$ and 4 and $\frac{4}{5}$, which make 11 and $\frac{1}{5}$. Detract it from 19 and $\frac{1}{5}$, 8 ounces is left, and eight ounces weighs that in middle. And it goes well, and thus all are done.

14.12 Uno homo è a Roma, et vole andare a Monpuleri. Et va'nne in 11 die né più né meno. Et uno altro homo è ad Monpuleri, et vole andare a Roma. Et va'nni in 9 dì né più né meno. Ora se partono ad una hora, et punto l'uno da Roma et l'altro da Monpuleri, et camminano l'uno inverso l'altro. Vo' sapere in quanti dì si troveranno inseme nel camino. Fa così et di', per che l'uno vene ad Roma in **(fol. 24')** 9 dì et l'altro va a Monpuleri in 11 dì, agiongi inseme 11 et 9, che fanno 20. Et questo è el partitore. Ora multiplica 9 via 11, fa 99. Parti in 20, che ne vene 4 et $\frac{19}{20}$. ⟨Et in 4 dì et $\frac{19}{20}$⟩ de dì se trovaranno inseme. Et sta bene. Et così se fanno le simili.

14.12 A man is in Rome, and wants to go to Montpellier. And he goes there in 11 days, neither more nor less. And another man is at Montpellier, and wants to go to Rome. And he goes there in 9 days, neither more nor less. Now they leave at one and the same hour, and precisely one from Rome and the other from Montpellier, and travel one toward the other. I want to know in how many days they will meet on their way. Do thus and say, because one comes to Rome in **(fol. 24')** 9 days and the other goes to Montpellier in 11 days, join together 11 and 9, which make 20. And this is the divisor. Now multiply 9 times 11, it makes 99. Divide in 20, from which results 4 and $\frac{19}{20}$. ⟨And in 4 days and $\frac{19}{20}$⟩ of day they will meet. And it goes well. And thus are done the similar.

14.13 Ancora diremo una altra simigliante ragione. Et diremo così, uno correro è ad Vignone et vole andare ad Tolosa, et toglie ad andare in 5 dì. Et un'altro è ad Tolosa et vole andare ad Vignone, et toglie andarvi in 4 dì. Ora se partono li correri ad una hora l'uno da Vigione et l'altro da Tolosa,

14.13 Further we shall say a similar computation. And we shall say thus, a courier is in Avignon and wants to go to Toulouse, and undertakes to go in 5 days. And another one is in Toulouse and wants to go to Avignon, and undertakes to go there in 4 days. Now the couriers leave at one and

dimme in quanti dì se trovaranno inseme. Fa così, et di', perché l'uno va in 5 dì, et l'altro in 4 dì, agiongi inseme 5 et 4, che fa 9, et questo è el partitore. Ora multiplica 4 via 5, fa 20, parti 20 in 9, che ne viene 2 et $\frac{2}{9}$. Et in 2 dì et $\frac{2}{9}$ si trovaranno inseme li dicti correri. Et sta bene. Et chosì fa tucte.

the same hour, one from Avignon and the other from Toulouse. Say me, in how many days they will meet. Do thus, and say, because one goes in 5 days, and the other in 4 days, join together 5 and 4, which makes 9, and this is the divisor. Now multiply 4 times 5, it makes 20, divide 20 in 9, from which results 2 and $\frac{2}{9}$. And in 2 days and $\frac{2}{9}$ will the said couriers meet. And it goes well. And do all thus.

14.14 Uno mercatante è oltramare con uno suo conpagno. Vogliono passare de qua da mare, et vengono al porto. Et trovano una nave insu la quale l'uno de costoro carcha 20 saccha de lana. Et l'altro conpagno ve ne carcha 24. Ora movono et vanno ad loro viagio. Et quando sonno gionti a porto discendono in terra. Et el patrone dela nave diceva, pagateme del nolo de questa lana che io v'ò arechata qui. Et li mercatanti dicono, noi non abiamo denari, ma togli di ciascheuno de noi uno saccho de lana, vendila, e pagati de quello che tu ài ad avere, et poi ce rende el resto. Et el patrone così fece, vendì la dicta lana, et pagòse del nolo, et poi rende al mercatante che aveva 24 sacche rechate insu la nave £ 6. Et a quello che aveva recati 20 saccha ne rende £ 8. Vo' sapere quanto

14.14 A merchant is overseas together with a partner of his. They want to come from there by sea, and come to the harbour. And they find a ship onto which one of them loads 20 sacks of wool. And the other partner loads 24 of it. Now they set out and go to their voyage. And when they have arrived to the harbour they go ashore. And the master of the ship said, pay me the freight charge for this wool which I have brought you here. And the merchants say, we have no money, but grasp from each of us a sack of wool, sell it, and pay yourself from that which you may get, and then give us the rest. And the master did thus, sold the said wool, and paid himself for the freight charge, and then gives back to the merchant who had 24 sacks brought on the ship *libre* 6. And to the one who had 20

vendì el saccho de questa lana, et quanto
tolse del nolo del saccho a ciascheuno de
questi merchatanti. Fa così, sappi primamente quanta lana ànno costoro l'uno più
che l'altro, che ne aveva saccha 4. Et ren-
dette el patrone all'uno più che all'altro £
2, che sonno ß 40. Ora parti ß 40 per 4, che
ne vene ß 10. Et ß 10 diremo che costui
glie tolse de nolo del saccho dela lana ad
ciascheuno de costoro. Et se voi provare se
sta bene, sappi quello che tolze al mercatante dele 24 saccha a ß 10 **(fol. 25ʳ)** el
saccho, sonno £ 12. Et 6 £ gle rende, ày £
18, et £ 18 vendì el saccho de questa lana.
Poi sappi quello che tolse al mercatante
dele 20 saccha a ß 10 el saccho, che tolse
£ 10, et 8 £ gle rende, ài £ 18. Et sta bene.
Et vedi che tanto vendì l'uno saccho quanto
l'altro de questa lana. Et sta bene, et vedi
che la ragione è provata, et tanto pagò de
nolo l'uno quanto l'altro merchatante. Et
a ciascheuno de loro rende quello glie
tocchava. Et così se fanno tucte le simigliante ragioni de quantunqua quantità de
lana o d'altra merchatantia che fusse.

sacks brought he gives back *libre* 8. I want
to know at how much he sold the sack of
this wool, and how much he took away for
the freight charge from the sack of each of
these merchants. Do thus, know firstly how
much wool one of these had more than the
other, and he had sacks 4. And the master
gave back to one *libre* 2 more than to the
other, which are *soldi* 40. Now divide *soldi*
40 by 4, from which results *soldi* 10. And
soldi we shall say that this one took away
in freight charge for a sack of wool from
each of them. And if you want to verify
whether it goes well, know that which he
took away from the merchant with the 24
sacks of wool at *soldi* 10 **(fol. 25ʳ)** the
sack, these are *libre* 12. And 6 *libre* he
gives him back, you have *libre* 18, and for
libre 18 he sold the sack of this wool. Then
know that which he took away from the
merchant withthe 20 sacks at *soldi* 10 the
sack, he took away *libre* 10, and 8 *libre* he
gives back, you have *libre* 18. And it goes
well. And you see that he sold one sack of
this wool for as much as the other. And it
goes well, and you see that the computation
is verified, and one merchant paid as much
in freight (rate) as the other. And to each
he gave what was due to him. And thus are
done all the similar computations for whatever quantity of wool or other merchandise
it might be.

14.15 Uno alboro è sotto terra el $\frac{1}{4}$ et $\frac{1}{5}$ de
tucto l'alboro. Et sopra terra n'à 20 braccia.
Vo' sapere quanto è longo tucto l'alboro.

14.15 A tree is underground the $\frac{1}{4}$ and $\frac{1}{5}$ of
the whole tree. And above ground it has 20
braccia. I want to know how long is the

Fa così et di', el $\frac{1}{4}$ e'l $\frac{1}{5}$ se trova in 20. Ora prendi el $\frac{1}{4}$ e'l $\frac{1}{5}$ di 20, che è 9. Infino in 20 si è 11, et questo è el partitore. Ora multiplica 20 via 20, che fa 400, et parti 400 in 11, che ne vene 36 et $\frac{4}{11}$. Et 36 b̃ʳ et $\frac{4}{11}$ de braccio è lungo tucto l'alboro. Et sta bene. Ora se la voi provare, prendi el $\frac{1}{4}$ di 36 et $\frac{4}{11}$, che fa 9 et $\frac{1}{11}$. Prendi el $\frac{1}{5}$ de 36 et $\frac{4}{11}$, che fa 7 et $\frac{3}{11}${Agiungi inseme.} Et giungi inseme, che fa 16 et $\frac{4}{11}$. Trailo di 36 et $\frac{4}{11}$. Resta 20. Et 20 b̃ʳ sonno quello che sonno fore dela terra. Et così se fanno le simili ragioni.

whole tree. Do thus and say, the $\frac{1}{4}$ and the $\frac{1}{5}$ one finds in 20. Now take the $\frac{1}{4}$ and the $\frac{1}{5}$ of 20, which is 9. Until 20 there is 11, and this is the divisor. Now multiply 20 times 20, which makes 400, and divide 400 in 11, from which results 36 and $\frac{4}{11}$. And 36 *braccia* and $\frac{4}{11}$ of *braccio* is the whole tree long. And it goes well. Now if you want to verify it, take the $\frac{1}{4}$ of 36 and $\frac{4}{11}$, which makes 9 and $\frac{1}{11}$. Take the $\frac{1}{5}$ of 36 and $\frac{4}{11}$, which makes 7 and $\frac{3}{11}$. {...} And join together, which makes 16 and $\frac{4}{11}$. Detract it from 36 and $\frac{4}{11}$. 20 is left. And 20 *braccia* are that which are outside the ground. And thus the similar computations are done.

14.16 E sonno tre conpagni che ànno a partire xx ß. Et l'uno de loro ne dè avere la mità. Et l'altro dè avere el terzo. Et l'altro dè avere el quarto. Ora dice el primo, io debo avere de questi ß 10, dameli.ᵃ Dice el secondo, io debo avere el terzo, dateme li ß 6 et deñ 8. Dice el terzo conpagno, io debo avere el $\frac{1}{4}$, dateme ß 5. Ora se ciascheuno de loro volesse questi denari, como chede ciascheuno, sì ce mancharebe ß 1 et deñ 8, et però vo' sapere in que modo si debono partire questi denari che ogni uno abia sua ragione. Fa così, di', $\frac{1}{2}$ et $\frac{1}{3}$ et $\frac{1}{4}$ se trova in 12. Ora prendi $\frac{1}{2}$ e'l $\frac{1}{3}$ e'l $\frac{1}{4}$ de 12, che gionti inseme sonno 13, et questo è el partitore. Ora multiplica per colui che dè avere ß 10. Multiplica 10 via 12 ß, che fa 120 ß, et partili in 13, che ne vene ß 9 et deñ 2 et $\frac{10}{13}$. Et cotanto dè avere colui che à ad avere la mita. Ora multiplica

14.16 And there are three fellows who have to divide xx *soldi*. And one of them shall have the half of it. And the other shall have the third. And the other shall have the fourth. Now the first says, I shall have 10 of these *soldi* 10, give them to me.ᵃ The second says, I shall have the third, give me the *soldi* 6 and *denari* 8. The third fellow says, I shall have the $\frac{1}{4}$, give me *soldi* 5. Now if each of them would have this money, as each (one) asks for, then *soldo* 1 and *denari* 8 would be lacking, and therefore I want to know in which way this money shall be divided so that everyone gets his share. Do thus, say, $\frac{1}{2}$ and $\frac{1}{3}$ and $\frac{1}{4}$ one finds in 12. Now take $\frac{1}{2}$ and the $\frac{1}{3}$ and the $\frac{1}{4}$ of 12, which joined together are 13, and this is the divisor. Now multiply for the one who shall have 10 *soldi*. Multiply 10 times 12 *soldi*, which makes 120 *soldi*, and

12 via ß 6 deñ 8, che fa 80. Parti in 13, che ne vene ß 6 **(fol. 25^r)** deñ 1 et $\frac{11}{13}$. Et cotanto ne dè avere el secondo conpagnio, cioè colui che à ad avere el terzo. Ora multiplica[b] per colui che à ad avere el $\frac{1}{4}$, et di', 12 via ß 5 fa 60, et parti in 13, che ne vene ß 4 δ 7 et $\frac{5}{13}$. Et cotanto dè avere el terzo, cioè, colui che domandava ß 5. Et è fatta. Et giongi inseme ß 9 δ 2 et $\frac{10}{13}$, et ß 6 δ 1 et $\frac{11}{13}$, et ß 4 δ 7 $\frac{5}{13}$. Fanno ß 20, et sta bene. Et così se fanno le simiglianti ragioni.

divide them in 13, from which results *soldi* 9 and *denari* 2 and $\frac{10}{13}$. And as much shall the one have who is to have the half. Now multiply 12 times *soldi* 6, *denari* 8, which makes 80. Divide in 13, from which results *soldi* 6, **(fol. 25^v)** *denari* 1 and $\frac{11}{13}$. And as much shall the second fellow have of it, that is, the one who is to have the third. Now multiply for the one who is to have the $\frac{1}{4}$, and say, 12 times *soldi* 5 makes 60, and divide in 13, from which results *soldi* 4, *denari* 7 and $\frac{5}{13}$. And as much shall the third have, that is, the one who asked for *soldi* 5. And it is done. And join together *soldi* 9, *denari* 2 and $\frac{10}{13}$, and *soldi* 6, *denaro* 1 and $\frac{11}{13}$, and *soldi* 4, *denari* 7$\frac{5}{13}$. They make *soldi* 20, and it goes well. And thus are done the similar computations.

[a] In the margin of this line, a different hand has written "compagnia".
[b] Here, p with a superscript vertical stroke is used for the first time in the abbreviation for "multiplicare"/"multipricare" and their kin. So far, and until fol. 26^r l. −2, all full writings are with "pli". Since the scribe has so far used the stroke for *ri* as well as *ni*, it cannot be excluded that he would also use it for *li*, even though most scribes would only use it for *ri*. However, the appearance of this contraction at approximately the same place as the full spelling with *pri* has made me prefer this expansion.

[14.17] Uno maestro[a] togle a morare, cioè a lavorare uno lavorezo in 30 dì, et ogni dì che lavora deba avere dal signore de cui è lo lavoro ß 5 el dì. Et ogni ^dì^ che non lavora debba dare al signore ß 7. Ora costui incomincia ad lavorare et lavora tanto in questi 30 dì, et tanto se sta che non lavora, che quando vene in capo de 30 dì, costui non dè avere né a dare alcuna cosa. Vo' sapere quanti dì costui à lavorato in questo lavoro, et quanti dì s'è stato che non à lavorato. Fa così, agiongi inseme ⌐[b] 7 e 5, cioè quello che dè avere et dare, che sonno

[14.17] A master undertakes to build, that is, to perform a piece of work in 30 days, and each day where he works he shall have from the gentleman for whom the work is done *soldi* 5 a day. And every day where he does not work he shall give the gentleman *soldi* 7. Now this one begins to build and works so much in these 30 days, and so much he stays away from working, that at the end of the 30 days, he shall neither have nor give anything. I want to know how many days this one has worked on this piece of work, and how many days he has

5 et 7, che fanno 12. Et questo è el partitore. Ora multiprica 5 via 30 dì, che fa 150,
et parti in 12, che ne vene dì 12 et $\frac{1}{2}$. Et
12 dì e $\frac{1}{2}$ stette che non lavorò. Che monta
quello che dè dare al signore a ß 7 el dì,
£ 4, ß 7 δ 6. Ora multiprica 7 via 30 dì,
che fa 210 dì, et parti in 12, che ne vene
17 dì et $\frac{1}{2}$. Et 17 dì et $\frac{1}{2}$ lavorò, che monta, a ß 5 el dì, £ 4 ß 7, δ 6. Et cotanto dè
avere dal signore. Et £ 4 ß 7 δ 6 dè dare
ad lui, siché a scontare l'uno per l'altro non
deve dare né avere da lui, como dicto è di
sopra. Et è fatta. Et così se fanno le simiglianti ragioni.

stayed away from working. Do thus, join
together {...} 7 and 5, that is that which he
shall have and give, which are 5 and 7,
which make 12. And this is the divisor.
Now multiply 5 times 30 days, which
makes 150, and divide in 12, from which
results days 12 and $\frac{1}{2}$. And 12 days and $\frac{1}{2}$
he stayed away from working. So that
which he shall give to the gentleman at
soldi 7 a day amounts to *libre* 4, *soldi* 7,
denari 6. Now multiply 7 times 30 days,
which makes 210 days, and divide in 12,
from which results 17 days and $\frac{1}{2}$. And 17
days and $\frac{1}{2}$ he worked, which amounts, at
soldi 5 a day, to *libre* 4, *soldi* 7, *denari* 6.
And as much shall he have from the gentleman. And *libre* 4, *soldi* 7, *denari* 6 shall
he give to him, so that, when one is de-
ducted from the other, he shall neither give
nor have from him, as is said above. And
it is done. And thus are done the similar
computations.

[a] Abbreviated "magͬo" (as "magistro" in the incipit) – thus another Latinism.
[b] Illegible sign, probably an erasure.

14.17A   [F.v.18]

14.17A   A cask has 3 taps and is full of wine
and, if I pulled one tap of it only, the cask
would be emptied in two days. And if I
pulled the other tap, it would be emptied
in three days. And if I pulled the third tap,
the cask would be emptied in 5 days. Now
it occurs that I pull all three taps at the
same time. Say me, in how many days the
said cask will be emptied. Do thus, say, a
half and a third and a fifth is found in 30.
Now take the half and the third and the
fifth of 30, it is 31. Now divide 30 in 31,
from which results $\frac{30}{31}$ of a day. And in so

much (time) will the cask be emptied, that is, in $\frac{30}{31}$ of a day.

$^{14.18}$ Una galea ^è^ a Genova et vole andare in Acquamorta. Et la dicta galea à doi vele, che coll'una vela ve andarebbe in 7 dì, et coll'altra vela ve andarebbe in 9 dì. Ora vene che costui vole rizzare suso ammedui le vele a un'ora. Vo' sapere in quanti dì questa galea averà facto suo corso, o suo viagio, cioè da Gienova a Acquamorta, hoperando ammendo queste due vele como te dicho. Fa così et di', l'una vela per se sola v'andarebbe in 7 dì, et l'altra in 9 dì. Et però agiongi inseme 9 et 7, che fanno 16. Et poi multiplica 7 via 9, fa 63. Et parti in 16, che ne vene 3 et $\frac{15}{16}$, et in cotanti dì arà conpiuti el suo viagio. Et è fatta. Et così se fanno **(fol. 26^r)** le simili ragioni.

$^{14.18}$ A galley is in Genoa and wants to go to Aigues-Mortes. And the said galley has two sails, with one of which sails it would go there in 7 days, and with the other sail it would go there in 9 days. Now it occurs that he wants to hoist up both sails at a time. I want to know in how many days this galley will have made its course, or its voyage, that is, from Genoa to Aigues-Mortes, operating both these sails as I say to you. Do thus and say, one sail by itself would go there in 7 days, and the other in 9 days. And therefore join together 9 and 7, which make 16. And then multiply 7 times 9, it makes 63. And divide in 16, from which results 3 and $\frac{15}{16}$, and in as many days will it have completed its voyage. And it is done. And thus are done **(fol. 26^r)** the similar computations.

$^{14.19}$ Uno à 400 pezze di drappi che ne vole fare 38 balle, et tale balla vole fare de 10 drappi, et tale de 11 per balla. Vo' sapere quante balle serano quelle de 10 drappi per balla, et quante seranno quelle de drappi 11. Fa così, multiplica 10 via 38, fa 380, et da 380 infino a 400 à 20. Et 20 balle dirremo che fusseno quelle de drappi 11 per balla. Et resto infino in 38 dirremo che seranno quelle de drappi 10 per balla, che sonno balle 18. Et è fatta. Et se la voli provare, di', 20 balle a 11 drappi per balla sonno drappi 220. Et balle 18 a drappi ^10^ per

$^{14.19}$ Somebody has 400 pieces of cloth of which he wants to makes 38 bales, and some bales he wants to make with 10 cloths, and some with 11 per bale. I want to know how many will be the bales with 10 cloths per bale, and how many will be those with cloths 11. Do thus, multiply 10 times 38, it makes 380, and from 380 until 400 there is 20. And 20 bales shall we say were those with cloths 11 per bale. And (the) rest until 38 we shall say will be those with cloths 10 per bale, which are bales 18. And it is done. And if you want to verify

balla sonno drappi 180. Agiongi inseme, fanno ~~bal~~ drappi 400. Et sta bene, et così se fanno le simili ragioni. Et se non te paresse tanto chiara questa ragione, sì te dico che ogni volta che te fosse data simile ragione, sappi primamente quante balle entrano nel numero de quelli drappi che tu ài a'mballare, cioè in balle intere. Come tu facesti de sopra, che sapisti che quelle 38 balle a 10 per balla montavano 380. Et restavati 20. Et 20 balle dicesti fossono l'altre. Et così serebbe venuto de quantunqua fusse stata la quantità de drappi. Et quelle che ne restano sonno l'altre balle.

it, say, 20 bales at 11 cloths per bale are cloths 220. And bales 18 at cloths 10 per bale are cloths 180. Join together, they make {...} cloths 400. And it goes well, and thus the similar computations are done. And if this computation should not seem too clear to you, then I say to you that every time that a similar computation should be given to you, know firstly how many bales enter in the number of those cloths that you have to put in bales, that is, in whole bales. As you did above, when you knew that these 38 bales at 10 per bale amounted to 380. And 20 were left for you. And 20 bales, you said, were the others. And thus would it have resulted whatever the quantity of cloths might have been. And those of them that are left are the other bales.

14.20 Uno presto a uno suo amicho una archa piena de biada. Et questa ^archa^ è per ogni verso 4 b̃ʳ, cioè lunga, alta et larga. Et quando venne in capo de uno tempo, et costui disse che revoliva la biava sua. Et costui che l'aviva ad rendare disse, io ò la biava, ma non ò archa grande como la tua, che tu me prestasti piena de biava. Ma io n'ò due altre arche che ciascheuna è per ogni verso 2 b̃ʳ. Vo' sapere se per due de queste arche costui è pagato, o quante volte gliele deba dare. Fa così, sappi primamente quante b̃ʳ quadre sonno l'archa magiore, cioè quella de b̃ʳ 4 per ogni verso. Et multiplica 4 via 4, cioè la lunghezza contra la larghezza, che fa 16. Et poi multiprica per l'altezza, cioè 4 via 16, che fa 64.

14.20 Somebody lends to a friend of his a chest full of feeding grain. And this chest is in all directions 4 braccia, that is, long, high and broad. And after a certain time had passed, and this one said that he wanted his grain back, and the one who had to give it back said, I have the feeding grain, but I do not have a chest as large as yours, which you lent me full of grain. But I have two other chests, each of which is 2 braccia in all directions. I want to know if this one is paid by two of these chests, or how many times he shall give him them. Do thus, know firstly how many square braccia are the larger chest, that is, the one of braccia 4 in each direction. And multiply 4 times 4, that is, the length against the

Et 64 b̃ʳ quadre è l'altezza l'archa magiore. **(fol. 26ʳ)** Ora arechamo a braccia quadre l'archetta minore, che è per ogni verso 2 b̃ʳ. Et multiplica 2 via 2, fa 4. Et poi 2 via 4, fa 8. Et 8 b̃ʳ quadre è l'archetta minore. Ora diciamo, se l'archa grande è 16 b̃ʳ et l'archetta piccola è 8 b̃ʳ, quante archetta entraranno in questa grande. Parti 16 per 2, che ne vene 8. Et octo volte gli à a rendere piena quella archetta minore per quella grande. Et sta bene. Et chosì se fanno le simili ragioni che tu avesse a'ffare.

breadth, which makes 16. And then multiply by the height, that is, 4 times 16, which makes 64. And 64 square *braccia* is {...} the larger chest. **(fol. 26ʳ)** Now we bring to square *braccia* the smaller chest, which in each direction is 2 *braccia*. And multiply 2 times 2, it makes 4. And then 2 times 4, it makes 8. And 8 square *braccia* is the smaller chest. Now we say, if the larger chest is 16 *braccia* and the smaller chest is 8 *braccia*, how many small chests enter in this large one. Divide 16 by 2, from which results 8. And eight times shall he give him back this smaller chest full for this large one. And it goes well. And thus the similar computations are done which you might have to do.

14.21 Uno homo se vole vestire de drappo. Et trova drappo che è largo, cioè che è alto palmi 3 et $\frac{1}{2}$. Et volne una robba b̃ʳ 11. Et ancora trova drappo che è largo palmi 5 $\frac{1}{2}$. Dimme quanto ne vorrá nella dicta robba de questo più largo. Fa così, la cosa che tu voli sapere si è quanto drappo ti bisogna de quello che è alto palmi 5 et $\frac{1}{2}$. Et la non simigliante cosa, si è che de quello che è alto palmi 3 et $\frac{1}{2}$ te ne bisogna b̃ʳ 11. Et però multiplica 11 via 3 et $\frac{1}{2}$, che fa 38 et $\frac{1}{2}$. Et parti in 5 et $\frac{1}{2}$. Et fa 2 via 5 et $\frac{1}{2}$, che fa 11 per partitore. E 2 via 38 et $\frac{1}{2}$ fa 77. Et parti 77 in 11, che ne vene 7. Et b̃ʳ 7 de quello che è alto palmi 5 et $\frac{1}{2}$ diremo che ne bisogna nella dicta roba. Et è fatta. Et sta bene ala simile ragione.

14.21 A man wants to dress in (woollen) cloth. And he finds cloth which is *palmi* 3 and $\frac{1}{2}$ broad, that is, high. And from this a dress requires *braccia* 11. And again he finds cloth which is *palmi* 5$\frac{1}{2}$ broad. Say me, how much will be required in the said dress of this broader (cloth). Do thus, the thing that you want to know is how much cloth you need of that which is *palmi* 5 and $\frac{1}{2}$ high. And the thing which is not similar, is that, of that (cloth) which is *palmi* 3 and $\frac{1}{2}$ high, you need *braccia* 11. And therefore multiply 11 times 3 and $\frac{1}{2}$, which makes 38 and $\frac{1}{2}$. And divide in 5 and $\frac{1}{2}$. And make 2 times 5 and $\frac{1}{2}$, which makes 11 as divisor. And 2 times 38 and $\frac{1}{2}$ makes 77. And divide 77 in 11, from which results 7. And *braccia* 7 of that which is *palmi* 5 and $\frac{1}{2}$

high we shall say is needed in the said dress. And it is done. And it goes well for the similar computation.

^{14.22} Una donna me manda a uno suo giardino a coglere melarancie, et diceme, cogline tante quante a te pare con questi pacti, che tu trovarai tre porte. Et al primo portannaro voglio che tu glie ne doni la mità de quelle che tu ài colte, et una più. Et al secondo portannaro voglio che tu doni la mità de quelle che te sonno remase, et una più. Et al terzo portannaro voglio che tu doni la mità de quelle che te **(fol. 27^r)** sono remase, e una più. Et facto questo voglio che tu me areche tre melarancie, et non più. Vo' sapere da te, quante melarancie gle convene coglere a ciò che non glie né remangheno né manchino più de tre. Fa così, noi diciamo che a ogni portanaro à ad lasciare la mità e una più. Et però noi diciamo che glie n'à ad avanzare 3, et una che gle ne dede più che la mità la sezza volte, siché adunqua pagato che ebbe, cioè dato che ebbe la mità al terzo portinnaro, glie n'era avanzate 4, siché a questo modo ne deveva avere 8. Ora ne poni 1 sopra a 8, cioè quella che dede a vantagio al secondo portinnaro, siché gle n'era avanzate 9, et nove ne dede al secondo portinnaro. Siché paghato el primo, glie ne campò 18. Ora poni uno sopra a 18, cioè quella che dede ⟨a⟩ vantagio al primo portinnaro, ài 19. Et 19 ne dede a'llui, siché ne venne a cogliere in tucto 38. Ora la prova. Al primo dede la mità, che sonno 19, et j più, ài 20.

^{14.22} A lady sends me to a garden of hers in order to pick oranges, and says to me, pick as many as you find fitting, with this agreement that you will find three doors. And to the first doorkeeper I want that you give him half of those that you have picked, and one more. And to the second doorkeeper I want that you give him the half of those that have remained for you, and one more. And to the third doorkeeper I want that you give him half of those that have **(fol. 27^r)** remained for you, and one more. And when this is done I want that you bring me three oranges, and no more. I want to know from you, how many oranges it suits him to pick so that no more than three remain, nor are lacking. Do thus, we say that to each doorkeeper he has to leave the half and one more. And then we say he shall leave over only 3, and one he gave more than the half the last time, so that when now he had paid, that is, when he had given the half to the third doorkeeper, 4 had been left over, so that in this way he shall have had 8. Now put 1 above 8, that is, the one he gave additionally to the second doorkeeper, so that 9 had been left over for him, and nine he gave to the second doorkeeper. So that when the first was paid, he saved 18. Now put one above 18, that is, the one which he gave additionally to the first doorkeeper, you get 19. And 19

Trai 20 de 38, resta 18. Ora dà la mità al secondo portinnaro, che sono 9, et j più, ài 10. Trailo de 18, resta 8. Ora dà la mità al terzo portinaro, che sono 4, et una in più, restò 3. Et 3 melarance glie remase. Et sta bene. Et così se fanno tucte le simili ragioni che te fossero date.

he gave to him, so that in all he came to have picked 38. Now the verification. To the first he gave the half, which are 19, and j more, you get 20. Detract 20 from 38, 18 is left. Now he gives the half to the second doorkeeper, which are 9, and j more, you get 10. Detract it from 18, 8 is left. Now he gives the half to the third doorkeeper, which are 4, and one more, 3 was left. And 3 oranges remained for him. And it goes well. And thus all the similar computations are done that might be given to you.

14.23 L'oncia dell'oro fino de 24 carrati vale £ 9 ß 7 δ 6. Dime quanto varranno le 125 oncie et 13 teri et 14 grani d'oro che sia ~~de~~ de carrati 22 et $\frac{1}{2}$ per oncia. Et sappi che 30 teri sonno una oncia, et 20 grani sonno uno teri. Fa così, sappi primamente quanto vale l'oncia dell'oro fino de 24 carrati, che vale £ 9 ß 7 δ 6. Ora multiplica 22 et $\frac{1}{2}$ via £ 9 ß 7 δ 6, che fanno £ 210 ß 18 δ 9. Et questo parti per 24, perché l'oro fino è de 24 carrati, che ne vene £ 8 ß 15 δ 9 et $\frac{3}{8}$ de denaro. Et cotanto vale l'oncia dell'oro de carrati 22 et $\frac{1}{2}$. Ora sappi quanto varranno le 125 oncie. Multiplica 125 via £ 8 ß 15 δ 9 $\frac{3}{8}$, che fanno £ 1098 ß 12 δ 7 et $\frac{7}{8}$. Ora sappiamo (fol. 27ᵛ) quanto o quelle che vagliono li 13 teri. Multiplica 13 via 8 £ et ß 15 deñ 9 et $\frac{3}{8}$, che fanno £ 114 ß 5 deñ $\frac{7}{8}$, et parti in 30 perché 30 tari o teri sonno una oncia, che ne vene £ 3 ß 16 δ 2 et $\frac{1}{30}$, ⟨e tanto vagliono li 13 teri⟩ᴹ d'oro de carrati 22 et $\frac{1}{2}$. Ora parti £ 8 ß 15 deñ 9 et $\frac{7}{8}$ per 600, cioè per 20, et per 30,

14.23 The ounce of fine gold of 24 carats is worth *libre* 9, *soldi* 7, *denari* 6. Say me, how much will 125 ounces and 13 *teri* and 14 grains of gold be worth which is of carats 22 and $\frac{1}{2}$ per ounce. And know that 30 *teri* are one ounce, and 20 grains are one *teri*. Do thus, know firstly how much is worth the ounce of fine gold of 24 carats, which is worth *libre* 9, *soldi* 7, *denari* 6. Now multiply 22 and $\frac{1}{2}$ times *libre* 9, *soldi* 7, *denari* 6, which make *libre* 210, *soldi* 18, *denari* 9. And divide this by 24, because fine gold is of 24 carats, from which results *libre* 8, *soldi* 15, *denari* 9 and $\frac{3}{8}$ of *denaro*. And as much is worth an ounce of gold of 22 carats and $\frac{1}{2}$. Now know how much will be worth the 125 ounces. Multiply 125 times *libre* 8, *soldi* 15, *denari* 9 $\frac{3}{8}$, which make *libre* 1098, *soldi* 12, *denari* 7 and $\frac{7}{8}$. Now let us know (fol. 27ᵛ) how much or what the 13 *teri* are worth. Multiply 13 times 8 *libre* and *soldi* 15, *denari* 9 and $\frac{3}{8}$, which make *libre* 114, *soldi* 5,

perché 30 tari sonno j oncia et 20 grani sono j tari, che ne vene deñ 3 et $\frac{1}{2}$. Et cotanto vale el grano. Dunqua vagliono gli 14 grani ß 4 deñ j. Ora micti inseme quello che vagliono 125 oncie et quello che vagliono li 13 teri et 14 grani, che sonno in tucto £ 1102 ß 13. Et è facta, et diremo che le 125 oncie et 13 tari et 14 grani d'oro de carrati 22 et $\frac{1}{2}$ cioè [*sic*, read "varrà"]$^{M+F}$ £ 1102 ß 13. Et così se fanno le simili ragioni.

denaro $\frac{7}{8}$, and divide in 30 because 30 *tari* or *teri* are one ounce, from which results *libre* 3, *soldi* 16, *denari* 2 and $\frac{1}{30}$ ⟨and as much are worth the 13 *teri*⟩^{M} of gold of 22 carats and $\frac{1}{2}$. Now divide *libre* 8, *soldi* 15, *denari* 9 and $\frac{7}{8}$ by 600, that is, by 20, and by 30, because 30 *tari* are j ounce and 20 grains are j *tari,* from which results *denari* 3 and $\frac{1}{2}$. And as much is worth the grain. Hence the 14 grains are worth *soldi* 4, *denaro* j. Now put together that which 125 ounces are worth and that which the 13 *teri* and 14 grains are worth, which are in all *libre* 1102, *soldi* 13. And it is done, and we shall say that the 125 ounces and 13 tari and 14 grains of gold of carats 22 and $\frac{1}{2}$ ⟨are worth⟩$^{M+F}$ *libre* 1102 *soldi* 13. And thus the similar computations are done.

14.24 Una borsia de 3 colori, cioè de seta biancha, de seta roscia, et de seta verde. La seta biancha pesa el $\frac{1}{4}$ de tucta la borsia. Et la seta rossia pesa el $\frac{1}{7}$ de tucta la borsia. Et la seta verde pesa oncie 2. Voglio sapere quanto pesa tucta la borsia. Fa così, di', $\frac{1}{4}$ et $\frac{1}{7}$ se trova in 28. El $\frac{1}{4}$ de 28 è 7, et el $\frac{1}{7}$ de 28 è 4. Agiongi inseme, fa 11. Et da 11 infino in 28 à 17, et questo è el partitore. Ora multiplica 28 via 2 oncie, che fa 56 oncie. Parti in 17, che ne vene 3 oncie et $\frac{5}{17}$. Et tanto pesa tucta la borsia. Et se la voi provare se sta bene, fa così, et di', toglie el $\frac{1}{4}$ de 3 et $\frac{5}{17}$, che è $\frac{14}{17}$. Et togli el $\frac{1}{7}$ de 3 et $\frac{5}{17}$, che è $\frac{8}{17}$. Ora a- giongi inseme $\frac{8}{17}$ et $\frac{14}{17}$, et 2 oncie che pesa la seta verde, fa in tucto oncie 3 et $\frac{5}{17}$. Et

14.24 A purse in 3 colours, that is, of white silk, of red silk, and of green silk. The white silk weighs the $\frac{1}{4}$ of the whole purse. And the red silk weighs the $\frac{1}{7}$ of the whole purse. And the green silk weighs ounces 2. I want to know how much weighs the whole purse. Do thus, say, $\frac{1}{4}$ and $\frac{1}{7}$ one finds in 28. The $\frac{1}{4}$ of 28 is 7, and the $\frac{1}{7}$ of 28 is 4. Join together, it makes 11. And from 11 until 28 there is 17, and this is the divisor. Now multiply 28 times 2 ounces, which makes 56 ounces. Divide in 17, from which results 3 ounces and $\frac{5}{17}$. And so much weighs the whole purse. And if you want to verify whether it goes well, do thus, and say, grasp the $\frac{1}{4}$ of 3 and $\frac{5}{17}$, which is $\frac{14}{17}$. And grasp the $\frac{1}{7}$ of 3 and $\frac{5}{17}$,

è facta, et così se fanno le simili ragioni.
Et ài provato che sta bene. Et ài che 2
oncie è la seta verde, $\frac{14}{17}$ d'oncia pesa la
seta biancha, et $\frac{8}{17}$ d'oncia pesa la seta
rossia.

which is $\frac{8}{17}$. Now join together $\frac{8}{17}$ and $\frac{14}{17}$,
and 2 ounces that the green silk weighs, it
makes in all 3 ounces $3\frac{5}{17}$. And it is done,
and thus the similar computations are done.
And you have verified that it goes well.
And you get that 2 ounces is the green silk,
the white silk weighs $\frac{14}{17}$ of ounce, and the
red silk weighs $\frac{8}{17}$ of ounce.

14.24A [F.v.26]

14.24A A piece of cloth, which is 7 palms
large and costs me 70 *soldi* for a rod
[*canna*], which is 8 palms. Another piece
of cloth, which is 5 palms large, costs me
for the said rod, which is 8 palms, 30 *soldi*.
Say me, which is cheaper and how much
is cheaper, the rod from one or from the
other. Do thus, know first how many square
palms is one piece of cloth and the other.
Multiply, for the first piece of cloth, 7
times 8, they make 56. And, for the second,
multiply 5 times 8, they make 40. And we
shall say that the first piece of cloth is
square palms 56, and the second is square
palms 40. Now divide *soldi* 70 by 56, from
which results *denari* 15. And as much is
the square palm worth of that which costs
soldi 70 the rod, that is, *denari* 15. Now
divide *soldi* 30 by 40, from which results
denari 9, and as much is the square palm
worth of that which costs *soldi* 30 the rod,
that is, *denari* 9. Now say thus: from 9
until 15 there are 6. Hence is 6 *denari*
cheaper the palm of that which costs *soldi*
30 the rod. Hence say, for 6 *denari* the
square palm are 56 palms worth 28 *soldi*,
and for 6 *denari* the square palm the 40

palms are worth 20 *soldi*. Now join 28 and 20, they make 48, and divide in half, from which results 24. And we shall say that the rod of the cloth that cost *soldi* 30 the rod will be 24 *soldi* cheaper than that which cost *soldi* 70. And it is done. Do all the similar thus.

^{14.25} Una coppa è de tre parti, cioè, el nappo una, el gambo un'altra, et el pede una altra. Il nappo pesa el $\frac{1}{4}$ de tucta la coppa. El pede pesa el $\frac{1}{6}$ de tucta la coppa. Et el gambo pesa oncie 5. Vo' sapere quanto pesa tucta la coppa. Et quanto pesa el nappo per se solo. Et quanto pesa el pede. Fa così et di', $\frac{1}{4}$ et $\frac{1}{6}$ se trova in 12. Ora prendi $\frac{1}{4}$ de 12, che è 3. Et prendi el $\frac{1}{6}$ de 12, che è 2. Agiongi inseme, fa 5. Et di', da 5 infino in 12 à 7, et questo è el partitore. Ora tu di' **(fol. 28ʳ)** che'l gambo pesa 5 oncie. Et però multiplica 12 via 5 oncie, che fa 60 oncie. Et parti in 7, che ne vene 8 et $\frac{4}{7}$. Et 8 oncie et $\frac{4}{7}$ d'oncia pesa tucta la coppa. Ora se voi sapere quanto pesa el nappo, che di' che pesa el $\frac{1}{4}$ de tucta, sì piglia $\frac{1}{4}$ de 8 et $\frac{4}{7}$, che è 2 et $\frac{1}{7}$. Et oncie 2 et $\frac{1}{7}$ d'oncia pesa el nappo. Ora se voi sapere quanto pesa el pede, che di' che pesa $\frac{1}{6}$ de tucta, prendi el $\frac{1}{6}$ de 8 et $\frac{4}{7}$, che ne vene j et $\frac{3}{7}$. Et oncie una et $\frac{3}{7}$ d'oncia diremo che pesa el pede. Et el gambo pesa oncie 5. ⟨Se la vuoi provare,⟩^{M+F} agiongi {f} inseme,

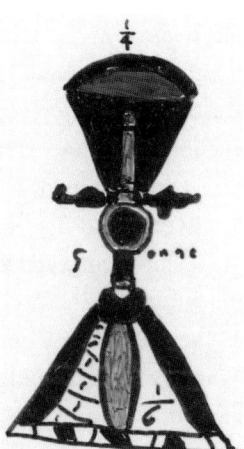

^{14.25} A goblet consists of three parts, that is, the cup one, the stem another, and the foot another. The cup weighs the $\frac{1}{4}$ of the whole goblet. The foot weighs the $\frac{1}{6}$ of the whole goblet. And the stem weighs ounces 5. I want to know how much weighs the whole goblet. And how much weighs the cup alone. And how much weighs the foot. Do thus and say, $\frac{1}{4}$ and $\frac{1}{6}$ one finds in 12. Now take $\frac{1}{4}$ of 12, which is 3. And take $\frac{1}{6}$ of 12, which is 2. Join together, it makes 5. And say, from 5 until 12 there is 7, and this is the divisor. Now you say **(fol. 28ʳ)** that the stem weighs 5 ounces. And therefore multiply 12 times 5 ounces, which makes 60 ounces. And divide in 7, from which results 8 and $\frac{4}{7}$. And the whole goblet weighs 8 ounces and $\frac{4}{7}$ of ounce. Now if you want to know how much the cup weighs, which you say weighs the $\frac{1}{4}$ of all, then seize $\frac{1}{4}$ of 8 and $\frac{4}{7}$, which is 2 and $\frac{1}{7}$. And the cup weighs ounces 2 and $\frac{1}{7}$ of ounce. Now if you want to know how much the foot weighs, which you say weighs $\frac{1}{6}$ of all,

fanno oncie 8 et $\frac{4}{7}$. Et è provata, et sta bene. Et così se fanno tucte le simili ragioni.

take the $\frac{1}{6}$ of 8 and $\frac{4}{7}$, from which results j and $\frac{3}{7}$. And ounces one and $\frac{3}{7}$ of ounce we shall say that the foot weighs. And the stem weighs ounces 5. ⟨And if you want to verify it⟩^M+F, join together, they make ounces 8 and $\frac{4}{7}$. And it is verified, and it goes well. And thus all the similar computations are done.

```
          36
    ┌──────────────┐
    │   quadro     │
    │  ┌────────┐  │
    │  │ br 4320│  │
    │  └────────┘  │        120
    │              │
    │   pietre     │
    │  ┌────────┐  │
    │  │ 34560  │  │
    │  └────────┘  │
    └──────────────┘
          36
```

14.26 Una sala, overo piazza, è lungha b̃ʳ 120, et largha b̃ʳ 36, né più né meno. Et io la voglio lastricare de lastre overo de pietre che sonno tucte de una grandezza. E ciascheuna pietra è lungha $\frac{1}{2}$ b̃ʳ et largha $\frac{1}{4}$. Vo' sapere quante pietre vorrà ad lastrecare la dicta sala. Fa così, primamente arrecha a b̃ʳ quadre tucta la dicta piazza. Et multiplica la lunghezza contra ala larghezza, cioè 120 via 36, che fa b̃ʳ 4320. Et simigliantemente recha a b̃ʳ quadre la pietra, et multiplica la lunghezza contra alla larghezza, cioè $\frac{1}{2}$ una [sic, read "via"]^M+F $\frac{1}{4}$, che fa $\frac{1}{8}$, siché la pietra è $\frac{1}{8}$ de braccio. Et perciò in ogni braccio andarano 8 pietre. Et tu di' che la piazza tucta è 4320 braccia quadre. Et però multiplica 8 via 4320 pietre, che fa 34560 pietre. Et diremo che in tucto quello terreno dela sala overo piazza entraranno 34560 pietre né più né meno. Et è facta, et così se fa l'altre. Et se la voli provare, fa così, et di' così, questa casa è lungha 120 braccia et è ampia 36 b̃ʳ. Et

14.26 A hall, or indeed piazza, is *braccia* 120 long, and *braccia* 36

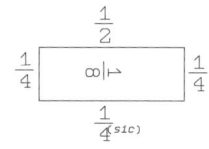

broad, neither more nor less. And I want to flag it with flags or slabs that are all of one and the same magnitude. And each slab is $\frac{1}{2}$ *braccia* long and $\frac{1}{4}$ broad. I want to know how many slabs are required to flag the said hall. Do thus, firstly bring to square *braccia* the whole said piazza. And multiply the length against the breadth, that is, 120 times 36, which makes *braccia* 4320. And similarly bring to square *braccia* the slab, and multiply the length against the breadth, that is, $\frac{1}{2}$ ⟨times⟩^M+F $\frac{1}{4}$, which makes $\frac{1}{8}$, so that the slab is $\frac{1}{8}$ of *braccio*. And so 8 slabs went into each *braccio*. And you say that the whole piazza is 4320 square *braccia*. And therefore multiply 8 times 4320 slabs, which makes 34560 slabs. And we shall say that in that whole terrain of the hall or indeed piazza 34560 slabs will enter, neither more nor less. And it is done, and thus the others are done. And if you want to verify it, do thus, and say thus, this house is 120 *braccia* long and is 36 *braccia* wide. And we say that the slabs

diciamo che le pietre con que le volemo lastricare, ogni una è lungha uno mezzo braccio et alta $\frac{1}{4}$. Dunqua se ogni pietra è lungha $\frac{1}{2}$ braccio, entraranno in tucte queste 120 braccia de lunghezza 240 pietre. Et per ampiezza entrano 4 pietre per braccio. Et però **(fol. 28ʳ)** multiplica 4 via 240 pietre, fanno 960 pietre. Et tante pietre entrano per altezza in ogni braccio. Et tu voi sapere quante n'entreranno in 36 braccia. Et però multiplica 36 via 960 pietre, fanno 34560. Et sta bene. Et vedi che venne al modo che l'avemo facta noi.

with which we wanted to flag, each one is half a *braccio* long and $\frac{1}{4}$ high. Hence, if each slab is $\frac{1}{2}$ braccio long, 240 slabs will enter in all these 120 *braccia* of length. And in width 4 slabs enter per *braccio*. And therefore **(fol. 28ʳ)** multiply 4 times 240 slabs, they make 960 slabs. And so many slabs enter in height in each *braccio*. And you want to know how many will enter in 36 *braccia*. And therefore multiply 36 times 960 slabs, they make 34560. And it goes well. And you see that it results in the way that we have done.

14.27 In Cicilia et nello Regno de Puglia si è tucto uno peso et una mesura, et uno conto. Et dovete sapere che in Cicilia et in Puglia se fanno tucti li pagamenti a oncie, et a tari et a grani. Et 20 grani sonno uno tari. Et 30 teri sonno una oncia. Et questo se intende in conto ma non a peso. El *f* d'oro de Firenze se conta tari 6 al conto, et *f* cinque sonno una oncia a conto. Et quattro carlini d'oro sonno una oncia de conto, e'l carlino d'oro se conta tari 7 et $\frac{1}{2}$. Et 2 carlini d'argento se chontano uno tari d'oro de conto et de pagamento. Et in ciò darremo per meglio intendare uno assempro. Et diremo così, el marcho dell' argiento, el quale è 8 oncie, vale 36 teri et ⟨1⟩3 grani. Dimme quanto varranno li 47 marchi et oncie $6\frac{1}{2}$ del dicto argento. Fa chosì, primamente multiprica 47 via 36 teri, et di' così per più breve, 47 via oncie una et 6 teri fa oncie 56 et 12 teri. Et multipricha 13 via 47 grani, fanno 30 teri et 11

14.27 In Sicily and in the Kingdom of Apulia there is everywhere the same weight and the same measure and the same accounting. And you shall know that in Sicily and in Apulia all payments are done in ounces, and in *tari* and in grains. And 20 grains are one *tari*. And 30 *teri* are one ounce. And this is understood on account but not in weight. The gold *fiorino* from Florence is counted *tari* 6 on account, and five *fiorini* are one ounce on account. And four gold *carlini* are one ounce on account, and the gold *carlino* is counted *tari* 7 and $\frac{1}{2}$. And 2 silver *carlini* are counted one *tari* of gold on account and in payment. And for better understanding we shall give an example to this. And we shall say thus, the mark of silver, which is 8 ounces, is worth 36 *teri* and ⟨1⟩3 grains. Say me, how much will be worth 47 mark and ounces $6\frac{1}{2}$ of the said silver. Do thus, firstly multiply 47 times 36 *teri*, and say thus for shortness, 47 times

grani. Et ài in tucto oncie 57 et 12 teri et 11 grani. Ora sappi quanto vagliono 6 oncie et $\frac{1}{2}$, che vagliono 29 teri et 16 grani et $\frac{9}{16}$. Et ài provato 58 oncie et 12 teri et 7 grani et $\frac{9}{16}$. Et è facta, et tanto vagliono li 47 marchi et 6 oncie et $\frac{1}{2}$ d'argento, cioè 58 oncie 12 teri et 7 grani et $\frac{9}{16}$.

ounces one and 6 *teri* makes ounces 56 and 12 *teri*. And multiply 13 times 47 grains, they make 30 *teri* and 11 grains. And you get in all ounces 57 and 12 *teri* and 11 grains. Now know how much are worth 6 ounces and $\frac{1}{2}$, and they are worth 29 *teri* and 16 grains and $\frac{9}{16}$. And you have verified 58 ounces and 12 *teri* and 7 grains and $\frac{9}{16}$. And it is done, and so much are worth the 47 mark and 6 ounces and $\frac{1}{2}$ of silver, that is, 58 ounces 12 *teri* and 7 grains and $\frac{9}{16}$.

14.27A–B   [F.v.30–31]

14.27A–B   In the fairs of Champagne purchases and sales and all payments are made in *provisini forti*, and *provisini* are sold the dozen. And of this we shall give an example. The dozen of *forti*, that is, 12 *libre*, is worth *libre* 37, *soldi* 10. Say me, how much will 1443 *libre* of *forti* be worth. Now know that you should do thus, and say thus, 1200 *libre* are one dozen of hundreds, hence the 1200 *libre* of *forti* will be worth *libre* 3750. Now *libre* 243 are saved for you, and the 240 *libre* are two dozens of tens, and each 120 *libre* of *provisini* are worth *libre* 375, hence two dozens are worth *libre* 750, an you gave in total *libre* 4500. And we have to make the 3 *libre*, which are worth the $\frac{1}{4}$ of *libre* 37, *soldi* 10, that is, *libre* 9, *soldi* 7, *denari* 6. And you have in all *libre* 4509 *soldi* 7 *denari* 6. And it is done, and we shall say that *libre* 1443 of *forti* are worth *libre* 4509, *soldi* 7, *denari* 6 of whatever money you posit at the rate of *libre* 37 and $\frac{1}{2}$ the dozen of *provi-*

sini.

And you should know, if 12 *forti* are worth *denari* 37 $\frac{1}{2}$ of *fiorini*, then 12 *soldi* of *forti* are worth *soldi* 37 and 1*overs*2 of *soldo*, that is, 6 *denari*. And 12 *libre* of *forti* will be worth *libre* 37 and *soldi* 10 of *fiorini*. And 120 *libre* of *forti* will be worth *libre* 370 and half a ten, that is, *libre* 5, hence they will be worth *libre* 375. And similarly 1200 *libre* of *forti* will be worth *libre* 3750 of *fiorini*. And by this rule of the dozens you can find the value of as many *forti* as might be said to you.

14.28 Uno merchatante conparò el quintale dela lana, el quale pesa £ 100, £ 10. Or vene in capo de uno tempo che questa lana se bagnò. Et el dicto mercatante la fece rasciucchare. Et quando fo rasciuccha trovò che ogni quintale scemò £ 5, **(fol. 29ʳ)** cioè ogni 100 £ tornò 95. Dime quanto glie convenne vendere el centinaro de questa lana a fare suo capitale. Fa così, et di', £ 100 de lana vagliono £ 10. Vo' sapere que varrà le 95 £. Et però multiplica 100 via ^10 £^, fa 1000 £. Et parti in 95, che ne vene 10 £ et 10 ß et 6 deñ et $\frac{6}{19}$ de deñ. Et è facta, et dirremo che gle convene vendere el centinaro de questa lana, a fare suo capitale, £ 10 ß 10 deñ 6 et $\frac{6}{19}$ de deñ. Et in questo modo se fanno tucte le simili ragioni. Et sta bene.

14.28 A merchant bought the quintal of wool, which weighs pounds 100, at *libre* 10. Now it occurred after a certain time that this wool became wet. And the said merchant had it dried. And when it was dried he found that each quintal had shrunk 5 pounds, **(fol. 29ʳ)** that is, each 100 *libre* had become 95. Say me at how much it suits him to sell the *centinaio* of this wool in order to get back his capital. Do thus, and say, pounds 100 of wool are worth *libre* 10. I want to know what the 95 pounds will be worth. And therefore multiply 100 times 10 *libre*, it makes 1000 *libre*. And divide in 95, from which results 10 *libre* and 10 *soldi* and 6 *denari* and $\frac{6}{19}$ of *denaro*. And it is done, and we shall say that it suits him to sell the *centinaio* of this wool, in order to get back his capital, *libre* 10, *soldi* 10, *denari* 6 and $\frac{6}{19}$ of *denaro*. And in this way all the similar computations are done. And it goes well.

14.29 Io ho al lato due borse, nele quali ho denari. Et nell'una borsa òe il $\frac{1}{2}$ e il $\frac{1}{3}$ de tucti li denari che sonno in ammedore le borse, et nell'altra borsa ò 13 deñ. Vo' sapere quanti deñ io ho in ammedore queste borse. Fa così, trova uno numero che abia $\frac{1}{2}$ et $\frac{1}{3}$. Et questo numero poi dire che sia 12. Et di', $\frac{1}{2}$ et $\frac{1}{3}$ se trova in 12. Et poi di', el $\frac{1}{2}$ de 12 si è 6. Et el $\frac{1}{3}$ de 12 si è 4. Agiongi inseme 4 et 6, che è el $\frac{1}{2}$ e'l $\frac{1}{3}$ de 12, fanno 10. Et poi di' così, da 10 infino in 12 ày 2. Et questo di' che sia el partitore. Et poi perché tu di' che nell'altra borsa ài 13 deñ, multiprica 12 via 13, che fa 156. Et parti in dui, che ne vene 78. Et è facta, et 78 deñ diremo che fosseno in queste borse. Et se la voi provare, fa così, prendi el $\frac{1}{2}$ de 78, che è 39, et prendi el $\frac{1}{3}$, che è 26. Ora agiongi inseme 39 et 26, che fa 65. Et 65 denari dirremo che costui avesse nell'una borsa. Et poi di', da 65 infino in 78 ày 13. Et tanto à nela seconda borsia, como tu di'. Et sta bene. Et ài la provata. Et similmente se fanno le simili ragioni.

14.29 I carry two purses, in which I have money. And in one of the purses I have the $\frac{1}{2}$ and the $\frac{1}{3}$ of all the money that is in both purses, and in the other purse I have 13 *denari*. I want to know how many *denari* I have in both these purses. Do thus, find a number that possesses $\frac{1}{2}$ and $\frac{1}{3}$. And you may say that this number is 12. And say, $\frac{1}{2}$ and $\frac{1}{3}$ one finds in 12. And then say, the $\frac{1}{2}$ of 12 is 6. And the $\frac{1}{3}$ of 12 is 4. Join together 4 and 6, which is the $\frac{1}{2}$ and the $\frac{1}{3}$ of 12, they make 10. And then say thus, from 10 until 12 there is 2. And say that this is the divisor. And then because you say that in the other purse you have 13 *denari*, multiply 12 times 13, which makes 156. And divide in two, from which results 78. And it is done, and 78 *denari* we shall say what there were in these purses. And if you want to verify it, do thus, take the $\frac{1}{2}$ of 78, which is 39, and take the $\frac{1}{3}$, which is 26. Now join together 39 and 26, which makes 65. And 65 *denari* we shall say that this one had in one purse. And then say, from 65 until 78 there is 13. And so much he has in the second purse, as you say. And it goes well. And you have verified it. And the similar computations are done similarly.

14.30 Egli è uno muro, el quale è lungho 12 braccia e alto sette. Et grosso uno et $\frac{1}{4}$. Et io l'ò tucto murato de quadroni che **(fol. 29^v)** sonno ciascheuno quadrone, cioè ogni uno lungho $\frac{1}{2}$ braccia et ampio $\frac{1}{3}$, et alto

14.30 There is a wall, which is 12 *braccia* long and seven high. And thick one and . And I have built it all from ashlars which **(fol. 29^v)** are each ashlar, that is, each one, long   *braccia* and wide   and high .[a]

$\frac{1}{4}$. Vo' sapere quanti quadroni sonno andati in tucto questo muro. Fa così, primamente sappi quante braccia è tucto el muro, et fa così, multiprica la lunghezza contra all'altezza, que fa 12 via 7, che fa 84. Et poi multiplica 84 via j b̃ʳ et $\frac{1}{4}$, per la grossezza del muro, che fanno b̃ʳ 105. Et cotante b̃ʳ quadre è tucto el muro prodotto. Ora arrchamo ad b̃ʳ quadre el quadrone, et multipricha la lunghezza contra all'altezza, cioè $\frac{1}{2}$ et $\frac{1}{3}$, fa $\frac{1}{6}$, et per la grossezza, multiprica $\frac{1}{6}$ via $\frac{1}{4}$, fa $\frac{1}{24}$. Et diremo che in ogni b̃ʳ de questo muro entrino 24 de questi quadroni. Et perché noi vogliamo sapere quanti n'entrano in tucto questo muro, sì multipricha 24 via 105, che fa 2520. Et è facta, et diremo che nel decto muro entrano 2520 quadroni, né più né meno. Et per questo modo fa quantunqua fusse la lunghezza del muro et del quadrone.

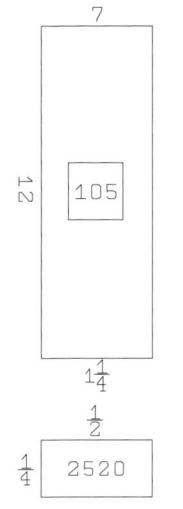

I want to know how many ashlars have gone into this whole wall. Do thus, firstly know how many *braccia* is the whole wall, and do thus, multiply the length against the height, which makes 12 times 7, which makes 84. And then multiply 84 times j *braccio* and $\frac{1}{4}$, for the thickness of the wall, which make *braccia* 105. And (in) as many square *braccia* is the whole wall constructed. Now we bring to square *braccia* the ashlar, and multiply the length against the height, that is, $\frac{1}{2}$ and $\frac{1}{3}$, it makes $\frac{1}{6}$, and by the thickness, multiply $\frac{1}{6}$ times $\frac{1}{4}$, it makes $\frac{1}{24}$. And we shall say that in each *braccio* of this wall enter 24 of these ashlars. And because we want to know how many enter in this whole wall, then multiply 24 times 105, which makes 2520. And it is done, and we shall say that in the said wall enter 2520 ashlars, neither more nor less. And do in this way whatever the length of the wall and the ashlar might be.

[a] Properly speaking, these stone bricks are thus not ashlars, which should be square-cut; but the same holds for *quadroni*.

14.31 La libra de zendadro me costa in Luccha 6 £ et 5 ß a *f*, et la £ de Luccha si è 12 oncie, la quale me torna in Monpuleri la £ oncie 15 et $\frac{1}{4}$ al peso de Monpuleri. Et el *f* d'oro de ~~Monpuleri~~ Firenze vale a Monpuleri ß 13 δ 4 de tornesi. Et in Luccha vale ß 29 a *f*. Vo' sapere per quanto porrò dare la £ a Monpuleri al peso de Monpuleri, cioè, per quanti ß de tornesi, a'cciò ^che io^ ne faccia mia capitale. Fa

14.31 The pound of *zendado* costs me in Lucca 6 *libre* and 5 *soldi* of *fiorini*, and the pound of Lucca is 12 ounces, which in Montpellier becomes for me for the (Lucca) pound ounces 15 and $\frac{1}{4}$ in the weight of Montpellier. And the gold *fiorino* from Florence is worth in Montpellier *soldi* 13, *denari* 4 of *tornesi*. And in Lucca it is worth *soldi* 29 of *fiorini*. I want to know for how much I shall be able to give the

così et di', oncie 15 et $\frac{1}{4}$ in Monpuleri mi costano a Luccha £ 6 et ß 5. Vo' sapere que varranno 12 oncie in Monpuleri. Multiprica 12 via 6 £ et 5 ß, che fanno 75 £, et parti in 15 et $\frac{1}{4}$, che ne vene 4 £ et 18 ß 4 deñ $\frac{20}{61}$. Et per cotanto po dare la £ in Monpuleri a f. Ora arechamo ad moneta tornesi. Noi diciamo che'l f vale in Luccha ß 29 a f. Et le £ 5 [sic] ß 18 deñ 4 $\frac{20}{61}$ de deñ a f sonno f 4 ß 2 deñ 4 $\frac{20}{61}$ de deñ a f. Et ß 2 deñ 4 $\frac{20}{61}$ de denaro a f possiamo dire che vagliono deñ 13 de tornesi però che ogni ß de tornesi vale ß 2 deñ 2 $\frac{1}{10}$ de deñ a f. Et se la voli provare, sì parti 29 ß a f per ß 13 deñ 4 de tornesi, como tu di' che vale el f a tornesi, et vin'ne como te dicho ß 2 deñ 2$\frac{1}{10}$. Siché como tu dedi (sic, read "vedi") poi dare la £ in Monpuleri per f 4 et ß j deñ j de tornesi. Et sta bene. Et se voi sapere per quanti ß de tornesi la poi dare, sì multiprica 4 via ß 13 (fol. 30^r)deñ ^4^, fa ß 53 deñ 4, et ponvi suso anco ß j deñ j, ài in tucto ß 54 deñ 5 de tornesi. Et è facta. Et così se fanno tucte le simigliante ragioni.

pound at Montpellier at the weight of Montpellier, that is, for how many *soldi* of *tornesi*, in order to get back from it my capital. Do thus and say, ounces 15 and $\frac{1}{4}$ in Montpellier cost me in Lucca *libre* 6 and *soldi* 5. I want to know what will be worth 12 ounces in Montpellier. Multiply 12 times 6 *libre* and 5 *soldi*, which make 75 *libre*, and divide in 15 and $\frac{1}{4}$, from which results 4 *libre* and 18 *soldi*, 4 *denari* $\frac{20}{61}$. And for as much can I give the *libre* in Montpellier in *fiorini*. Now we bring to *tornesi* coin. We say that the *fiorino* is worth in Lucca 29 *soldi* in *fiorini*. And the *libre* ⟨4⟩, *soldi* 18, *denari* 4, $\frac{20}{61}$ of *denaro* in *fiorini* are *fiorini* 4 *soldi* 2 *denari* 4, $\frac{20}{61}$ of *denaro* in *fiorini*. And we can say that *soldi* 2, *denari* 4, $\frac{20}{61}$ of *denaro* in *fiorini* are worth *denari* 13 of *tornesi* because each *soldo* of *tornesi* is worth *soldi* 2, *denari* 2, $\frac{1}{10}$ of *denaro* in *fiorini*. And if you want to verify it, then divide 29 *soldi* of *fiorini* by *soldi* 13, *denari* 4 of *tornesi*, as you say that the *fiorino* is worth in *tornesi*, and from this results, as I say to you, *soldi* 2, *denari* 2$\frac{1}{10}$. So that, as you {see}, you can give the pound in Montpellier for *fiorini* 4 and *soldo* j, *denaro* j of *tornesi*. And it goes well. And if you want to know for how many *soldi* of *tornesi* you may give it, then multiply 4 times *soldi* 13, (fol. 30^r) *denari* 4, it makes *soldi* 53 *denari* 4, and put on top there also *soldo* j, *denaro* j, you have in all *soldi* 54 *denari* 5 of *tornesi*. And it is done. And thus all the similar computations are done.

14.31A   [F.v.36]

14.31A   The rod [*canna*] of cloth, which is 4 *braccia* in Florence and is palms 8 and $\frac{3}{4}$ in the measure of Nîmes. And it costs me *soldi* 43 of *fiorini*, say me for how many *soldi* of *tornesi* I can give the rod of Nîmes, which is 8 palms. And the gold *fiorino* is worth in Florence *soldi* 29 of *fiorini*, and in Nîmes it is worth *soldi* 13, *denari* 4 of *tornesi*. Do thus, say, palms 8 and $\frac{3}{4}$ of Florence (*sic*) are worth *soldi* 43 of *fiorini*; how much are 8 palms of Nîmes worth? Multiply *soldi* 43 by 8 and $\frac{3}{4}$, which makes *libre* 18, *soldi* 16, *denari* 3. And divide *libre* 18, *soldi* 16, *denari* 3 by 8, from which results *soldi* 47 and $\frac{3}{8}$ of one *denaro*. And as much will it suit him to sell the rod of Nîmes, that is, *soldi* 47 and $\frac{3}{8}$ of a *denaro* of *fiorini*. Now let us bring them to *tornesi*, and say: soldi 43 $\frac{1}{2}$, that is, one gold *fiorino* and a half, are worth *soldi* 20 of *tornesi*; *soldi* 3, *denari* 6 of *fiorini* are saved, which in *tornesi* are *denari* 21, and you have in all *soldi* 21, *denari* 9. And as will it suit him to sell the rod of cloth in Nîmes, that is, *soldi* 21 and *denari* 9 of *tornesi*, and it is done.

14.32   Sonno tre conpagni che ciascheuno ànno denari in borsa. Dice l'uno agli altri due, io ho in borsa el $\frac{1}{4}$ de tucti denari che noi abiamo tra tucti noi. Dice l'altro conpagnio, et io ho in borsa l'octava. Dice el terzo conpagnio, et io ho in borsa uno denaro et non più. Vo' sapere quanti deñ costoro ànno tra tucti et tre, et quanti n'ànno per uno. Fa così et di', $\frac{1}{2}$ et $\frac{1}{8}$ se trova

14.32   There are three fellows who, each, have money in their purse. One says to the other two, I have in (my) purse $\frac{1}{4}$ of all the money that we all have together. The other fellow says, and I have in (my) purse the eighth. The third fellow says, and I have in (my) purse one *denaro* and no more. I want to know how many *denari* they have all three together, and how many (each) one

in otto. Ora prendi el $\frac{1}{2}$ de octo, che è 4. Et prendi l'octavo de otto, che è j. Agiongi inseme 4 et 1, fa 5, et questo è el partitore. Ora perché tu di' che el terzo conpagnio à j deñ et non più, sì multipricha 8 via uno denaro, fa 8 deñ. Et parti in 5, che ne vene uno denaro et $\frac{3}{5}$. Et cotanti deñ ànno tra tucti e tre costoro. Ora se voi sapere quanti deñ à ciascheuno de loro, sì togli $\frac{1}{4}$ de uno deñ et $\frac{3}{5}$, che è $\frac{2}{5}$ de uno deñ. Et cotanto à colui che dice che à $\frac{1}{4}$ de tucti. Et poi togli $\frac{1}{8}$ de uno denaro et $\frac{3}{5}$, che è uno quinto. Et cotanto à collui che dice che à l'octavo. Ora trai $\frac{2}{5}$ et $\frac{1}{5}$ de j denaro et $\frac{3}{5}$, resta appunto uno denaro intero. Et cotanto à el terzo, cioè uno denaro come tu di'. Et è facta et provata. Et sta bene. Et così se fanno tucte simili ragioni.

has. Do thus and say, $\frac{1}{2}$ and $\frac{1}{8}$ one finds in eight. Now take the $\frac{1}{2}$ of eight, which is 4. And take the eighth of eight, which is j. Join together 4 and 1, it makes 5, and this is the divisor. Now because you say that the third partner has j *denaro* and no more, then multiply 8 times one *denaro*, it makes 8 *denari*. And divide in 5, from which results one *denaro* and $\frac{3}{5}$. And as many *denari* do all those three have together. Now if you want to know how many *denari* each of them has, then grasp $\frac{1}{4}$ of one *denaro* and $\frac{3}{5}$, which is $\frac{2}{5}$ of one *denaro*. And as much has the one who says he has $\frac{1}{4}$ of all. And then grasp $\frac{1}{8}$ of one *denaro* and $\frac{3}{5}$, which is one fifth. And as much has the one who says he has the eighth. Now detract $\frac{2}{5}$ and $\frac{1}{5}$ of j *denaro* and $\frac{3}{5}$, precisely one whole *denaro* is left. And as much has the third, that is, one *denaro* as you say. And it is done and verified. And it goes well. And thus are done all similar computations.

[15. Practical Geometry, with Approximate Computation of Square Roots]

15.1 *In nomine Domini amen. Qui appresso incominciaremo, et dirremo de tucte maniere de mesure. Et primamente dirremo del tundo ad conpasso. Et in ciò mostraremo l'assemplo per propria regola.*

15.1 *In the name of God, amen. Next to here we shall begin, and speak of all modes of measurement. And firstly we shall speak of the compass-made round. And about this we shall show the example by proper rule.*

(fol. 30ᵛ)

15.2 **Egl'è** uno terreno, el quale è tucto rotondo como tu vedi de rinpetto, el quale

15.2 There is a terrain, which is all round, as you see opposite, which goes around in

gira dintorno braccia 44. Vo'
sapere quanto è el suo diametro,
cioè, per lo diricto de mezzo.
Dè se fare così, et questa è la
sua propria et legitima regola.
Sempre fa che quando tu sai la
sua circumferentia dintorno, cioè
la sua mesura, et tu voi sapere quanto è el
suo diricto de mezzo, sì parti la circumferentia sua per 3 e $\frac{1}{7}$. Et quello che ne
vene, tanto serà el suo diametro, cioè, el
dericto de mezzo. Et similmente quando tu
sai el dericto de mezzo de una circumferentia et tu voli sapere quanto gira dintorno,
sì multiplica el dericto de mezzo per 3 e $\frac{1}{7}$,
et cotanto quanto farrà, tanto gira dintorno
el dicto tundo. Et se volissi sapere per che
cagione parti et moltipriche per 3 e $\frac{1}{7}$, sì
te dico che la ragione è perché ogni tundo
de qualunqua mesura se sia è intorno {in-
torno} 3 volte et $\frac{1}{7}$ quanto è el suo diametro, cioè el diricto de mezzo. Et per
questa cagione ày a moltiplichare et partire
como io t'ò dicto de sopra. Dunqua, como
dice la nostra regola, abbiamo a partire la
circumferentia del tundo sopradicto, che è
44 braccia, per 3 e $\frac{1}{7}$, che ne vene 14
appunto. Et 14 braccia diremo che sia el
suo diametro. Cioè el diricto de mezzo. Et
è facta. Et così se fanno tucte le simili
ragioni.

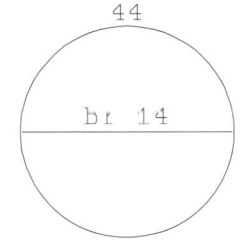

44

b r 14

braccia 44. I want to know how
much is its diameter, that is,
(how much it is) by the straight
in middle. One shall do thus,
and this is its proper and legitimate rule. Always do, that
when you know its circumfer-
ence around, that is, its measure, and you
want to know how much is its straight in
middle, then divide its circumference by 3
and $\frac{1}{7}$. And that which results from it, so
much will its diameter be, that is, the
straight in middle. And similarly when you
know the straight in middle of a circumference and you want to know in how much
it goes around, then multiply the straight
in middle by 3 and $\frac{1}{7}$, and as much as it
makes, in so much does the said round go
around. And if you should want to know
for which cause you divide and multiply by
3 and $\frac{1}{7}$, then I say to you that the reason
is that every round of whatever measure it
might be is around {...} 3 times and $\frac{1}{7}$ as
much as is its diameter, that is, the straight
in middle. And for this cause you have to
multiply and divide as I have said to you
above. Hence, as our rule says, we have to
divide the circumference of the round
spoken of above, which is 44 braccia, by
3 and $\frac{1}{7}$, from which results precisely 14.
And 14 braccia we shall say that its diameter is. That is, the straight in middle.
And it is done. And thus all the similar
computations are done.

15.3 **Ora** diremo un'altra assempro del

15.3 Now we shall say another example of

tondo, et disegnaremo de rinpetto. Et diremo così, egli è uno tondo ad conpasso, el quale è per lo diricto de mezzo braccia 19. Vo' sapere quanto gira tucto dintorno. Fa così como dice la nostra re-gola. Multiprica 19 via 3 e $\frac{1}{7}$, che fa 59 et $\frac{5}{7}$. Et 59 b̃ʳ et $\frac{5}{7}$ de braccio diremo che sia la sua circumferentia dintorno. Et è facta. Et sta bene.

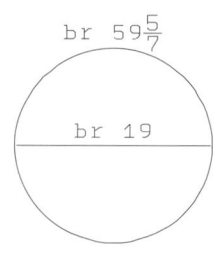

br $59\frac{5}{7}$

br 19

the round, and we shall draw (it) opposite. And we shall say thus, there is a compass-made round, which by the straight in middle is *braccia* 19. I want to know how much it turns around in all. Do thus as our rule says. Multiply 19 times 3 and $\frac{1}{7}$, which makes 59 and $\frac{5}{7}$. And 59 *braccia* and $\frac{5}{7}$ of *braccio* we shall say that its circumference around is. And it is done. And it goes well.

15.4 Uno terreno, el quale è tucto ritondo a conpasso, et gira din-torno b̃ʳ 22, como tu vedi de rinpetto designato. Vo' sapere quante b̃ʳ quadre serà tucto que-sto terreno dentro da questo circhio. Fa così, parti 22 per 3 e $\frac{1}{7}$, che ne vene 7 braccia. Et 7 braccia è el suo diametro di mezzo. Ora multiprica 7 via 22, **(fol. 31ʳ)** che fa 154, parti 154 in 4, che ne vene 38 b̃ʳ e $\frac{1}{2}$. Et 38 braccia et $\frac{1}{2}$ quadre è tucto questo terreno, sicomo te mostro designato de rimpetto. Et è facta.

br 22

serra quadro

br $38\frac{1}{2}$

15.4 A terrain, which is wholly compass-made round, and goes around in *braccia* 22, as you see drawn opposite. I want to know how many square *braccia* this whole terrain within this circle will be. Do thus, divide 22 by 3 and $\frac{1}{7}$, from which results 7 *braccia*. And 7 *braccia* is its diameter in middle. Now multiply 7 times 22, **(fol. 31ʳ)** which makes 154, divide 154 in 4, from which results 38 *braccia* and $\frac{1}{2}$. And 38 square *braccia* and $\frac{1}{2}$ is this whole terrain, as I show you drawn opposite. And it is done.

15.5 Uno terreno à tre canti, si-como tu vidi designato de rin-petto. Et dui canti diricti, et l'altra faccia scuadrata. Et el lato diricto, et minore, è b̃ʳ 30. Et l'altro lato è b̃ʳ 40. Voglio sapere quanto serà l'altra faccia, cioè la squadrante dall'una punta del terreno all'altra. Fa così, multiprica primamente

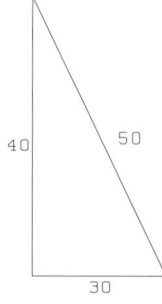

40   50

30

15.5 A terrain has three edges, as you see drawn opposite. And two straight edges, and the other face skew. And the straight side, and smaller, is *braccia* 30. And the other side is *braccia* 40. I want to know how much the other face will be, that is, the skew one from one tip of the terrain to the other.

ogni faccia per se medesima, de quelle che tu sai la mesura. Cioè, 40 via 40, fa 1600, et poi multiprica 30 via 30, fa 900. Agiongi inseme 1600 et 900, che fanno 2500. Ora trova la sua radice de 2500, che è 50, però che 50 via 50 fa 2500. Et 50 $\tilde{b}^r$ dirremo che à la terza faccia de questo terreno. Et è facta. Et così se fanno le simili ragioni.

Do thus, multiply firstly each face by itself, of these of which you know the measure. That is, 40 times 40, it makes 1600, and then multiply 30 times 30, it makes 900. Join together 1600 and 900, which make 2500. Now find the root of 2500, which is 50, because 50 times 50 makes 2500. And 50 *braccia* shall we say that the third face of this terrain is. And it is done. And thus the similar computations are done.

15.6 Uno terreno è quadro, cioè con quattro faccie, come vedi figurato de rimpetto. El quale è lungo per ogni faccia 10 $\tilde{b}^r$. Voglio sapere quanto serà dall'una punta del terreno ad l'altra, misurando ad traverso. Fa così, multiprica l'una faccia del terreno contra ad l'altra, cioè 10 via 10, fa 100. Et ancora multiprica le altre dui, 10 via 10, fa 100. Ora agiongi inseme, fa 200. Et de questo trova la sua radice, che è 14 e $\frac{1}{7}$. Cioè la più pressa, però che {appunto} appunto non si può trovare. Et cotanto diremo che serà dall'una punta all'altra del terreno, cioè $\tilde{b}^r$ 14 e $\frac{1}{7}$. Ed è facta. Et così se fanno le simili ragioni.

15.6 A terrain is square, that is, with four faces, as you see in figure opposite. Which is 10 *braccia* long by each face. I want to know how much there will be from one tip of the terrain to the other, measuring across. Do thus, multiply one face of the terrain against the other, that is, 10 times 10, it makes 100. And further multiply the other two, 10 times 10, it makes 100. Now join together, it makes 200. And of this find its root, which is 14 and $\frac{1}{7}$. That is, the closest, because {...} precisely it cannot be found. And as much we shall say that there will be from one tip to the other of the terrain, that is, *braccia* 14 and $\frac{1}{7}$. And it is done. And thus the similar computations are done.

15.7 Una torre, come tu vederai designata qui, è alta 30 $\tilde{b}^r$. Et a pede dela dicta torre è una serpe, che vole salire in suso la torre. Et ogni dì sale suso uno terzo de $\tilde{b}^r$. Et la nocte scende $\frac{1}{4}$. Vo' sapere in quanto

15.7 A tower, as you will see drawn here, is 30 *braccia* high. And at the foot of the said tower there is a serpent, which wants to climb to the top of the tower. And each day it climbs toward the top one third of

tempo serà salita insuso la dicta torre. Fa così, et di', $\frac{1}{3}$ et $\frac{1}{4}$ se trova in 12. Et poi di', el $\frac{1}{3}$ de 12 si è 4, cioè che sonno $\frac{4}{12}$. Et poi di', el $\frac{1}{4}$ de 12 si è 3, cioè $\frac{3}{12}$. Siché, como **(fol. 31ʳ)** tu vedi, la serpe sale el dì $\frac{4}{12}$ de b̃ʳ. Et la nocte discende $\frac{3}{12}$, siché in tucto viene ad acquistare tra el dì et la nocte $\frac{1}{12}$ de braccio. Et in 12 dì colle nocte viene a salire uno b̃ʳ. Ora se voli sapere in quanto serà in cima dela torre, che diciamo che è alta b̃ʳ 30, sì multiprica 12 via 30, che fa 360. Et in 360 dì tra dì et nocte serà salita in suso la torre. Et è facta et sta bene. Et così se fanno le simili ragioni.

braccio. And in the night it descends $\frac{1}{4}$. I want to know in how much time it will have climbed to the top of the said tower. Do thus, and say, $\frac{1}{3}$ and $\frac{1}{4}$ one finds in 12. And then say, the $\frac{1}{3}$ of 12 is 4, that is, that they are $\frac{4}{12}$. And then say, the $\frac{1}{4}$ of 12 is 3, that is, $\frac{3}{12}$. So that, as **(fol. 31ʳ)** you see, the serpent climbs the day $\frac{4}{12}$ of braccio. And in the night it descends $\frac{3}{12}$, so that in all it comes to acquire in day and night together $\frac{1}{12}$ of braccio. And in 12 days with the nights it comes to climb one braccio. Now if you want to know in how much (time) it will be on the peak of the tower, which we said to be braccia 30 high, then multiply 12 times 30, which makes 360. And in 360 days, days and nights together it will have climbed on top of the tower. And it is done and it goes well. And thus the similar computations are done.

¹⁵·⁸ Uno terreno à 4 faccie, le quale sonno le due faccie magiore che l'altre due, como tu vedi de rinpetto designato. Le quale due faccie magiori sonno lunghe ogni una b̃ʳ 60. Et le altre due sonno b̃ʳ 17. Dime quante b̃ʳ quadre serà tucto questo torreno. Fa così, et questa è la sua regola. Multiprica una dele faccie minori contra una dele magiori, cioè 17 via 60, che fa ~~ehe fa~~ 1020. Et 1020 b̃ʳ

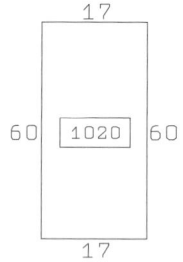

¹⁵·⁸ A terrain has 4 faces, two of which are larger than the other two, as you see drawn opposite. Each of which two larger faces are long braccia 60. And the other two are braccia 17. Say me, how many square braccia will this whole terrain be. Do thus, and this is its rule. Multiply one of the smaller faces against one of the larger, that is, 17 times 60, which makes 1020. And 1020 square

quadre dirremo che serà tucto questo terreno. Et è facta, et sta bene. Et così se fanno le simili ragioni.

braccia shall we say that this whole terrain will be. And it is done, and it goes well. And thus the similar computations are done.

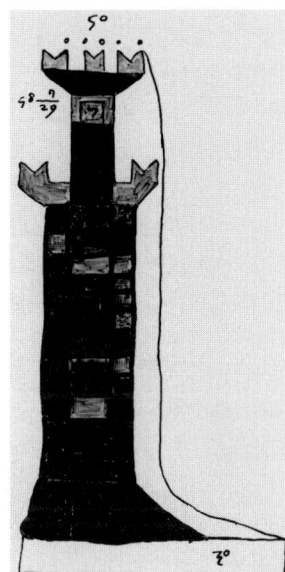

15.9 Una torre è alta 50 braccia sicomo tu vedi designata de rinpetto. Et a'ppe dela dicta torre è uno fosso, el quale è largho 30 b̃ʳ. Ora io voglio portare una fune dala cima dela torre infino all'orlo del fosso. Vo' sapere quanto serà lungha la dicta fune. Fa chosì, multiprica l'altezza dela torre per se medesema, cioè 50 via 50, che tu di' che è alta, che fa 2500. Et ancora multipricha per se medesemo la larghezza del fosso, che è largho 30 b̃ʳ. Et di', 30 via 30, che fa 900. Ora agiongi inseme 2500 et 900, che fanno 3400. Ora trova la sua radice, cioè de 3400, la quale è 58 e $\frac{9}{29}$, cioè la più presso, et più presso non si po trovare. Però che 58 e $\frac{9}{29}$ via 58 e $\frac{9}{29}$ fa 3400 et $\frac{81}{841}$, et appunto non si trova. Siché noi possiamo dire che la fune che noi vogliamo porre dala sponda del fosso infino ala cima dela torre serrà lungha b̃ʳ 58 e $\frac{9}{29}$ de braccio. Et è facta. Et chosì se fanno tucte le simili ragioni.

15.9 A tower is 50 *braccia* high as you see drawn opposite. And at the foot of the said tower there is a moat, which is 30 *braccia* broad. Now I want to carry a rope from the peak of the tower until the border of the moat. I want to know how much the said rope will be long. Do thus, multiply the height of the tower by itself, that is, 50 times 50, which you say it is high, which makes 2500. And further multiply by itself the breadth of the moat, which is 30 *braccia* broad. And say, 30 times 30, which makes 900. Now join together 2500 and 900, which make 3400. Now find its root, that is, of 3400, which is 58 and $\frac{9}{29}$, that is, the closest, and closer one cannot find. Because 58 and $\frac{9}{29}$ times 58 and $\frac{9}{29}$ makes 3400 and $\frac{81}{841}$, and precisely it is not found. So that we can say that the rope that we want to carry from the bank of the moat until the peak of the tower will be *braccia* 58 and $\frac{9}{29}$ of *braccio* long. And it is done. And thus all the similar computations are done.

15.10 Una torre è alta 40 braccia, como tu

15.10 A tower is 40 *braccia* high, as you see

vidi designata de rimpetto. Et a'ppe della torre si è uno fosso. Et io pongho una fune dala cima dela torre ala sponda del dicto fosso. La quale fune è lungha 50 b̃ʳ, né più né meno. Vo' sapere quanto è **(fol. 32ʳ)** largho el dicto fosso. Fa così, multiprica l'altezza del^a^ dicta ~~fosse~~ torre per se medesema, chome tu facesti de sopra all' altra torre. Et di', 40 via 40 fa 1600. Et ancora multiprica la lunghezza dela fune, cioè 50 via 50, fa 2500. Ora non si vuole agiungere inseme come tu facesti quella de sopra. Ancho se vole trare l'una multipricatione dell'altra, cioè, 1600 de 2500, che resta 900. Et de questo se vole trovare la sua radice, che è 30, però che 30 via 30 fa 900. Et è facta, et dirremo che el dicto fosso è largho b̃ʳ 30, né più né meno. Et così se fanno le simiglianti ragioni.

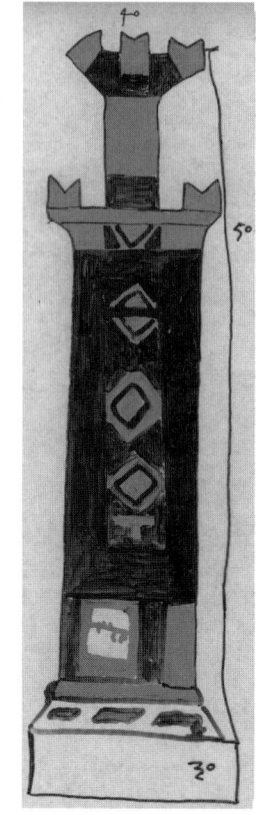

drawn opposite. And at the foot of the tower there is a moat. And I stretch a rope from the peak of the tower to the bank of the said moat. Which rope is 50 *braccia* long, neither more nor less. I want to know how much **(fol. 32ʳ)** the said moat is broad. Do thus, multiply the height of the said tower by itself, as you did above with the other tower. And say, 40 times 40 makes 1600. And further multiply the length of the rope, that is, 50 times 50, it makes 2500. Now one shall not join together as you made the one above. Instead one shall detract one multiplication from the other, that is, 1600 from 2500, and 900 is left. And from this one shall find its root, which is 30, because 30 times 30 makes 900. And it is done, and we shall say that the said moat is *braccia* 30 broad, neither more nor less. And thus are done the similar computations.

15.11 Uno tundo ad conpasso como tu vidi designato de rimpetto gira dintorno 100 b̃ʳ. Vo' sapere quanto serà el diametro suo, cioè el dericto de mezzo. Fa così, io t'ò anche dicto de sopra, ogni tondo, a volere sapere quanto è el suo diametro, si vole partire per 3 e $\frac{1}{7}$. Et però parti 100 per

15.11 A compass-made round as you see drawn opposite goes around in 100 *braccia*. I want to know how much its diameter will be, that is, the straight in middle. Do thus, I have also said it to you above, (for) every round, if one wants to know how much is its diameter, one

3 e $\frac{1}{7}$, et quello che ne vene, tanto è'l suo diametro, cioè, tante braccia. Et se non sapissi partire per 3 e $\frac{1}{7}$, sì te insegnarò. Fo così, ogni volta che tu averai a partire per numero sano ^et rotto^, si vole arrechare a rotti tucto el numero, cioè, quelli rotti che tu ài a partire. Et similemente anchora quello numero che tu ài a partire. Et fa così per meglio intendare. Io te dicho che tu parte 100 b̃ᵣ per 3 et $\frac{1}{7}$. Et però arrecha a settimi tucto el partitore. Et di' così, 7 via 3 e $\frac{1}{7}$ fa 22, et questo è el partitore. Et poi arrecha a settimi le 100 braccia, che sonno 7 via 100, fa 700 settimi. Ora parti 700 settimi in 22, che ne vene 31 et $\frac{9}{11}$. Et 31 braccio et $\frac{9}{11}$ de braccio dirremo che serrà el diametro de questo tundo. Et è facta, et sta bene. Et per altro modo che io te dica non se po partire, a volere appunto la ragione, et così se fanno tucte le altre.ᵃ

shall divide by 3 and $\frac{1}{7}$. And therefore divide 100 by 3 and $\frac{1}{7}$, and that which results from it, so much is its diameter, that is, so many *braccia*. And if you should not know how to divide by 3 and $\frac{1}{7}$, then I shall teach you. Do thus, each time you shall have to divide by an integer and broken number, one shall bring the full number to fractions, that is, these fractions (by) which you have to divide. And similarly also that number which you have to divide. And do thus for understanding better. I say to you to divide 100 *braccia* by 3 and $\frac{1}{7}$. And therefore bring to sevenths the full divisor. And say thus, 7 times 3 and $\frac{1}{7}$ makes 22, and this is the divisor. And then bring to sevenths the 100 *braccia*, which are 7 times 100, it makes 700 sevenths. Now divide 700 sevenths in 22, from which results 31 and $\frac{9}{11}$. And 31 *braccio* and $\frac{9}{11}$ of *braccio* shall we say that the diameter of this round will be. And it is done, and it goes well. And in another way than what I say one cannot divide if one wants the computation to be precise, and thus are done all the others.ᵃ

ᵃIn translation, the corresponding problem in **M+F** runs as follows (a dittography in **F** is left out): A compass-made round, which goes around in 100 *braccia*, say me, how much is its diameter, that is, the straight in middle. Do thus, and this is its proper rule. Divide 100 by 3 and ¹⁄₇ in this way. Say, 7 times 3 and ¹⁄₇ make 22. And say, 7 times 100 make 700, and as much is to divide 700 by twenty-two as 100 by 3 and ¹⁄₇, from which results 31 and ⁹⁄₁₁. And as much is this round by the straight in middle, that is, *braccia* 31 and ⁹⁄₁₁ of *braccio*, such as I show you in the drawn diagram [*forma*].

15.12 **Questa** è una regola, la quale se insegna trovare radice **(fol. 32ᵛ)** a ogni numero, cioè la più pressa radice che ⟨abbia⟩ quello cotale numero del quale noi voles-

15.12 This is a rule which teaches to find the root **(fol. 32ᵛ)** to every number, that is, the closest root that such a number ⟨has⟩ whose root we wanted to find. Because every

semo trovare la sua radice. Però che ogni numero non à radice appunto. Et principalmente diciamo così per assempro ala dicta regola per meglio intendare. Et questa serà la sua propria regola.[a]

La radice de 4 si è

2 però che 2 via 2 fa 4. Et la radice de 9 si è
3 però che 3 via 3 fa 9. Et la radice de 16 si è
4 però che 4 via 4 fa 16. Et la radice de 25 si è
5 però che 5 via 5 fa 25. Et la radice de 36 si è
6 però che 6 via 6 fa 36. Et la radice de 49 si è
7 però che 7 via 7 fa 49. Et la radice de 64 si è
8 però che 8 via 8 fa 64. Et la radice de 81 si è
9 però che 9 via 9 fa 81. Et la radice de 100 si è
10 però che 10 via 10 fa 100. Et la radice de 121 si è
11 però che 11 via 11 fa 121. Et così poteremo andare a questo modo infino a 100. Et 100 via 100 è la radice de 10000. Et così addevene de ogni altro numero che è multiplicato in se medesimo. Quello medesimo numero è radice dela sua multipricatione. Sicomo tu vedi (sic, read "vedi" per assempro neli soprascripti.

number does not have root precisely. And firstly we say thus as example to the said rule for understanding better. And this will be its proper rule.[a]

The root of 4 is

2 because 2 times 2 makes 4. And the root of 9 is
3 because 3 times 3 makes 9. And the root of 16 is
4 because 4 times 4 makes 16. And the root of 25 is
5 because 5 times 5 makes 25. And the root of 36 is
6 because 6 times 6 makes 36. And the root of 49 is
7 because 7 times 7 makes 49. And the root of 64 is
8 because 8 times 8 makes 64. And the root of 81 is
9 because 9 times 9 makes 81. And the root of 100 is
10 because 10 times 10 makes 100. And the root of 121 is 11 because 11 times 11 makes 121. And thus we could go in this way until 100. And 100 times 100 is the root of 10000. And thus happens with every other number which is multiplied in itself. This same number is the root of its multiplication. So as you {see} by example in what was written above.

[a] In translation, the introduction to the corresponding passage in M.15.12=F.unnumbered runs as follows:

> This is a rule which shows us how to find the root of every number of which one can find the root, or indeed the closest root that one can find. And this we shall show by proper rule. First we say thus, as example:

In the following list, **M+F** gives the true roots of 4, 9, 16, 100, 169 and 10000. The closing remarks are almost the same as in that of **V**.

15.13 **Ora** dirremo in qual modo se trova radice a ogni numero, cioè a quello che non l'ànno appunto, la più pressa. Sappi che tu dì fare così, ogni volta che tu volissi trovare radice a uno numero che non l'avesse over che non la sapessi, se vole trovare uno numero che multipricato per se medesimo sia più presso a quello numero del quale

15.13 Now we shall say in which way the root is found to every number, that is, to those which do not have it precisely, the closest. Know that you shall do thus, each time that you might want to find root to a number that does not have it or indeed that you do not know it, one shall find a number which multiplied by itself is closer to

voli trovare la radice che nisi-
uno altro numero, et poi partire et [*sic*, read "el"][M+F] remanente che fusse da quello
insuso per lo duppio de quillo
cotal numero che tu avissi
multipricato. Et quello che ne
vene po⟨n⟩i sopra al numero
(fol. 33ʳ) che tu multipricasti,
et quella serà la più pressa radice.[a]

that number of which you
want to find the root than any
other number, and then divide
⟨the⟩[M+F] remainder that there
might be from this upwards by
the double of that number
which you had multiplied. And
that which results from it, put
above the number **(fol. 33ʳ)**
which you multiplied, and this
will be the closest root.[a]

[a] In translation, the corresponding
text in M.15.13=F.unnumbered
runs as follows:

> Now we shall say in which
> way the root can be found
> for every number for which
> it can be found, or indeed the
> closest root. Know that you
> shall do thus. You shall find
> a number which, when multi-
> plied by itself, is closer to
> the number of which you
> want to find root than any
> other number. And them
> divide the remainder by the
double of that number which you multiplied. And in this way one finds true or closest root.

[15.14] Et in ciò dirremo l'assempro a'cciò
che tu intende meglio. Et dirremo così,
trovami radice de 10. Et 10 è uno de quelli
numeri che non la po avere appunto. Ma
a volere trovare la più pressa, fa chosì
chomo io t'ò dicto de sopra, trova uno
numero che multipricato per se medesemo
sià più presso a 10. Et questo numero è 3,
perché 3 via 3 fa 9. Ora te dicho che quello
numero che tu multipriche si vole raddop-
piare et in quello se vole partire quello che

[15.14] And to this we shall say the example
in order that you understand better. And we
shall say thus, find me the root of 10. And
10 is one of these numbers that cannot have
it precisely. But wanting to find the closest,
do thus as I have said to you above, find
a number that multiplied by itself will be
closest to 10. And this number is 3, because 3 times 3 makes 9. Now I say to you
that this number which you multiply shall
be doubled, and in this shall be divided that

te avanza quando l'ai multipricato, da indi in su perfino a quello numero de que voi trovare la radice. Ora tu di' che voi trovare radice de 10, et ày multipricato 3 via 3, fa 9, per infino in 10 à uno. Ora parti j in dui cotanti che el numero che tu multipricasti, cioè nel duppio de 3, che è 6. Parti j in 6, che ne vene $\frac{1}{6}$. Ora poni $\frac{1}{6}$ sopra al numero che tu multiprichasti, cioè a 3, et ày 3 e $\frac{1}{6}$. *Et 3 et $\frac{1}{6}$ dirremo* dirremo che sia la più pressa radice che abbia 10. Et questa se chiama la sua radice, ma non appunto. Però che 3 e $\frac{1}{6}$ via 3 e $\frac{1}{6}$ fa 10 e $\frac{1}{36}$. Et per questo modo poi trovare radice a ogni numero, chomo io t'ò dicto.[a]

which is left over for you when you have multiplied it, from there upward until that number of which you want to find the root. Now you say that you want to find the root of 10, and you have multiplied 3 times 3, it makes 9, until 10 there is one. Now divide j in two (times) as many as the number that you multiplied, that is, in the double of 3, which is 6. Divide j in 6, from which results $\frac{1}{6}$. Now put $\frac{1}{6}$ above the number that you multiplied, that is, above 3, and you have 3 and $\frac{1}{6}$. And 3 and $\frac{1}{6}$ we shall say to be the closest root that 10 possesses. And this is called its root, but not precisely. Because 3 and $\frac{1}{6}$ times 3 and $\frac{1}{6}$ makes 10 and $\frac{1}{36}$. And in this way you can find root to every number, as I have said to you.[a]

[a] In translation, the corresponding text in **M+F** runs like this:
And in this we shall say the example. And say thus, find me the root of 10. Do thus, say, 3 times 3 make 9. And say, from 9 until 10 there is 1. Now divide 1 by the double of 3, that is, by 6, from which results $\frac{1}{6}$. And join $\frac{1}{6}$ above 3, they are 3 and $\frac{1}{6}$. And we shall say that the root of 10 is 3 and $\frac{1}{6}$, that is, the closest root than one can find. And in this way and by this rule you can find root to every number, or indeed the closest root that can be found, by the rule stated above.

15.15 Ancora dirremo uno altro assempro per che meglio intende. Et diremo chosì, trova radice de 67. Fa così et di', 8 è quello numero che multiprichato per se medesimo è più presso a 67 che nigiuno altro, però che non se vole multipricare se non numero intero. Et però di' chosì, 8 via 8 fa 64. Et similemente di', 8 et 8, 16, et di', da 64 infino in 67 à 3. Ora parti 3 in 16, che ne vene $\frac{3}{16}$. Ora poni $\frac{3}{16}$ sopra a 8, ày 8 et $\frac{3}{16}$. Et dirremo che 8 et $\frac{3}{16}$ via 8 et $\frac{3}{16}$ sia radice de 67. Ma como io t'ò dicto de sopra non è appunto, però che non l'à, ma questa

15.15 Further we shall say another example so that you understand better. And we shall say thus, find root of 67. Do thus and say, 8 is that number which multiplied by itself is closer to 67 than any other, because one does not want to multiply but whole numbers. And therefore say thus, 8 times 8 makes 64. And similarly say, 8 and 8, 16, and say, from 64 until 67 there is 3. Now divide 3 in 16, from which results $\frac{3}{16}$. Now put $\frac{3}{16}$ above 8, you have 8 and $\frac{3}{16}$. And we shall say that 8 and $\frac{3}{16}$ times 8 and $\frac{3}{16}$ is the root of 67. But as I have said to you

è la sua propria regola.

above not precisely, because it does not have that, but this is its proper rule.

(fol. 33ʳ)

15.16 **Anchora te ne vo' dire un'altra, per** che tu intende meglio. Et vo' dire chosì, dimi quanto è la radice de 82. Fa così chomo io t'ò dicto, quale è quello numero che è più presso a 82, multiprichato per se medesimo che nigiuno altro. È 9, però che 9 via 9 fa 81. Ora di', ^da^ 81 infino in 82 si à uno. Ora radoppia quello numero che tu multipricasti, cioè 9, che fa 18, et parti uno in 18, che ne vene $\frac{1}{18}$. Pollo sopra a 9, ày 9 et $\frac{1}{18}$. Et ⟨diremo che⟩ᴹ⁺ᶠ 9$\frac{1}{18}$ via 9 e $\frac{1}{18}$ sia la radice de 82. Et in questo modo fa de ciascheuno numero de que voli trovare la sua radice. Et questo baste sopra a questa materia. Et tornamo ale misure.

15.16 Further I want to say you another one, so that you understand better. And I want to say thus, say me how much is the root of 82. Do thus as I have said to you. Which is that number which is closer to 82 (when) multiplied by itself than any other. It is 9, because 9 times 9 makes 81. Now say, from 81 until 82 there is one. Now double this number that you multiplied, that is, 9, which makes 18, and divide one in 18, from which results $\frac{1}{18}$. Put it above 9, you get 9 and $\frac{1}{18}$. And ⟨we shall say that⟩ᴹ⁺ᶠ 9$\frac{1}{18}$ times 9 and $\frac{1}{18}$ is the root of 82. And in this way do with every number of which you want to find its root. And this will be enough about this matter. And we turn back to the measures.

15.17 **Uno terreno è lungho b̃ʳ 567,** et largho braccia 31, sicomo te mostro de rinpetto designato. Ora io voglio accasare tucto questo terreno, et voglio far'vi dentro chase, che ciascheuna casa sia lungha b̃ʳ 11, et largha b̃ʳ 7, né più né meno. Vo' sapere quante case ve posso fare dentro ad em-piere tucto questo terreno, et che non me remangha punto de voto. Fa così, et questa à la sua legi-tima regola, che tu arreche a b̃ʳ quadre

15.17 An area is *braccia* 567 long, and *braccia* 31 broad, as I show you drawn opposite. Now I want to build on this whole area, and I want to make houses inside, so that each house is *braccia* 11 long, and *braccia* 7 broad, nei-ther more nor less. I want to know how many houses I can make inside so as to fill this whole area, and so than no empty remains for me at all. Do thus, and this is its legitimate rule, that you

primamente tucto questo terreno. Et fa così, multipricha la sua lunghezza contra la sua larghezza, cioè 31 via 567, 17577. Et cotante $\tilde{b}^r$ quadre è tucto questo terreno. Et anchora arecha a $\tilde{b}^r$ quadre la chasa che tu vi voi fare, et multipricha 7 via 11, fa 77. Et cotante braccia quadre è ogni una casa che tu voi fare. Ora se voi sapere quante ve ne poi fare dentro, sì parti 17577 in 77, che ne vene 228 et $\frac{3}{11}$. Et 228 case et $\frac{3}{11}$ de casa poi fare in questo terreno. Et sta bene.

firstly bring to square *braccia* this whole area. And do thus, multiply its length against its breadth, that is, 31 times 567, 17577. And as many square *braccia* is this whole area. And further bring to square *braccia* the house that you want to make there, and multiply 7 times 11, it makes 77. And as many square *braccia* is each single house that you want to make. Now if you want to know how many you can make of them there inside, then divide 17577 in 77, from which results 228 and $\frac{3}{11}$. And 228 case and $\frac{3}{11}$ of house can you make in this area. And it goes well.

^{15.18} **Uno** pozzo è quadro, con quattro faccie, el quale è per ogni ~~quadro~~ faccia 2 $\tilde{b}^r$. Et è cupo $\tilde{b}^r$ 50, et è pieno d'acqua infino suso razente l'urla chomo tu vedi designato **(fol. 34ʳ)** de rinpecto. Ora vene per caso che vi cade entro una colonna la quale è lungha $\tilde{b}^r$ 25 et è quadra con quattro faccie, et per ogni faccia uno $\tilde{b}^r$. Vo' sapere quanta acqua è ussita fore del dicto pozzo per la caduta dela dicta colonna. Fa così, arecha a braccia quadre el pozzo, et multiprica primamente per la larghezza, cioè 2 via 2, fa 4. Ora multiplica contra ala chupezza, cioè 4 via 50, che fa 200. E ducento $\tilde{b}^r$ quadre è tucto questo pozzo. Et simigliantemente arecha a $\tilde{b}^r$ quadre la colonna, et multiprica

^{15.18} A well is square, with four faces, and is 2 *braccia* by each face. And it is *braccia* 50 deep, and is so full of water that it touches the brim as you see drawn **(fol. 34ʳ)** opposite. Now it occurs by chance that a column falls into it which is *braccia* 25 long and is square with four faces, and by each face one *braccio*. I want to know how much water has flown out from the said well by the fall of the said column. Do thus, bring the well to square *braccia*, and multiply firstly for the

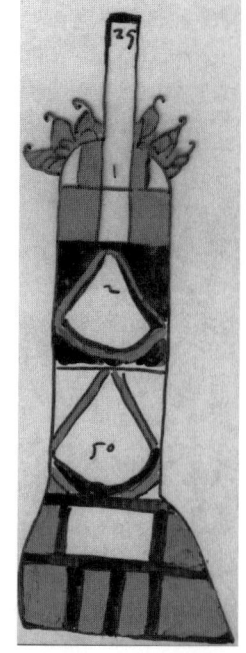

al modo facesti de sopra, cioè, 1 via 1 fa uno, et questa è la larghezza. Ora multiprica contra ala lunghezza, cioè uno via 25, fa 25. Et 25 b̃ʳ quadre è la colonna. Ora se voi sapere quanta acqua n'è ussita del dicto pozzo, parti 200 b̃ʳ del puzzo in 25 b̃ʳ che è la colonna, che ne vene octo. Et 8 braccia quadre d'acqua è ussita del pozzo per la caduta dela dicta colonna. Et è facta, et sta bene. Et così se fanno lo simili ragioni.

breadth (that is, the cross-section or base), that is, 2 times 2, it makes 4. Now multiply against the depth, that is, 4 times 50, which makes 200. And two hundred square *braccia* is this whole well. And similarly bring to square *braccia* the column, and multiply in the way you did above, that is, 1 times 1 makes one, and this is the breadth. Now multiply against the length, that is, one times 25, it makes 25. And 25 square *braccia* is the column. Now if you want to know how much water has flown out from the said well, divide 200 *braccia* of the well in 25 *braccia* which is the column, from which results eight. And 8 square *braccia* of water has flown out from the well by the fall of the said column. And it is done, and it goes well. And thus are done the similar computations.

15.19 Uno terreno como tu vedi designato de rimpecto à 5 faccie eguali, como tu vidi, el quale se chiama el pertecone,ᵃ et è per ogni faccia 8 b̃ʳ. Vo' sapere quante b̃ʳ quadre è tucto questo terreno. Fa così, et questa è la sua regola. Multipricha l'una dele faccie in se medesema, cioè 8 via 8, che fa 64. Ora multiprica per le altre tre faccie, 3 via 64, fa 192. Cavane l'una delle faccie, cioè 8. Resta 184, ed è facta. Et dirremo che tucto quello terreno è b̃ʳ 184 quadre.

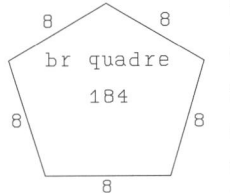

15.19 A terrain as you see drawn opposite has 5 equal faces, as you see, which is called the pentagon, and it is by each face 8 *braccia*. I want to know how many square *braccia* this whole terrain is. Do thus, and this is its rule. Multiply one of the faces by itself, that is, 8 times 8, which makes 64. Now multiply by the other three faces, 3 times 64, it makes 192. Remove from it one of the faces, that is, 8. 184 is left, and it is done. And we shall say that this whole terrain is square *braccia* 184.

ᵃ Written ptecone; I have noticed no other place where "p" does not stand for "per" or "par". However, an intended "pentecone" cannot be ruled out.

^{15.20} Uno schodo, cioè uno triangholo, como tu vedi designato qui de rempetto. E el suo dericto de mezzo è braccia 5. Vo' sapere quanto serà el dicto triangholo per ogni faccia. Fa cossì, multipricha el de-ricto de mezzo per se medesemo, cioè 5 b̃ᵣ via **(fol. 34ᵛ)** 5 b̃ᵣ, fa 25 b̃ᵣ. Ora parti per 3, che ne vene 8 e $\frac{1}{3}$. Ora aggiongi sopra a 25, fa 33 e $\frac{1}{3}$.ᵃ Ora trova la sua radice, cioè de 33 e $\frac{1}{3}$, che ne vene a essere 5 e $\frac{7}{9}$ meno $\frac{4}{18}$. Et 5 b̃ᵣ $\frac{7}{9}$ meno $\frac{4}{18}$ de b̃ᵣ diremo che questo trianglo serà per ogni faccia. Et è facta, et sta bene. Et così se fanno le simili ragioni.

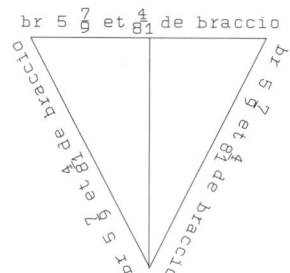

br 5 $\frac{7}{9}$ et $\frac{4}{81}$ de braccio

^{15.20} A shield, that is, a triangle, as you see drawn here opposite. And its straight in middle is *braccia* 5. I want to know how much the said triangle will be by each face. Do thus, multiply the straight in middle by itself, that is, 5 *braccia* times **(fol. 34ᵛ)** 5 *braccia*, it makes 25 *braccia*. Now divide by 3, from which results 8 and $\frac{1}{3}$. Now join above 25, it makes 33 and $\frac{1}{3}$.ᵃ Now find its root, that is, of 33 and $\frac{1}{3}$, which comes to be 5 and $\frac{7}{9}$ less $\frac{4}{18}$. And 5 *braccia* $\frac{7}{9}$ less $\frac{4}{18}$ of *braccio* we shall say that this triangle will be by each face. And it is done, and it goes well. And thus the similar computations are done.

ᵃ From this point onward, the text in **M+F** is rewritten, and runs as follows:
  And know that the root of 33 and ⅓ will the said triangle be by face. The root you shall find according to the rule we have said, which root we say will be 5 and ⅚ less ¹⁷⁄₅₄ not precisely. And as much will the shield be by face. I show you the diagram for understanding better. Thus do all the similar. And this is understood about a shield which has faces of equal measure.

^{15.21} Uno padiglione como tu vedi designato qui da pede. E el ferristo suo di mezzo, cioè la colonna che'l sostene, è alta 40 braccia. Et el panno del padiglione è lung-ho, dala cima del ferristo infino all'urlo del padiglione, b̃ᵣ 50. Va [*sic*, read "Vo'"] sapere quanto è tucto questo panno, et quanta terra posscede socto se el dicto padi-glione quando è teso. Fa così, perché el panno è lungo 50 b̃ᵣ, sì multiprica 50 via 50, fa 2500. Et perché el ferristo è lungo 40 b̃ᵣ, sì multipricha 40 via 40, fa 1600. Ora tray 1600 de 2500, resta 900. Trova la radice de 900, che è 30. Ora raddoppia 30,

^{15.21} A pavilion as you see drawn here at bottom. And its mid-pole, that is, the post that supports it, is 40 *braccia* high. And the cloth of the pavilion, from the peak of the pole until the border of the pavilion, is *braccia* 50 long. ⟨I want to⟩ know how much is all this cloth, and how much ground the said pavilion occupies when it is pitched. Do thus, because the cloth is 50 *braccia* long, then multiply 50 times 50, it makes 2500. And because the pole is 40 *braccia* long, then multiply 40 times 40, it makes 1600. Now detract 1600 from 2500, 900 is left. Find the root of 900, which is

fa 60. Et cotanto è largho el padiglione per lo dericto de mezzo, cioè cotanto è el suo diametro. Ora multipricha 60 via 3 et $\frac{1}{7}$, che fa 188 et $\frac{4}{7}$. Et cotanto serà dintorno el dicto padiglione. Ora se voli sapere quanta terra possciede socto se, sì parti per mezzo el dericto de mezzo del padiglione, cioè 60, che è 30, et similemente parti el circhio del padiglione per mezzo, che ne vene 94 et $\frac{2}{7}$. Ora multipricha 30 via 94 et $\frac{2}{7}$, che fa 2828 $\frac{4}{7}$. Et b̃^{r} 2828 e $\frac{4}{7}$ quadre $\langle...\rangle$. Et cotanta terra posscede socto se el dicto padiglione. Ora se voli sapere quanto è tucto el panno, parti el diametro, cioè 60, per mezzo, che ne vene 30. Multipricha 30 via 50, fa 1500. Et cotante b̃^{r} quadre è tucto el padiglione. Et sta bene et è facta.

30. Now double 30, it makes 60. And as much is broad the pavilion by the straight in middle, that is, as much is its diameter. Now multiply 60 times 3 and $\frac{1}{7}$, which makes 188 and $\frac{4}{7}$. And as much will the said pavilion be around. Now if you want to know how much ground it occupies, then divide in half the straight in middle of the pavilion, that is, 60, which is 30, and similarly divide the circle of the pavilion in half, from which results 94 and $\frac{2}{7}$. Now multiply 30 times 94 and $\frac{2}{7}$, which makes 2828 $\frac{4}{7}$. And square *braccia* 2828 and $\frac{4}{7}$ $\langle...\rangle$. And as much ground occupies the said pavilion. Now if you want to know how much is all the cloth, divide the diameter, that is, 60, in half, from which results 30. Multiply 30 times 50, it makes 1500. And as many square *braccia* is the whole pavilion. And it goes well and it is done.

(fol. 35ʳ)

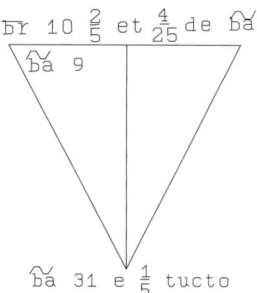

[15.22] Uno schudo over triangolo como tu vedi designato
de rimpetto, è tucto oguale,
tanto per l'una faccia quanto
per l'altra. Et el suo dericto
de mezzo è 9 braccia. Vo'
sapere quanto sonno tucte et
tre le faccie, cioè, quante b̃ʳ
è lungha ogni faccia.[a] Fa
così, multipricha 9 via 9, fa 81. Et parti 81
in 3, che ne vene 27. Agiongi 27 sopra a
81, fa 108. Et de questo numero trova la
sua radice, cioè, de 108, che è 10 et $\frac{2}{5}$
meno $\frac{4}{25}$. Et tante b̃ʳ serà per faccia el dicto
triangholo. Et se voli sapere quanto è tucto
dintorno, multipricha 3 via 10 et $\frac{2}{5}$ meno
$\frac{4}{25}$, che fa 31 e $\frac{1}{5}$. Et sta bene. Et così fa
tucte le simile ragioni.

[15.22] A shield or triangle as
you see drawn opposite, is
all equal, so much by one
face as by the other. And its
straight in middle is 9 *braccia*. I want to know how
much are all three faces, that
is, how many *braccia* each
face is long.[a] Do thus, multiply 9 times 9, it makes 81. And divide 81
in 3, from which results 27. Join 27 above
81, it makes 108. And of this number find
its root, that is, of 108, which is 10 and $\frac{2}{5}$
less $\frac{4}{25}$. And so many *braccia* will the said
triangle be by face. And if you want to
know how much it is all around, multiply
3 times 10 and $\frac{2}{5}$ meno $\frac{4}{25}$, which makes
31 and $\frac{1}{5}$. And it goes well. And do thus
all similar computations.

[a] In **M+F**, the statement is replaced by this:
Further we shall say another example of the triangle in order to show it more clearly, and we
shall say thus: a triangle which is equal by face, that is, as much by one face as by the other,
and it is from the point below to the face above, that is, by the straight in middle, 9 *braccia*,
say me how much is all three faces, that is, how much is it by each side.

^{15.23} **Due** lancie, como tu vidi
designato de rinpetto, sonno fitte
in terra, cioè in uno piano. Et
l'una lancia è lungha b̃ᵣ 17, et
l'altra b̃ᵣ 10. Et dall'una lancia ad
l'altra si à 20 b̃ᵣ. Vo' sapere
quante b̃ᵣ arà dall'una dele punte
all'altra dele dicte lancie. Fa così,
tray l'una lunghezza dele dicte
lancie dell'altra, cioè, trai 10 de
17, remane 7. Ora multipricha 7
via 7, fa 49. Et similmente multipricha la lunghessa che è dall'una
all'altra, cioè 20 via 20, che fa
400. Agiongi inseme, che fa 449.
Et de questo numero trova la
radice sua, che è 21 et $\frac{4}{21}$. Et è
facta. Et braccia 21 e $\frac{4}{21}$ de b̃ᵣ
dirremo che arà dall'una punta
all'altra dele dicte lance. Et così
fa l'altre simili ragioni.

^{15.23} Two lances, as you see
drawn opposite, are stuck in
ground, that is, in one plane. And
one lance is *braccia* 17 long, and
the other *braccia* 10. And from
one lance to the other there is 20
braccia. I want to know how
many *braccia* there will be from
one of the points to the other of
the said lances. Do thus, detract
one length of the said lances from
the other, that is, detract 10 from
17, 7 remains. Now multiply 7
times 7, it makes 49. And similarly multiply the length that there
is from one to the other, that is,
20 times 20, which makes 400.
Join together, which makes 449.
And of this number find its root,
which is 21 and $\frac{4}{21}$. And it is
done. And *braccia* 21 and $\frac{4}{21}$ of *braccio*
shall we say that there will be from one
point to the other of the said lances. And
do thus the other similar computations.

^{15.24} **E** sonno due torri in uno piano, come
per figura vederai designate de socto in
questa carta. ⟨De⟩le quali l'una torre è alta
25 b̃ᵣ, et l'altra 20 b̃ᵣ. Et nel mezzo de queste due torri si è una coppa piena d'acqua.
La quale è appunto nel mezzo, che tanto
è appresso all'una torre quanto all'altra. Et
in su ogni una de queste torri si è una
colonba, le quale vogliono andare a bere in
quella coppa. Et dall'una torre all'altra si
à 100 b̃ᵣ. Et ciascheuno de quelle colonbe

^{15.24} And there are two towers in one plane,
as you will see drawn in figure at bottom
of this page. ⟨Of⟩ which one tower is 25
braccia high, and the other 20 *braccia*. And
in the middle between these two towers
there is a goblet full of water. Which is
precisely in the middle, so that it is as close
to one tower as to the other. And on the top
of each one of these towers there is a dove,
and these want to go drink from this goblet.
And from one tower to the other there is

se movono a una hora, et volano ogual-
mente, cioè che tanto vola l'una quanto l'
altra. Voglio sapere in quanto serà più tosto
l'una che l'altra a bevere in quella coppa.
Fa così et di', perché dall'una torre all'altra
(fol. 35ʳ) à 100 b̃ʳ, sì parti 100 per mezzo,
che ne vene 50. Et 50 braccia à da ogni una
dele torri ala coppa. Ora multipricha 50 via
50, fa 2500. Et perché una dele torri è alta
25 b̃ʳ, sì multipricha 25 via 25, che fa 625.
Agiongi sopra a 2500, fa 3125. Et de que-
sto ⟨trova⟩^(M+F) la sua radice, che è 55 et $\frac{10}{11}$.
Et 55 b̃ʳ et $\frac{10}{11}$ de braccio ⟨diremo che ...⟩.
Et se voli sapere in quanto vi serà l'altra,
sì multipricha al simil modo et di', 20 via

100 *braccia*. And each of these doves set
out at one and the same hour, and fly
equally, that one flies as much as the other.
I want to know how much earlier one will
be there than the other to drink from that
goblet. Do thus and say, because from one
tower to the other **(fol. 35ᵛ)** there is 100
braccia, then divide 100 in half, from
which results 50. And 50 *braccia* there is
from one of the towers to the goblet. Now
multiply 50 times 50, it makes 2500. And
because one of the towers is 25 *braccia*
high, then multiply 25 times 25, which
makes 625. Join above 2500, it makes
3125. And of this ⟨find⟩^(M+F) its root, which

20 fa 400. Agiongi sopra a 2500, fa 2900. Et de questo trova la sua radice, che è 53 et $\frac{9}{106}$.[a] Et in cotante b̃ʳ vi serà l'altra. Ora se voli sapere in quanto vi serà più tosto l'una che l'altra, sì{a} tray 53 e $\frac{9}{106}$ de 55 e $\frac{10}{11}$, che resta 2 e $\frac{87}{106}$ e uno pocho pocho più.[b] Ma non se po vedere appunto. Et cotanto vi serà più tosto l'una che l'altra. Cioè, 2 b̃ʳ e $\frac{87}{106}$ de braccio, et como dico uno pocho pocho più. Et è facta. E così se fanno le simili ragioni.

is 55 and $\frac{10}{11}$. And 55 *braccia* and $\frac{10}{11}$ of *braccio* ⟨we shall say that ...⟩. And if you want to know when the other will be there, then multiply in a similar way, and say, 20 times 20 makes 400. Join above 2500, it makes 2900. And of this find its root, which is 53 and $\frac{9}{106}$.[a] And in as many *braccia* will the other be there. Now if you want to know how much earlier one will be there than the other, then {...} detract 53 and $\frac{9}{106}$ from 55 and $\frac{10}{11}$, and 2 and $\frac{87}{106}$ and a little bit more is left.[b] But it cannot be seen precisely. And as much earlier will one be there before the other. That is, 2 *braccia* and $\frac{87}{106}$ of *braccio*, and as I say a little bit more. And it is done. And thus the similar computations are done.

[a] Error for 53⁹¹/₁₀₆.

[b] This "pocho pocho più" is $\frac{4/11}{106}$.

(fol. 36ʳ)

15.25 Uno cictadino vole fare over à facto fare uno palagio como tu vedi qui da pede designato, el quale è quadro con quattro faccie oguali. Et è lungo per ogni faccia 40 b̃ʳ. Ora costui vole fare porre suso el tecto a doi piovetoi et non più. Et vole che'l dicto tetto sia

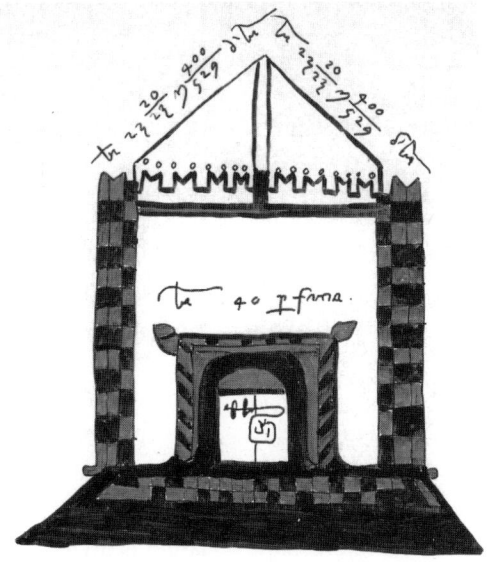

15.25 A townsman wants to make or indeed has had a mansion made as you see drawn here at bottom, which is square with four equal faces. And it is 40 *braccia* long by each face. Now he wants to have two and no more gutters put on top of the roof. And he wants that the said

alto nel colmigno b̃ʳ 13. Vo' sapere quanto vogliono essere lunghi li dicorrenti, cioè, quante b̃ʳ li quali li ponghono dal colmigno in su'l muro. Fa così, daché'l palagio è alto [*sic*, read "ampio"] 40 braccia, et vole fare doy piovetoi, sì dividi 40 per mezzo, che ne vene 20. Et multipricha 20 via 20, che fa 400. Ora per che el tecto vole essere alto 13 b̃ʳ, sì multipricha 13 via 13, che fa 169. Ora agiongi inseme con 400, che fa 569, et de questo numero trova la sua radice, che è 23 et $\frac{20}{23}$ meno $\frac{400}{529}$. Et cotanto dirremo che vogliono essere lunghi li dicorrenti che ponghono desu el muro al colmiglio del tetto, cioè b̃ʳ 23 e $\frac{20}{23}$ de b̃ʳ meno $\frac{400}{529}$ de braccio. Et è facta, et così se fanno le simili ragioni che ti fossono date intorno a questa materia.

roof may be *braccia* 13 high at the roof-ridge. I want to know how long the drains shall be, that is, how many *braccia* they lay them from the roof-ridge to the top of the wall. Do thus, given that the mansion is 40 *braccia* high [*sic*, error for "wide"], and he wants to make two gutters, 40 is split in half, from which results 20. And multiply 20 times 20, which makes 400. Now because the roof shall be 13 *braccia* high, then multiply 13 times 13, which makes 169. Now join together with 400, which makes 569, and of this number find its root, which is 23 and $\frac{20}{23}$ less $\frac{400}{529}$. And as much shall we say that the drains shall be long which they lay from the top of the wall to the roof-ridge, that is, *braccia* 23 and $\frac{20}{23}$ of *braccio* less $\frac{400}{529}$ of *braccio*. And it is done, and thus the similar computations are done which might be given to you about this matter.

(fol. 36ᵛ)

[16. Rules and Examples for Algebra until the Second Degree]

16.1 **Q**uando le cose sonno eguali al numero, si vole partire el numero nelle cose, et quello che ne vene si è numero. Et cotanto vale la cosa.

16.1 When the things are equal to the number, one shall divide the number in the things, and that which results from it is number. And as much is worth the thing.

16.2 **P**ongoti assempro ala dicta ragione. Et vo' dire così, fammi de 10 doy parti, che partita la magiore nella minore ne vengha 100. Fa così, poni che la magiore parte fosse una cosa. Adunqua la minore serà lo

16.2 I propose to you an example to the said computation. And I want to say thus, make two parts of 10 for me, so that when the larger is divided in the smaller, 100 results from it. Do thus, posit that the larger part

rimanente infino in 10 che serà 10 meno una cosa. Et così abiamo facto de dece doy parti, che la magiore sia una cosa, et la minore sia 10 meno una cosa. Ora si vole partire la magiore nella minore, cioè una cosa in 10 meno una cosa, che ne dè venire 100. Et però dè multipricare 100 via 10 meno una cosa. Fa 1000 meno 100 cose, che s'aoguaglino a una cosa. Ora ristora ciascheuna parte, cioè dè giungere 100 cose che sonno meno a ciascheuna parte. Arai che 101 cosa sonno iguali a 1000 numeri. Et però se vole partire li numeri nelle cose, cioè 1000 numeri in 101 cosa, che ne vene 9 et $\frac{92}{101}$ [*sic*, read "$\frac{91}{101}$"], et cotanto vale la cosa. Et noi porremo che la magiore parte fusse una cosa, dunqua vale, et dirremo che la magiore parte de 10 sia 9 et $\frac{91}{101}$.[a] Et la seconda serà el resto infino in 10, che serà $\frac{10}{101}$. Et abiamo che la magior parte de 10 serà 9 e $\frac{91}{101}$, et la minore $\frac{10}{101}$. Ora parti 9 et $\frac{91}{101}$ in $\frac{10}{101}$, che ne vene appunto 100. Et sta bene. Et così se fa le simili ragioni.

was a thing. Hence the smaller will be the remainder until 10, which will be 10 less a thing. And thus we have made two parts of ten, of which the larger is a thing, and the smaller is 10 less a thing. Now one shall divide the larger in the smaller, that is, a thing in 10 less a thing, from which shall result 100. And therefore one shall multiply 100 times 10 less a thing. It makes 1000 less 100 things, which equal one thing. Now restore each part, that is, you shall join 100 things which are less to each part. You will get that 101 thing are equal 1000 numbers. And therefore one shall divide the numbers in the things, that is, 1000 numbers in 101 thing, from which results 9 and $\frac{\langle 91 \rangle}{101}$, and as much is worth the thing. And we posited that the larger part was a thing, {...} and we shall say that the larger part of 10 is 9 and $\frac{91}{101}$.[a] And the second will be the rest until 10, which will be $\frac{10}{101}$. And we get that the larger part of 10 will be 9 and $\frac{91}{101}$, and the smaller $\frac{10}{101}$. Now divide 9 and $\frac{91}{101}$ in $\frac{10}{101}$, from which results precisely 100. And it goes well. And thus one makes the similar computations.

[a] To be understood either as "{dunqua vale,} et dirremo che la magiore parte de 10 sia 9 et $^{91}/_{101}$," or as "dunqua vale ⟨9 et $^{91}/_{101}$⟩, et dirremo che la magiore parte de 10 sia 9 et $\frac{91}{101}$". The translation follows the first conjecture, which agrees well with a version of the problem in the *Libro di molti ragioni* from c. 1330 [ed. Arrighi 1973: 195].

16.3 Anco' ti voglio porre uno altro assempro. Et vo' dire chosì, e sonno tre conpagni che ànno guadagnato 30 libre. El primo conpagnio misse 10 £. El secondo misse 20 £. El terzo misse tanto che de questo guadagnio gle tocchò 15 £. Vo' sapere quanto misse el terzo conpagnio, et quanto tocca

16.3 Again, I want to propose to you another example, and I want to say thus, there are three partners who have gained 30 *libre*. The first partner put in 10 *libre*. The second put in 20 *libre*. The third put in so much that 15 *libre* of this gain was due to him. I want to know how much the third partner

per uno de guadagnio de quelli altri doy conpagni. Fa così, se noi vogliamo sapere quanto misse el terzo conpagnio, poni che el terzo mettesse una chosa. Appresso se vole raccoglere quello che mise el primo et el secondo, cioè £ 10 et £ 20, che sonno 30. Et arai che sonno tre conpagni, che el primo mette in conpagnia 10 £. El secondo mette 20 £. El terzo mette una chosa. Siché el corpo dela conpagnia è 30 £ et una cosa. Et ànno guadagnato 30 £. (fol. 37ʳ) Ora se noi vogliamo sapere quanto toccha al terzo conpagnio de questo guadagnio, che abbiamo posto che mettesse una cosa, sì'tti convene multipricare una chosa via quello che egli ànno guadagniato, et partire in tucto el corpo dela conpagnia. Et però abbiamo a multiprichare 30 via una cosa. Fa 30 cose, le quale te convene partire nel corpo dela conpagnia, cioè per 30 et una cosa, et quello che ne vene cotanto toccha al terzo conpagnio. Et questo non ci fa bisogno partire perché noi sappiamo che glie ne toccha 15 £. Et però multipricha 15 via 30 et una cosa. Fanno 450 et 15 cose. Dunqua 450 numeri et 15 cose s'aoguagliano a 30 cose. Ristora ciascheuna parte, cioè dè cavare de ciascheuna parte 15 cose. Et arai che 15 cose se aoguagliano a 450 numeri. Et però devi partire li numeri nelle cose, cioè 450 in 15, che ne vene 30. Et cotanto vale la chosa. Et noi ponemo che el terzo conpagnio mettesse una cosa, siché vene ad avere messo 30 £. El secondo 20 £. El primo 10 £. Et se volesse sapere quanto ne toccha al primo et al secondo, sì cava di 30 £ 15 che'nne toccha al terzo.

put in, and how much gain is due to (each) one of those two other partners. Do thus, if we want to know how much the third partner put in, posit that the third put in a thing. Next one shall aggregate that which the first and the second put in, that is, *libre* 10 and *libre* 20, which are 30. And you will get that there are three partners, and that the first puts in the partnership 10 *libre*. The second puts in 20 *libre*. The third puts in a thing. So that the principal of the partnership is 30 *libre* and a thing. And they have gained 30 *libre*. (fol. 37ʳ) Now if we want to know how much of this gain is due to the third partner, when we have posited that he put in a thing, then it suits you to multiply a thing times that which they have gained, and divide in the total principal of the partnership. And therefore we have to multiply 30 times a thing. It makes 30 things, which it suits you to divide in the principal of the partnership, that is, by 30 and a thing, and that which results from it, as much is due to the third partner. And this we do not need to divide, because we know that 15 *libre* of it is due to him. And therefore multiply 15 times 30 and a thing. It makes 450 and 15 things. Hence 450 numbers and 15 things equal 30 things. Restore each part, that is, you shall remove from each part 15 things. And you will get that 15 things equal 450 numbers. And therefore you shall divide the numbers in the things, that is, 450 in 15, from which results 30. And as much is worth the thing. And we posited that the third partner put in a thing, so that he comes to have put in

Restano 15 £. Et dirrai che sonno 2 conpagni che ànno guadagnato 15 £. Et el
primo misse 10 £. Et el secondo misse 20
£. Quanto ne toccha per uno. Fa così et di',
20 £ et 10 £ sonno 30 £, et questo è el
corpo dela conpagnia. Ora multipricha per
lu primo, che mise 10 £, 10 via 15 che
ànno guadagniato. Fanno 150. Parti in 30,
che ne vene 5 £. Et cotanto ne toccha al
primo. Et poi per lo secondo multipricha
20 via 15, che fa 300 £. Parti in 30 che ne
vene 10 £, et cotanto toccha al secondo
conpagnio. Et è facta, et sta bene. Et così
se fanno le simili ragioni.

30 *libre*. The second 20 *libre*. The first
10 *libre*. And if you should want to know
how much of it is due to the first and to the
second, then remove from 30 *libre* 15 of
them which are due to the third. 15 *libre*
are left. And you will say that there are 2
partners who have gained 15 *libre*. And the
first put in 10 *libre*. And the second put in
20 *libre*. How much of it is due to (each)
one. Do thus, and say, 20 *libre* and 10 *libre*
are 30 *libre*, and this is the principal of the
partnership. Now multiply for the first, who
put in 10 *libre*, 10 times 15 which they
have gained. It makes 150. Divide in 30,
from which results 5 *libre*. And as much
is due to the first. And then for the second,
multiply 20 times 15, which makes 300
libre. Divide in 30, from which results 10
libre, and as much is due to the second
partner. And it is done, and it goes well.
And thus the similar computations are done.

16.4 **Quando** li censi sonno oguali al numero, si vole partire el numero per li censi. Et
la radice de quello che ne vene vale la
cosa.

16.4 When the *censi* are equal to the number, one shall divide the number by the
censi. And the root of that which results
from it is worth the thing.

(fol. 37ᵛ)

16.5 **Assemplo** ala dicta regola. Et vo' dire
chosì, trovame doi numeri che siano in
propositione sicome è 2 de 3. Et multipricato ciascheuno per se medesimo, et tracta
l'una multipricatione dell'altra, remangha
20. Vo' sapere qual' numeri sonno questi.
Fo così, et poni che l'uno numero fosse 2

16.5 Example to the said rule. And I want
to say thus, find me two numbers that are
in proportion as is 2 of 3 and when each
(of them) is multiplied by itself, and one
multiplication is detracted from the other,
20 remains. I want to know which are these
numbers. Do thus, and posit that one num-

chose et l'altro fosse 3 cose. Et bene sonno in propositione sicome sonno 2 et 3. Appresso si vole multipricare li numeri, ciascheuno per se medesemo, et cavare l'una multipricatione dell'altra. Et deve remanere 20. Et però multipricha ciascheuno per se, et di', duo cose via 2 cose fanno 4 censi. Et tre cose via 3 cose fanno 9 censi. Ora cava l'una multipricatione dell'altra, cioè 4 de 9. Resta 5 censi, i quali s'aoguagliano a 20 numeri. Et noi diciamo che se voli ^partire^ li numeri nelli censi, siché se vole partire 20 numeri in 5 censi. Che ne vene 4 numeri, et cotanto vale la cosa, cioè la sua radice che è 2θ. Dicemo che fosse el primo numero 2 cose et el secondo 3 cose. Però vedi chiaro che 2 cose vagliono 4 numeri. Et 3 cose 6 numeri. Et così te dicho che questi numeri sonno l'uno 4 et l'altro 6. Et tal parte è 4 de 6 qual 2 de 3. Ora se la voi provare, multipricha 6 via 6, fa 36. Et multipricha 4 via 4, fa 16. Tray 16 de 36. Resta 20, et sta bene. Et chosì se fano tucte le simiglianti ragioni, cioè secondo questa regola.

ber was 2 things and the other was 3 things. And they are well in proportion as are 2 and 3. Next one shall multiply the numbers, each (one) by itself. And remove one multiplication from the other. And 20 shall remain. And therefore multiply each (one) by itself, and say, two things times 2 things make 4 *censi*. And three things times 3 things make 9 *censi*. Now remove one multiplication from the other, that is, 4 from 9. 5 *censi* is left, which equal 20 numbers. And we say that one shall divide the numbers in the *censi*, so that one shall divide 20 numbers in 5 *censi*. From which results 4 numbers, and as much is worth the thing, that is, its root, which is 2. We said that the first number was 2 things and the second 3 things. Therefore you see clearly that 2 things are 4 numbers. And three things 6 numbers. And thus I say to you that these numbers are 4, one, and 6, the other. And such part is 4 of 6 as 2 of 3. Now if you want to verify it, multiply 6 times 6, it makes 36. And multiply 4 times 4, it makes 16. Detract 16 from 36. 20 is left, and it goes well. And thus all the similar computations are done, that is, according to this rule.

16.6 **Q**uando li censi sonno oguali ale chose, se vole partire le cose per li censi, et quello che ne vene si è numero. Et cotanto vale la cosa.

16.6 When the *censi* are equal to the things, one shall divide the things by the *censi*, and that which results from it is number. And as much is worth the thing.

16.7 Assemplo ala dicta regola. Trovami 2 numeri che siano in propositione sicomo è 4 de 9. Et multiprichato l'uno contra l'altro faccia quanto ragionti inseme. **(fol. 38ʳ)** Vo' sapere qual' numeri sonno questi. Fa così, poni che l'uno numero sia 4 cose. Et l'altro numero sia 9 chose. Et bene è in proposi-tione come è 4 a 9. Adunqua l'uno numero è 4 chose. Et l'altro è 9 chose. Et noi di-ciamo cho vogliamo fare tanto multiprichati l'uno contra al'altro quanto raggionti in-seme. Et però multipricha 4 cose via 9 cose, fanno 36 censi. Et aggiongi inseme 4 e 9 cose, fanno 13 cose, et ài che 36 censi sonno oguali a 13 cose. Et però parti 13 cose in 36 numeri. Che ne vene $\frac{13}{36}$ de numero, et cotanto vale la cosa. Ora noi ponemo che l'uno numero fusse 4 cose. Però multipricha 4 via {erasure} $\frac{13}{36}$, che fa $\frac{52}{36}$, che sonno j e $\frac{4}{9}$. Et cotanto è l'uno numero. Et ponemo che l'altro numero fusse 9. Però multipricha 9 via $\frac{13}{36}$, che fa $\frac{117}{36}$, che sonno 3 et $\frac{1}{4}$. Et cotanto è'll'altro numero. Ora se la voli provare, sì multi-pricha j et $\frac{4}{9}$ via 3 et $\frac{1}{4}$, che fanno 4 et $\frac{25}{36}$. Ora agiongi inseme li dicti numeri, che fanno quello medesimo. Et sta bene. Et così se fanno le simili ragioni.

16.7 Example to the said rule: Find me 2 numbers that are in proportion as is 4 of 9. And when one is multiplied against the other, it makes as much as when they are joined together. **(fol. 38ʳ)** I want to know which are these numbers. Do thus, posit that one number is 4 things. And the other number is 9 things. And they are well in proportion as is 4 to 9. Hence one number is 4 things. And the other is 9 things. And we say that we want them to make as much when they are multiplied one against the other as when they are joined together. And therefore multiply 4 things times 9 things, it makes 36 *censi*. And join together 4 and 9 things, they make 13 things, and you get that 36 *censi* are equal to 13 things. And therefore divide 13 things in 36 numbers. From which results $\frac{13}{36}$ in number, and as much is worth the thing. Now we posited that one number was 4 things. Therefore multiply 4 times $\frac{13}{36}$, which makes $\frac{52}{36}$, which are j and $\frac{4}{9}$. And as much is one number. And we posited that the other number was 9. Therefore multiply 9 times $\frac{13}{36}$, which makes $\frac{117}{36}$, which are 3 and $\frac{1}{4}$. And as much is the other number. Now if you want to verify it, then multiply j and $\frac{4}{9}$ times 3 and $\frac{1}{4}$, which make 4 and $\frac{35}{36}$. Now join together the said numbers, which make the same. And it goes well. And thus the similar computations are done.

16.8 **Quando** li censi et le cose sonno oguali al numero se vole partire neli censi, et poi demezzare le cose et multiprichare per se

16.8 **When** the *censi* and the things are equal to the number, one shall divide in the *censi*, and then halve the things and multiply by

medesimo et giungere sopra al numero. Et la radice dela somma meno el dimezzamento dele cose vale la cosa.

16.9 **Assemplo**[a] ala dicta regola. Et vo' dire chosì, uno prestò a un'altro 100 £ al termine de 2 anni a fare capo d'anno. Et quando vene ala fine de 2 anni et quegli glie rendì £ 150. Vo' sapere ad **(fol. 38ᵛ)** que ragione fo pres⟨ta⟩ta la £ el mese. Fa così, pone che fusse prestata a una cosa el mese de deñ, siché vene a valere l'anno ^la £^ 12 cose de deñ, che 12 cose de deñ sonno el vintesimo de una £, siché la £ vale l'anno $\frac{1}{20}$ de una £ ⟨de cosa⟩[A]. Et però di' così, se la £ vale ~~la li~~ l'anno $\frac{1}{20}$ de una £, que varranno 100 £. Multipricha 100 via $\frac{1}{20}$. Fa $\frac{100}{20}$, che sonno 5 cose. Agiongi ^sopra a^ 100 £, fanno 100 £ e 5 cose per uno anno. Ora se voli sapere per lo secondo anno, multipricha 100 £ et 5 cose via $\frac{1}{20}$ de cosa. Fanno 5 cosa et $\frac{1}{4}$ censo, le quali se vogliono agiongere a 100 £ et 5 cose, che fanno 100 £ e 10 cose et $\frac{1}{4}$ censo. Et cotanto sonno le 100 £ in 2 anni, tra merito et capitale, et essendo prestata la £ el mese a una cosa. Et noi sappiamo de certo che le 100 £ ànno guadagniato in 2 anni 50 £. Siché le 150 £ vagliono le 100 £ e 10 cose et $\frac{1}{4}$ censo. Siché le 100 £, 10 cose et $\frac{1}{4}$ censo sonno oguali a 150 £. Ristora ciascheuna parte, cioè ⟨dè⟩ cavare 100 £ de ogni parte, et arai che 10 cose et $\frac{1}{4}$ censo sonno oguali a 50. Ora fa sicomo dice la nostra regola, cioè dè arrechare a uno censo, cioè dè partire in $\frac{1}{4}$ censo, et arai che

itself and join above the number. And the root of the total less the halving of the things is worth the thing.

16.9 Example to the said rule. And I want to say thus, someone lent to another 100 *libre* at the term of 2 years, to make (up at) the end of year. And when it came to the end of the two years, then that one gave back to him *libre* 150. I want to know at **(fol. 38ᵛ)** which rate the *libra* was lent a month. Do thus, posit that it was lent at one thing in *denaro* a month, so that the *libra* comes to be worth 12 things of *denaro* a year, which 12 things of *denaro* are the twentieth of one *libra*, so that the *libra* is worth the year $\frac{1}{20}$ of one *libra* ⟨of thing⟩[A]. And therefore say thus, if the *libra* is worth $\frac{1}{20}$ of one *libra* a year, what will 100 *libre* be worth. Multiply 100 times $\frac{1}{20}$. It makes $\frac{100}{20}$, which are 5 things. Join above 100 *libre*, they make 100 *libre* and 5 things for one year. Now if you want to know for the second year, multiply 100 *libre* and 5 things times $\frac{1}{20}$ of thing. They make 5 thing and $\frac{1}{4}$ *censo*, which shall be joined to 100 *libre* and 5 things, which make 100 *libre* and 10 things and $\frac{1}{4}$ *censo*. And as much are the 100 *libre* in 2 years, interest and capital together, and being lent the *libra* at one thing a month. And we know for sure that the 100 *libre* have gained 50 *libre* in 2 years. So that the 150 *libre* are the 100 *libre* and 10 things and $\frac{1}{4}$ *censo*. So that the 100 *libre*, 10 things and $\frac{1}{4}$ *censo* are equal to 150 *libre*. Restore each

j censo et 40 cose sonno oguali a 200 numeri. Ora demezza le cose, sonno 20. Multipricha per se medesemo, fa 400. Aggiongi sopra li numeri, fanno 600. Trova la sua radice, la quale è sorda, cioè, che è manifisto, de non avere radice appunto, et cotanto dirremo che vaglia la cosa, cioè la radice di 600 meno 20, cioè el dimezzamento dele cose. Et noi ponemo che fusse prestata la £ el mese a una cosa de denaro, dunqua dirremo che fusse prestata la £ el mese a denari, la radice di 600 meno 20 denari. Et sta bene. Et così se fanno le simiglianti ragioni.

part, that is, ⟨you shall⟩ remove 100 *libre* from each part, and you will get that 10 things and $\frac{1}{4}$ *censo* are equal to 50. Now do so as our rule says, that is, you shall bring to one *censo*, that is, you shall divide in $\frac{1}{4}$ *censo*, and you will get that j *censo* and 40 things are equal to 200 numbers. Now halve the things. They are 20. Multiply by itself, it makes 400. Join above the numbers, they make 600. Find its root, which is surd, that is, as it is manifest, to have no precise root, and as much will we say that the thing is, that is, the root of 600 less 20, that is, the halving of the things. And we posited that the *libra* was lent at one thing of *denaro* a month, hence we shall say that the libra was lent at *denari,* the root of 600 less 20 *denari* a month. And it goes well. And thus the similar computations are done.

ª Corrected from "assempro" – unless the correction went the other way, the ink is the same. In any case the copyist is seen to be conscious of orthography, and probably to try to follow his original.

(fol. 39ʳ)

16.10 E sonno dui homini che ànno denari. Dice el primo al secondo, se tu me dessi 14 de toi denari, che io li racchozzasse co' mey, io arei 4 cotanti de te. Dice el secondo al primo, se tu me desse la radice de toy denari, io arei deñ 30. Vo' sapere quanto aveva ciascheuno homo. Fa chosì, poniamo che'l primo homo avesse j censo. Et egli adimanda 14 al secondo, siché verrà ad avere j censo e 14. Et dice de avere 4 cotanti de lui. Dunqua convene che rimangha al secondo el $\frac{1}{4}$ ⟨censo⟩ᴬ e $3\frac{1}{2}$. Dunqua

16.10 And there are two men that have *denari*. The first says to the second, if you gave me 14 of your *denari*, and I threw them together with mine, I should have 4 times as much as you. The second says to the first: if you gave me the root of your *denari*, I should have 30 *denari*. I want to know how much each man had. Do thus, let us posit that the first man had j *censo*. And he asks for 14 from the second, so that he will come to have j *censo* and 14. And he says to have 4 times as much as him.

nanzi che 'l secondo desse nulla al primo sì n'aveva egli $17\frac{1}{2}$ et $\frac{1}{4}$ censo. Et così abiamo che 'l primo vene ad avere uno censo. Et el secondo $17\frac{1}{2}$ et $\frac{1}{4}$ censo. Et poi domanda el secondo al primo la radice de soi deñ, cioè de j censo, che è una chosa, la quale se vole agiongere a $\frac{1}{4}$ censo e $17\frac{1}{2}$. Et in verità fa $\frac{1}{4}$ censo et una chosa et $17\frac{1}{2}$, et con questo dice che dè avere 30. Adunqua abiamo che $\frac{1}{4}$ censo et una chosa et $17\frac{1}{2}$ numeri sonno oguali a 30. Ristora ciascheuna parte, cioè tray $17\frac{1}{2}$ de ogni una parte. Et arai che $\frac{1}{4}$ censo et una chosa sonno oguali a $12\frac{1}{2}$ numero. Dei partire per j [*sic*, it means "$\frac{1}{4}$"] censo et arai che uno censo e 4 chose sonno oguali a 50. Ora demezza le cose, sonno 2. Multipricha per se medesimo, fa 4. Aggiongi sopra ai numeri, à 54, et de questo trova la sua radice, et cotanto vale la cosa meno el dimezzamento dele cose, cioè 2. Et noi ponemo el primo avesse uno censo. Et però ti convene sapere que vale el censo. Et però dì multiprichare radice de 54 meno 2 via radice de 54 meno 2. Et cotanto varrà el censo. Che in verità, radice de 54 meno 2 via radice de 54 meno 2, fa 58 meno radice de [a] et abbiamo che vale el censo 58 meno radice   . Et noi ponemo avesse el primo uno censo. Dunqua vene ad avere 58 meno radice de   . ⟨Ora sappi el secondo, che ponesti ch'avesse $\frac{1}{4}$ censo e $17\frac{1}{2}$ numeri. Adunqua piglia el $\frac{1}{4}$ de 58 meno radice de   ⟩[A] ch'è $14\frac{1}{2}$ meno radice de 54, sopra el quale vi giongi $17\frac{1}{2}$. Fanno 32 mino la radice de 54. Et così abiamo che el primo à 58 meno

Hence it suits that $\frac{1}{4}$ ⟨*censo*⟩[A] and $3\frac{1}{2}$ remain for the second. Hence, before the second gave anything to the first, he had $17\frac{1}{2}$ and $\frac{1}{4}$ *censo*. And thus we have that the first comes to have one *censo*. And the second $17\frac{1}{2}$ and $\frac{1}{4}$ *censo*. And then the second asks the first for the root of his *denari*, that is, of j *censo*, which is a thing, which one shall join to $\frac{1}{4}$ *censo* and $17\frac{1}{2}$. And truly it makes $\frac{1}{4}$ *censo* and one thing and $17\frac{1}{2}$, and with this he says that he shall have 30. Hence we get that $\frac{1}{4}$ *censo* and one thing and $17\frac{1}{2}$ numbers are equal to 30. Restore each part, that is, detract $17\frac{1}{2}$ from each part. And you will get that $\frac{1}{4}$ *censo* and one thing are equal to $12\frac{1}{2}$ number. You shall divide by ⟨$\frac{1}{4}$⟩ *censo* and you will get that one *censo* and 4 things are equal to 50. Now halve the things, they are 2. Multiply by itself, it makes 4. Join above the numbers, one has 54, and of this find its root, and as much is worth the thing less the halving of the things, that is, 2. And we posited that the first had a *censo*. And therefore it suits you to know what the *censo* is, and therefore you shall multiply root of 54 less 2 times root of 54 less 2. And as much will the *censo* be. And truly, root of 54 less 2 times root of 54 less 2, makes 58 less root of [   ],[a] and we get that the *censo* is 58 less root [   ]. And we posited the first had a *censo*. Hence he comes to have 58 less root of [   ]. ⟨Now know the second, of whom you posited that he had $\frac{1}{4}$ *censo* and $17\frac{1}{2}$ numbers. Hence seize $\frac{1}{4}$ of 58 less root of   ⟩[A], which is $14\frac{1}{2}$ less

radice de   . Et el secondo homo
à 32 meno radice de 54. Et è facta. Et così
se fanno le simiglianti ragioni.

root of 54, above which join $17\frac{1}{2}$. They
make 32 less the root of 54. And so we get
that the first has 58 less the root of
[   ]. And the second man has 32
less root of 54. And it is done. And thus
the similar computations are done.

ª Instead of "864", the manuscript leaves open c. 2 cm. In the margin the copyist writes the
commentary "così stava nel'originale spatii". The corresponding problem in **A** writes "radici di
864" instead of having spaces.

(fol. 39ᵛ)

16.11 Quando le cose sonno oguali ali censi
et al numero, se vole partire nelli censi, et
poi dimezzare le cose et multiprichare per
se medesimo et cavare el numero, et la
radice de quello che romane, et poi el
dimezzamento dele cose vale la cosa. Overo
el dimezzamento dele chose meno la radice
de quello che remane.

16.11 When the things are equal to the *censi*
and to the number, one shall divide in the
censi, and then halve the things and multiply by itself and remove the number, and
the root of that which remains and then the
halving of the things is worth the thing. Or
indeed the halving of the things less the
root of that which remains.

16.12 Asemplo ala dicta regola. Et vo' dire
chosì, fammi de 10 dui parti, che multipricata la magiore contra la minore faccia 20.
Adimando quanto serà ciascheuna parte. Fa
chosì, poni la minore parte fosse una chosa.
Dunqua la magiore serà rimanente infino
in 10, che serà 10 meno una chosa. Appresso si vole multiprichare la minore, che
è una cosa, via la magiore, che è 10 meno
una cosa. Et diciamo che vole fare 20. Et
però multipricha una cosa via 10 meno una
cosa. Fa 10 cose meno uno censo, la quale
multiprichatione è oguale a 20. Ristora
ciascheuna parte, cioè dè aggiongere uno
censo a ciascheuna parte, et arai che 10
cose sonno oguali a uno censo et 20 numeri. Arrecha a uno censo, et poi dimezza

16.12 Example to the said rule. And I want
to say thus, make two parts of 10 for me,
so that when the larger is multiplied against
the smaller, it shall make 20. I ask how
much each part will be. Do thus, posit that
the smaller part was a thing. Hence the
larger will be the remainder until 10, which
will be 10 less a thing. Next one shall
multiply the smaller, which is a thing, by
the larger, which is 10 less a thing. And we
say that it will make 20. And therefore
multiply a thing times 10 less a thing. It
makes 10 things less one *censo*, which
multiplication is equal to 20. Restore each
part, that is, you shall join one *censo* to
each part, and you will get that 10 things
are equal to one *censo* and 20 numbers.

le cose, ve ne vienne 5. Multipricha per se medesimo, fa 25. Cavane el numero, che è 20, remane 5, del quale piglia la sua radice, la quale è manifesta che non l'à apponto. Adunqua vale la cosa 5, cioè el dimezzamento meno radice de 5. Et noi ponemo che la parte, cioè la minore, fosse una chosa. Adunqua è 5 meno radice de 5. Et la seconda è rimanente infino in 10, che è 5 et più radice de 5. Et sta bene.

16.13 Uno fa doi viaggi, et al primo viagio guadagna 12. Et al secondo viagio guadagna a quella medesema ragione che fece nel primo. Et quando ebe conpiuti li soi viaggi et egli se trovò tra guadagniati et capitale 54. Vo' sapere con quanti se mosse. Poni che se movesse con una chosa, et nel primo viaggio guadagniò 12. Dunqua conpiuto el primo viaggio si trovò una cosa et 12. Adunqua manifestamente vedi che de ogni una cosa nel primo viaggio fa una chosa e 12. Quanto serrà a quella medesema ragione nel secondo viaggio. Convienti multiprichare una cosa et 12 via **(fol. 40ʳ)** una cosa et 12, che fa uno censo et 24 cose e 144 numeri, li quali sicondo che dice la regola si vole partire in una cosa, et dè ne venire 54. Et però multipricha 54 via una cosa. Fa 54 cose, le quali se oguagliano a uno censo et 24 cose e 144 numeri. Ristora ciascheuna parte, cioè dè cavare 24 cose de ciascheuna parte. Et arai che 30 cose sonno oguali a uno censo et 144 nu-

Bring it to one *censo*, and then halve the things, from which 5 results. Multiply by itself, it makes 25. Remove from it the number, which is 20, 5 remains, of which seize the root, which it is manifest that it does not have precisely. Hence the thing is 5, that is, the halving less root of five. And we posited that the part, that is, the smaller, was a thing. Hence it is 5 less root of 5. And the second is the remainder until 10, which is 5 and added root of 5. And it goes well.

16.13 Somebody makes two voyages, and in the first voyage he gains 12. And in the second voyage he gains at that same rate as he did in the first. And when he had completed his voyages he found himself with 54, gains and capital together. I want to know with how much he set out. Posit that he set out with one thing, and in the first voyage he gained 12. Hence, when the first voyage was completed, he found himself with one thing and 12. Hence you see manifestly that from each one thing in the first voyage he makes a thing and 12. How much will it be at that same rate in the second voyage. It suits you to multiply a thing and 12 times **(fol. 40ʳ)** a thing and 12, which makes one *censo* and 24 things and 144 numbers, which, according to what the rule says, one shall divide in a thing, and 54 shall result from it. And therefore multiply 54 times a thing. It makes 54 things, which equal one *censo* and 24 things and 144 numbers. Restore each part, that is, you

meri. Parti in uno censo, vene quello medesemo. Dimezza le cose, remanghono 15.
Multipricha per se medesemo, fanno 225.
Traine li numeri, che sonno 144, resta 81.
Trova la sua radice, che è 9. Trailo del
dimezzamento dele cose, cioè de 15. Resta
6, et cotanto vale la chosa. Et noi dicemmo
che se movesse con una chosa. Dunqua
vedi manifestamente che se mosse con 6.
Et se la voi provare, fa così. Tu di' che nel
primo viaggio guadagnio 12 et con 6 se
mosse, à 18. Siché nel primo viaggio se
trovò 18. E peró di' così, se de 6 io fo 18,
que farò de 18 a quella medesema ragione.
Multipricha 18 via 18, fa 324. Parti in 6,
che ne vene 54, et sta bene. Et così se
fanno le simili ragioni.

shall remove 24 things from each part. And
you will get that 30 things are equal to one
censo and 144 numbers. Divide in one
censo, the same results. Halve the things,
15 remain. Multiply by itself, they make
225. Detract from it the numbers, which are
144, 81 is left. Find its root, which is 9.
Detract it from the halving of the things,
that is, from 15. 6 is left, and as much is
worth the thing. And we said that he set out
with one thing. Hence you see manifestly
that he set out with 6. And if you want to
verify it, do thus. You say that in the first
voyage he gained 12, and with 6 he set out,
one has 18. So that in the first voyage he
found himself with 18. And therefore say
thus, if from 6 I make 18, what will I make
from 18 at that same rate. Multiply 18
times 18, it makes 324. Divide in 6, and 54
results from it, and it goes well. And thus
the similar computations are done.

16.14 Ancora si poterebbe dire che si movesse colla radice de rimanente et più el dimezzamento dele cose, cioè cola radice de
81, che è 9. Pollo sopra a 15, fa 24. Et
cossì sta bene nell'uno modo como nell'altro. Et eccho la prova. Noi abbiamo facta
all'altro modo che se movesse con 6. Et
abbiamo facto ragione che, conpiuti i viaggi, si trovò 54 chomo noi diciamo. Ora
faciamo ragione che se movesse con 24, et
diciamo che nel primo viaggio guadagnò
12. Siché se trovò 36. Ora di' chosì, se con
24 io fo 36, que farrò con 36. Multiplica
36 via 36, fa 1296, et parti in 24, che ne

16.14 Again one could say that he set out
with the root of the remainder and added
the halving of the things, that is, with the
root of 81, which is 9. Put it above 15. It
makes 24. And so it goes well in one way
as well as the other. And here is the verification. We have made in the other way that
he set out with 6. And we have computed
that, when the voyages were completed, he
found himself with 54, as we say. Now we
make the computation that he set out with
24, and we say that in the first voyage he
gained 12. So that he found himself with
36. Now do thus, if with 24 I make 36,

vene 54, et sta bene. Siché tu vedi che all'uno modo et all'altro sta bene. Et però quella così facta regola è molto da lodare, che ce dà doi responsioni et così sta bene all'una come all'altro. Ma abbi a mente che tucte le ragioni che reduchono a questa regola non si possono **(fol. 40ᵛ)** respondere per doi responsioni se non ad certe. Et tali sonno che te conviene pigliare l'una responsione, et tale l'altra. Cioè a dire che a tali ragioni te converà rispondere che vaglia la cosa el dimezzamento dele cose meno la radice de rimanente. Et a tale te converrà dire la radice de remanente e più el dimezzamento dele cose. Onde ogni volta che te venisse questo co'tale raoguaglamento, trova in prima l'una responsione. Et se non te venisse vera, de certo si piglia l'altra senza dubio. Et averai la vera responsione. Et abi a mente questa regola. In bona verità vorrebbe una grande despositione. Ma non mi distendo troppo però che me pare stendere et scrivere in vile cosa. Ma questo baste qui et in più dire sopra ciò non mi vo' stendere.

what will I make with 36. Multiply 36 times 36, it makes 1296, and divide in 24, and 54 results from it, and it goes well. So that you see that it goes well in one way as well as the other. And therefore the rule so constituted is much to be praised, which gives us two answers, and it goes well thus, one as well as the other. But keep in mind that not all computations that lead back to this rule can **(fol. 40ʳ)** be answered with two answers, but only some of them. And there are some for which it suits you to seize one answer, and some, the other. That is to say that to some computations it suits you to answer that the thing is the halving of the things less the root of the remainder. And to some it suits you to say the root of the remainder and added the halving of the things. Wherefore every time that you are asked with such an equation, find first one answer. And if it does not turn out true for you, certainly the other is seized without doubt. And you will get the true answer. And keep in mind this rule. Verily, a vast exposition would be needed. But I will not enlarge too much, because I seem to expand and write about a base thing. But this will be enough here, and I will not expand to say more about it.

16.15 **Pongoti** assemplo a quello che abbiamo dicto denanzi, et dichò chosì, fami de 10 dui parti, che multipricata l'una contra l'altra et sopra la dicta multiplichatione giontovi la differentia che à dall'una parte all'altra faccia 22. Adimando, quanto

16.15 I propose to you an example to that which we have said before, and I say thus, make two parts of 10 for me, so that when one is multiplied against the other and above the said multiplication is joined the difference which there is from one part to

serrà ciascheuna parte. Fa chosì, poni che l'una parte fusse una cosa. Dunqua l'altra parte serrà lo rimanente infino in 10, che serà 10 meno una cosa. Appresso multipricha l'una contra all'altra, cioè una cosa via 10 meno una cosa, che fa 10 cose meno uno censo. Appresso sopra a questa multipricatione poni la differenza che è da una cosa a 10 meno una cosa, che è 10 meno ij cose, le quali differenze se vole giungere a 10 cose meno uno censo, et arai che fanno 10 numeri e otto cose meno uno censo, le quali se aggiongono a 22. Ristora ciascheuna parte, cioè dè cavare 10 numeri de ciascheuna parte. Et arai che 8 cose meno uno censo sonno oguali a 12 numeri. Dà uno censo a ogni parte, et arai che 8 cose sonno oguali a 12 numeri et uno censo. **(fol. 41ʳ)** Parti nelli censi, vene quello medesimo. Dimezza le cose, sonno 4. Multipricha per se medesimo, sonno 16. Cavane li numeri, che sonno 12, remane 4. E piglia la sua radice et più el dimezzamento dele cose. Et cotanto vale la cosa. La radice de 4 è 2. Et più el dimezzamento dele cose, che sonno 4, et 2, ày 6, et cotanto vale la chosa. Et noi dicemo che l'una parte fosse una cosa. Dunqua vene ad essere 6. Et la seconda parte l'avanzo infino in 10, che è 4. Provala, et multipricha 4 via 6, fa 24. Giongi suso la differenza che è dall'una all'altra, che è 2, ài 26. Et noi vogliamo 22. Siché vedi manifestamente che non sta bene. Però che in questa ragione la cosa non vale la radice de quello che remane et più el dimezzamente dele cose. Adunqua abiamo provata questa, et

the other, it makes 22. I ask, how much will each part be. Do thus, posit that one part was a thing. Hence the other part will be the remainder until 10, which will be 10 less a thing. Next multiply one against the other, that is, a thing times 10 less a thing, which makes 10 things less one *censo*. Next, above this multiplication put the difference which there is from a thing to 10 less a thing, which is 10 less ij things, which difference one shall join to 10 things less a *censo*, and you will get that they make 10 numbers and 8 things less one *censo*, which join together to 22. Restore each part, that is, you shall remove 10 numbers from each part. And you will get that 8 things less one *censo* are equal to 12 numbers. Give one *censo* to each part, and you will get that 8 things are equal to 12 numbers and one *censo*. **(fol. 41ʳ)** Divide in the *censi*, the same results. Halve the things, they are 4. Multiply by itself, they are 16. Remove from them the numbers, which are 12, 4 remains. And seize its root, and added the halving of the things. And as much is worth the thing. The root of 4 is 2. And added the halving of the things, which are 4, and 2, you get 6, and as much is worth the thing. And we said that one part was a thing. Hence it comes to be 6. And the second part the excess until 10, which is 4. Verify it, and multiply 4 times 6, it makes 24. Join on top the difference which there is from one to the other, which is 2, you get 26. And we want 22. So that you see manifestly that it does not go well. Because in this computation the thing is not

non ce vene bene de certo. L'altra provamo
e de certo verrà bene, cioè dè pigliare el
dimezzamento dele cose meno la radice de
rimanente. E'l dimezzamento dele cose è
4. La radice de rimanente è 2, però che
como tu sai ce remase 4, et la sua radice
è 2. Cava 2 de 4, remane 2. Et cotanto vale
la cosa. Et l'altra parte serrà rimanente
infino in 10, che è 8. Et sta bene. Et provala. Multiplicha 2 via 8, fa 16. Poni suso la
differenza ch'è da 2 a 8, che è[a] 6, à 22. Et
sta bene. Et così se fanno le simiglianti
ragioni.

worth the root of that which remains and
added the halving of the things. Hence we
have verified this one, and it certainly did
not result well for us. We verify the other,
and it will certainly result well, that is, you
shall seize the halving of the things less the
root of the remainder. And the halving of
the things is 4. The root of the remainder
is 2, because as you know 4 remained for
us, and its root is 2. Remove 2 from 4, 2
remain. And as much is worth the thing.
And the other part will be the remainder
until 10, which is 8, and it goes well. And
verify it: multiply 2 times 8, it makes 16.
Put on top the difference which there is
from 2 to 8, one has 22. And it goes well.
And thus the similar computations are done.

[a] Corrected, perhaps from "fa".

16.16   Quando li censi sonno oguali alle cose
et al numero, se vole partire nelli censi, et
poi dimezzare le cose, et multiplicare per
se medesmo et giongere al (fol. 41') numero. Et la radice dela summa più el dimezzamento dele cose vale la cosa.

16.16   When the *censi* are equal to the things
and to the number, one shall divide in the
censi, and then halve the things, and multiply by itself and join to the (fol. 41') number. And the root of the total plus the halving of the things is worth the thing.

16.17   Assemplo ala dicta regola. Et vo' dire
così, uno à 4̶0̶ 40 *f* d'oro et canbiòli a
venetiani. Et poi de quelli venetiani tolse
60 et recanbiòli a *f* d'oro a uno venetiano
più per *f* che meli cambiò in prima. Et
quando à così cambiato et quello trovò, che
tra venetiani che glie rimaseno quando ne
trasse 60, et li fiorini che ebe de 60 venetiani, gionti inseme fece 100. Vo' sapere
quanto valze el *f* a venetiani. Di' così,
pognamo che 'l *f* valesse una cosa. Dunqua

16.17   Example to the said rule. And I want
to say thus, somebody has 40 gold *fiorini*
and changed them to *venetiani*. And then
from those *venetiani* he grasped 60 and
changed them back into gold *fiorini* at one
venetiano more per *fiorino* than he changed
them at first for me. And when he has
changed thus, that one found that the *venetiani* which remained with him when he
detracted 60, and the *fiorini* he got for the
60 *venetiani*, joined together made 100. I

40 *f* vagliono 40 chose de venetiani. E poi
ne tolse 60 de quelli venetiani, et cambiòli
a *f* d'oro a uno venetiano più el *f*. Adunqua cava 60 venetiani de 40 cose de venetiani. Remangono 40 cose meno 60 venetiani. Et questi venetiani che glie sono
remasti, raggionti co' *f* che egli ebe de 60
venetiani, fanno 100. Dunqua se noi traes-
semo 40 cose meno 60 venetiani de 100,
remarracte [*sic*, read "remarran'te"?] quello
che vagliono li 60 venetiani a *f* d'oro.
Adunqua trai 40 cose meno 60 de 100,
rimane 160 meno 40 cose. Et dunqua li *f*
che egli ebe de 60 venetiani forono 160
meno 40 cose. Et quando egli recambiò 60
venetiani a *f* d'oro si cambiò a uno venetiano più el *f* che prima. Dunqua li 60
venetiani cambiò a una cosa et uno venetiano. Et noi abbiamo che 60 venetiani
vagliono a fiorini d'oro 160 meno 40 chose.
Dunqua dobbiamo sapere se 160 meno 40
chose fiorini d'oro, a avalere el *f* una cosa
et uno venetiano, se vale 60 venetiani.
Adunqua multipricha 160 meno 40 cose via
una cosa et uno, fanno 120 cose meno 40
censi et più 160 numeri. **(fol. 42ʳ)** Sonno
oguali a 60 venetiani. Et così abiamo che
120 cose meno 40 censi et più 160 numeri
sonno oguali a 60. Ristora ciascheuna parte,
arai che 40 censi sonno oguali a 120 cose
et 100 numeri. Parti nelli censi, arai che
uno censo sia oguali a 3 chose e dui numeri
et mezzo. Dimezza le chose, j $\frac{1}{2}$ ᵃ. Multipricha per se medesimo, fa 2 et $\frac{1}{4}$. Giungi
sopra al numero, fa 4 et $\frac{3}{4}$, et abbiamo che
la chosa vale la radice de 4 et $\frac{3}{4}$ et più el
dimezzamento dele chose, che fo uno ⟨e⟩

want to know how much was worth the
fiorino in *venetiani*. Say thus, let us posit
that the *fiorini* was worth one thing. Hence
40 *fiorini* are worth 40 things of *venetiani*.
And then he grasped 60 of these *venetiani*,
and changed them to *fiorini* at one *venetiano* more the *fiorino*. Hence remove 60
venetiani from 40 things of *venetiani*. 40
things less 60 *venetiani* remain. And these
venetiani which remained with him, joined
with the *fiorini* which he got from 60 *venetiani*, make 100. Hence, if we detracted 40
things less 60 *venetiani* from 100, will(?)
remain for you that which the 60 *venetiani*
are worth in gold *fiorini*. Hence detract 40
things less 60 from 100, 160 less 40 things
remain. And hence the *fiorini* which he got
from 60 *venetiani* were 160 less 40 things.
And when he changed back 60 *venetiani*
into gold *fiorini* they were changed at one
venetiano more the *fiorino* than before.
Hence he changed the 60 *venetiani* at one
thing and a *venetiano*. And we have that 60
venetiani are worth in gold *fiorini* 160 less
40 things. Hence we shall know whether
160 less 40 things gold *fiorini*, the *fiorino*
being one thing and a *venetiano*, is worth
60 *venetiani*. Hence multiply 160 less 40
things times one thing and one, they make
120 things less 40 *censi* and added 160
numbers. **(fol. 42ʳ)** They are equal to 60
venetiani. And thus we get that 120 things
less 40 *censi* and added 160 numbers are
equal to 60. Restore each part, you will get
that 40 *censi* are equal to 120 things and
100 numbers. Divide in the *censi*, you will
get that one *censo* is equal to 3 things and

mezzo. Et noi ponemo che'l f valesse una chosa, dunqua valse la radice de 4 et $\frac{3}{4}$ et più el dimezzamento dele cose, che è j$\frac{1}{2}$. Et è facta.

two numbers and a half. Halve the things, j$\frac{1}{2}$. Multiply by itself, it makes 2 and $\frac{1}{4}$. Join above the number, it makes 4 and $\frac{3}{4}$, and we get that the thing is the root of 4 and $\frac{3}{4}$ and added the halving of the things, which makes one ⟨and a⟩ half. And we posited that the *fiorino* was worth one thing, hence it was worth the root of 4 and $\frac{3}{4}$ and added the halving of the things, which is j$\frac{1}{2}$. And it is done.

ᵃ The same writing of $\frac{1}{2}$ is used in 22.30, fol. 59ʳ.

[17. Rules without Examples for Reducible Third- and Fourth-Degree Equations]

¹⁷·¹ *Qui finischo le sey regole conposte con alquanti assempri. Et incomincia l'altre regole che sequitano le sopradicte sey como vederete.*

¹⁷·¹ *Here I end the six rules combined with various examples. And begins the other rules that follow the six told above, as you will see.*

¹⁷·² **Quando** li Censi [*sic*, read "cubi"] sonno oguali al numero, si vole partire el numero per li chubi, et la radice chubicha de quello che ne vene vale la cosa.

¹⁷·² When the ⟨*cubi*⟩ are equal to the number, one shall divide the number by the *cubi*, and the cube root of that which results from it is worth the thing.

¹⁷·³ **Quando** li chubi sonno oguali alle cose, si vole partire le cose per li chubi, et la radice de quello che ne vene vale la cosa.

¹⁷·³ When the *cubi* are equal to the things, one shall divide the things by the *cubi*, and the root of that which results from it is worth the thing.

¹⁷·⁴ ⟨Q⟩uando li chusi [*sic*, read "chubi"] sonno oguali a li censi, si vole partire li censi per li chubi. Et quello che ne vene si è numero, et cotanto vale la cosa.

¹⁷·⁴ When the *cubi* are equal to the *censi*, one shall divide the *censi* by the *cubi*. And that which results from it is number, and as much is worth the thing.

17.5 **Q**uando li chubi et li censi sonno oguali alle chose, se vole partire nelli chubi, et
poi dimezzare li censi et multipricare per
se medesimo et giongerlo ale cose. Et la
radice dela somma meno el dimezzamento
de' censi vale la cosa.

17.5 When the *cubi* and the *censi* are equal
to the things, one shall divide in the *cubi*,
and then halve the *censi* and multiply by
itself and join to the things. And the root
of the total less the halving of the *censi* is
worth the thing.

17.6 **Q**uando li censi sonno oguali alli chubi
et alle cose, **(fol. 42ᵛ)** devi partire nelli
chubi et poi dimezzare li censi et multiplicare per se medesimo et cavarne le cose,
et la radice de quello ⟨che⟩ rimane più el
dimezzamento deli censi vale la cosa.
Overo el dimezzamento de' censi meno la
radice de rimanente.

17.6 When the *censi* are equal to the *cubi*
and to the things, **(fol. 42ᵛ)** you shall divide
in the *cubi* and then halve the *censi* and
multiply by itself and remove from it the
things, and the root of that ⟨which⟩ remains
plus the halving of the *censi* is worth the
thing. Or indeed the halving of the *censi*
less the root of the remainder.

17.7 **Q**uando li chubi sonno oguali alli censi
et alle cose, dei partire ⟨ne⟩li chubi et poi
dimezzare li censi, et multiprichare per se
medesimo et agiungere alle cose, et la
radice dela summa più el dimezzamento de'
censi vale la chosa.

17.7 When the *cubi* are equal to the *censi*
and to the things, you shall divide ⟨in⟩ the
cubi and then halve the *censi*, and multiply
by itself and join to the things, and the root
of the total plus the halving of the *censi* is
worth the thing.

17.8 **Q**uando li censi de censi sonno oguali
al numero, se vole partire el numero nelli
censi de censi. Et la radice ⟨della radice⟩
de quello che ne vene vale la cosa.

17.8 When the *censi* of *censi* are equal to
the number, one shall divide the number in
the *censi* of *censi*. And the root ⟨of the
root⟩ of that which results from it is worth
the thing.

17.9 **Q**uando li censi de censi sonno oguali
alle cose se vole partire le cose per li censi
de censi, et la radice chubicha de quello
vale la cosa.

17.9 When the *censi* of *censi* are equal to
the things one shall divide the things by the
censi of *censi*, and the cube root of that is
worth the thing.

^{17.10} **Q**uando li censi de censi sonno oguali a censi, se vole partire li censi per li censi de censi, et la radice de quello che ne vene vale la chosa.

^{17.10} When the *censi* of *censi* are equal to the *censi*, one shall divide the *censi* by the *censi* of *censi*, and the root of that which results from it is worth the thing.

^{17.11} **Q**uando li censi de censi sonno oguali ali chubi, se vole partire li chubi per li censi de censi. Et quello che ne vene si è numero, et cotanto vale la cosa.

^{17.11} When the *censi* of *censi* are equal to the *cubi*, one shall divide the *cubi* by the *censi* of *censi*. And that which results from it is number, and as much is worth the thing.

^{17.12} **Q**uando li censi de censi et li chubi sonno oguali ali censi, si vole partire nelli censi de censi, et poi dimezzare li chubi et multiprichare per se medesimo, et agiungere alli censi. Et la radice dela summa meno el dimezzamento de' chubi vale la cosa.

^{17.12} When the *censi* of *censi* and the *cubi* are equal to the *censi*, one shall divide in the *censi* of *censi*, and then halve the *cubi* and multiply by itself, and join to the *censi*. And the root of the total less the halving of the *cubi* is worth the thing.

^{17.13} **Q**uando li chubi sonno oguali alli censi de censi et {d}a censi, si vole partire nelli censi de censi, et poi dimezzare li chubi, et multiprichare per se medesimo, et cavarne li censi, et la radice dela summa^a et el dimezzamento de' chubi vale la chosa. Overo el dimezzamento de' chubi meno la radice de quello che remane.

^{17.13} When the *cubi* are equal to the *censi* of *censi* and to *censi*, one shall divide in the *censi* of *censi*, and then halve the *cubi*, and multiply by itself, and remove from it the *censi*, and the root of the total^a and the halving of the *cubi* is worth the thing. Or indeed the halving of the *cubi* less the root of that which remains.

^a As in an account, the *summa* is any kind of total, not just an arithmetical sum. Elsewhere it may result from a multiplication or an (imagined) measurement.

(fol. 43ʳ)

^{17.14} **Q**uando li censi de censi sonno oguali a chubi et a censi, vole partire nelli censi de censi, et poi dimezzare li chubi, et multiprichare per se medesimo, et giungere alli

^{17.14} When the *censi* of *censi* are equal to *cubi* and to *censi*, one shall divide in the *censi* of *censi*, and then halve the *cubi*, and multiply by itself, and join to the *censi*.

censi. Et la radice dela summa più el dimezzamento de' chubi vale la cosa.

And the root of the total plus the halving of the *cubi* is worth the thing.

17.15 **Q**uando li censi de censi et li censi sonno oguali al numero, se vole partire nelli censi de censi, et poi dimezzare li censi ⟨et multiplichare per se medesimo⟩[a] et agiungere al numero. Et la radice dela radice dela summa et meno el dimezzamento de' censi vale la cosa.

[a] The same lacuna is found in **A**.

17.15 When the *censi* of *censi* and the *censi* are equal to the number, one shall divide in the *censi* of *censi*, and then halve the *censi* ⟨and multiply by itself⟩[a] and join to the number. And the root of the root of the total and less the halving of the *censi* is worth the thing.

17.16 **Q**ui finischono le xv regole sopradicte senza niuna dispositione, le qual' cose como io t'ò dicto se reduchono alle sey regole de prima.

17.16 Here end the xv rules spoken of above without any exposition, which things as I have said to you lead back to the six rules from before.

[18. A Grain Problem of Alloying Type]

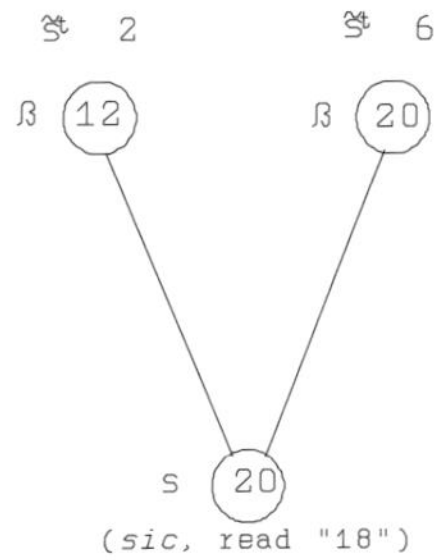

18.1 **U**no homo à 100 staia de grano che vale ß 20 lo s̃t, et grano che vale ß 12 lo s̃t. Ora vene per caso che costui vole mettere, de quello che vale ß 12 lo s̃t, sopra a quello che vale ß 20 lo s̃t, tanto che così mescolato vaglia ß 18 lo s̃t. Vo' sapere quanto ve ne mettarà. Fa così, pogniamo che li ponggi cossì ordinati, et di' così, da ß 12 lo s̃t infino a quello da ß 18 si à ß vj. Et poni 6 sopra a quello da 20 lo s̃t. Et poi di' così, da ß 20 infino in ß 18 si à 2. Et poni 2 sopra a quello de ß 12 lo s̃t. Ora di' così, quando tolgho s̃t 6 de quello che vale lo s̃t ß 20, sì tolgo s̃t 2 de quello cha vale ß 12. Vo' sapere, quando io torrò s̃t 100 de quello

18.1 A man has 100 *staia* of grain that is worth *soldi* 20 the *staio*, and grain that is worth *soldi* 12 the *staio*. Now it occurs by chance that he wants to put, of that which is worth *soldi* 12 the *staio*, so much above that which is worth *soldi* 20 the *staio* that, thus blended, it is worth *soldi* 18 the *staio*. I want to know how much of it he will put to it. Do thus, let us posit that he posits them thus ordered. And say thus, from *soldi* 12 the *staio* until that of *soldi* 18 there is *soldi* vj. And put 6 above that of 20 the *staio*. And then say thus, from *soldi* 20 until *soldi* 18 there is 2. And put 2 above that of *soldi* 12 the *staio*. Now say

de ß 20 lo š^t, quanto torrò de quello de ß 12. Et però, 100 via 2 š^t de ß 12 lo š^t fa š^t 200. Et parti in 6 che ne vene š^t 33 e $\frac{1}{3}$ š^t.^{a} Siché se tu metterai š^t 33 e $\frac{1}{3}$ de quello che vale ß 12 lo š^t sopra ad š^t 100 de ß 20 lo š^t, arai in tucto š^t 133 e $\frac{1}{3}$ de grano de ß 20 et de ß 12 lo š^t. Ora la prova se sta bene. Tu di' che volini potere dare per ß 18 lo š^t così miscolato. Sappi prima che vagliono š^t 100 de ß 20 lo š^t, che vale £ 100. Ora sappi quello che vale š^t 33 e $\frac{1}{3}$ de ß 12 lo š^t, che vale £ 20. Poni sopra a **(fol. 43^v)** 100 £ et ài 120 £. Et cotanto vale così miscolato le 133 š^t e $\frac{1}{3}$ per ß 18 lo š^t, che vagliono appunto £ 120. Et sta bene, et è facta. Et così se fanno tucte le simili ragioni.

thus, when I grasp *staia* 6 from that which is worth *soldi* 20 the *staio*, *staia* 2 are grasped from that which is worth 12 *soldi*. I want to know, when I shall grasp *staia* 100 from that of *soldi* the *staio*, how much will I grasp from that of *soldi* 12. And therefore, 100 times 2 *staia* of *soldi* 12 makes *staia* 200. And divide in 6, and *staia* 33 and $\frac{1}{3}$ result from it.^{a} So that if you will put *staia* 33 and $\frac{1}{3}$ of that which is worth *soldi* 12 the *staio*, above *staia* 100 of *soldi* 20 the *staio*, you will get in total 133 *staia* and $\frac{1}{3}$ of grain of *soldi* 20 and of *soldi* 12 the *staio*. Now the verification whether it goes well: You say that they will be able to give for *soldi* 18 the *staio* thus blended. Know first what are worth *staia* 100 of *soldi* 20 the *staio*, which is worth *libre* 100. Now know that which is worth *staia* 33 and $\frac{1}{3}$ of *soldi* 12 the *staio*, which is worth *libre* 20. Put it above **(fol. 43^v)** 100 *libre* and you will get 120 *libre*. And as much is worth, thus blended, the 133 *staio* and $\frac{1}{3}$ of *soldi* 18 the *staio*, which are worth precisely *libre* 120, and it goes well, and it is done. And thus all the similar computations are done.

^{a} The margin summarizes "Staia 33 $\frac{1}{3}$ staio".

[19. Second- and Third-Degree Problems about Continued Proportions Dressed as Wage Problems and Solved without the Use of *cosa-census* Algebra]

$^{19.1}$ Uno sta a uno fondacho 3 anni, et à de salario tra'l primo anno e'l terzo 20 f. El secondo anno à 8 f. Vo' sapere que glie

$^{19.1}$ Somebody stays in a warehouse 3 years, and in the first and third year together he gets in salary 20 *fiorini*. The

venne el primo anno et que el terzo preci-
samente, ogni uno per se solo. Fa così, et
questo te sia sempre a mente, che tanto
vole fare multiprichato el secondo anno per
se medesimo quanto el primo nel terzo. Et
fa così, multipricha el secondo per se medesimo che di' che ebe 8 f. Multipricha 8 via
8, fa 64 f. Ora te convene fare de 20 f, che
tu di' che ebbe tra'l primo e'l terzo anno,
{tra} 2 parti che moltipricha⟨ta⟩ l'una
contra l'altra faccia 64 f. Et farrai così,
cioè che sempre dimezze quello che à nelli
2 anni. Cioè, dimezza 20, venne 10. Moltipricha l'uno contra all'altro, fa 100. Cavane
la multiprichatione facta del secondo anno
che è 64, resta 36. Et de questo trova la sua
radice, et dirrai che l'una parte serà 10,
cioè el primo anno [*sic*, this word order]
meno radice de 36. Et l'altra parte, cioè el
secondo anno, serà 8 f. Et la terza serà da
10 meno radice de 36 infino in 20 f, che
sonno f 10 et più radice de 36. Et se la voli
provare, fa così et di', el primo anno à 10
f meno radice de 36, che è 6. Tray 6 de 10,
resta 4 f. Et 4 f ebbe el primo anno. Et el
secondo ebe 8 f. Et el terzo ebbe f 10 et
più radice de 36, che è 6. Ora poni 6 f
sopra a 10 f, arai 16 f. Et tanto ebe el
terzo anno. Et sta bene. Et tanto fa multiprichato el primo contra al terzo quanto el
secondo per se medesimo. Et tal parte è el
secondo del terzo quale el primo del
secondo. Et è fatta.

second year he gets 8 *fiorini*. I want to
know what he received accurately the first
year and the third year, each one by itself.
Do thus, and let this always be in your
mind, that the second year multiplied by
itself will make as much as the first in the
third. And do thus, multiply the second by
itself, in which you say that he got 8 *fiorini*. Multiply 8 times 8, it makes 64 *fiorini*.
Now it suits you to make of 20 *fiorini*,
which you say he got in the first and third
year together, {...} two parts which when
multipli⟨ed⟩ one against the other makes 64
fiorini. And you will do thus, that is that
you always halve that which he got in the
two years. That is, halve 20, 10 result.
Multiply the one against the other, it makes
100. remove from it the multiplication
made from the second year which is 64, 36
is left. And of this find its root, and you
will say that one part, that is, the first year,[a]
will be 10 less root of 36. And the other
part, that is, the second year, will be 8
fiorini. And the third will be from 10 less
root of 36 until 20 *fiorini*, which are *fiorini*
10 and added root of 36. And if you want
to verify it, do thus and say: the first year
he gets 10 *fiorini* less root of 36, which is
6. Detract 6 from 10, 4 *fiorini* is left. And
4 *fiorini* he got the first year. And the
second year he got 8 *fiorini*. And the third
he got *fiorini* 10 and added root of 36,
which is 6. Now put 6 *fiorini* above 10
fiorini, you will get 16 *fiorini*. And so
much he got the third year. And it goes
well. And the first multiplied against the
third makes as much as the second by

itself. And such a part is the second of the third as the first of the second. And it is done.

(fol. 44ʳ)

^{19.2} Uno sta a uno fondicho 4 anni, et el primo anno ebe 15 *f* d'oro. El quarto ebe 60 *f*. Vo' sapere quanto ebe el secondo anno e'l terzo a quella medesima ragione. Fa così, che tu parte quello che egli ebbe el quarto anno in quello che ebbe el primo anno. Et dirai che quello che ne vene sia radice chubicha. Ora ài a partire 60 *f* in 15, che ne vene 4 *f*. Et questo 4 si è ℞ᵃ chubicha. Et sempre piglia el partitore et arrechalo a radice, cioè arrecha 15 a radice. Et di' chosì, multipricha 15 via 15, fa 225 [corrected from "125"]. Ora multipricha 15 via 225, che fa 3375. Ora multipricha la radice chubica, cioè 4, che è radice chubicha, contra ala radice chubicha {contra ala radice chubicha} de 3375, che fa radice chubicha de 13500. Et cotanto ebbe el secondo anno. Ora facciamo per lo terzo anno et multipricha 4, che è dicto de sopra, contra a radice chubicha de 13500, che fa radice chubicha 54000, et cotanto ebbe el terzo anno ad quella medesema ragione che ebbe el primo e'l quarto anno. Siché noi dirremo che costui avesse el primo anno *f* 15. El secondo anno ebbe radice chubica de 13500 *f* d'oro. El terzo anno ebbe radice chubicha de *f* 54000, et el quarto anno ebe *f* 60 d'oro. Et sta bene.

^{19.2} Somebody stays in a warehouse 4 years, and in the first year he got 15 gold *fiorini*. The fourth he got 60 *fiorini*. I want to know how much he got the second year and the third at that same rate. Do thus, that you divide that which he got in the fourth year in that which he got in the first year. And you will say that what results from it is cube root. Now you have to divide 60 *fiorini* in 15, from which results 4 *fiorini*. And this 4 is cube root. And always seize the divisor and bring it to root, that is, bring 15 to root, and say thus, multiply 15 times 15, it makes 225. Now multiply 15 times 225, which makes 3375. Now multiply the cube root, that is, 4, which is cube root, against the cube root {...} of 3375, which makes cube root of 13500. And as much he got the second year. Now let us do for the third year and multiply 4, which is said above, against cube root of 13500, which makes cube root ⟨of⟩ 54000, and as much he got the third year at that same rate as he had the first and the fourth year. So that we will say that he got the first year, *fiorini* 15. The second year he got cube root of 13500 gold *fiorini*. The third year he got cube root of *fiorini* 54000, and the fourth year gold *fiorini* 60. And it goes well.

^{19.3} Uno sta a uno fondicho 4 anni. Et tra'l primo anno e'l quarto ebe 90 ƒ d'oro. Et tra'l secondo anno e'l terzo ebbe 60 ƒ d'oro. Vo' sapere que gli venne ogni uno per se solo. Et siano in propositione et sia tal parte el primo del secondo come el secondo del terzo, et come el terzo del quarto. Et sempre te stia a mente questo, che tanto fa a multiprichare el primo anno nel quarto quanto el secondo anno nel terzo. Et tanto fa a partire el quarto anno nel secondo quanto el terzo anno nel primo. Ora fa così, che sempre tu arreche quello che egli à tra'l secondo e'l terzo anno a radice chubicha. Et poi multipricha quello **(fol. 44ᵛ)** che egli à tra'l secondo e'l terzo anno per 3. Et sopra aquello giongi quello che gl'à tra'l primo e'l quarto anno. Et questo è el partitore. Et ài a partire la radice chubicha sopradicta. Et per che tu intende meglio, fa così, multipricha 60 via 60, fa 3600. Et 60 via 3600 fa 216000, et ài a partire in quello che fa 3 via 60 giontovi suso 90, che fa 270. Et questo è el partitore. Parti 216000 in 270, che ne vene 800. Et tanto fa multiprichato el primo anno nel quarto. Et multiprichato el secondo nel terzo fa ancho 800. Siché te convene fare de 90 doi parti, che multiprichata l'una contra l'altra faccia 800. E però fa così, dimezza 90, venne 45. Moltiprichalo per se medesimo, fa 2025. Cavane 800, resta 1225. Et dirai che l'una parte, cioè el primo anno, avesse ƒ 45 meno radice de 1225 ƒ. Et el quarto anno lo resto infine in 90 ƒ che è ƒ 45 et più radice de 1225 ƒ. Et afacto [*sic*, read "è facto"] pe'l primo e'l quarto

^{19.3} Somebody stays in a warehouse 4 years. And in the first year and the fourth together he got 90 gold *fiorini*. And in the second year and the third together he got 60 gold *fiorini*. I want to know what resulted for him, each one by itself. And let them be in proportion and let the first be such part of the second as the second of the third, and as the third of the fourth. And let it always stay in your mind this, that to multiply the first year in the fourth makes as much as the second year in the third. And it makes as much to divide the fourth year in the second as the third year in the first. Now do thus, that you always bring to cube root that which he gets in the second and third year together. And then multiply that **(fol. 44ᵛ)** which he gets in the second and third year together by 3. And above this you join that which he gets in the first and fourth year together. And this is the divisor. And you have to divide the cube root spoken of above. And in order that you understand better, do thus, multiply 60 times 60, it makes 3600. And 60 times 3600 makes 216000, and you have to divide in that which 3 times 60 makes when 90 is joined on top, which makes 270. And this is the divisor. Divide 216000 in 270, from which results 800. And so much makes the first year multiplied in the fourth. And the second multiplied in the third also makes 800. So that it suits you to make from 90 two parts, of which one, multiplied against the other, makes 800. And therefore do thus, halve 90, 45 result. Multiply it by itself, it makes 2025.

anno. Et per che tu intende meglio questo numero, cioè 1225, la sua radice si è 35, però che fa 35 via 35 1225. Siché el primo anno di' che ebbe *f* 45 meno 35, resta 10 *f*. Et *f* 10 ebe el primo anno. El quarto anno ebe *f* 45 di et [*sic*, this word order] più radice de 1225, che è *f* 35. Poni sopra a 45, fa 80. Et *f* 80 ebbe el quarto anno. Ora facciamo per lo secondo et terzo anno, et fa in simile modo, che tu faccia de 60ta 2 parti che moltipricha^ta^ l'una contra all'altra faccia 800. Et però fa così, dimezza 60, venne 30. Multipricha per se medesimo, fa 900. Traine 800, resta 100. Et dirai che'l secondo anno avesse *f* 30 meno radice de 100. Et el terzo anno el resto infino ~~el secondo~~ in 60, che è 30 et più radice de 100. Et la radice de 100 si è 10, sicomo tu sai 10 via 10 fa 100. Et però, perché tu di' che'l secondo anno à *f* 30 meno {meno} radice de 100 *f*, che sonno *f* 10, trai 10 de 30, resta 20. Et *f* 20 ebbe el secondo anno. Et el terzo ebbe *f* 30 et più radice de *f* 100 che è 10. Poni 10 sopra a 30, fa 40, et 40 *f* ebe el terzo anno. Et è facta, et bene vedi chiaro che ciascheuno de questi numeri sonno in propositione. Et tal parte **(fol. 45ʳ)** è el primo del secondo quale el secondo del terzo et quale el terzo del quarto. Ciascheuno è la mità. Et anchora vedi chiaro che tanto fa multiplicato el primo contra al quarto, che fa tanto quanto multiprichato el secondo contra al terzo. Et tante ne vene a partire el quarto nel secondo quanto vene a partire el terzo nel primo. Siché vedi chiaro che la allegatione sta bene. Et è facta apponto. Et così se fanno le simiglianti

Remove 800 from it, 1225 is left. And you will say that one part, that is, the first year, he got 45 *fiorini* less root of 1225 *fiorini*. And the fourth year the rest until 90 *fiorini*, which is 45 *fiorini* and added root of 1225 *fiorini*. And ⟨it is⟩ done for the first and the fourth year. And in order that you understand better this number, that is, 1225, its root is 35 because 35 times 35 makes 1225. So that the first year you say that he got *fiorini* 45 less 35, 10 *fiorini* is left. And *fiorini* 10 he got the first year. The fourth year he got *fiorini* 45 and added root of 1225, which is *fiorini* 35. Put it above 45, it makes 80. And *fiorini* 80 he got the fourth year. Now let us do for the second and third year and do in a similar way, that you make from 60 2 parts so that, when one is multiplied against the other, it makes 800. And therefore do thus, halve 60, 30 results from it. Multiply by itself, it makes 900. Detract from it 800, 100 is left. And you will say that the second year he had *fiorini* 30 less root of 100, And the third year the rest until 60, which is 30 and added root of 100. And the root of 100 is 10, since as you know 10 times 10 makes 100. And therefore, because you say that the second year he gets *fiorini* 30 less {...} root of 100 *fiorini*, which are *fiorini* 10, detract 10 from 30, 20 is left. And *fiorini* 20 he got the second year. And the third he got *fiorini* 30 and added root of *fiorini* 100, which is 10. Put 10 above 30, it makes 40, and 40 *fiorini* he got the third year. And it is done, and you see well clearly that each of these numbers are in proportion.

ragioni.

And such part **(fol. 45ʳ)** is the first of the second as the second of the third, and as the third of the fourth: each is the half. And again you see clearly that the first multiplied against the fourth makes as much as the second makes when multiplied against the third. And as much results from it when the fourth is divided in the second as results when the third is divided in the first. So that you see clearly that the composition goes well. And it is done precisely. And thus the similar computations are done.

El primo	anno	ebbe como ài veduto	f 10 d'oro appunto.
El secondo	anno	ebbe ———————	f 20 d'oro appunto.
El terzo	anno	ebbe ———————	f 40 d'oro appunto.
El quarto	anno	ebbe ———————	f 80 d'oro appunto.

The first	year	he got, as you have seen,	precisely gold *fiorini* 10.
The second	year	he got ———————	precisely gold *fiorini* 20 o.
The third	year	he got ———————	precisely gold *fiorini* 40.
The fourth	year	he got ———————	precisely gold *fiorini* 80.

19.4 Uno sta a uno fundecho 4 anni. Et tra'l primo anno e'l terzo ebe f 20 d'oro. Et tra'l secondo e'l quarto anno ebbe f 30 {f} d'oro. Vo' sapere que glie toccho el primo anno e'l secondo e'l terzo e'l quarto. Et che tal parte sia el primo del secondo quale è el terzo del quarto. Fa così, et questo abbi sempre per regola, che tu parte sempre quello ch'egl'à tra'l secondo et quarto anno in quello ch'egl'à tra'l primo e'l terzo. Et ciò che ne vene multiprichalo per se medesimo. Et sopra quello che fa, sempre poni uno per regola, et quello che fa si è el partitore. Et in quello ài a partire amendori li salarii, cioè quello che egli à in questi 4

19.4 Somebody stays in a warehouse 4 years. And in the first year and the third together he got gold *fiorini* 20. And in the second and the fourth year he got gold *fiorini* 30 {...}. I want to know what was due to him the first year and the second and the third and the fourth. And that the first be such part of the second as the third is of the fourth. Do thus, and have this always as a rule, that you always divide that which he has in the second and fourth year together in that which he had in the first and third together. And that which results from it, multiply it by itself. And above that which it makes, always put one

anni, ciascheuno salario de per se. Et per che tu intende meglio, fa così como di sopra abiamo dicto, che tu parti quello ch'egl'à tra'l secondo et quarto anno in quello che egli à tra'l primo e'l terzo. Et però fa così, parti f 30 in f 20, che ne vene f j$\frac{1}{2}$. Multiprichalo per se medesimo, fa 2 et $\frac{1}{4}$. Ponvi suso uno, como dice la regola, fa 3 e $\frac{1}{4}$. Et questo è el partitore. Ora parti 20 f, che egli à tra'l primo e'l terzo anno in 3 e $\frac{1}{4}$, che ne vene f 6 e $\frac{2}{13}$ de f. Et tanto glie toccha **(fol. 45ʳ)** el primo anno. Et el terzo anno el resto infino in 20 f che è f 13 e $\frac{11}{13}$ de fiorino. Ora parti f 30 in 3 e $\frac{1}{4}$, venne f 9 e $\frac{3}{13}$ de f. Et tanto gle toccha el secondo anno. Et el resto infino in 30 f gle toccha el quarto anno, che è f 20 e $\frac{10}{13}$ de f. Et è facta, et vedi che sonno li salarii in propositione, che tal parte è el primo del secondo quale ~~el secondo de~~ el terzo del quarto. Et tal parte è el primo del secondo quale el secondo del terzo, et quale è el terzo del quarto, che ciascheuno numero è $\frac{2}{3}$ dell'altro. Et sta bene. Et così se fanno tucte le simigliante ragioni.

by rule, and that which it makes is the divisor. And in this you have to divide both salaries, that is, that which he got in these 4 years, each salary by itself. And in order that you understand better, do thus as we have said above, that you divide that which he got in the second and fourth year together in that which he got in the first and third together. And therefore do thus, divide *fiorini* 30 in *fiorini* 20, from which results *fiorini* j$\frac{1}{2}$. Multiply it by itself, it makes 2 and $\frac{1}{4}$. Put one on top there, as the rule says, it makes 3 and $\frac{1}{4}$. And this is the divisor. Now divide 20 *fiorini*, which he gets in the first and third year together, in 3 and $\frac{1}{4}$, from which results *fiorini* 6 and $\frac{2}{13}$ of *fiorino*. And so much is due to him **(fol. 45ʳ)** the first year. And the third year the rest until 20 *fiorini*, which is *fiorini* 13 and $\frac{11}{13}$ of *fiorino*. Now divide *fiorini* 30 in 3 and $\frac{1}{4}$, *fiorini* 9 and $\frac{3}{13}$ of *fiorino* results. And so much is due to him the second year. And the rest until 30 *fiorini* is due to him the fourth year, which is *fiorini* 20 and $\frac{10}{13}$. And it is done, and you see that the salaries are in proportion, that such part is the first of the second as the third of the fourth. And such part is the first of the second as the second of the third, and as the third of the fourth, that each number is $\frac{2}{3}$ of the other. And it goes well. And thus all the similar computations are done.

| El primo | anno | ebbe —— | f 6 | e $\frac{2}{13}$ de f |
| El secondo | anno | ebbe —— | f 9 | e $\frac{3}{13}$ de f |

| El terzo | anno | ebbe —— | f 13 | e $\frac{11}{13}$ de f |
| El quarto | anno | ebbe —— | f 20 | e $\frac{10}{13}$ de f |

The first	year	he had	*fiorini* 6	and $\frac{2}{3}$ of *fiorino*
The second	year	he had	*fiorini* 9	and $\frac{3}{13}$ of *fiorino*
The third	year	he had	*fiorini* 13	and $\frac{11}{13}$ of *fiorino*
The fourth	year	he had	*fiorini* 20	and $\frac{10}{13}$ of *fiorino*

[20. Tabulated Degrees of Fineness of Coins]

20.1 *In Christi nomine amen. Qui sonno sotto scripte tucte maniere de leghe de monete. Et similmente tucti allegamenti de oro, argento et ramo, como se legano l'una moneta et l'altra, over lo lighare o d'oro in verghe, o argento de tucte ragioni.*

20.1 *In the name of Christ amen. Here are written below all modes of alloys of coins. And similarly all alloyings of gold, silver and copper, how one coin and another are alloyed, or indeed the alloying either of gold in ingots, or silver of all rates.*

20.2 Incomenciaremo a dire così, dovete sapere che una oncia de oro fino si è 24 charrati d'oro. Et quanto l'oro è peggiore, meno carrati n'à nell'oncia. Et simigliante-mente vene dell'argento, che'sse allegha a oncie, overo a denari pesi. Et l'argento che tene 12 oncie per £ è argento fino e bono e puro.

20.2 We shall begin by saying thus, you shall know that one ounce of fine gold is 24 carats of gold. And the baser the gold, the less carats of that it has in the ounce. And similarly happens with silver, whether it is alloyed at ounces, or indeed at *denari* of weight. And the silver that holds 12 ounces per pound is fine silver and good and pure.

20.3 Fiorini d'oro de Firenze sonno a'llegha de carrati . . . . 24 per oncia

Agostani d'oro sonno a carrati . . . . 20$\frac{1}{2}$ per oncia

Perperi pagli a dochati[a] sonno a charrati . . . . 15 per oncia

Dobbole dela mirra —— sonno a charrati . . . . 23$\frac{1}{2}$ per oncia

(fol. 46ʳ)

Dobbole de rascetto sonno a charrati . . . . 23$\frac{1}{4}$ per oncia

Castellani d'oro sonno a charrati . . . . 23$\frac{1}{2}$ per oncia

Anfogiani d'oro sonno a charrati . . . . 20$\frac{1}{2}$ per oncia

Tornesi d'oro sonno a charati . . . . $23\frac{3}{4}$ per oncia

Bisanti vecchi d'oro sonno a charati . . . . 24 per oncia

Perperi vecchi d'oro comunali et mezzani sonno a charrati . . . . . . 17 per oncia

Bisanti saracinati d'oro, che ne vanno 12 per oncia, sonno a charati 15 per oncia

Lucchesi d'oro a cavallo sonno a charati . . . . 18 per oncia

Lucchesi d'oro a'ppede sonno a charati . . . . 23 per oncia

Perperi novi sonno a charati . . . . 14 per oncia

Genovini d'oro a chavallo sonno a charati . . . . 24 meno[b] $\frac{1}{15}$ per oncia

Genovini d'oro a'ppede sonno a charati . . . . $23\frac{1}{4}$ per oncia

Carlini d'oro sonno a charati . . . . 24 per oncia

Pezzicti de bisanti sonno a charati . . . . 12 meno $\frac{1}{4}$ per oncia

Romani d'oro sonno a charati . . . . 24 meno $\frac{1}{18}$ per oncia

Parigini d'oro a chavallo sonno a charati . . . . 24 meno $\frac{1}{4}$ per oncia

Duchati d'oro venetiani sonno a charati . . . . 24 scarsi per oncia

Ragonisi d'oro sonno a charati . . . . 24 meno $\frac{1}{4}$ per oncia

Bisanti d'Acri colla croce sonno a charati . . . . $16\frac{1}{3}$ per oncia

Santoline fine sonno a charati . . . . 21 per oncia

Maraboctini d'oro sonno a charati . . . . 21 per oncia

Medaglie[c] Massamutine sonno a charati . . . . 24 per oncia

Oro de paglola secondo chomo te⟨ne⟩[M+F], el migliore si è a carati 22,
    El communale è a charati 20 infino in 21 . . . . per oncia

Bisanti vecchi de Alexandria sono a charati . . . . 24 per oncia

Nota che 30 teri sonno una oncia. Et 20 grani sonno uno teri, chomo altresi a te de sopra è dicto.

20.3 *Fiorini* of gold from Florence are alloyed at carats . . . . 24 per ounce

Augustales of gold are at carats . . . . $20\frac{1}{2}$ per ounce

Perperi pagliolati[a] are at carats . . . . 15 per ounce

Dobbole della mirra — — are at carats . . . . $23\frac{1}{2}$ per ounce

(fol. 46ʳ)

Dobbole de rascetto are at carats . . . . $23\frac{1}{4}$ per ounce

Castellani of gold are at carats . . . . $23\frac{1}{2}$ per ounce

Anfogiani of gold are at carats . . . . $20\frac{1}{2}$ per ounce

Tornesi of gold are at carats . . . . $23\frac{3}{4}$ per ounce

Old *Bizanti* of gold are at carats . . . . 24 per ounce

Old communal and intermediate *Perperi* of gold are at carats . . . . 17 per ounce

Saracen *bizanti* of gold, of which 12 go per ounce, are at carats . . 15 per ounce

Lucchesi of gold "(Saint Vultus as crusader) on horseback"
                                         are at carats . . . . 18 per ounce

Lucchesi of gold "(Saint Vultus as crusader) on foot" are at carats . 23 per ounce

New *Perperi* are at carats . . . . 14 per ounce

Genovini of gold "on horseback" are at carats . . . . 24 less[b] $\frac{1}{15}$ per ounce

Genovini of gold "on foot" are at carats . . . . 23$\frac{1}{4}$ per ounce

Carlini of gold are at carats . . . . 24 per ounce

Pezzicti of *bizanti* are at carats . . . . 12 less $\frac{1}{4}$ per ounce

Romani of gold are at carats . . . . 24 less $\frac{1}{18}$ per ounce

Parigini of gold "on horseback" are at carats . . . . 24 less $\frac{1}{4}$ per ounce

Venetian *Ducati* of gold are at carats . . . . almost 24 per ounce

Ragonisi of gold are at carats . . . . 24 less $\frac{1}{4}$ per ounce

Bizanti from Accra "with the cross" are at carats . . . . 16$\frac{1}{3}$ per ounce

Fine *santoline* are at carats . . . . 21 per ounce

Marabottini of gold are at carats . . . . 21 per ounce

Massamutine medals[c] are at carats . . . . 24 per ounce

Pagliola gold [a gold quality] according to what is contains. The best is
        at 22 carats. The ordinary is at carats from 20 until 21 . . . . per ounce

Old *Bizanti* from Alexandria are at carats . . . . 24 per ounce

Note that 30 *teri* are one ounce. And 20 grains are one *teri*, as it was also said to
you above.

[a] **M+F** has the somewhat better "paglialocati" (referring to the Byzantine *Palaeologoi*); not
understanding, the copyist of **V** or his predecessor has introduced a *Verschlimmbesserung*.
[b] *Meno* is abbreviated ⓜ here and elsewhere in the list. The sign is actually written much like capital
"G", but it is known from elsewhere to represent an encircled "m", see [Vogel 1977: 11]. The sign
is conspicuously absent from the mathematical text proper.
[c] *Medaglia/*"medal", from *medius>medialia*, is mostly a half-*denaro*, here however the half of a
more valuable unit.

(fol. 46ᵛ)

Qui sono scripte le leghe de monete piccioli. Et nota per errore trapassai la regola dele
monete de argento, como tu vedi de rinpecto nel sequento foglio a questo segnio.[a]

Parigini primeri sonno a	denari 5 et grani 18 de legha[b]
Parigini sechondi sonno a	deñ 4 grani 16 de legha
Parigini terzi sonno a	deñ 3 g̃ʳ 14 de legha
Tolosini vecchi dala croce sonno a	deñ 6 g̃ʳ 18 de legha[c]
Murlani sonno a	deñ 7 g̃ʳ 7 de legha
Reali primeri sonno a	deñ 4 g̃ʳ 18 de legha

Reali secondi sonno a	deñ 3 g̃ʳ 18 de legha
Reali terzi sonno a	deñ 3 de legha
Ternali sonno a	deñ 3 g̃ʳ 14 de legha
Medaglie ternali sonno a	deñ 3 g̃ʳ 3 de legha
Coronati de Re Carlo primeri a	deñ 4 de legha
Coronati secondi sonno a	deñ 3 g̃ʳ 18 de legha
Coronati terzi sonno a	deñ 3 de legha
Rinfazzati sonno a	deñ 3 grani 15 de legha
Reali de Marsilia sonno a	deñ 3 grani 18 de legha
Margonesi valenzani et capo de Re sonno a	deñ 3 $\frac{1}{2}$ de legha
Coronati vecchi sonno a	deñ 2 grani 18 de legha
Caorsini sonno a	deñ 3 de legha

Vaselamento[d] de Parigi et de Torso et de
Monpoleri sonno a oncie 11 $\frac{3}{4}$ per libra

✝ Vaselamento de Marsilia sonno a oncie 11 $\frac{1}{2}$ per libra

20.5 ✝ Here are written the alloyings of small coins. And note that by error I passed over the rule of the silver coins, as you see opposite in the next sheet at this sign.[a]

Parigini of first class are at	*denari* 5 and grains 18 of alloy[b]
Parigini of second class are at	*denari* 4 grains 16 of alloy
Parigini of third class are at	*denari* 3 grains 14 of alloy
Old *Tolosini* "with the cross" are at	*denari* 6 grains 18 of alloy[c]
Murlani are at	*denari* 7 grains 7 of alloy
Reali of first class are at	*denari* 4 grains 18 of alloy
Reali of second class are at	*denari* 3 grains 18 of alloy
Reali of third class are at	*denari* 3 of alloy
Ternali are at	*denari* 3 grains 14 of alloy
Ternali medals are at	*denari* 3 grains 3 of alloy
Coronati of King Carlo of first class at	*denari* 4 of alloy
Coronati of second class are at	*denari* 3 grains 18 of alloy
Coronati of third class are at	*denari* 3 of alloy
Rinfazzati are at	*denari* 3 grains 15 of alloy
Reali from Marseille are at	*denari* 3 grains 18 of alloy
Margonesi from Valence and "head of King" are at	*denari* 3 $\frac{1}{2}$ of alloy
Old *coronati* are at	*denari* 2 grains 18 of alloy
Caorsini are at	*denari* 3 of alloy

Vasellamento[d] from Paris and from Tours and
   from Montpellier are at ounces 11 $\frac{3}{4}$ per pound

Vasellamento from Marseilles are at ounces $11\frac{1}{2}$ per pound

[a] Since there was space for this observation (three lines in the manuscript), the omission of the section on silver coins and the subsequent insertion on the following page cannot be due to the ultimate copyist, who instead has carefully copied the order and the explanation of his original – which itself will therefore have been a copy.

The same conclusion follows from the observation that the words "questo segnio" are not followed by any sign, nor is any found on top of the following page where the omitted section is inserted.

[b] Comparison with the other coin lists shows that *denari* in the present list of small coins is to be understood as *ounces*, and *grani* as *denari*. This explains that no fineness goes beyond 12.

[c] After this line, **M+F** has "Tolosany ala fiore sono a δ 7 grani 4 di legha"/"*Tolosani* with a flower are at *denari* 7 grains 4 of alloy".

[d] *Vasellamento* is unminted silver.

(fol. 47ʳ)

20.4 **Qui** sonno scripte tucte tenute de monete de argento.

Tornesi grossi . . . . sonno a oncie	$11\frac{1}{2}$ per £
Et intendesi che la £ sia oncie 12 de argento fino in tucti allegamenti.

Medaglie de Torre primere sonno a oncie . . . . 11 per £[a]

Carlini et Merchoresi et Barzellonesi sonno a oncie $11\frac{1}{4}$ per £

Starlini . . . . sonno a oncie . . . . 11 δ 2 per £

Venetiani da Venegia . . . . . . sonno a oncie . . . . 11 $\frac{3}{4}$ per £

Popolini da Firenze et da Siena et da
        Pisa sonno comunamente a oncie 11 δ 15 . . . . per £

Aquilani vecchi da Pisa . . . . sonno a oncie . . . . 11 per £

Bolognini grossi . . . . sonno a oncie . . . . 9 δ 21 per £

Astegiani . . . . sonno a oncie . . . . 8 δ 18 per £

Imperiali et Piacentini . . . . . . sonno a oncie . . . . 9 per £

Romani de peso del tornese . . sonno a oncie . . . . 11 δ 8 per £

Genovini {sonno} . . . . sonno a oncie . . . . 11 δ 12 per £

Baldacchini dela guglia . . . . . sonno a oncie . . . . 11 δ 8 per £

Fresarchisi d'Aquilea cola ghuglia et
        dela torre et del giglio et dela luna sonno a oncie . . . . 8 δ $10\frac{1}{2}$ per £

Angontani grossi . . . . sonno a oncie . . . . 10 δ 5 per £[b]

Senesi vecchi . . . . sonno a oncie . . . . 11 δ 6 per £

Volterani grossi . . . . sonno a oncie . . . . 9 — per £

Et nota che se intende 12 oncie la £ et 24 denari de peso per oncia.

Reguarda qui denanzi a questa prima faccia, nella quale sonno scripte le leghe de monete picciole. El quale trapassamento fo facto per errore. Cioè a questo segnio.[c]

20.4 Here are written what all silver coins contain.

Tornesi grossi .... are at ounces .... $11\frac{1}{2}$ per pound

And it is to be understood that the *pound* is of 12 ounces of fine silver in all allegations

Medals from Tours, first class, are at ounces .... 11 per pound[a]

Carlini and *mercoresi* and *barzellonesi* are at ounces $11\frac{1}{4}$ per pound

Sterlini .... are at ounces .... 11 *denari* 2 per pound

Venetiani from Venice ...... are at ounces .... $11\frac{3}{4}$ per pound

Popolini from Florence and from Siena and from
Pisa are generally at ounces 11 *denari* 15 .... per pound

Old *Aquilani* from Pisa ..... are at ounces .... 11 per pound

Bolognini grossi .... are at ounces .... 9 *denari* 21 per pound

Astegiani .... are at ounces .... 8 *denari* 18 per pound

Imperiali and *Piacentini* .... are at ounces .... 9 per pound

Romani of the weight of the *tornese* are at ounces ..... 11 *denari* 8 per pound

Genovini {...} .... are at ounces ...... 11 *denari* 12 per pound

Baldacchini "with the pinnacle" are at ounces .... 11 *denari* 8 per pound

Fresciacchini d'Aquilea "with the pinnacle" and "with the tower" and "with the
lily" and "with the moon" are at ounces .... 8 *denari* $10\frac{1}{2}$ per pound

Angontani grossi .... are at ounces .... 10 *denari* 5 per pound[b]

Old *senesi* .... are at ounces .... 11 *denari* 6 per pound

Volterrani grossi .... are at ounces .... 9 — per pound

And note that 12 ounces the pound and 24 *denari* of weight per ounce is understood. See before in the previous page, in which are written the alloys of small coins. Which were passed over by error. That is, at this sign.[c]

---

[a] After this line, **M+F** has "**M**edaglie terzeriole sono a oncie 11 per libra"/"*Medaglie terzeriole* [third of *tornese*] are at ounces 11 per pound".
[b] After this line, **MF** has "**R**avingniany grossi sono a oncie 10 δ 12 per libra"/"*Ravignani grossi* are at ounces 10, *denari* 12 per pound".
[c] The sign is actually missing in the present place; the reference is obviously to the sign at top of the previous page.

(fol. 47ᵛ)

[21. Alloying Problems]

^{21.1} **Qui** sonno finite tucti allegamenti de monete. Ora incomenciamo a fare alchuna ragione de allegamenti.

^{21.2} **Io** ho oncie 60 d'oro che tene charati 16 per oncia, et vogliolo mettere al focho et affinarlo tanto che torni a charati 21 per oncia. Vo' sapere quanto tornaranno queste 60 oncie a peso tractolo del focho, che sia de charati 21, chomo io t'ò dicto, né più né meno. Fa così, sappi primamente quanti charati d'oro ài nelle dicte 60 oncie che tu di' che tene charati 16 per oncia. Moltipricha 60 via 16 charrati, fa 960. Et 960 carrati d'oro à nelle dicte 60 oncie che tu voli mettere a'ffocho. Ora se tu voli sapere quanto tornerà a a peso le dicte 60 oncie quando serà affinato de charrati 21 per oncia, sì parti 960 charrati in 21, che ne vene oncie 45 et $\frac{5}{7}$ d'oncia. Et cotante oncie diremo che sia, et così è vero. Torneranno le dicte 60 oncie quando serà affinato como io t'ò dicto, cioè oncie 45 et $\frac{5}{7}$ d'oncia d'oro de carrati 21 per oncia. Et è facta, et sta bene. Et così se fanno tucte le simiglianti ragioni, che ogni volta {che} ^se^ fanno per questa regola, de ogni quantità che tu volessi affinare, et a qualunqua legha le volissi fare.

^{21.1} Here end all the alloys of coins. Now we begin to make some computations of alloying.

^{21.2} I have ounces 60 of gold which contains carats 16 per ounce, and I want put it in fire and refine it so much that it becomes of carats 21 per ounce. I want to know how much how much these 60 ounces will become in weight, taken out of the fire when it is of carats 21, as I have said to you, neither more nor less. Do thus, know firstly how many carats of gold you have in the said 60 ounces of which you say that it contains 16 carats per ounce. Multiply 60 times 16 carats, it makes 960. And 960 carats of gold there is in the said 60 ounces that you want to put in fire. Now if you want to know how much will become in weight the said 60 ounces when it will be refined to carats 21 per ounce, then divide 960 carats in 21, from which results ounces 45 and $\frac{5}{7}$ of ounce. And as many ounces we shall say that it is, and thus it is true. The said 60 ounces when refined will become as I have said to you, that is, ounces 45 and $\frac{5}{7}$ of ounce of gold of carats 21 per ounce. And it is done, and it goes well. And thus are done all the similar computations, and every time {...} they are done by this rule, about any quantity that you might want to refine, and at

whatever alloying you might want to make it.

21.2A [F.vii.2]

I have 18 ounces of gold, which is at $20\frac{1}{2}$ carats per ounce, and I want put it in fire and refine it so much that it becomes pure gold, that is, at 24 carats per ounce. Say me how much the said 18 ounces will be of weight when taken out of the fire and being of 24 carats. Know first how many carats of gold there are in the said 18 ounces which you put in fire, and multiply 18 by $20\frac{1}{2}$, they make 369, and so many carats of gold are there in the said ounces, that is, carats 369. Now, if you want to know how much it becomes in weight, then divide 369 by 24, because you want that it becomes 24 carats per ounce, from which results 15 and $\frac{3}{8}$. And it is done. And we shall say that the said 18 ounces, which you put in fire at $20\frac{1}{2}$ carats per ounce, when taken from the fire becomes in weight ounces 15 and $\frac{3}{8}$ at carats 24 per ounce.

21.3 Io ho oncie 7 d'oro, el quale è a charrati $19\frac{1}{2}$ per oncia. Et ò oncie 9 d'oro che è de charrati 20 e $\frac{1}{4}$. Et ho oncie 16 d'oro de charrati 21 e $\frac{2}{7}$ per oncia. Et ò oncie 20 d'oro de charrati $23\frac{3}{4}$ per oncia. Io voglio tucti questi 4 ori fondere inseme e farne una vergha. Vo' sapere quanto serà tucta questa vergha a peso, de quanti charati d'oro serà per oncia appunto. Fa così, e questa è la sua legitima regola. Primamente sappi quanti charati d'oro ài nelle prime 7 oncie. Et moltipricha 7 via 19 charati e $\frac{1}{2}$,

21.3 I have ounces 7 of gold, which is at carats $19\frac{1}{2}$ per ounce. And I have ounces 9 of gold which is of carats 20 and $\frac{1}{4}$. And I have ounces 16 of gold of carats 21 and $\frac{2}{7}$ per ounce. And I have ounces 20 of gold of carats $23\frac{3}{4}$ per ounce. I want to fuse all these 4 (types of) gold together and make an ingot of them. I want to know how much this whole ingot will be in weight, and of how many carats of gold per ounce it will be precisely. Do thus, and this is its legitimate rule. Firstly know how many

che fa 136 ⟨e $\frac{1}{2}$⟩. Et cotanto oro ài in queste 7 oncie. Ora sappi quanto n'ài nelle 9 oncie che tene 20 $\frac{1}{4}$ per oncie. Et moltipricha 9 via 20$\frac{1}{4}$, che fa 182 charati e $\frac{1}{4}$. Et cotanto n'à nelle 9 oncie. Ora sappi quanto n'à nelle 16 oncie che tene charati 21 e $\frac{2}{3}$ per oncia. Multipricha 16 via 21 charrato **(fol. 48ʳ)** e $\frac{2}{3}$, che fa 346 charrati e $\frac{2}{3}$. Et cotanto n'à nelle 16 oncie. Ora sappi quanto n'à nelle 20 oncie che tene charrati 23 e $\frac{3}{4}$ per oncia. Et multipricha 20 via 23 charati et $\frac{3}{4}$, che fa 475 charati. Et tanto n'è nelle 20 oncie. Ora giungi inseme tucti questi charrati, che sonno in tucto charrati 1140 e $\frac{5}{12}$. Ora giongi inseme 7 oncie et 9 oncie et 16 oncie et 20 oncie, che sonno in tucto oncie 52. Et in 52 ài a partire 1140 charati e $\frac{5}{12}$, che ne vene charati 21 e $\frac{581}{624}$ de charato. *Et è facta, et sta bene, et diremo che la dicta verghetta fonduta {che serà} serà oncie 52 d'oro de charati 21 e $\frac{581}{624}$ de charato.*ᵃ Et così se fanno tucte le simigliante ragioni. Et se volesse fondere inseme de 100 ragioni d'oro et de diversi ragioni, sì fa sempre per questa reghola. Et non poi errare.

carats of gold you have in the first 7 ounces. And multiply 7 times 19 carats and $\frac{1}{2}$, which makes 136 ⟨and $\frac{1}{2}$⟩. And as much gold is there in these 7 ounces. Now know how much there is in the 9 ounces that contains 20$\frac{1}{4}$ per ounces. And multiply 9 times 20$\frac{1}{4}$, which makes 182 carats and $\frac{1}{4}$. And as much is there in the 9 ounces. Now know how much there is in the 16 ounces that contains carats 21 and $\frac{2}{3}$ per ounce. Multiply 16 times 21 carat **(fol. 48ʳ)** and $\frac{2}{3}$, which makes 346 carats and $\frac{2}{3}$. And as much is there in the 16 ounces. Now know how much there is in the 20 ounces that contains carats 23 and $\frac{3}{4}$ per ounce. And multiply 20 times 23 carats and $\frac{3}{4}$, which makes 475 carats. And so much is there in the 20 ounces. Now join together all these carats, which are in all carats 1140 and $\frac{5}{12}$. Now join together 7 ounces and 9 ounces and 16 ounces and 20 ounces, which are in all 52 ounces. And in 52 you have to divide 1140 carats and $\frac{5}{12}$, from which results carats 21 and $\frac{581}{624}$ of carat. And it is done, and it goes well, and we shall say that the said small fused ingot {...} will be ounces 52 of gold of carats 21 and $\frac{581}{624}$ of carat. And thus all the similar computations are done. And if you might want to fuse together of 100 rates of gold and of different rates, then do always by this rule. And you cannot go wrong.

ᵃ This passage in the margin is written with a finer pen, and thus later, by a similar but probably different hand – not the one which added the marginal comment "compagnia" on fol. 25ʳ, but possibly the same as made the initials.

²¹·⁴ **Io** ò de una ragione bologni, el quali tene denari 11 de legha. Et ò de una altra

²¹·⁴ I have *bologni* of one rate, which contains *denari* 11 of alloy. And I have *bo-*

ragione bologni da deñ 4 de legha. Ora io voglio fare una moneta che sia a deñ 7 de legha. Et voglio alleghare 100 marchi. Vo' sapere quanto me convene torre de ciascheuno de questi 2 bologni a fare 100 marchi de deñ 7 de legha app-unto. Fa così, e questa è la sua propria regola. Et di' così, la moneta che io voglio fare si è che sia a deñ 7 de legha. Et el biglione che io ho el più alto si è a deñ 11. Et però di' così, da 7 a 11 si à 4. Et prendi marchi 4 del contrario, cioè

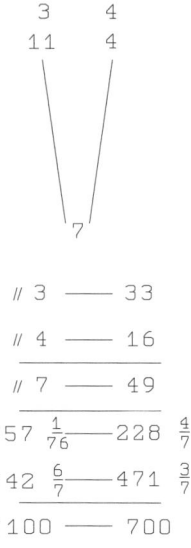

$$3 \qquad 4$$
$$11 \qquad 4$$
$$\diagdown \; \diagup$$
$$7$$

$$/\!/\,3 \;\text{——}\; 33$$
$$/\!/\,4 \;\text{——}\; 16$$
$$/\!/\,7 \;\text{——}\; 49$$
$$/\!/\,57\tfrac{1}{76}\;\text{——}\;228\tfrac{4}{7}$$
$$/\!/\,42\tfrac{6}{7}\;\text{——}\;471\tfrac{3}{7}$$
$$/\!/\,100 \;\text{——}\; 700$$

del più basso, cioè de quello che tene denari 4. Et poi di', da 4 a 7 si à 3. Et togli 3 marchi de quello che à deñ 11 de legha. Ora ài allegato marchi 7 a deñ 7 de legha. Et ài vi messo marchi 4 de quello che tene deñ 4 de legha, et marchi 3 de quello che tene deñ 11 de legha. Ora prova se sta bene. Marchi ⟨3⟩ de deñ 11 si sonno denari 33. Et marchi 4 de denari 4 sonno denari 16. Aggiongi inseme, fa 49. Et li marchi che tu ài alleghati a denari 7 l'uno sonno ancho 49. Et sta bene. Ora tu di' che voli alleghare 100 marchi de denari 7 de legha l'uno. Et però fa così, et reduci questa ragione al modo che se fossono doi con-pagni che facesseno conpagnia asseme. Et **(fol. 48ᵛ)** l'uno conpagnio mettesse 4, et l'altro 3. Et arebbono ^7^ in conpagnia tra amedoi. Et ànno guadagniato 100. Que toccherà per uno. Et però fa così, multi-pricha per lo primo che mette 4, et di', 4

logni of another rate at *denari* 4 of alloy. Now I want to make a coin that is at *denari* 7 of alloy. And I want to alloy 100 mark. I want to know how much it suits me to grasp of each of these two (types of) *bologni* for making 100 mark of precisely *denari* 7 of alloy. Do thus, and this is its proper rule. And say thus, the coin that I want to make shall be at *denari* 7 of alloy. And the highest bullion that I have is at *denari* 11. And therefore say thus, from 7 to 11 there is 4. And take 4 mark of the opposite, that is, of the lowest, that is, of that which contains *denari* 4. And then say, from 4 to 7 there is 3. And grasp 3 mark of that which has *denari* 11 of alloy. Now you have alloyed mark 7 at *denari* 7 of alloy. And you have put into it mark 4 of that which contains *denari* 4 of alloy, and mark 3 of that which contains *denari* 11 of alloy. Now verify whether it goes well. Mark ⟨3⟩ of *denari* 11 are *denari* 33. And mark 4 of *denari* 4 are *denari* 16. Join together, it makes 49. And the mark that you have alloyed at *denari* 7 (each) one are also 49. And it goes well. Now you say that you want to alloy 100 mark of *denari* 7 of alloy (each) one. And therefore do thus, and lead this computation back to the mode as if there were two partners who made partnership together. And **(fol. 48ᵛ)** the one partner put in 4, and the other 3. And they had 7 in

via 100 fa 400. Ora parti nel corpo dela conpagnia, cioè in 7, che ne vene 57 e $\frac{1}{7}$. Et 57 marchi e $\frac{1}{7}$ de biglone de quello che à deñ 4 de legha metterai ne' 100 marchi che tu voi fare. Ora multipricha per l'altra parte, 3 via 100, fa 300. Et parti in 7, che ne vene 42 e $\frac{6}{7}$. Et marchi $42\frac{6}{7}$ torrai et metterai de quello biglione che è a deñ 11 de legha ne' 100 marchi che tu voli alleghare. Et è fatta, et ày alleghati appunto 100 marchi. Ora però che tu sì ⟨vedi⟩ bene chiaro, voglio che noi la provamo se sta bene. Noi abbiamo alleghati 100 marchi. Et diciamo che è a denari 7 de legha. Ora sappiamo ⟨se⟩ le doi ragioni del biglione che noi abbiamo messo dele doi leghe ànno de legha 700 deñ. Et fa così, sappi quanti deñ ànno de legha prima li 42 marchi $\frac{6}{7}$ che tu di' che vi mecti de quello che tene deñ 11 de legha. Multipricha 11 via 42 e $\frac{6}{7}$, che fanno denari 471, $\frac{3}{7}$ de denaro. Ora multipricha l'altra parte che di' che vi metti de quello che à deñ 4 de legha, marchi 57 e $\frac{1}{7}$. Et però multipricha 4 via $57\frac{1}{7}$, fanno denari 228 e $\frac{4}{7}$ de denaro. Ora aggiongi inseme 57 marchi e $\frac{1}{7}$, et 42 marchi e $\frac{6}{7}$, che fanno appunto 100 marchi. Ora agiungi inseme 471 deñ e $\frac{3}{7}$, et 228 denari et $\frac{4}{7}$, che fanno appunto 700 denari. Siché vedi che la ragione sta bene, et è bene alleghato. Et così se fanno le simili ragioni de quantunqua ragioni fosseno el biglioni. Et de rimpetto è posto, como tu vedi, nela forma che se fa la dicta ragione.

partnership both two together. And they have earned 100. What will be due to (each) one. And therefore do thus, multiply for the first who puts in 4, and say, 4 times 100 makes 400. Now divide in the principal of the partnership, that is, in 7, from which results 57 and $\frac{1}{7}$. And 57 mark and $\frac{1}{7}$ of bullion of that which has *denari* 4 of alloy will you put in the 100 mark that you want to make. Now multiply for the other part, 3 times 100, it makes 300. And divide in 7, from which results 42 and $\frac{6}{7}$. And mark $42\frac{6}{7}$ will you grasp and put in of that bullion which is at *denari* 11 of alloy, in the 100 mark that you want to alloy. And it is done, and you have alloyed precisely 100 mark. Now therefore that you ⟨may see⟩ quite clearly, I want that we verify whether it goes well. We have alloyed 100 mark. And we say that it is at *denari* 7 of alloy. Now let us know ⟨whether⟩ the two rates of bullion that we have put in of the two alloys have 700 *denari* of alloy. And do thus, know first how many *denari* of alloy have the 42 mark $\frac{6}{7}$ that you say that you put in of that which contains *denari* 11 of alloy. Multiply 11 times 42 and $\frac{6}{7}$, which make *denari* 471, $\frac{3}{7}$ of *denaro*. Now multiply the other part which you say that you put in, of that which has *denari* 4 of alloy, mark 57 and $\frac{1}{7}$. And therefore multiply 4 times $57\frac{1}{7}$, they make *denari* 228 and $\frac{4}{7}$ of *denaro*. Now join together 57 mark and $\frac{1}{7}$, and 42 mark and $\frac{6}{7}$, which make precisely 100 mark. Now join together 471 *denari* and $\frac{3}{7}$, and 228 *denari* and $\frac{4}{7}$, which make precisely 700 *denari*. So that

you see that the computation goes well, and it is well alloyed. And thus the similar computations are done, whatever the rates of the bullions might be. And opposite it is shown, as you see, in the diagram in which the said computation is made.

^{21.5} **Io** ho marchi 5 de argento, el quale è a oncie 9 per £. Et ò marchi 8, el quale è a oncie $10\frac{1}{4}$ per libra. Et ò marchi 2 de rame pretto. Ora io voglio fare fondare inseme tucto questo argento et rame, et farne **(fol. 49ʳ)** fare una vergha. Vo' sapere quanto tornerà a'ppeso tucto, et ad que legha serà tucto questo argento mescolato co rame. Fo così, giongi prima asseme tucti questi 3 pesi che tu voli fondere inseme, cioè 5 marchi et otto marchi de argento et 2 marchi de rame, che in tucto sonno marchi 15. Et questo è el partitore. Ora sappi quanti oncie à nelli 5 marchi, che v'è n'à oncie 45. Et poi sappi quanta n'à nelli 8 marchi, che ne vene 82. Ora sappi quanti n'à nelli 2 marchi de rame, che non ve'n'è punto, però che è pretto rame. Ora agiongi inseme 45 oncie et 82 oncie, che fa 127. Ora parti 127 in 15, che ne vene oncie $8\frac{7}{15}$ d'oncia. Et è facta, et diremo che tucto questo argento et rame mescholato inseme serranno marchi 15 ad oncie $8\frac{7}{15}$ d'oncia per £. Et sta bene. Et così se fanno le simili ragioni.

^{21.5} I have mark 5 of silver, which is at ounces 9 per pound. And I have mark 8, which is at ounces $10\frac{1}{4}$ per pound. And I have mark 2 of pure copper. Now I want to have fused together all this silver and copper together, and have **(fol. 49ʳ)** an ingot made from it. I want to know how much all will become in weight, and of which alloy all this silver mixed with copper will be. Do thus, join together first all these 3 weights that you want to fuse together, that is, 5 mark and eight mark of silver and 2 mark of copper, which in all are mark 15. And this is the divisor. Now know how many ounces there are in the 5 mark, and there are ounces 45. And then know how many there are in the 8 mark, from which results 82. Now know how many there are in the 2 mark of copper, where there is nothing at all, because it is pure copper. Now join together 45 ounces and 82 ounces, which makes 127. Now divide 127 in 15, from which results ounces $8\frac{7}{15}$ of ounce. And it is done, and we shall say that all this silver and copper mixed together will be mark 15 at ounces $8\frac{7}{15}$ of ounce per pound. And it goes well. And thus the similar computations are done.

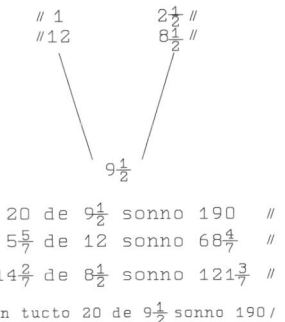

21.6 Io ho de una ragione argento che è fino che tene oncie 12 per £. Et ho de un'altra ragione argento più basso che tene oncie $8\frac{1}{2}$ per £. Ora io voglio alle-ghare 20 marchi de una moneta che tenga oncie $9\frac{1}{2}$ per £. Vo' sapere quanto me bisogna torre de cia-scheuno de questi 2 argenti a ciò che me vengha bene alleghato. Fa così, et questa se fa propriamente come quella che tu ài nanzi a quella ragione de sopra a questa, et in questa forma. Et però non me stendarò in si longho dire como feci in quella. Et però di' così, la legha che io voglio fare si è a oncie $9\frac{1}{2}$ per £. Et però di' così, da $9\frac{1}{2}$ infino in 12 si à $2\frac{1}{2}$. Et marchi $2\frac{1}{2}$ de argento torrai de quello che tene oncie $8\frac{1}{2}$ per £. Et poi di', da $9\frac{1}{2}$ infino in $8\frac{1}{2}$ à uno, et marcho uno torrai del più fino argento, cioè, de 12 oncie per £. Et ài allegato in tucto marchi $3\frac{1}{2}$, che voli che sia a oncie $9\frac{1}{2}$ per £, et così è. Ora tu di' che voli alleghare 20 marchi. Et però fa così, multi-pricha 20 via uno marcho d'argento del più fino, fa 20 **(fol. 49ᵛ)** marchi d'argento. Ora ài a partire in $3\frac{1}{2}$, che ne vene marchi $5\frac{5}{7}$. Et marchi $5\frac{5}{7}$ d'argento metterai de quello che tene 12 oncie per £ nelli 20 marchi che tu voli alleghare. Ora multipricha per l'al-tro, cioè 20 via marchi $2\frac{1}{2}$ d'argento, fanno marchi 50 d'argento, et parti in $3\frac{1}{2}$, che ne vene marchi $14\frac{2}{7}$. Et cotanto torrai del più basso argento. Ora prova se sta bene, et di' così, 20 marchi che noi abbiamo allegati

```
// 1          2½ //
//12          8½ //
```

21.6 I have silver of one rate which is fine and contains ounces 12 per pound. And I have silver of another, lower rate which contains ounces $8\frac{1}{2}$ per pound. Now I want to alloy 20 mark of a coin that shall contain ounces $9\frac{1}{2}$ per pound. I want to know how much I need to grasp of each of these 2 (types of) silver so that it results well alloyed for me. Do thus, and this is properly done as the one you have before the computation above the present one, and in this (same) diagram. And therefore I shall not expand in as long speech as I did in that one. And therefore say thus, the alloy that I want to make is at ounces $9\frac{1}{2}$ per pound. And therefore say thus, from $9\frac{1}{2}$ until 12 there is $2\frac{1}{2}$. And mark $2\frac{1}{2}$ of silver you will grasp of that which contains ounces $8\frac{1}{2}$ per pound. And then say, from $9\frac{1}{2}$ until $8\frac{1}{2}$ there is one, and mark one you will grasp of the finest silver, that is, that of 12 ounces per pound. And you have alloyed in all mark $3\frac{1}{2}$, which you want to be at ounces $9\frac{1}{2}$ per pound, and thus it is. Now you say that you want to alloy 20 mark. And therefore do thus, multiply 20 times one mark of the finest silver, it makes 20 **(fol. 49ᵛ)** mark of silver. Now you have to divide in $3\frac{1}{2}$, from which results mark $5\frac{5}{7}$. And mark $5\frac{5}{7}$ of silver will you put in of that which contains 12 ounces per pound in the 20 mark that you want to alloy. Now multiply for the other, that is,

voglamo che tengano oncie $9\frac{1}{2}$ per £. Siché li 20 marchi degono tenere in tucto oncie 190, però che 20 via $9\frac{1}{2}$ fa 190. Ora sappi che tu ve ài messo dentro apponto argento tra dell'una et dell'altra legha, che siano apponto oncie 190. Et di' così, li marchi $5\frac{5}{7}$ diciamo che sia de quello che tene oncie 12 per £, siché nelli dicti 5 marchi $\frac{5}{7}$ ne vene ad avere oncie 68 et $\frac{4}{7}$, però che 12 via $5\frac{5}{7}$ fanno appunto $68\frac{4}{7}$. Et marchi $14\frac{2}{7}$ che tu v'ài messo de quello argento che tene oncie $8\frac{1}{2}$ per £ sonno apponto oncie $121\frac{3}{7}$, però che $14\frac{2}{7}$ via $8\frac{1}{2}$ fa $121\frac{3}{7}$. Ora aggiongi inseme $121\frac{3}{7}$ et $68\frac{4}{7}$, fanno appunto oncie 190. Siché sta bene, et abiamo bene allegato. Et pongote de rimpetto la forma. In questo modo se fa la dicta reghola. Et così se fanno le simiglianti ragioni.

21.7 **Questo** è uno ⟨allegamento⟩$^{M+F}$ spitiale o generale, et dirrò meglio allegamento de 4 ragioni biglioni. Et per lo dicto modo poteremo allegare oro et argento et ramo de qualunqua tenuta se fossono o de quantunqua ragioni de biglioni volessi fare la legha. Et ciò scriveremo qui appresso. Et simile⟨mente⟩$^{M+F}$ porremo la figura. Nel modo si fa como abbiamo facto de sopra nell'altra ragione. Et così se pongono et

20 times mark $2\frac{1}{2}$ of silver, they make mark 50 of silver, and divide in $3\frac{1}{2}$, from which results mark $14\frac{2}{7}$. And as much will you grasp of the baser silver. Now verify whether it goes well, and say thus, we want that the 20 mark that we have alloyed shall contain ounces $9\frac{1}{2}$ per pound. So that the 20 mark shall contain in all ounces 190, because 20 times $9\frac{1}{2}$ makes 190. Now know that you have precisely put in silver of one and the other alloy together, which are precisely ounces 190. And say thus, the mark $5\frac{5}{7}$ we say to be of that which contains ounces 12 per pound, so that in the said mark $5\frac{5}{7}$ there results to be ounces 68 and $\frac{4}{7}$, because 12 times $5\frac{5}{7}$ make precisely $68\frac{4}{7}$. And mark $14\frac{2}{7}$ that you have put in of that silver that contains ounces $8\frac{1}{2}$ per pound are precisely ounces $121\frac{3}{7}$, because $14\frac{2}{7}$ times $8\frac{1}{2}$ makes $121\frac{3}{7}$. Now join together $121\frac{3}{7}$ and $68\frac{4}{7}$, they make precisely ounces 190. So that it goes well, and we have alloyed well. And I show you the diagram opposite. In this way the said rule is done. And thus are done the similar computations.

21.7 This is a special or general ⟨alloying⟩, and I'd better say alloying of bullions of 4 rates. And in the said way we may alloy gold and silver and copper of whatever contents they might be or of whatever rates of bullions one might want to make the alloy. And this we shall write next to here. And similarly we shall show the figure. It is done in the way as we have done above in the other computation. And thus the

fanno le simiglianti ragioni. Ora io vo' dire chosì.

similar computations are proposed and done. Now I want to say thus.

(fol. 50ʳ)

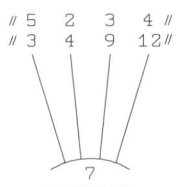

```
// 5   2   3   4 //
// 3   4   9  12 //
```

```
         7

//  30 de   7 sonno 210  //
// 10 5/7 de  3 sonno  32 1/7 //
//  8 4/7 de 12 sonno 102 6/7 //
//  4 2/7 de  4 sonno  17 1/7 //
//  6 3/7 de  9 sonno  57 6/7 //
//In tucto 30 de 7 sonno 210//
```

²¹·⁸ Io ho de 4 ragioni biglione, cioè de 4 leghe. Cioè biglione che tengha denari 3 de legha. Et biglione che à deñ 4 de legha. Et biglione che à deñ 9. Et biglione che à deñ 12. Ora io voglio fare una moneta che sia a deñ 7 de legha, et voglio allegare 30 marchi de questi 4 biglioni. Vo' sapere quanto me convene torre de ciascheuno. Fa così come io te disse ancho nell'altra ragione. Et di' così, el biglione overo moneta che io voglio fare si è che sia a deñ 7 de legha. Et el più fino biglione si è a deñ che io ho si è a deñ 12 de legha. Et però di', da 7 infino a 12 à cinque, et togli marchi 5 del più basso, cioè de quello de deñ 3. Et poi di', da 7 infino in 3 menoma 4, et 4 marchi torrai del più fino, cioè de deñ 12. Et poi di', da 7 infino in 9 à 2, et marchi 2 torrai dell'altro, cioè de quello de deñ 4. Et poi di', da 7 infino in 4 mancha 3, et marchi 3 torrai de quello de deñ 9. Ora vedi che noi abbiamo alleghato in tucto marchi 14 de questi 4 biglioni. Ora prova se sta bene, et di' chosì, noi abbiamo alleghati 14 marchi, et vogliamo che siano a deñ 7 de legha l'uno. Adunqua 14 marchi sonno a denari 98 in tucto, però che 7 via 14 fa 98. Ora sappi se tu hai messo apponto deñ 98 alla

²¹·⁸ I have bullion of 4 rates, that is, of 4 alloys. That is, bullion that contains *denari* 3 of alloy. And bullion that has *denari* 4 of alloy. And bullion that has *denari* 9. And bullion that has *denari* 12. Now I want to make a coin that is to be at 7 *denari* of alloy, and I want to alloy 30 mark of these 4 bullions. I want to know how much it suits me to grasp of each. Do thus as I also said to you in the other computation. And say thus, the bullion or indeed coin that I want to make is to be at *denari* 7 of alloy. And the finest bullion that I have is at *denari* 12 of alloy. And therefore say, from 7 until 12 there is five, and grasp mark 5 of the basest, that is, of that of *denari* 3. And then say, from 7 until 3 lessens 4, and 4 mark will you grasp of the finest, that is, of *denari* 12. And then say, from 7 to 9 there is 2, and mark 2 will you grasp of the other, that is, of that of *denari* 4. And then say, from 7 until 4, 3 are lacking, and 3 mark will you grasp of that of *denari* 9. Now you see that we have alloyed in all 14 mark of these 4 bullions. Now verify whether it goes well, and say thus, we have alloyed 14 mark, and we want that they shall be at *denari* 7 of alloy (each) one. Hence 14 mark are at

ragione che noi abbiamo facta de questi quattro bigloni. Et di' così, marchi 5 che noi abiamo messo de deñ 3 l'uno de legha sonno denari 15. Et marchi 4 de denari 12 l'uno sonno deñ 48. Et marchi 2 de deñ 4 l'uno sonno deñ 8. Et marchi 3 de deñ 9 l'uno sonno deñ 27. Ora aggiongi inseme 15, 48, 8 et 27, fanno appunto 98. Siché sta bene, et abiamo bene alleghato questi 14 marchi. Ora tu di' che voi allegare in tucto marchi 30, però fa come t'ò dicto io in altra ragione adietro. Cioè questa *(fol. 50ʳ)* ragione se po dirizzare a 4 homini che facesseno conpagnia. Et l'uno homo mettesse 5. Et l'altro 4. Et l'altro 2. Et l'altro 3. Che in tucto mettono in conpagnia 14. Ora ànno guadagniato 30. Vo' sapere que toccha per uno. Et però multipricha per colui che mette 5, 5 via 30 fa 150, et parti in 14, che ne vene $10\frac{5}{7}$. Et marchi $10\frac{5}{7}$ torrai del biglione de deñ 3 de legha. Et poi multipricha per lo secondo, che mette quattro, 4 via 30 fa 120, et parti in 14, che ne vene 8 e $\frac{4}{7}$, et marchi $8\frac{4}{7}$ torrai del biglone de deñ 12 de legha. Et poi multipricha per lo terzo che mette doi, 2 via 30 fa 60, parti in 14, che ne vene 4 e $\frac{2}{7}$, et marchi $4\frac{2}{7}$ torrai del biglone de denari 4 de legha. Et poi multipricha per lo quarto, che mette tre, 3 via 30 fa 90, parti in 14, che ne ⟨vene 6 e $\frac{3}{7}$, et marchi⟩ $6\frac{3}{7}$ torrai del biglone de deñ 9 de lega. Ora aggiongi inseme marchi $10\frac{5}{7}$ et $8\frac{4}{7}$ et $4\frac{2}{7}$ et $6\frac{3}{7}$, fa appunto marchi 30 de deñ 7 de legha. Et è facta et sta bene. Et se la provi, troverai che sta così como io {t'ò} te dicho. Et così se fanno le similianti ragioni. Ora te pongho de rempetto,

denari 98 in all, because 7 times 14 makes 98. Now know whether you have put in precisely *denari* 98 at the computation that we have done of these four bullions. And say thus, mark 5 that we have put in of *denari* 3 of alloy (each) one are *denari* 15. And mark 4 of *denari* 12 one are *denari* 48. And mark 2 of *denari* 4 one are *denari* 8. And mark 3 of *denari* 9 one are *denari* 27. Now join together 15, 48, 8 and 27, they make precisely 98. So that it goes well, and we have alloyed well these 14 mark. Now you say that you want to alloy in all mark 30, therefore do as I have said to you in another computation above. That is, this *(fol. 50ᵛ)* computation can be referred to 4 men who make a partnership. And one man put in 5. And the other 4. And the other 2. And the other 3. So that in all they put into the partnership 14. Now they have earned 30. I want to know what is due to (each) one. And therefore multiply for the one who puts in 5, 5 times 30 makes 150, and divide in 14, from which results $10\frac{5}{7}$. And mark $10\frac{5}{7}$ will you grasp of the bullion of *denari* 3 of alloy. And then multiply for the second, who puts in four, 4 times 30 makes 120, and divide in 14, from which results 8 and $\frac{4}{7}$, and mark $8\frac{4}{7}$ will you grasp of the bullion of *denari* 12 of alloy. And then multiply for the third who puts in two, 2 times 30 makes 60, divide in 14, from which results 4 and $\frac{2}{7}$, and mark $4\frac{2}{7}$ will you grasp of the bullion of *denari* 4 of alloy. And then multiply for the fourth, who puts in three, 3 times 30 makes 90, divide in 14, from which ⟨results 6 and

como se ancora per figura a volere fare la dicta reghola.

$\frac{3}{7}$, and mark⟩ $6\frac{3}{7}$ will you grasp of the bullion of *denari* 9 of alloy. Now join together mark $10\frac{5}{7}$ and $8\frac{4}{7}$ and $4\frac{2}{7}$ and $6\frac{3}{7}$, it makes precisely mark 30 of *denari* 7 of alloy. And it is done and it goes well. And if you verify it, will you find that it goes thus as I {...} say to you. And thus are done the similar computations. Now I show you opposite how, (one can also) if wanted, do the said rule by figure.

[22. Further Mixed Problems, Including Practical Geometry]

22.1 Uno homo toglie una boctegha a'ppeggione, et venni a stare dentro in kalende gienaro. Ora viene un altro, acconpagnasse colui in kalende aprile. Viene un altro, acconpagnase coloro kalende luglo. Viene un altro, acconpagnase coloro in kalende ottobre. El primo mette in conpagnia, cioè mise en la boctegha el primo dì che la tolze a pegione £ 100. El secondo mise el dì che se acconpagniò con loro £ 200. El terzo mise £ 300. El quarto mise £ 400. Et così stanno tucti et quattro inseme infino in kalende gienaro. Et in capo dell'anno elli vegono loro conto. Et trovasi guadagnato £ 100. Adomandoti como (fol. 51ʳ) s⟨er⟩à a partire questo guadagno, et quello ⟨che⟩ toccharà per uno. Devi fare così, merita ciascheuno li soi deñ per lo tempo che egli è stato nela conpangia, a 2 deñ per £ el mese. Et diciamo così, el primo è stato in conpagnia uno anno, et misse £ 100, che dè avere de merito £ 10. Et colui che mise 200, cioè el secondo, è stato in conpagnia

22.1 A man rents a shop, and comes to stay there the first of January. Now comes another one, enters in partnership with him the first of April. Another one comes, and enters in partnership with them the first of July. Another one comes, and enters in partnership with them the first of October. The first puts in partnership, that is, puts in the shop the first day he rented it, *libre* 100. The second, on the day he entered partnership with them, put *libre* 200. The third put *libre* 300. The fourth put *libre* 400. And thus all four stay together until the first of January. And at the end of the year they inspect their accounts. And the gain is found to be *libre* 100. I ask you how (fol. 51ʳ) this gain shall be divided, and that ⟨which⟩ is due to one. You shall do thus, each (one) puts on interest his money for the time he has been in the partnership, at 2 *denari* the *libra* the month. And we say thus, the first has been in partnership for a year, and put *libre* 100,

mesi 9, che dè avere de merito £ 15. El terzo, che mise £ 300, è stato in conpagnia mesi 6, dè avere de merito £ 15. El quarto, che mise in conpagnia £ 400, dè avere de merito per tre mesi £ 10. Ora di' così, e sonno 4 conpagni che ànno facto conpagnia inseme. Et l'uno mette in conpagnia £ 10. Et l'altro £ 15. Et l'altro £ 15. Et l'altro £ 10. Et ànno guadagnato £ 100. Que toccharà per uno. Fa così, raccogli inseme tucto quello che ànno messo in conpagnia, che sonno £ 50, et questo è'l corpo dela conpagnia. Ora multipricha per lo primo, che mise £ 10, et di', 10 via 100 ^£^ fa 1000. Parti in 50, che ne vene 20 £. Et tanto toccha al primo. Ora multipricha per lo secondo. 15 via 100 fa 1500 £. Parti in 50, che ne vene £ 30. Et tanto toccha al secondo. Ora multipricha per lo terzo. 15 via 100 £ fa 1500, che fa ancho 1500 £. Parti in 50, anco' ne vene 30 £. Et tanto toccha al terzo. Ora multipricha per lo quarto. 10 via 100 £ fa 1000 £. Parti in 50, che ne vene 20 £. Et tanto toccha al quarto. Et è facta. Et così se fanno le simiglianti ragioni.

and shall have in interest *libre* 10. And he who put 200, that it the second, has been in partnership months 9, and shall have in interest *libre* 15. The third, who put *libre* 300, has been in partnership months 6, he shall have in interest *libre* 15. The fourth, who put *libre* 400 in partnership, shall have in interest for three months *libre* 10. Now say thus, and there are 4 partners, who have made partnership together. And one puts in partnership *libre* 10. And the other *libre* 15. And the other 15. And the other *libre* 10. And they have gained *libre* 100. What is due to (each) one. Do thus, aggregate together all that which they have put in partnership, which are *libre* 50, and this is the principal of the partnership. Now multiply for the first, who put in 10, and say, 10 times 100 *libre* makes 1000. Divide in 50, from which results 20 *libre*. And so much is due to the first. Now multiply for the second. 15 times 100 makes 1500 *libre*. Divide in 50, from which results *libre* 30. And so much is due to the second. Now multiply for the third. 15 times 100 *libre*, which again makes 1500 *libre*. Divide in 50, from which again results 30 *libre*. And so much is due to the third. Now multiply for the fourth. 10 times 100 *libre* makes 1000 *libre*. Divide in 50, from which results 20 *libre*. And so much is due to the fourth. And it is done. And thus the similar computations are done.

[22.2] **Se** noi volessemo sapere la mesura de una torre o de uno arboro senza salirne

[22.2] If we should want to know the measure of a tower or of a tree without climbing to

suso, fa così, togli una mazza, et ficcala in terra al lato ala cosa de que voli sapere la sua mesura de l'altezza. Et fa che dela dicta mazza avanzi sopra a terra doi b̃ʳ o iij o iv secondo che te vene tolta. Et poi guarda el dì, quando el sole percote nela mazza et nella torre overo albore. Et mesura apponto la ombria dela mazza et la ombria dela torre overo arboro. Et poi multiprica la mesura dela mazza **(fol. 51ᵛ)** contra la mesura della ombria dela torre overo arboro. Et parti nela mesura dell'onbria dela mazza. Et quello che ne vene, tanto è alta quella cosa che tu voli sapere. Et a'cciò che tu intende meglio te voglio dare l'assemplo.

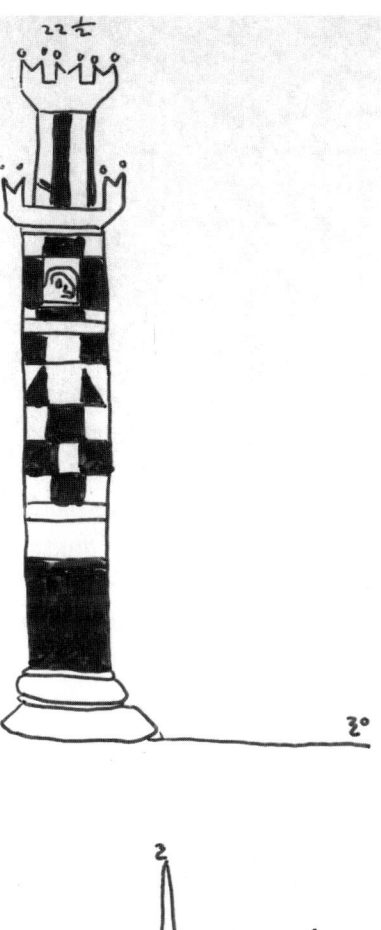

its top, do thus, cut off a stick, and drive it into the ground beside the thing of which you want to know the measure of the height. And do so that of the said stick is left over above the ground two *braccia* or iij or iv according to how it happens to be cut off for you. And then look during the day, when the sun hits the stick and the tower or tree. And measure precisely the shadow of the stick and the shadow of the tower or tree. And then multiply the measure of the stick **(fol. 51ᵛ)** against the measure of the shadow of the tower or tree. And divide in the measure of the shadow of the stick. And that which results

Togli una mazza, et ficchala in terra, et fa che sopra a terra avanzi 3 braccia, et diciamo che la ombra sua sia 4 braccia. Et la ombr^i^a dela torre overo arboro diciamo che sia braccia 30. Ora multipricha 3 via 30 braccia, fa braccia 90, et parti in 4, che ne vene vinti dui et mezzo. Et b̃ʳ 22$\frac{1}{2}$ è alta quella torre overo alboro che tu voi mesurare. Et è facta, et così se fanno le similianti ragioni.

from it, so much is high that thing which you want to know. And in order that you understand better I want to give you the example. Cut off a stick, and drive it into the ground, and do so that it leaves over 3 *braccia* above the ground, and let us say that its shadow is 4 *braccia*. And let us say that the shadow of the tower or tree is 30 *braccia*. Now multiply 3 times 30 *braccia*, it makes *braccia* 90, and divide in 4, from which results twenty-two and a half. And *braccia* $22\frac{1}{2}$ is high that tower or tree which you want to measure. And it is done, and thus the similar computations are done.

---

²²·³ Una torre è digno de fora de 50 braccia et à grosso el muro b̃ʳ ij$\frac{1}{4}$. Vo' sapere quanto serà de giro dal lato dentro. Fa così, sempre per reghola multipricha la grossezza del muro per octo et di' così, 8 via $2\frac{1}{4}$ fa 18 b̃ʳ, trai 18 de 50, resta 32, et cotante braccia gira la torre dentro. Et simiglante-mente se dice'ssi una torre gira dentro b̃ʳ cotante, et el muro è grosso cotanto, sì multipricha la grossezza del muro per 8 et agiongi sopra esso, et tanto gira de fore la torre. Et è facta, et sta bene.

²²·³ A tower is *braccia* 50 gross outside and has the wall *braccia* ij$\frac{1}{4}$ thick. I want to know how much it will be all around on the inside. Do thus, by rule multiply always the thickness of the wall by eight and say thus, 8 times $2\frac{1}{4}$ makes 18 *braccia*, detract 18 from 50, 32 is left, and as many *braccia* is the tower all around inside. And similarly is you should say, a tower is so and so many *braccia* all around inside, and the wall is thick so and so much, then multiply the thickness of the wall by 8 and join above that, and so much is the tower all around outside. And it is done, and it goes well.

---

²²·⁴ Una coppa d'argento è in tre pezzi, o in tri parti. Cioè el gambo, el nappo, e'l coperchio. El nappo pesa $\frac{1}{3}$ e'l $\frac{1}{4}$ de se medesemo et del gambo. El coperchio pesa el$\frac{1}{4}$ e'l $\frac{1}{5}$ de se medesemo et del nap-po. Et pesa el coperchio oncie 6. Adomandote que pesa el gambo et que pesa el nappo per se mede-simo, et que pesa tucta la coppa. Fa così, tu di' che'l coperchio pesa el $\frac{1}{4}$ e'l $\frac{1}{5}$ de se et del nappo, et pesa oncie 6. Et però trova uno numero che'l $\frac{1}{4}$ e'l $\frac{1}{4}$ [*sic*, read "$\frac{1}{5}$"] sia 6. Et questo te convene fare pro [*sic*, read "per"] positione. Et però fa così,

²²·⁴ A goblet of silver consists of three pieces, or three parts. That is, the stem, the cup, and the lid. The cup weighs $\frac{1}{3}$ and the $\frac{1}{4}$ of itself and of the stem. The lid weighs the $\frac{1}{4}$ and the $\frac{1}{5}$ of itself and of the cup. And the lid weighs ounces 6. I ask you what the stem weighs and what the cup weighs by itself, and what all the goblet weighs. Do thus, you say that the lid weighs the $\frac{1}{4}$ and the $\frac{1}{5}$ of itself and of the cup, and weighs ounces 6. And therefore find a number of which the $\frac{1}{4}$ and the ⟨$\frac{1}{5}$⟩ is 6. And this it suits you to do ⟨by⟩ position. And therefore do

trova uno numero che abia $\frac{1}{4}$ e $\frac{1}{5}$, che è 20. Ora piglia el $\frac{1}{4}$ e'l $\frac{1}{5}$ de 20, che è 9. Et tu di' che vorresti 6. Et però di' chosì, per 20 che io me appongho me vene 9, et io voglio 6. Et multipricha 6 via 20, fa 120, et parti in 9, che ne vene $13\frac{1}{3}$, et questo è el numero de que 6 è $\frac{1}{4}$ e $\frac{1}{5}$. Ora trai 6 de $13\frac{1}{3}$, resta {resta} $7\frac{1}{3}$. Et $7\frac{1}{3}$ oncia pesa el nappo. Ora tu di' che'l nappo pesa **(fol. 52ʳ)** el $\frac{1}{3}$ e'l $\frac{1}{4}$ de se medesemo et del gambo. Et però te convene trovare uno numero che $7\frac{1}{3}$ sia el $\frac{1}{3}$ e'l $\frac{1}{4}$, et conventi fare como facesti l'altra de sopra, per positione. Trova uno numero che abia el $\frac{1}{3}$ e'l $\frac{1}{4}$, che è 12. E'l $\frac{1}{3}$ e'l $\frac{1}{4}$ de 12 si è 7. Et tu voli $7\frac{1}{3}$. Et però di' così, per 12 che io me appongno me viene 7. Et io voglio $7\frac{1}{3}$. Multipricha 12 via $7\frac{1}{3}$, fa 88, et parti in 7, che ne vene 12 e $\frac{4}{7}$. Et 12 e $\frac{4}{7}$ è el numero de que $7\frac{1}{3}$ è el $\frac{1}{3}$ e'l $\frac{1}{4}$. Or⟨a tr⟩ai $7\frac{1}{3}$ che pesa el nappo, resta oncie 5 e $\frac{5}{21}$, et oncie 5 e $\frac{5}{21}$ pesa el gambo. Siché tu ài che'l nappo pesa oncie $7\frac{1}{3}$, e'l gambo pesa oncie 5 e $\frac{5}{21}$. E'l corpo [*sic*, read "coperchio"] pesa oncie 6. Ora se voli sapere quanto pesa la coppa, aggiongi inseme questi numeri, fanno in tucto oncie $18\frac{4}{7}$. Et tanto pesa la coppa. Et è facta. Et chosì se fanno le simili ragioni.

thus, find a number that possesses $\frac{1}{4}$ and $\frac{1}{5}$, which is 20. Now seize the $\frac{1}{4}$ and the $\frac{1}{5}$ of 20, which is 9. And you say that you wanted 6. And therefore say thus, for 20 that I posit 9 results for me, and I want 6. And multiply 6 times 20, it makes 120, and divide in 9, from which results $13\frac{1}{3}$, and is the number of which 6 is $\frac{1}{4}$ and $\frac{1}{5}$. Now detract 6 from $13\frac{1}{3}$, $7\frac{1}{3}$ is left {...}. And $7\frac{1}{3}$ ounce weighs the cup. Now you say that the cup weighs **(fol. 52ʳ)** the $\frac{1}{3}$ and the $\frac{1}{4}$ of itself and of the stem. And therefore it suits you to find a number of which $7\frac{1}{3}$ is the $\frac{1}{3}$ and the $\frac{1}{4}$, and it suits you to do as you did the other above, by position. Find a number that has the $\frac{1}{3}$ and the $\frac{1}{4}$, which is 12. And the $\frac{1}{3}$ and the $\frac{1}{4}$ of 12 is 7. And you want $7\frac{1}{3}$. And therefore say thus, for 12 that I posit 7 results for me. And I want $7\frac{1}{3}$. Multiply 12 times $7\frac{1}{3}$, it makes 88, and divide in 7, from which results 12 and $\frac{4}{7}$. And 12 and $\frac{4}{7}$ is the number of which $7\frac{1}{3}$ is the $\frac{1}{3}$ and the $\frac{1}{4}$. No⟨w de⟩tract $7\frac{1}{3}$ that the cup weighs, ounces 5 and $\frac{5}{21}$ is left, and ounces 5 and $\frac{5}{21}$ weighs the stem. So that you have that the cup weighs ounces $7\frac{1}{3}$, and the stem weighs ounces 5 and $\frac{5}{21}$. And the ⟨lid⟩ weighs ounces 6. Now if you want to know how much the goblet weighs, join together these numbers, they make in all ounces $18\frac{4}{7}$. And so much weighs the goblet. And it is done. And thus the similar computations are done.

22.5 Se te fosse dicto, egli è uno giro tundo a sesto tale che'l dericto de mezzo è 14 b̃ʳ.

22.5 If it was said to you, there is an ambit-round orbit such that the straight in middle

Vo' sapere quante braccia quadre è egli sanza spiare el crichulazio [*sic*, read "circhulazio"] dintorno. Sì se fa a questo modo, multiprica sempre la summo dele b̃^{r} del dericto de mezzo, et fa'nne numero.

Et de quello numero trai el septimo et la mità del septimo. E quello che rimane, tante braccia quadre è tucto el giro. Ora la prova. Tu di' che'l dericto de mezzo è 14 b̃^{r}. Multipricha 14 via 14, fa 196. Traine el septimo, che è 28, et la mità del septimo, che è 14, ài 42. Trailo de 196, resta 154. Et tante b̃^{r} quadre è quillo giro. Et ⟨è⟩ facta. Et così se fanno le simili ragioni.

is 14 *braccia*. I want to know how many square *braccia* it is without espying the circulation around. It is made in this way, always multiply (by itself) the total of the *braccia* of the straight in middle, and make it number. And from this number detract the seventh and the half of the seventh. And that which remains, so many square *braccia* is all the orbit. Now the verification. You say that the straight in middle is 14 *braccia*. Multiply 14 times 14, it makes 196. Detract the seventh, which is 28, and the half of the seventh, which is 14, you get 42. Detract it from 196, 154 is left. And so many square *braccia* is that orbit. And ⟨it is⟩ done. And thus the similar computations are done.

22.6 Uno giro tundo a sesto è tale che'l dericto suo de mezzo è 14 b̃^{r}. Vo' sapere quanto gira dintorno, et quante b̃^{r} quadre è tucto. Fa così como io t'ò dicto ancho de sopra, multipricha la summa del dericto di mezzo per $3\frac{1}{7}$. Et 14 via 3 e $\frac{1}{7}$ fa 44, et 44 b̃^{r} gira dintorno questo tundo. Ora se voli sapere quante b̃^{r} quadre è egli, sì multipricha 14 via 44, che fa 616, et parti in 4, che ne vene 154, et tante b̃^{r} quadre è tucto el giro. Et vedi che torna como quella de sopra, che la facciamo senza sapere el circulario dintorno, el quale è ancho b̃^{r} 44, et tornano a uno modo. Et però ho facta questa al lato a quella, che tu intende bene l'una et l'altra, et che l'una

22.6 An ambit-round orbit is such that its straight in middle is 14 *braccia*. I want to know in how much it goes around, and how many square *braccia* it is in all. Do thus as I have also said to you above, multiply the total of the straight in middle by $3\frac{1}{7}$. And 14 times 3 and $\frac{1}{7}$ makes 44, and in 44 *braccia* this round goes around. Now if you want to know how many square *braccia* it is, then multiply 14 times 44, which makes 616, and divide in 4, from which results 154, and so many square *braccia* the whole orbit. And you see that it becomes as the one above, which we make without knowing the circulation around, which is also *braccia* 44, and

et {la} l'altra è bona reghola. Et stanno bene. Et così se fanno le simili ragioni.

they become the same. And therefore I have made this beside that, so that you understand well one as well as the other, and that one as well as {...} the other is a valid rule. And they go well. And thus the similar computations are done.

(fol. 52ᵛ)

22.7 **Trova** uno numero che, tractone el $\frac{1}{2}$ e'l $\frac{1}{4}$ e'l $\frac{1}{6}$, e lo remanente multiprichato per se medesimo, faccia quello medesimo numero. Fa cosa [sic, read "così"] ogni volta che te sonno date de simili ragioni. Ove sieno questi rocti, te convene sempre trovare uno numero nel quale se trovino tucti questi rocti. Altramente serrebe inpossibile a'ffare. Et però trova ala ragioni proposta uno numero in que se trovino $\frac{1}{2}$ $\frac{1}{4}$ $\frac{1}{6}$, et questo numero el più presso che tu abia si è 12, che tal te fa a torre uno quanto un altro. Ma toglamo questo, che è più presso. Ora trai de 12 el $\frac{1}{2}$, $\frac{1}{4}$ e $\frac{1}{6}$, resta uno. Ora multipricha quello uno per se medesimo, fa pur j. Et tu di' di' sopra che voi che faccia quello medesimo numero. Cioè che vorresti facesse 12, como tu trovasti. Et però di' così, per 12 che io me appongho me remane uno, et io voglio me remanga 12. Multipricha 12 via 12, fa 144, parti in j, venne quello medesimo. Siché el numero che noi vogliamo si è 144. Ora la prova, tray de 144 el $\frac{1}{2}$, che è 72, resta 72. Trai de 144 el $\frac{1}{4}$, che è 36, traylo de 72 ~~che è~~ ^resta^ 36. Trai de 144 el $\frac{1}{6}$, che è 24, trailo de 36, resta 12. Ora ài che tracto de 144 el $\frac{1}{2}$ e'l $\frac{1}{4}$ e'l $\frac{1}{6}$, resta 12. Et tu di'

22.7 Find a number which, when the $\frac{1}{2}$ and the $\frac{1}{4}$ and the $\frac{1}{6}$ are detracted from it, and the remainder multiplied by itself, makes this same number. Do ⟨thus⟩ each time that similar computations are given to you. Where there are these fractions, it always suits you to find a number in which all these fractions are found. Otherwise it would be impossible to make it. And therefore find to the proposed computation a number in which $\frac{1}{2}$ $\frac{1}{4}$ $\frac{1}{6}$ are found, and this number, the closest that you have, is 12, as you may as well grasp one as the other. But let us grasp this, which is the closest. Now detract from 12 the $\frac{1}{2}$, $\frac{1}{4}$ and $\frac{1}{6}$, one is left. Now multiply this one by itself, it makes j once again. And you say above that you want to it to be that same number. That is, that you want it should make 12, as you found. And therefore say thus, for 12 that I posit one remains for me, and I want that 12 remain. Multiply 12 times 12, it makes 144, divide in j, that same results. So that the number that we want is 144. Now the verification, detract of 144 the $\frac{1}{2}$, which is 72, 72 is left. Detract of 144 the $\frac{1}{4}$, which is 36, detract it from 72, 36 {...} is left. Detract of 144 the $\frac{1}{6}$, which is 24,

che voi che quello che te resta multiprichato per se medesimo faccia quello medesimo numero. Et però multipricha 12 via
12, fa 144. Et sta bene, che fa quello medesimo. Et è bene facta et provata. Et così se
fanno le simigliante ragioni.

detract it from 36, 12 is left. Now you have
that when from 144 the $\frac{1}{2}$ and the $\frac{1}{4}$ and
the $\frac{1}{6}$ have been detracted, 12 is left. And
you say that you want that that which is left
for you multiplied by itself make that same
number. And therefore multiply 12 times
12, it makes 144. And it goes well, and it
makes that same. And is well done and
verified. And thus the similar computations
are done.

$^{22.8}$ **Trova** uno numero che, tractone el $\frac{1}{2}$
e'l $\frac{1}{3}$, lo rimanente sia 24. Questa se fa
ancho como io t'ò dicto de sopra. Trova
uno numero che abia $\frac{1}{3}$ e $\frac{1}{2}$, che è 6. El $\frac{1}{2}$
e'l $\frac{1}{3}$ de 6 si è 5. Trailo de 6, resta j. Et tu
di', voli te resti 24. Et però di' così, per 6
che io me appongho me vene j, et io voglio
24. Multipricha 6 via 24, fa 144, parti in
j, viene 144. Et è facta che tractone el $\frac{1}{2}$
e'l $\frac{1}{3}$ de 144 che è 120, resta apponto 24.
Et sta bene. Et così se fanno le simigliante
ragioni.

$^{22.8}$ Find a number so that, when the $\frac{1}{2}$ and
the $\frac{1}{3}$ are detracted from it, the remainder
is 24. This is also done as I have said to
you above. Find a number that possesses $\frac{1}{3}$
and $\frac{1}{2}$, which is 6. The $\frac{1}{2}$ and the $\frac{1}{3}$ of 6
is 5. Detract it from 6, j is left. And you
say, you want that 24 shall be left for you.
And therefore say thus, for 6 that I posit
j results for me, and I want 24. Multiply 6
times 24, it makes 144, divide in j, 144
results. And it is done that, when the $\frac{1}{2}$ and
the $\frac{1}{3}$ of 144 are detracted, which is 120,
precisely 24 is left. And it goes well. And
thus the similar computations are done.

(fol. 53^{r})

$^{22.9}$ **Uno** homo à una nave con doi vele. Et
vole andare al suo viagio. Et coll'una vela
fa el viagio suo in tre dì. Et coll'altra vela
el fa in 4 dì. Vo' sapere in quanti dì el
farebbe rizzando ammedore le vele a uno
tracto et navighando con esse doy. Fa così,
l'una vela fa el viagio in 3 dì, et l'altra in
4 dì. Et però multipricha 3 via 4, fa 12. Et
agiongi inseme 3 et 4, fa 7. Parti 12 in 7,

$^{22.9}$ A man has a ship with two sails. And
he wants to go to his voyage. And with one
sail he makes his voyage in three days. And
with the other he makes it in 4 days. I want
to know in how many days he would make
it hoisting up both sails at one time and
sailing with these two. Do thus, one sail
makes the voyage in 3 days, and the other
in 4 days. And therefore multiply 3 times

che ne vene uno e $\frac{5}{7}$. Et in uno dì et $\frac{5}{7}$ de dì fa uno viagio con ammedoro. Et è facta. Et così se fanno le simili ragioni.

4, it makes 12. And join together 3 and 4, it makes 7. Divide 12 in 7, from which results one and $\frac{5}{7}$. And in one day and $\frac{5}{7}$ of day he makes a voyage with both. And it is done. And thus the similar computations are done.

---

^{22.10} Uno campo de terra. È lungho b̃^r 25, et è largho braccia 16. Ora io el voglio quadrare. Vo' sapere quante b̃^r serà per ogni faccia. Fa così, multipricha la lunghezza contra ala larghezza, cioè 16 via 25, che fa 400. Ora trova la sua radice, che è 20. Et 20 b̃^r serà per ogni faccia. Et è facta. Et così se fanno le simiglianti ragioni.

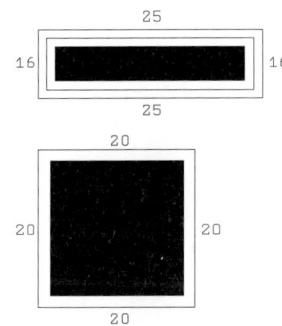

^{22.10} A field. It is 25 *braccia* long, and it is 16 *braccia* broad. Now I want to square it. I want to know how many *braccia* it will be by each face. Do thus, multiply the length against the breadth, that is, 16 times 25, which makes 400. Now find its root, which is 20. And 20 *braccia* will it be by each face. And it is done. And thus are done the similar computations.

---

^{22.11} Una torre, como tu vedi designata de rimpetto. Et è alta braccia 40. Et a'ppe dela torre si è uno fosso, che è ampio 30 b̃^r. Ora io voglio porre una fune che agiungha dall'urlo del fosso infino ala cima dela torre. Vo' sapere, quanto vole essere lungha, ponendola all'orlo del fosso de fuori. Fa così, multipricha l'altezza dela tore per se medesima, et di', 40 via 40 fa 1600. Ora multipricha per la larghezza del

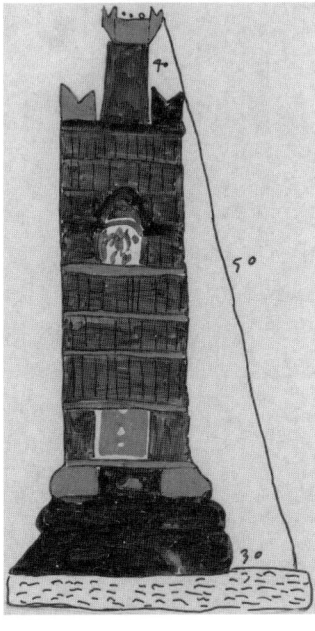

^{22.11} A tower, as you see drawn opposite. And it is 40 *braccia* high. And at the foot of the tower there is a moat, which is 30 *braccia* wide. Now I want to stretch a rope that reaches from the border of the moat until the peak of the tower. I want to know, how much it has to be long, stretching it to the outer border of the moat. Do thus, multiply the height of the tower by itself, and say, 40 times 40 makes 1600. Now multiply for the

fosso, et di', 30 via 30 fa 900. Ora agiongi inseme, fa 2500. De questo numero trova la sua radice, che è 50. Et 50 braccia vole esser longha la dicta fune. Et è facta, et sta bene. Et così se fanno le simili ragioni.

breadth of the moat, and say, 30 times 30 makes 900. Now join together, it makes 2500. Of this number find its root, which is 50. And 50 *braccia* has the said rope to be long. And it is done, and it goes well. And thus the similar computations are done.

(fol. 53ᵛ)

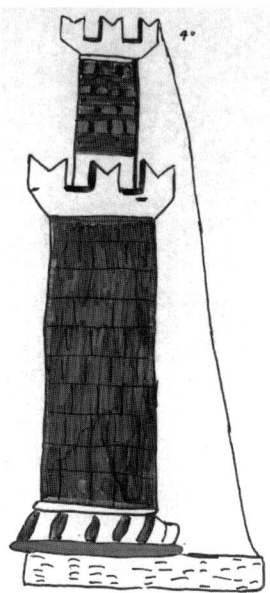

22.12 Una torre chomo tu vedi designata de rinpecto. È alta ma non so quanto. Ma a'ppe dela torre si è uno fosso, el quale è largho 30 b̃ʳ. Et io ho posta una fune dala sponda del fosso dal canto de fuori, la quale aggiongie infino ala cima dela torre. Et è lungha la dicta fune b̃ʳ 50. Vo' sapere, quanto vene a essere alta la dicta torre. Fa così, multipricha la lunghezza dela fune per se medesima, et di', 50 via 50 fa 2500. Ora multipricha la larghezza del fosso per se medesimo, che è largho 30 b̃ʳ, et di', 30 via 30 fa 900. *Ora trai de 2500 900,* che resta 1600. Ora trova la radice de 1600, che è 40, però che 40 via 40 fa 1600. Et 40 b̃ʳ vene a essere la dicta torre. Et sta bene ala simile ragione. Et è facta. Et così se fanno le simili ragioni.

22.12 A tower as you see drawn opposite. It is high but I do not know how much. But at the foot of the tower there is a moat, which is 30 *braccia* broad. And I have stretched a rope from the bank of the moat on the outer edge, which reaches until the peak of the tower. And the said rope is *braccia* 50 long. I want to know how high the said tower results to be. Do thus, multiply the length of the rope by itself, and say, 50 times 50 makes 2500. Now multiply the breadth of the moat by itself, which is 30 *braccia* broad, and say, 30 times 30 makes 900. Now detract 900 from 2500, and 1600 is left. Now find the root of 1600, which is 40, because 40 times 40 makes 1600. And the said tower results to be 40 *braccia*. And it goes well to the similar computation, and it is done. And thus the similar computations are done.

22.13 Uno vinaio^a murato ampio per ogni faccia 30 b̃ʳ. Et è alto assai. Et l'acqua v'è alta dentro 3 b̃ʳ. Ora vene per caso che vi cade dentro una pietra quadra per ogni faccia 20 b̃ʳ et grossa b̃ʳ 3. Vo' sapere quanto crebbe

l'acqua nel vinaro per questa pietra. Fa così, multipricha 30 via 30, fa 900. Cotante b̃ʳ quadre posscede el vinaro. Ora multipricha 3 via 900, fa 2700. Ora ài che l'acqua è 2700 b̃ʳ quadre. Ora sappi quanto è la pietra, et multipricha 20 via 20, fa 400. Et multipricha 3 via 400, fa 1200. Agiongi inseme 2700 e 1200, fa 3900 braccia tra acqua e pietra. Ora se voli sapere quanto l'acqua cresce, fa così, lo vinaro possc⟨e⟩de in fundo 900 b̃ʳ quadre. Et però parti 3900 per 900, che ne vene 4$\frac{1}{3}$. Et b̃ʳ 4$\frac{1}{3}$ è alta l'acqua nel vinaro cola pietra, et la pietra è grossa 3 b̃ʳ. Dunqua crebbe l'acqua nel vinaro per cagione dela pietra braccio j e $\frac{1}{3}$. La pietra andò al fundo, et à di sopra a'sse b̃ʳ j e $\frac{1}{3}$ d'acqua. Et cotanto crebbe, et è facta. Et chosì se fanno tucte le simiglianti ragioni.

^a Error for *vivaio*.

22.13 A walled basin 30 *braccia* wide by each face. And it is sufficiently high. And the water there inside is 3 *braccia* high. Now it occurs by chance that a square stone falls into it, by each face 20 *braccia* and thick *braccia* 3. I want to know how much the water rose in the basin because of this stone. Do thus, multiply 30 times 30, it makes 900. As many square *braccia* does the basin occupy. Now multiply 3 times 900, it makes 2700. Now you have that the water is 2700 square *braccia*. Now know how much the stone is, and multiply 20 times 20, it makes 400. And multiply 3 times 400, it makes 1200. Join together 2700 and 1200, it makes 3900 *braccia*, water and stone together. Now if you want to know how much the water rose, do thus, the basin occupies as bottom 900 square *braccia*. And therefore divide 3900 by 900, from which results 4$\frac{1}{3}$. And *braccia* 4$\frac{1}{3}$ is the water high in the basin with the stone, and the stone is 3 *braccia* thick. Hence the water rose *braccio* j and $\frac{1}{3}$ in the basin by reason of the stone. The stone went to the bottom, and above itself has *braccio* j and $\frac{1}{3}$ of water. And as much it rose, and it is done. And thus are done all the similar computations.

(fol. 54ʳ)

22.14 Uno vinaio ascesto gira intorno 44 b̃ʳ et è vi alta dentro l'acqua 5 braccia. Vene per caso che vi cade entro una pietra tonda, et de giro de b̃ʳ 22, et è grossa 6 b̃ʳ. Vo'

22.14 An ambit-made basin goes around in 44 *braccia* and the water is 5 *braccia* high inside it. It occurs by chance that a round stone falls into it, and all around of *braccia*

sapere quanto crebbe l'acqua entro nel dicto vinaro per cagione dela dicta pietra. Fa così, prima sappi quanto terreno possede el vinaro, et fa secondo la regola. Parti 44 b̃ʳ che gira dintorno el vinaro in 3 e $\frac{1}{7}$, che ne vene 14. Et 14

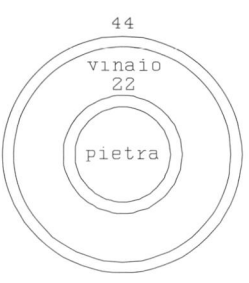

b̃ʳ è el dicto vinaro per lo mino longho suo. Ora multipricha 14 via 44, che fa 616, parti in 4, che ne vene 154. Et cotante b̃ʳ quadre è lo spatio del vinaro. Ora tu di' che l'acqua è alta 5 b̃ʳ, multipricha 5 via 154, che fa 770. Et cotante b̃ʳ quadre è l'acqua nel vinaro. Ora sappi quante b̃ʳ è la pietra, la quale tu di' che gira 22 b̃ʳ. Parti 22 in 3$\frac{1}{7}$, che ne vene 7. Et 7 b̃ʳ è la pietra per lo milongho. Ora multipricha 7 via 22, che fa 154. Parti in 4, che ne vene 38$\frac{1}{2}$. Et cotante b̃ʳ quadre è lo spatio dela pietra. Ora tu di' che ella è grossa 6 b̃ʳ. Et però multipricha 6 via 38$\frac{1}{2}$, che fa 231. Et cotante b̃ʳ quadre è tucta la pietra. Ora agiongi inseme 770 b̃ʳ, che è alta l'acqua nel vinaio, et 231 b̃ʳ, che è la pietra, fanno b̃ʳ 1001. Et cotante b̃ʳ quadre è tra l'acqua e la pietra. Et lo spatio del vinaro è b̃ʳ 154. Ora parti 1001 in 154, che ne vene b̃ʳ 6 e $\frac{1}{2}$. Et cotanto torna l'acqua alta nel vinaio. Et noi diciamo che prima vi cadesse la pietra, v'era alta entro l'acqua b̃ʳ 5. Siché per cagione dela pietra è cressiuta l'acqua nel dicto vinaro b̃ʳ j e $\frac{1}{2}$. Et così sta secondo la ragione. Et è facta, et sta bene. Et così se fanno tucte le simigliante ragioni.

22, and it is 6 *braccia* thick. I want to know how much the water rose inside the said basin by reason of the said stone. Do thus, first know how much terrain the basin occupies, and do according to the rule. Divide 44 *braccia,* in which the basin goes around, in 3 and $\frac{1}{7}$, from which results 14. And 14 *braccia* is the said basin by its middle. Now multiply 14 times 44, which makes 616, divide in 4, from which results 154. And as many square *braccia* is the space of the basin. Now you say that the water is 5 *braccia* high, multiply 5 times 154, which makes 770. And as many square *braccia* is the water in the basin. Now know how many *braccia* the stone is, which you say goes around in 22 *braccia*. Divide 22 in 3$\frac{1}{7}$, from which results 7. And 7 *braccia* is the stone by the middle. Now multiply 7 times 22, which makes 154. Divide in 4, from which results 38$\frac{1}{2}$. And as many square *braccia* is the space of the stone. Now you say that it is 6 *braccia* thick. And therefore multiply 6 times 38$\frac{1}{2}$, which makes 231. And as many square *braccia* is the whole stone. Now join together 770 *braccia*, which the water is high in the basin, and 231 *braccia*, which the stone is, they make 1001 *braccia*. And as many square *braccia* is water and stone together. And the space of the basin is 154 *braccia*. Now divide 1001 in 154, from which results *braccia* 6 and $\frac{1}{2}$. And as much will the water become high in the basin. And we say that before the stone fell

into it, the water was *braccia* 5 high inside it. So that by reason of the stone the water has risen *braccio* j and $\frac{1}{2}$ in the said basin. And thus it is according to the computation. And it is done, and it goes well. And thus all the similar computations are done.

22.15 Uno homo toglie uno puzzo a cavare tondo, **(fol. 54ᵛ)** che sia largho 3 braccia et copo 40, del quale dè avere £ 12. Ora dice colui de chi è el puzzo, cavalo tanto più largho che sia 3 b̃ʳ e $\frac{1}{2}$, et io te pagarò ala simile ragione. Vo' sapere, quanto dè avere più per cavatura del dicto pozzo. Fa così, multipricha questa mesura per se medesima, cioè la prima mesura. Et di' così, 3 via 3 fa 9. Ora trai secondo che dice la regola de 9 el septimo et la metà del septimo, che resta 7 e $\frac{1}{14}$. Ora multipricha 40 via 7 e $\frac{1}{14}$, che fa 228 e $\frac{6}{7}$. Et cotante b̃ʳ dovea essere el puzzo del quale dovea avere 12 £. Ora ello dice el vole cavato b̃ʳ 3$\frac{1}{2}$ per largho, siché sappia quante braccia debba tornare cavato. Moltipricha 3$\frac{1}{2}$ via 3 e $\frac{1}{2}$, fa 12$\frac{1}{4}$. Ora cavane como de sopra el $\frac{1}{7}$ et la mità del $\frac{1}{7}$, che resta 9 e $\frac{5}{8}$. Ora multiplicha 40 via 9 e $\frac{5}{8}$, che fa 385. Et cotante b̃ʳ quadre torna cavato el puzzo de b̃ʳ 3 e $\frac{1}{2}$ largho. Siché tu ài che vene a essere cavato più che prima b̃ʳ 102 e $\frac{1}{7}$. Ora, di' così, uno dè avere ⟨per⟩ cavatura de uno puzzo de b̃ʳ 282 e $\frac{6}{7}$ £ 12. Vo' sapere che dè avere de b̃ʳ 102 e $\frac{1}{7}$. Et

22.15 A man undertakes to dig a round well, **(fol. 54ᵛ)** which is to be 3 *braccia* broad and 40 deep, for which he shall get 12 *libra*. Now the one whose well it is says, dig it so much broader that it is 3 *braccia* and $\frac{1}{2}$, and I shall pay you in similar rate. I want to know how much more he shall get for the digging of the said well. Do thus, multiply this measure by itself, that is, the first measure. And say thus, 3 times 3 makes 9. Now detract according to what the rule says from 9 the seventh and the half of the seventh, and 7 and $\frac{1}{14}$ is left. Now multiply 40 times 7 and $\frac{1}{14}$, which makes 228 and $\frac{6}{7}$. And as many *braccia* should the well be for which he should get 12 *libre*. Now that one says that he wants it dug 3$\frac{1}{2}$ *braccia* in breadth, and therefore know how many *braccia* have to be dug out. Multiply 3$\frac{1}{2}$ times 3 and $\frac{1}{2}$, it makes 12$\frac{1}{4}$. Now remove as above $\frac{1}{7}$ and the half of $\frac{1}{7}$, and 9 and $\frac{5}{8}$ is left. Now multiply 40 times 9 and $\frac{5}{8}$, which makes 385. And as many square *braccia* are to be dug out for the well which is *braccia* 3 and $\frac{1}{2}$ broad. So that it results for you that *braccia* 102

multipricha 102 $\tilde{b}^r$ e $\frac{1}{7}$ via 12 £, ⟨fa⟩ 1225 e $\frac{5}{7}$. Et parti per 282 $\frac{6}{7}$, che ne vene 4 £ et 6 ß et deñ 8. Et cotanto dè avere sopra alle 12 £ colui che à cavato el puzzo de $\tilde{b}^r$ 3 e $\frac{1}{2}$ largho, che sonno in tucto £ 16 ß 6 deñ 8. Et è facta, et sta bene. Et così se fanno le simiglianti ragioni.

and $\frac{1}{7}$ more than before have to be dug out. Now say thus, one shall get ⟨for the⟩ digging of a well of *braccia* 282 and $\frac{6}{7}$ *libra* 12. I want to know what he shall get for *braccia* 102 and $\frac{1}{7}$. And multiply 102 *braccia* and $\frac{1}{7}$ times 12 *libre*, ⟨it makes⟩ 1225 and $\frac{5}{7}$. And divide by 282 $\frac{6}{7}$, from which results 4 *libre* and 6 *soldi* and *denari* 8. And as much shall this one get above the 12 *libre* who dug the well *braccia* 3 and $\frac{1}{2}$ broad, which are in all *libre* 16, *soldi* 6, *denari* 8. And it is done, and it goes well. And thus are done the similar computations.

22.16 **I**o vo a uno giardino, et giongho a' ppede de una melarancia. Et coglione una. Et poi coglio el decimo de rimanente. Poi vene un altro dopo me, et coglene doy, et anchora el decimo de rimanente. Poi vene un altro et coglene 3, et anchora el decimo de rimanente. Poi vene un altro et coglene 4 et el decimo de rimanente. Et così venghono molti. Poi quello che vene da sezzo, cioè dercto, coglie tucte quelle che retrova. Et non ve ne trova né più né meno che abiamo auti li altri. Et tanto ne colze l'uno **(fol. 55ʳ)** quante l'altro. Et tanti homini quanti erano, tante melarancie ebbe per uno. Vo' sapere quanti homini forono, et quante melarancie colseno per uno, et quante ne colzeno fra tucti quanti. Fa così, tray uno de 10, resta 9, et 9 homini forono, et 9 melarancie colseno per uno. Et colzero in tucto 81 melarancie. Et se la voli provare, fa così.

22.16 I go to a garden, and come to the foot of an orange. And I pick one of them. And then I pick the tenth of the remainder. Then comes another after me, and picks two of them, and again the tenth of the remainder. Then comes another and picks 3 of them, 3, and again the tenth of the remainder. Then comes another and picks 4 of them and the tenth of the remainder. And thus come many. Then the one who comes last, that is, behind, picks all that which he finds left. And finds by this neither more nor less than all the others got. And one picked as much **(fol. 55ʳ)** as the other. And as many men as there were, so many oranges were there for (each) one. I want to know how many men there were, and how many oranges they picked (each) one, and how many they picked all together. Do thus, detract one from 10, 9 is left, and there were 9 men, and 9 oranges (each) one picked. And they picked in all 81 oranges.

And if you want to verify it, do thus,

El primo ne colze j, restano

80.	El decimo	è octo, et ày che illo n'ebbe 9. Restano
72.	El sccondo	2, rcstano 70, cl dccimo è 7, et ebe ne 9, restano
63.	El terzo	3, restano 60, el decimo è 6, et ebe e 9, restano
54.	El quarto	4, restano 50, el decimo è 5, et ebe ne 9, restano
45.	El quinto	5, restano 40, el decimo è 4, et ebe ne 9, restano
36.	El sexto	6, restano 30, el decimo è 3, et ebe ne 9, restano
27.	El sectimo	7, restano 20, el decimo è 2, et ebe ne 9, restano
18.	Ell'octavo	8, restano 10, el decimo è 1, et ebe ne 9, restano
9.	El nono, cioè quello da sezzo, colze quelle 9, né più né meno, che non ve n'erano più. Siché vedi che ella è bene facta. Et sta bene. Et così se fano le simiglianti ragioni.	

the first picked j of them, left

80.	The tenth	is eight, and you have that this one got 9. Left
72.	The second	2, left 70, the tenth is 7, and he got 9, left
63.	The third	3, left 60, the tenth is 6, and he got 9, left
54.	The fourth	4, left 50, the tenth is 5, and he got 9, left
45.	The fifth	5, left 40, the tenth is 4, and he got 9, left
36.	The sixth	6, left 30, the tenth is 3, and he got 9, left
27.	The seventh	7, left 20, the tenth is 2, and he got 9, left
18.	The eighth	8, left 10, the tenth is 1, and he got 9, left
9.	The ninth, that is, the last one, picked these 9, neither more nor less, as there were no more. So that you see that it is well done. And it goes well. And thus are done the similar computations.	

22.17 Uno tundo asexto como vedi designato de rinpetto[a] gira dintorno 22 b̃ʳ. Vo' sapere quante braccia debba essere el mino longho suo. Et quante b̃ʳ quadre è'l dicto tondo. Fa così como dice la regola, parti 22 per 3 e $\frac{1}{7}$, che ne vene 7. Et cotante b̃ʳ è el dericto di mezzo del mino longho. Ora se voli sapere quante braccia quadre è, multiplicha 7 via 22, fa 154, parti in 4, ven'ne $38\frac{1}{2}$. Et cotante

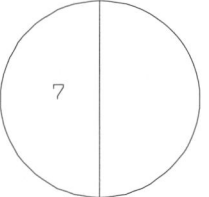

22.17 An ambit-made round as you see drawn opposite[a] goes around in 22 *braccia*. I want to know how many *braccia* its middle shall be. And how many square *braccia* the said round is. Do thus as the rule says, divide 22 by 3 and $\frac{1}{7}$, from which results 7. And as many *braccia* is the straight in middle of the middle.[b] Now if you want to know how many square *braccia* it is, multiply 7 times

braccia quadre serrà. Et è facta. Et così se fanno le simiglianti ragioni.

22, it makes 154, divide in 4, $38\frac{1}{2}$ results from it. And as many square *braccia* will it be. And it is done. And thus are done the similar computations.

ª In the manuscript, the diagram in question is misplaced, and found on fol. 57ᵛ.
ᵇ This translation of "el dericto di mezzo del mino longho" is certainly not elegant; but the original is hardly better.

(fol. 55ᵛ)

22.18 Una torre, chomo tu vedi designata de rimpetto, è alta b̃ʳ 40. Et nel fondo dela torre si è una gacta che vole salire suso. Et ogni ^dì^ acquista $\frac{1}{4}$ de braccio, et la nocte perde $\frac{1}{5}$ de b̃ʳ. Et insu la cima del torre si è uno topo che vole discendere giù dela torre. Et ogni dì discende $\frac{1}{3}$ de b̃ʳ, et la nocte perde $\frac{1}{4}$ de b̃ʳ. Vo' sapere in quanto tempo raccozzaranno inseme l'uno coll'altro, et in quanto tempo seranno l'uno fora et l'altro giunto giò nel fondo dela torre. Fa così, sappi prima quanto acquistano ogni uno tra dì et nocte, scontando quello che perdono la nocte. Et fa così, prima per la gatta, la quale acquista el dì $\frac{1}{4}$ et la nocte perde $\frac{1}{5}$. Et di', $\frac{1}{4}$ e $\frac{1}{5}$ se trova in 20, et di', el $\frac{1}{4}$ de 20 si è 5, et el $\frac{1}{5}$ de 20 si è 4. Cava 4 de 5, resta j, et questo è $\frac{1}{20}$. Siché tra dì et nocte, la gatta acquista $\frac{1}{20}$ de b̃ʳ. Et in 20 dì acquista j b̃ʳ. Et multipricha

22.18 A tower, as you see drawn opposite, is *braccia* 40 high. And in the basement of the tower there is a cat that wants to climb to the top. And each ^day^ it acquires $\frac{1}{4}$ of *braccio*, and in the night it loses $\frac{1}{5}$ of *braccio*. And on top of the tower there is rat that wants to descend the tower. And each day it descends $\frac{1}{3}$ of *braccio*, and in the night it loses $\frac{1}{4}$ of *braccio*. I want to know in how much time they will run into each other, and in how much time they will be, the one out and the other down in the basement of the tower. Do thus, know first how much each one acquires in day and night together, discounting that which they lose in the night. And do thus, first for the cat, which acquires $\frac{1}{4}$ in the day and loses $\frac{1}{5}$ in the night. And say, $\frac{1}{4}$ and $\frac{1}{5}$ is found in 20, and say, the $\frac{1}{4}$ of 20 is 5, and

20 via 40, fa 800, siché in 800 dì la gatta serà insu la cima dela torre. Ora sappi in quanti dì serà disceso el topo giù, el quale acquista el dì $\frac{1}{3}$ de b^r, et la nocte perde $\frac{1}{4}$ de b̃^{r}. Et di', $\frac{1}{3}$ e $\frac{1}{4}$ se trova in 12. E'l $\frac{1}{3}$ de 12 è 4, e'l $\frac{1}{4}$ de 12 è 3. Cava 3 de 4, resta j, et questo è $\frac{1}{\langle1\rangle2}$, siché el topo acquista tra dì et nocte, schonto quello che perde, $\frac{1}{12}$ de braccio, et in 12 dì acquista j b̃^{r}. Multipricha 12 via 40, fa 480, et in 480 dì serà disceso el topo giù. Ora se voli sapere in quanto se raccozzaranno inseme, sì multipricha l'uno di contra all'altro, cioè 480 via 800, cha fa 384000. Et poi agiongi inseme 480 et 800, che fa 1280. Ora parti 384000 in 1280, che ne vene 300, et in 300 dì si racozzaranno inseme. Et in 300 dì serà disceso el topo 25 b̃^{r}, et la gatta serà salita 15 b̃^{r}. Et el topo à ancora a discendere giù 15 b̃^{r}, et à tempo ancora da 300 dì infino in 480, che sonno 180. Et noi diciamo che egli acquista tra dì et nocte $\frac{1}{12}$ de b̃^{r}, siché in 180 dì sonno $\frac{180}{12}$ de b̃^{r} **(fol. 56^r)** che sonno braccia 15. Et sta bene, et è facta. Et la gacta à a salire anchora 25 braccia, et à tempo ancora 500 dì. Et noi diciamo che acquista tra dì et nocte $\frac{1}{20}$ de b̃^{r}. Siché in 500 dì salirà $\frac{500}{20}$ de b̃^{r}, che sonno b̃^{r} 25. Et sta bene. Et così se fanno le simili ragioni.

the $\frac{1}{5}$ of 20 is 4. remove 4 from 5, j is left, and this is $\frac{1}{20}$. So that in day and night together, the cat acquires $\frac{1}{20}$ of *braccio*. And in 20 day it acquires j *braccio*. And multiply 20 times 40, it makes 800, so that in 800 the cat will be on top of the peak of the tower. Now know in how many days the rat will have descended, which acquires in the day $\frac{1}{3}$ of *braccio*, and in the night loses $\frac{1}{4}$ of *braccio*. And say, $\frac{1}{3}$ and $\frac{1}{4}$ one finds in 12. And the $\frac{1}{3}$ of 12 is 4, and the $\frac{1}{4}$ of 12 is 3. remove 3 from 4, j is left, and this is $\frac{1}{\langle1\rangle2}$, so that the rat acquires in day and night together, discounted what it loses, $\frac{1}{12}$ of *braccio*, and in 12 days it acquires j *braccio*. Multiply 12 times 40, it makes 480, and in 480 days will the rat have descended. Now if you want to know in how much they will run into each other, then multiply the one against the other, that is, 480 times 800, which makes 384000. And then join together 480 and 800, which makes 1280. Now divide 384000 in 1280, from which results 300, and in 300 days will they run into each other. And in 300 days will the rat have descended 25 *braccia*, and the cat will have climbed 15 *braccia*. And the rat has still to descend 15 *braccia*, and still has time from 300 days until 480, which are 180. And we say that it acquires in day and night together $\frac{1}{12}$ of *braccio*, so that in 180 days these are $\frac{180}{12}$ of *braccio* **(fol. 56^r)** which are *braccia* 15. And it goes well. And the cat still has to climb 25 *braccia*, and still has time for 500 days. And we say that it acquires in day and night together $\frac{1}{20}$ of *braccio*. So that in 500 days it will climb $\frac{500}{20}$ of *braccio*, which are *braccia* 25. And it goes well, and it is done. And thus the similar computations are done.

22.19 Chi te dicesse overo che volesse {sapere} arechare a braccia quadre lo scudo oguale el quale è cotante $\tilde{b}^r$ per faccia, fa così, ponte al mezzo dell'una dele faccie, et sappi quanto è da ivi al canto che v'è de rinpotto. Et saputo quanto è quella lunghezza, allora multipricha lunghezza contra ala lunghezza del mino longho delo schudo. Cioè contra ala lunghezza de mezzo. Et quello che fa, chotante $\tilde{b}^r$ quadre è tucto lo schudo. Et poni bene mente a queste ragioni, che elle sonno belle ragioni.

22.19 ⟨If⟩ somebody should indeed say to you that he wanted to bring to square *braccia* an equal(-sided) shield which is so and so many *braccia* by face, do thus, put yourself in the middle of one of the faces, and know how much there is from there to the opposite corner. And when you have known how much this length is, then multiply (this) length against the length of the middle of the shield. That is, against length in middle. And that which it makes, as many square *braccia* is the whole shield. And keep well in mind these computations, as they are pleasant computations.

22.20 Et quando te fosse dicto, uno schudo è cotante braccia per ogni faccia, vo' sapere quante $\tilde{b}^r$ serà el dericto de mezzo per lo mino longho suo, sì se fa a questo modo che tu multipriche l'una dele faccie per se medesima, et dela somma abacti el quarto. Et delo remanente trova la sua radice. Et la radice che ne vene, chotante $\tilde{b}^r$ serà per lo dericto de mezzo, cioè, el mino longho. Et se altri dicesse, uno schudo è tale che mino longho suo è chotante $\tilde{b}^r$, vo' sapere quanto serà lo schudo per ogni faccia, sì multipricha quella lunghezza per se medesima, et fa numero. Et de quello numero piglia el terzo, et pollo sopra a esso, et de

22.20 And when it was said to you, a shield is so and so many *braccia* by each face, I want to know how many *braccia* the straight mid(-line) in its middle will be, it is made in this way that you multiply one of the faces by itself, and from the total strike off the fourth. And of the remainder find its root. And the root that results from it, as many *braccia* will it be by the straight in middle, that is, the middle. And if others should say, a shield is such that its middle is so and so many *braccia*, I want to know how much the shield will be by each face, then multiply that length by itself, and make number. And of this number seize the

tucta la summa piglia la sua radice. Et quello che te vene per radice, chotante b̃ʳ serrà lo schudo per ogni faccia. Et tucte le simigliante ragioni se fanno per questo modo et non altramente.

third, and put it above that one, and of the whole total seize its root. And that which results from it as root, as many *braccia* will the shield be by each face. And all the similar computations are done in this way and not otherwise.

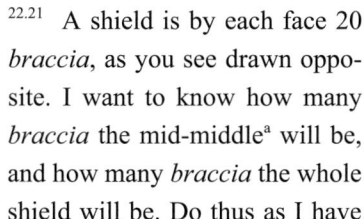

22.21 Uno schudo è per ogni faccia 20 b̃ʳ, chomo tu vedi designato de rinpetto. Vo' sapere quante braccia serà per lo mino longho de mezzo, et quante b̃ʳ serà tucto lo schudo. Fa così como io t'ò dicto de sopra, multipricha l'una dele faccie per se medesima, et di', 20 via 20 fa 400, e'l quarto de 400 è 100. Trailo de 400, **(fol. 56ʳ)** resta 300. Trova la radice de 300, che è 17 e $\frac{3}{10}$ et uno pocho più, tanto quanto 17 e $\frac{3}{10}$ via 17 e $\frac{3}{10}$ fa 299 e $\frac{29}{100}$. Ma'nnon se po trovare apponto. Et cotante braccia serà lo schudo per lo dericto di mezzo. Cioè 17 e $\frac{3}{10}$, et uno pocho pocho più, chomo dico de sopra. Ora se voli sapere quante b̃ʳ quadre è tucto lo schudo, fa como dice la regola de sopra. Cioè che tu pigli la metà dell'una faccia, che è 10 b̃ʳ, et multipricha 10 via 17 e $\frac{3}{10}$ et uno pocho più, che fa 173 et uno pocho più ⟨b̃ʳ⟩ quadre. Et è facta. Ma chomo io dicho de sopra non appunto, perché non si po fare, perché radice non si po trovare appunto de 300. Et così se fanno tucte le simigliante ragioni.

22.21 A shield is by each face 20 *braccia*, as you see drawn opposite. I want to know how many *braccia* the mid-middle[a] will be, and how many *braccia* the whole shield will be. Do thus as I have said to you above, multiply one of the faces by itself, and say, 20 times 20 makes 400 and the fourth of 400 is 100. Detract it from 400, **(fol. 56ʳ)** 300 is left. Find the root of 300, which is 17 and $\frac{3}{10}$ and a little more, inasmuch as 17 and $\frac{3}{10}$ times 17 and $\frac{3}{10}$ makes 299 and $\frac{29}{100}$. But it cannot be found precisely. And as many *braccia* will the shield be by the straight in middle. That is, 17 and $\frac{3}{10}$, and a little bit more, as I say above. Now if you want to know how many square *braccia* the whole shield is, do as the above rule says. That is, that you seize the half of one face, which is 10 *braccia*, and multiply 10 times 17 and $\frac{3}{10}$ and a little more, which makes 173 square *braccia* and a little more. And it is done. But as I say above not precisely, because it cannot be done, because the root of 300 cannot be found precisely. And thus all similar computations are done.

---

[a] This clumsy translation (corresponding to another combination of *dericto de mezzo* and *mino longo*, cf. note [b] to 22.17) reflects the groping for a vocabulary for the mid-line of various symmetric figures.

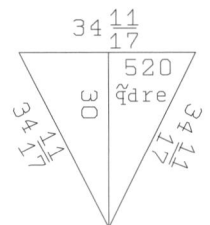

^{22.22} Uno schudo, como tu vedi designato de rinpecto, è tale che el dericto del mino lungho suo è 30 b̃^r. Voglio sapere quante b̃^r serà per ogni faccia, et quante b̃^r quadre serà tucto lo schudo. Fa così, multipricha como dice la regola, questa mesura per se medesima. Cioè 30 via 30, che fa 900. Ora piglia el terzo de 900, che è 300. Agiongi inseme, fa 1200. Et de questo numero trova la sua radice, che è 34 e $\frac{11}{17}$, et uno pocho meno, tanto quanto 34 e $\frac{11}{17}$ via 34 e $\frac{11}{17}$ fanno 1200 e $\frac{121}{289}$. Et 34 ^b̃^r^ e $\frac{11}{17}$ de braccio dirremo che el dicto schudo serrà per ogni faccia, che possiamo dire che sia 34 b̃^r e $\frac{2}{3}$ o pocho meno. Ora se voi sapere quante b̃^r quadre serà tucto lo schudo, sì piglia la mità dell'una dele faccie, che possiamo dire che sia 17 e $\frac{1}{3}$. Et multipricha contra ala lunghezzo del mino longho, cioè contra a 30. Et multipricha 30 via 17 e $\frac{1}{3}$, che fa 520. Et 520 b̃^r quadre dirremo che sia tucto lo schudo. Ma non appunto chomo dicho de sopra, perché non si po fare per la cagione sopradicta, che la radice non si po trovare appunto. Et per questa via et regola se fanno tucte le simigliante ragioni. Como de sopra io ho dicto.

^{22.22} A shield, as you see drawn opposite, is so that the straight (line) of its middle is 30 *braccia*. I want to know how many *braccia* it will be by each face, and how many square *braccia* the whole shield will be. Do thus, multiply, as the rule says, this measure by itself. That is, 30 times 30, which makes 900. Now seize the third of 900, which is 300. Join together, it makes 1200. And of this number find its root, which is 34 and $\frac{11}{17}$, and a little less, inasmuch as 34 and $\frac{11}{17}$ times 34 and $\frac{11}{17}$ make 1200 and $\frac{121}{289}$. And 34 braccia and $\frac{11}{17}$ of *braccio* shall we say that the said shield will be by each face, and we can say that it is 34 *braccia* and $\frac{2}{3}$ or a little less. Now if you want to know how many square *braccia* the whole shield will be, then seize the half of one of the faces, and we can say that it is 17 and $\frac{1}{3}$. And multiply against the length of the middle, that is, against 30. And multiply 30 times 17 and $\frac{1}{3}$, which makes 520. And 520 square *braccia* shall we say that the whole shield will be. But not precisely as I say above, because it cannot be done fare for the reason told above, that the root cannot be found precisely. And by this course and rule all the similar computations are done, as I have said above.

(fol. 57^r)

^{22.23} Se tu voi arrogere al tondo et farne quadrato et sapere quanto torna per faccia, et quante b̃^r quadre è tucto quello quadro,

^{22.23} If you want to adjoin to the round and make a squared (shape) out of it and know how much it becomes by face, and how

fa così. Quello tondo che è de rinpecto è tale che el dericto de mezzo del mino longho suo è 12 b̃ʳ, quante braccia serà ⟨el⟩ quadro. Respondo e dicho che se'l mino longho è 12 b̃ʳ, dico che vene a essere per ogni faccia 12 b̃ʳ. Et però multiplicha 12 via 12, fa 144. Et 144 braccia è tucto quello terreno. Et è facta, et sta bene.

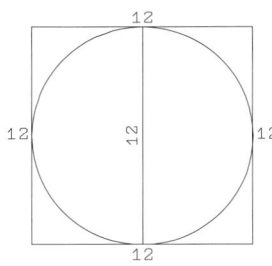

many square *braccia* this whole square is, do thus. That round which is opposite is so that the straight in middle of its middle is 12 *braccia*, how many *braccia* will ⟨the⟩ square be. I answer and say, that if the middle is 12 *braccia*, I say that it comes to be 12 *braccia* by each face. And therefore multiply 12 times 12, it makes 144. And 144 *braccia* is that whole terrain. And it is done, and it goes well.

22.24 Se tu volessi arrogere al quadro e farne uno tundo como tu vedi designato de rinpecto, et sapere quanto gira d'intorno, et quante b̃ʳ sia quadre, sì te'l mosterò testé. Io dico che quello quadro si è per ogni faccia 10 braccia. Vo' sapere quante b̃ʳ è el circhio dintorno, et quante b̃ʳ quadre fia tucto. Fa così, multipricha 10 via 10, fa 100. Ora radoppia, fa 200. Trova la sua radice de 200, che è 14 e $\frac{1}{7}$. Ma non è apponto ma uno pocho pocho meno, tanto quanto 14 e $\frac{1}{7}$ via 14 e $\frac{1}{7}$ fa 200 e $\frac{1}{49}$. Ora ài che la corda del quadro dall'uno canto al'altro è 14 b̃ʳ e $\frac{1}{7}$. Et altretanto è el dericto di mezzo di quel tondo. Ora se voli sapere quanto gira, sì multipricha 3 $\frac{1}{7}$ via {v̄} 14 e $\frac{1}{7}$, che fa 44 e $\frac{22}{49}$. Et cotanto gira dintorno el tondo. Ora se voli sapere quante b̃ʳ quadre è tucto, sì multipricha 14 e $\frac{1}{7}$ per se medesimo, fa 200 et uno pocho poco

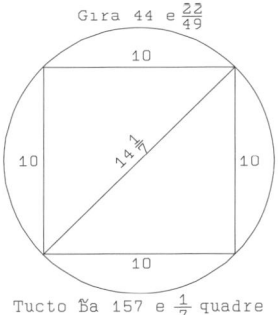

22.24 If you should want to adjoin to the square and make a round out of it as you see drawn opposite, and know how much it goes around, and how many square *braccia* it is, then I shall show you just now. I say that that square is 10 *braccia* by each face. I want to know how many *braccia* the circle around is, and how many square *braccia* the whole makes. Do thus, multiply 10 times 10, it makes 100. Now double, it makes 200. Find of 200 its root, which is 14 and $\frac{1}{7}$. But it is not precise but a little bit less, inasmuch as 14 and $\frac{1}{7}$ times 14 and $\frac{1}{7}$ makes 200 and $\frac{1}{49}$. Now you have that the cord of the square from one corner to the other is 14 *braccia* and $\frac{1}{7}$. And the same is the straight in middle of that round. Now if you want to know in how much it goes around, then multiply 3 $\frac{1}{7}$ times 14 and $\frac{1}{7}$, which makes 44 and $\frac{22}{49}$.

più. Ora traine el $\frac{1}{7}$ et la mità del $\frac{1}{7}$, che è 42 e $\frac{6}{7}$. Resta 157 e $\frac{1}{7}$. Et cotante $\tilde{b}^r$ quadre è tucto quello tundo. Et è facta. Et così se fanno le simigliante ragione.

And in as much goes around the round. Now if you want to know how many square *braccia* the whole is, then multiply 14 and $\frac{1}{7}$ by itself, it makes 200 and a little bit more. Now detract the $\frac{1}{7}$ and the half of the $\frac{1}{7}$, which is 42 and $\frac{6}{7}$. 157 and $\frac{1}{7}$ is left. And as many square *braccia* is that whole round. And it is done. And thus the similar computations are done.

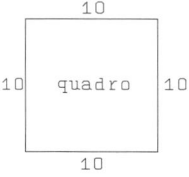

22.25 Se tu voli scemare del quadro, et farne tundo. Et de quello {quadro} tundo sapere quante $\tilde{b}^r$ gira dintorno. Et ancora quante $\tilde{b}^r$ quadre è tucto el dicto tundo, sì ti mostro la regola qui appresso. Et vo' dire chosì, uno quadro como tu vedi designato de rinpetto, è tale che per ogni faccia è 10 $\tilde{b}^r$. Ora io voglio del dicto quadro fare uno tondo, come tu vederai ancora designato nell'altro lato. Voglio sapere quante **(fol. 57ᵛ)** $\tilde{b}^r$ serà dintorno el dicto tondo. Et ancora quante $\tilde{b}^r$ quadre serà tucto. Fa così, tu di' che'l quadro è per ogni verso over faccia 10 $\tilde{b}^r$. Adunqua el dericto del tondo vene a essere ancho 10 $\tilde{b}^r$. Et però multipricha, como dice la regola, 10 via 3 e $\frac{1}{7}$, che fa 31 e $\frac{3}{7}$. Et $\tilde{b}^r$ 31 e $\frac{3}{7}$ gira dintorno el dicto tundo facto nel quadro. Ora se tu voli sapere quante $\tilde{b}^r$ quadre è egli, sì multipricha 10 per se medesimo, che fa 10 via 10, 100. Traine el $\frac{1}{7}$ et la mità del $\frac{1}{7}$, che è in tucto 21 e $\frac{3}{7}$. Trailo de 100, resta 78 e $\frac{4}{7}$. Et è facta. Et $\tilde{b}^r$ 78

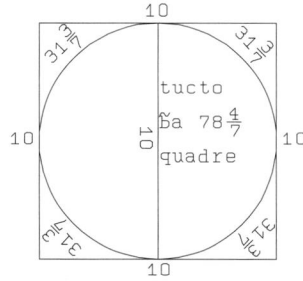

22.25 If you want to cut away from the square, and make from it a round. And about this round know in how many *braccia* it goes around. And further how many square *braccia* the whole said round is, then I show you the rule next to here. And I want to say thus, a square as you see drawn opposite, is such that by each face it is 10 *braccia*. Now I want of the said square to make a round, as you see again drawn to the other side. I want to know how many **(fol. 57ᵛ)** *braccia* the said round will be around. And further how many square *braccia* the whole will be. Do thus, you say that the square is in all directions or faces 10 *braccia*. Hence the straight of the round also comes to be 10 *braccia*. And therefore multiply, as the rule says, 10 times 3 and $\frac{1}{7}$, which makes 31 and $\frac{3}{7}$. And in *braccia* 31 and $\frac{3}{7}$ goes around the said round made

e $\frac{4}{7}$ quadre possiede el dicto tundo. Et così se fanno le simili ragioni.

from the square. Now if you want to know how many square *braccia* it is, then multiply 10 by itself, which makes 10 times 10, 100. Detract from it the $\frac{1}{7}$ and the half of the $\frac{1}{7}$, which in all is 21 and $\frac{3}{7}$. Detract it from 100, 78 and $\frac{4}{7}$ is left. And it is done. And 78 square *braccia* and $\frac{4}{7}$ does the said round occupy. And thus the similar computations are done.

22.26 Se tu volexi scemare del tundo e farne quadro, et sapere quante b̃ᵣ el dicto quadro torna per faccia, sì te'l mostrarò testé. Et vo' dire chosì, uno tundo como tu vedi designato de rinpetto è tale che'l

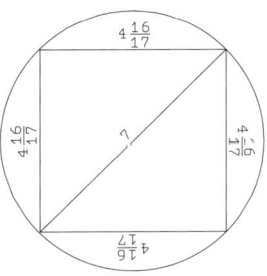

dericto de mezzo è 7 b̃ᵣ. Vo' sapere quante b̃ᵣ serà per faccia quello quadro che io vo' fare nel dicto tundo, chome tu vedi designato anchora da piede.[a] Fa così, tu di' che'l dericto de mezzo del tondo è 7 b̃ᵣ. Et però multipricha 7 via 7, fa 49. Dividilo per mezzo, che è $24\frac{1}{2}$. Ora trova la radice de $24\frac{1}{2}$, che è 4 e $\frac{16}{17}$ et uno poco pocho più, tanto quanto {è} 4 e $\frac{16}{17}$ via 4 e $\frac{16}{17}$ fa 24 et $\frac{120}{289}$. Et tu voli $24\frac{1}{2}$, che vi mancha $\frac{24\,e\,\frac{1}{2}}{289}$. Ma non si po trovare appunto. Et cotanto dirremo che'l dicto quadro torna per faccia. Cioè b̃ᵣ 4 e $\frac{16}{17}$. Et è facta. Et chosì se fanno le simiglianti ragioni.

22.26 If you should want to cut away from the round and make a square from it, and know how many *braccia* the said square becomes by face, then I shall show you just now. And I want to say thus, a round as you see drawn opposite is such that the straight in middle is 7 *braccia*. I want to know how many *braccia* by face that square will be that I want make in the said round, as you see drawn again at bottom.[a] Do thus, you say that the straight in middle of the round is 7 *braccia*. And therefore multiply 7 times 7, it makes 49. Split it in half, which is $24\frac{1}{2}$. Now find the root of $24\frac{1}{2}$, which is 4 and $\frac{16}{17}$ and a little bit more, inasmuch as {...} 4 and $\frac{16}{17}$ times 4 and $\frac{16}{17}$ makes 24 and $\frac{120}{289}$. And you want $24\frac{1}{2}$, to which $\frac{24\,and\,\frac{1}{2}}{289}$ is lacking. But it cannot be found precisely. And as much shall we say that the said square becomes by face. That is, *braccia* 4 and $\frac{16}{17}$. And it is done. And thus the similar computations are done.

[a] Actually, the diagram is in the right and not the lower margin. The formulation must refer to the organization of a parent manuscript.

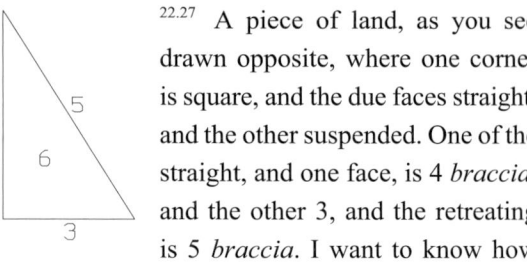

22.27 Uno pezzo de terra, como tu vedi designata de rinpetto, che l'uno canto è quadro, et le due faccie diricte, et l'altra pende. L'uno diricto, et l'una faccia è 4 b̃ʳ, et l'altra 3, et lo schifo è 5 b̃ʳ. Vo' sapere quante b̃ʳ è questo terreno in summa. Fa così, deli doi lati dericti multipricha **(fol. 58ʳ)** l'uno contra all'altro, et fa numero. Et quello numero dividi per mezzo. Et averai quante b̃ʳ quadre serà. Ora fa così, multipricha 3 via 4, fa 12. Dividi per mezzo, che ne vene 6. Et cotante b̃ʳ quadre è el terreno. Et è facto. Et così se fanno le similiante ragioni.

22.27 A piece of land, as you see drawn opposite, where one corner is square, and the due faces straight, and the other suspended. One of the straight, and one face, is 4 *braccia*, and the other 3, and the retreating is 5 *braccia*. I want to know how many *braccia* this area is in total. Do thus, of the two straight sides multiply **(fol. 58ʳ)** one against the other, and make number. And this number split in half. And you will get how many square *braccia* it will be. Now do thus, multiply 3 times 4, it makes 12. Split in half, from which results 6. And as many square *braccia* is the area. And it is done. And thus are done the similar computations.

22.28 Uno pozzo como tu vedi designato de rinpecto. È tondo a sesto, et è largho per lo mino longho suo b̃ʳ iij ½. Et è chupo b̃ʳ 32. El quale è pieno d'acqua. Et uno homo l'à tolto ad votare, et mettere l'acqua che ne cava in uno vinaio murato, el quale è quadro, per ogni faccia b̃ʳ 6. Et è alto assai tanto che l'acqua de quello pozzo non l'empie. Et de ogni b̃ʳ che illo cresce nel vinaio dè avere ß 40 per voitatura del pozzo. Vo' sapere quanti denari dè avere ⟨per⟩ votatura de quello pozzo. Fa chosì, sappi prima quante b̃ʳ gira dintorno el pozzo. Multiprica como dice la regola 3 1/7 via {ⅴ} 3 ½, che fa 11. Et 11 b̃ʳ gira d'intorno el pozzo. Ora sappi quante b̃ʳ è egli in tucto, che tu di' che egli è cupo b̃ʳ 32. Multiplica 11 via 3 ½, fa 38 e ½. Parti in 4, che ne

22.28 A well as you see drawn opposite. It is ambit-round, and it is *braccia* iij   broad by its middle. And it is *braccia* 32 deep. Which is full of water. And a man has undertaken to empty it, and to put the water which he removes in a walled basin, which is square, by each face *braccia* 6. And is sufficiently high so that the water from that well does not fill it up. And for each *braccio* that he raises (the water) in the basin he shall have *soldi* 40 for the emptying of the well. I want to know how much money he shall get ⟨for⟩ the emptying of that well. Do thus, know first in how many *braccia* the well goes around. Multiply as the rule says 3   times 3 , which makes 11. And in 11 *braccia* does the well go around. Now know how many *braccia* it is in all,

vene $9\frac{5}{8}$. Et b̃ʳ 9 e $\frac{5}{8}$ quadre
è el pozzo, cioè possiede. Ora
se voi sapere quante b̃ʳ è tucto,
sì multipricha 9 e $\frac{55}{88}$ via 32,
che fa 308. Et 308 b̃ʳ quadre
è tucto el pozzo. Et 308 b̃ʳ è
l'acqua che è nel pozzo. Ora
costui à voto el pozzo e messa
l'acqua nel vinaro. Vo' sapere
quante b̃ʳ è alta l'acqua dentro.
Sappi quante b̃ʳ è tucto el
vinaro, cioè per la larghezza e
per la lunghezza, el qual'è per
ogni faccia 6 b̃ʳ. Multipricha 6
via 6, fa 36. Et 36 b̃ʳ quadre
possciede el vinaro dintorno.
Ora parti 308 in 36, che ne
vene 8 e $\frac{5}{9}$. Et 8 b̃ʳ e $\frac{5}{9}$ de b̃ʳ
alza tucta questa acqua nel vi-
naro. Ora se voli sapere quanti
denari dè avere ad ragione de
ß 40 el braccio, multipricha
(fol. 58ᵛ) 8 e $\frac{5}{9}$ via 40 ß, che
fa 342 ß et 2 deñ et $\frac{2}{3}$, che sonno £ 17 ß
2 deñ 2 e $\frac{2}{3}$. Et cotanti denari dè avere
colui ⟨per⟩ votatura del dicto pozzo, a
metterlo nel dicto vinaio. Et è facta, et sta
bene. Et così se fanno le simili ragioni.

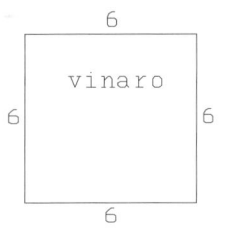

as you say that it is *braccia* 32
deep. Multiply 11 times $3\frac{1}{2}$, it
makes 38 and $\frac{1}{2}$. Divide in 4,
from which results $9\frac{5}{8}$. And
square *braccia* 9 and $\frac{5}{8}$ is the
well, that is, does it occupy.
Now if you want to know how
many *braccia* it is in all, then
multiply 9 and $\frac{5}{8}$ times 32,
which makes 308. And 308
square *braccia* is the whole
well. And 308 *braccia* is the
water that is in the well. Now
he has emptied the well and
put the water in the basin. I
want to know how many *braccia* the water is high inside it.
Know how many *braccia* the
whole basin is, that is, by the
breadth and by the length,
which by each face ⟨is⟩ 6
braccia. Multiply 6 times 6, it
makes 36. And 36 square
braccia does the basin occupy inside. Now
divide 308 in 36, from which results 8 and
$\frac{5}{9}$. And 8 *braccia* and $\frac{5}{9}$ of *braccio* does
all this water raise in the basin. Now if you
want to know how much money he shall
get at the rate of *soldi* 40 the *braccio*,
multiply *(fol. 58ᵛ)* 8 and $\frac{5}{9}$ times 40 *soldi*,
which makes 342 *soldi* and 2 *denari* and
$\frac{2}{3}$, which are *libre* 17, *soldi* 2, *denari* 2 and
$\frac{2}{3}$. And as much money shall he get ⟨for⟩
the emptying of the said well, and putting
it in the said basin. And it is done, and it
goes well. And thus the similar computations are done.

^{22.29} **E**gli è uno che fa conpagnia con un altro. Et costui mette in su la bottegha una quantità di denari. Et quando vengono in capo dell'anno costui se trova avere guadagnato el terzo de quello che mise de capitale. Et ancora lo mise in conpagnia sopra al capitale primo. Et poi in capo del secondo anno se trova avere guadagnato el quarto de ogni cosa, cioè che à in su la bottegha. Et anchora questo mette in su la bottegha sopra alli altri. Et poi in capo del terzo anno se trova avere guadagnato el quinto de ciò che à in su la bottegha. Et tra quello che vi misse de primo capitale et el guadagnio facto se trova avere in tucto in su la bottega f 1200. Vo' sapere quanti denari misse de prima in su la bottegha. Fa chosì chomo in molte altre ragioni adietro abbiamo facto. Questa conviene se faccia per positione. Cioè che trove uno numero nel quale sia $\frac{1}{3}$ e $\frac{1}{4}$ e $\frac{1}{5}$, et questo numero è 60. Ora tu di' che'l primo anno guadagnò el $\frac{1}{3}$ de quello che vi mise. Siché noi diremo vi mettesse f 60, guadagna el $\frac{1}{3}$, che è 20. Pollo sopra a esso, fa 80. Siché el secondo anno mette in conpagnia 80. Et tu di' che guadagnia el $\frac{1}{4}$, che vene a guadagnare 20. Pollo sopra a esso, fa 100. Siché'l terzo anno mette in conpagnia 100. Et tu di' che guadagnia el quinto, che vene a guadagnare ancho 20. Pollo sopra a esso, fa 120. Siché a questo modo se trovarebbe tra'l capitale e'l guadagno in capo de tre anni 120 f. Et noi diciamo che se trovò avere guadagniato f 1200. Et però diremo così, per 60 f che io me appongo me viene

^{22.29} There is one who makes partnership with another. And he puts into the shop an amount of money. And when they come to the end of the year, he finds to have gained the third of what he put in as capital. And again he put it in partnership in addition to the first capital. And then in the end of the second year he finds to have gained the fourth of everything, that is, of that which he put into the shop. And this again he puts into the shop in addition to the other (contributions). And then in the end of the third year he finds to have gained the fifth of that which he put into the shop. And with that which he put into it as the first capital and the gains he made he finds to have in all in the shop 1200 *fiorini*. I want to know how much money he put into the shop at first. Do thus as we have done in many other computations before. This one it suits to do by position. That is, that you find a number in which there is $\frac{1}{3}$ and $\frac{1}{4}$ and a $\frac{1}{5}$, and this number is 60. Now you say that in the first year he gained the $\frac{1}{3}$ of that which he put in. So that we shall say that he put in 60 *fiorini*, and gains the $\frac{1}{3}$, which is 20. Put it above that (amount), it makes 80. So that the second year he puts into partnership 80. And you say that he gains the $\frac{1}{4}$, then he comes to gain 20. Put it above that (amount), it makes 100. So that the third year he puts into partnership 100. And you say that he gains the fifth, then he comes to gain again 20. Put it above that (amount), it makes 120. So that in this way capital and gains together in the end of

f 120. Et io voglio me vengha f 1200. Et però multipricha 60 via 1200, che fa 72000 de fiorini, parti in 120, che ne vene f 600. Et cotanto **(fol. 59^r)** mise di primo capitale in su la bottegha. Provala. El primo anno guadagniò el $\frac{1}{3}$, sonno 200 f. Ài che ebe f 800. El secondo guadagniò el $\frac{1}{4}$. Sonno ancho 200 f, ài che ebe f 1000. El terzo anno guadagnò el $\frac{1}{5}$, sonno ancho f 200. Ài che ebe in tucto f 1200. Et sta bene et è facta. Et così se fanno le simili ragioni.

three years would be found to be 120 *fiorini*. And we say that he found to have gained *fiorini* 1200. And therefore we shall say thus, for 60 *fiorini* which I posit, *fiorini* 120 results for me. And I want that 1200 *fiorini* result for me. And therefore multiply 60 times 1200, which makes 72000 of *fiorini*, divide in 120, from which *fiorini* 600 results. And as much **(fol. 59^r)** he put as first capital into the shop. Verify it: The first year he gained $\frac{1}{3}$, (which) are 200 *fiorini*. You get that he had *fiorini* 800. The second year he gained $\frac{1}{4}$. These are again 200 *fiorini*, you get that he had *fiorini* 1000. The third year he gained $\frac{1}{5}$, these are again *fiorini* 200, you get that he had in all *fiorini* 1200. And it goes well, and it is done. And thus the similar computations are done.

22.30 **I**o vo a uno mercato et porto certi deñ in borsa, et quando sonno gionto conpro alchuna cosa, et poi la rivendo. Et radoppio li mey deñ. Et do'nne per l'amore de Dio deñ 12. Et poi spendo quelli che me sonno remasi che ne conpro ancho alcuna cosa, et poi la revendo. Et ancho radoppio li mey deñ. Et poi ne do ancho per l'amore de Dio deñ 12. Et poi de quello che me remane conpro ancho alcuna cosa, et poi la rivendo. Et ancho radoppio li mey deñ. Et ancho do per l'amore de Dio deñ 12. Et non me rimane nulla. Vo' sapere con quanti deñ me partì da casa quando andai al merchato. Fa così, tu di' che la terza volta desti per l'amore de Dio xij deñ, che tu di' trovasti

22.30 I go to a market and bring certain *denari* in my purse, and when I have arrived I buy some things, and then I resell them. And I double my *denari*. And for love of God I give away of them *denari* 12. And then I spend those that have remained for me, buying again some things, and then reselling them. And again I double my *denari*. And then again for love of God I give away *denari* 12 of them. And then for those that remain for me I buy again some things, and then resell them. And again I double my *denari*. And again I give away for love of God *denari* 12. And nothing remains for me. I want to know with how many *denari* I left home when I went to the

che avivi radoppiato dela merchantia che tu conparasti. Et però, quando tu conp⟨r⟩asti la merchantia, la terza volta, avivi tu 6 deñ et non più. Et 12 desti per l'amore de Dio ancho la seconda volta. Siché tu ti trovasti quando avisti venduta la merchantia la seconda volta deñ 18, né più né meno. Et di' che venivi ad radoppiare quello che tu avivi speso. Siché quando la conparasti, spendesti deñ 9. Et 12 n'avivi dati per l'amore de Dio la prima volta, fa 21. Et 21 denaro pigliasti dela merchantia che tu conparasti la prima volta. Et di' che tu radoppiavi li deñ che tu portasti, siché non venisti a portare altro che deñ $10\frac{1}{2}$. Et così sta. Et se la voli provare, sì fa così, tu di' che la prima volta spendesti quelli che tu portasti, che forono deñ $10\frac{.}{.}$,[a] et poi revendesti quello ⟨che⟩ conparasti, et radoppiasti li deñ. Siché tu vendesti quella cosa deñ 21. Et de questi desti per l'amore de Dio deñ 12, restatene 9. Et poi la seconda volta spendesti quelli 9, et poi revendesti et radoppiasti, siché tu te trovasti deñ 18. Ed de quelli desti per Dio deñ 12, restòtene 6. Et quelli spendesti la terza volta. Et poi revendesti et radoppiasti, si**(fol. 59ᵛ)**ché tu te trovasti la terza volta deñ 12. Et quelli 12 desti per l'amore de Dio. Et non te campò nulla, como tu di'. Siché la dicta ragione sta bene. Et così se fanno le simili ragioni.

market. Do thus, you say that the third time you gave away for love of God xij *denari*, which you say that you found to have doubled by the goods that you bought. And therefore, when you bought these goods, the third time, you had 6 *denari* and no more. And 12 you gave away for love of God also the second time. So that you found yourself when you had sold the goods the second time with *denari* 18, neither more nor less. And you say that you had come to double that which you had spent. So that when you bought it, you spent *denari* 9. And 12 you had given for love of God the first time, it makes 21. And 21 *denaro* you seized from the goods that you bought the first time. And you say that you doubled the *denari* that you brought, so that you brought nothing different from *denari* $10\frac{1}{2}$. And thus it goes. And if you want to verify it, then do thus, you say that the first time you spent that which you brought, which were $10⟨\frac{1}{2}⟩$ *denari*, and then resold that ⟨which⟩ you bought, and doubled the *denari*. So that you sold this thing at *denari* 21. And of these you gave away for love of God *denari* 12, 9 being left for you. And then the second time you spent these 9, and then resold and doubled, so that you found yourself with *denari* 18. And of these you gave away for God *denari* 12, 6 were left for you. And these you spent the third time. And then you resold and doubled, so **(fol. 59ʳ)** that you found yourself the third time with *denari* 12. And those 12 you gave away for love of God. And you saved nothing, as you say. So that the said computation goes well. And thus the similar computations are done.

ᵃ The same writing of ½ is used in 16.17, fol. 42ʳ.

22.31 Trovame uno numero che'l $\frac{1}{3}$ e'l$\frac{1}{4}$ e'l $\frac{1}{6}$ sia 18. Fa chosì como in molte altre adietro abiamo dicto et facto. Et secondo che dice la regola, trova uno numero ⟨che abia⟩ $\frac{1}{3}$ $\frac{1}{4}$ $\frac{1}{6}$ ne numeri interi. Et de questi ce sonno assai, come sonno 12, 24, 36, 48 et 60, et molti altri. Ma togliamo pure el minore, per fare minore multiprichatione, et diciamo che sia 12. Ora piglia el $\frac{1}{3}$ de 12, che è 4, e'l $\frac{1}{4}$ de 12, che è 3, e'l $\frac{1}{6}$ de 12, che è 2. Agiongi inseme, fanno 9. Et tu voli faccia 18. Et però di' così, per 12 io me appongho me vene 9, et io voglio 18. Et però multipricha 12 via 18, che fa 216. Parti in 9, che ne vene 24. Et è facta. Et 24 è quello numero che'l $\frac{1}{3}$ è 8 e'l $\frac{1}{4}$ è 6 e'l$\frac{1}{6}$ è 4. Agiongi inseme, fa 18. Et sta bene. Et così se fanno le simigliante ragioni.

22.31 Find me a number of which the 1/3 and the $\frac{1}{4}$ and the $\frac{1}{6}$ is 18. Do thus as we have said and done earlier in many other (cases). And according to what the rule says, find a number ⟨which possesses⟩ $\frac{1}{3}$ $\frac{1}{4}$ $\frac{1}{6}$ in whole numbers. And of these there are sufficiently, as there are 12, 24, 36, 48 and 60, and many others. But let us once again grasp the smallest, so as to have a smaller multiplication, and let us say that it is 12. Now seize the $\frac{1}{3}$ of 12, which is 4, and the $\frac{1}{4}$ of 12, which is 3, and the$\frac{1}{6}$ of 12, which is 2. Join together, they make 9. And you want it to make 18. And therefore say thus, for 12 I posit 9 results for me, and I want 18. And therefore multiply 12 times 18, which makes 216. Divide in 9, from which 24 results. And it is done. And 24 is that number of which the $\frac{1}{3}$ is 8 and the $\frac{1}{4}$ is 6 and the $\frac{1}{6}$ is 4. Join together, it makes 18. And it goes well. And thus the similar computations are done.

22.32 Trovami uno numero che'l terzo e'l quarto multiprichato per 5 faccia 25. Fa così, ancora te convene fare como facesti de sopra. Trova uno numero che abia $\frac{1}{3}$ e $\frac{1}{4}$. Et questo numero anco è 12. Togli el $\frac{1}{3}$, che è 4, e'l $\frac{1}{4}$, che è 3. Agiongi inseme, fa 7. Multipricha per 5, fa 35. Et tu voli faccia 25. Et però di' così, per 12 che io me appongho me vene 7. Et io voglio 5. Multipricha 5 via 12, fa 60. Parti in 7, che

22.32 Find me a number of which the third and the fourth multiplied by 5 makes 25. Do thus, again it suits you to do as you did above, find a number that possesses $\frac{1}{3}$ and $\frac{1}{4}$. And this number again is 12. Grasp the $\frac{1}{3}$, which is 4, and the $\frac{1}{4}$, which is 3. Join together, it makes 7. Multiply by 5, it makes 35. And you want it to make 25. And therefore say thus, for 12 that I posit 7 results for me. And I want 5. Multiply 5

ne vene 8 e $\frac{4}{7}$. Et 8 e $\frac{4}{7}$ dirremo ⟨è quello numero⟩. Et de quello numero el terzo e'l quarto multiprichato per 5 fa 25. Provala. Togli el $\frac{1}{3}$ de 8 e $\frac{4}{7}$, che è 2 e $\frac{6}{7}$. Et togli el $\frac{1}{4}$, che è 2 e $\frac{1}{7}$. Agiongi inseme 2 e $\frac{6}{7}$, e 2 e $\frac{1}{7}$, fa 5 interi. Multipricha per 5, fa 25. Et sta bene. Et così se fanno le simigliante ragioni.

times 12, it makes 60. Divide in 7, from which results 8 and $\frac{4}{7}$. And 8 and $\frac{4}{7}$ we shall say ⟨that number to be⟩. And of that number the third and the fourth multiplied by 5 makes 25. Verify it. Grasp the $\frac{1}{3}$ of 8 and $\frac{4}{7}$, which is 2 and $\frac{6}{7}$. And grasp the $\frac{1}{4}$, which is 2 and $\frac{1}{7}$. Join together 2 and $\frac{6}{7}$, and 2 and $\frac{1}{7}$, it makes 5 integer. Multiply by 5, it makes 25. And it goes well. And thus are done the similar computations.

APPENDIX
THE REVISED VERSION, MILAN AND FLORENCE

Introduction

Since Annalisa Simi's transcription of **F** – published in a *Rapporto matematico* from the Dipartimento di Matematica of the Università di Siena – is accessible only to the happy few, I have prepared the following transcription of **M** with indication of all except the merely orthographic deviations of **F** from the wording of **M**.[366] Since I compare **M** not with the manuscript of **F** but with Simi's transcription, this can only be a "semi-critical" and not a critical edition in the strictest sense. However, comparison of Simi's transcription with a facsimile of one page from the manuscript that is included in her publication shows the transcription to be precise.[367] For all practical purposes, the semi-critical edition should thus be indistinguishable from a critical edition.

Since only two manuscripts are involved, I have preferred to avoid an unwieldy apparatus. Instead I use the following typographical keys in the text itself:

- Normal roman type indicates that the two manuscripts are identical.
- A word or passage in contour type followed by a subscript word or passage means that the former is in **M** and the latter in **F**, and that there is no reason to prefer the reading of **F** over that of **M** (nor, in most cases, that of **M** over that of **F**).[368]
- A superscript word or passage (normally enclosed in { } because it is superfluous) followed by a word or passage in italics means that the former is in **M** and the latter in **F**, and that the reading of **F** is to be preferred – either for internal reasons or, rarely, because a parallel passage in **V** shows this to be the original formulation. *Neglecting*

---

[366] I count as "merely orthographic" also seemingly conjugational differences in verbal forms (singular/plural, indicative/subjunctive) but *not* differences between finite and infinite verbal forms. The former are "seemingly conjugational" in the sense that one manuscript may contain an apparent singular and the other an apparent plural of the verb, whereas the subject is, for instance, plural in both. The scribe thus does not seem to have felt a real conjugational difference.

[367] I have only one doubt. In the (admittedly not very clear) facsimile, it is impossible to distinguish what Simi reads as *z* from a *c* (in the words *iscienzia, scienza, grazia, conoscenza* and *terzo*). I suppose this interpretation corresponds to a systematic choice on her part – which I find acceptable but would not share, given the vacillation in the orthography of the time between *c*, *ç* and *z*, and in view of the development of spellings like *science, ciencia, tierce, gracia* and *graça* in later French, Castilian, Catalan and Portuguese. Admittedly, splitting the same medieval grapheme into two letters *c* and *z* is not different in principle from the customary splitting of one medieval letter into *u* and *v* (which I *do* follow).

[368] Both manuscripts contain errors, for instance, skipped passages caused by repeated phrases. Above the level of orthography, however, **M** seems to be somewhat better than **F**. It thus seems reasonable to use **M** as exemplar.

all superscript and subscript, one thus essentially gets a text which is close to the common archetype for the two manuscripts.

– In order to give an impression of the level of orthographic discrepancy between the two manuscripts, the transcription of the first two folios indicate all orthographic variants, underlining words in **M** whose spelling differ from that of **F** and still giving the corresponding text of **F** in subsequent subscript. In the numerous cases where a geminated consonant in **F** corresponds to a single consonant in **M** I have merely underlined this consonant; an underlined word *e* means that the corresponding form in **F** is *et*. From fol. 3 onward, merely orthographic variants are no longer indicated.

– Both when referring to **M** and to **F**, "=" means that the corresponding manuscript has no counterpart of a word or a passage found in the other manuscript.

– "•" indicates that an *e* or *et* in **F** has no counterpart in **M** (this happens 131 times, whereas only 31 instances of *e* or *et* in **M** have no counterpart in **F**).

– When paragraphs start in **M** with a decorated initial or similar paragraph indication, the first letter is boldfaced.

For the numbering of paragraphs in **M**, I use those of my transcription of **V**; this should facilitate a comparison of these two manuscripts. Paragraphs that have no counterpart in **V** are assigned the number of the previous paragraph with an added letter A (and B if necessary); paragraphs that are displaced in **M** with respect to **V** are treated similarly, but the corresponding number in **V** is added in parenthesis.[369] For **F**, I indicate Simi's numbering. The former occurs as superscript before the paragraph, the latter as subscript, both preceded by the corresponding siglum. None of them are in the manuscripts. * after a paragraph number for **M** indicates that there is no corresponding paragraph separation in **V** (M.4.4+M.4.4* thus correspond to V.4.4). F.— means that the passage is present in **F** but not as a separate paragraph.

As in the transcription of **V**, I have retained the abbreviations for monetary and metrological units wherever such are used in used in **M**. £, ß and δ still stand for *lire*, *soldi* and *danari*, *f* stands for *fiorini*, Ǧ for *gienovini*, δᶻ for *onzie*, g̃ʳ for *grani*, mᵒ for *marchi* (or the corresponding singular forms); the *staio* and the alternative abbreviation for *denari* do not appear in **M+F**, and the *braccio* is not abbreviated in **M**. Since Simi's transcription expands all such symbols, I cannot indicate differences on this account; all cases where one manuscript has the unit before the numeral and the other the opposite order are pointed out, however. Even in this respect, the manuscripts are erratic.

Other abbreviations are expanded, word separations are normalized, and punctuation, accents and apostrophes are added as in my transcription of **V**. In cases where the manuscript adds *e* to a final vowel as indication of stress, this is however taken to replace

---

[369] That is, M.24.27A has no counterpart in **V** but follows after M.14.27, and is followed by M.14.27B. M.15.21A(20) follows after M.15.21 but corresponds to M.15.20.

the accent (thus *ae*, not *àe*, where modern Italian has *ha*, and *prestoe*, "lent", not *prestòe)*.

Some numerical tables and methods that might be obscure to a modern reader are explained, and errors in the numerical tables are indicated. For the rest, the commentary is integrated with the discussion of the contents of manuscript **V**, pp. 45–128.

All diagrams and illustrations are redrawn by me, following **M**. They are very similar to those of **F**.

Two observations can be made specifically about **M**. Firstly, this manuscript is written in Genoa, which explains its vacillation between Tuscan orthography – in all probability that of the shared archetype for **F** and **V** – and northern spellings, in particular a strongly reduced number of geminations.[370] It also explains the insertion of supplementary coin lists with Genovese and Lombardian coin – in part coins not known from other sources, in part even from mints that are otherwise unknown! (Lucia Travaini, personal communication).

Secondly, the pseudo-gemination in *cossa/chossa* and the spelling *onzia* (visible also in the abbreviation) shows clearly that **M**, rather than the Tuscan area, represents the region where the German *Rechenmeister*, with their *coß* and *unze*, took their inspiration.

---

[370] The almost constant spelling *cossa/chossa/cosse* (18 occurrences of 20) does not contradict this observation. According to [Rohlfs 1966: I, 284], the spelling of this word with -ss- was used in the northern areas in order to distinguish the unvoiced intervocalic *s* from the voiced sound, not in order to render phonetic gemination.

The Text

[1. Incipit and General Introduction]

(fol. 1ʳ)

M.1.1 [F.—] Incipit tractatus algorismi. Huius autem artis novem sunt species, silicet numeratio, additio, subtractio, mediatio, duplatio, multipricatio(multiplicatio), divisio, progresio et radicum extracio(extractio). Compilatus a magistro Jacobo de Florentia apud montem Pesulanum anno domini millessimo(millesimo) trecientessimo(trecentesimo) septimo in mense setembris(septembris).

M.1.2 [F.—] Con ciò sia chossa(cosa) che tute quele chosse(cose) le quali l'umana generatione di questo sechulo(secolo) sanno o posono sapere si sanno per due principali vie, le quali vie son(sono) queste: la prima si è senno et la seconda si è scientia(iscienzia). Et ciaschuna(ciascuna) di queste due via(vie) si à con(com) secho(seco) due gentili e nobili conpagnie:(compagnie:) l'una si è gratia(grazia) da Dio et l'altra si è conosentia(conoscenza) per rasone(ragione), e le compagnie de la scientia(scienza) si è l'una l'amarstramento(l'amaestramento) dele scriture, et l'altra si è intendimento chon(con) bono(buono) ingenio(ingegno). E secondamente che dicie(dice) la santa scritura, lo senno è'l(è il) più nobile tesoro(thesoro) che sia al mondo, e dovete sapere che Salamone, che fue(fu) quasi il più savio huomo di tuto il mondo, sì domandò in sua gioventudine al nostro signore(segnore) Gieso(Giesus) Cristo che'lli(ke li) desse senno, e'l nostro signore(segnore) sì =li disse che'l suo domando fue il più alto adomandamento che gli potesse avere domandato. Onde gli diede il terzo del senno d'Adamo, e questo senno fue per gratia(grazia) di Dio. Ancora dicie(dice) la santa scritura che tutti gli omini(huomini) che anche furo(fuoro) ad Dio non domandaro niuno più bello né più alto domandamento di quello, per cioe che tuti li boni(buoni) e li perfeti doni disciesono(discesono) da quelo domandamento. E vera chossa(cosa) è che l'uomo puote nominare lo senno e la scienzia(scienza), l'uno= senno naturale et l'altro scienzia accidentale. E dovete sapere che tuto cioe che gli omini(huomini) sanno naturalmente *e accidentalmente* si è che'l nostro padre à concieduto(conceduto) a(ad) sapere per la sua santissima **(fol. 1ᵛ)** vertù(virtù) e gratia(grazia) e misericordia. Dunque siamo noi debiti di rendere grazia a lui ch'è si dolcie(dolce) padre et signore(segnore) che ci à dato a conosciere(conoscere) chotanta(cotanta) sientia(scientia) a nostra utilitade. E noi al suo santissimo honore e ala sua santissima misericordia si cominciamo lo(il) nostro trattato, lo quall'è(quale è) detto algorismus. E sapiate che noi il chiamiamo algorismus perchè questa scientia(scienza) fu principalmente fatta in Arabia, e quelli che la trovoe(trovò) fu simigliantemente arabo. E l'arte è deta(detto) in lingua arabia algo e'l numero è deto rismus$_{\text{risimus}}$ e percioe$_{\text{perciò}}$ è deto algorismus. Lo quale algorismus distringnie$_{\text{distrigne}}$ cinque capitoli, li quali vi mostremo$_{\text{mosterremo}}$ manifestamente nel nostro tratato ordinatamente, seguiendo$_{\text{seguendo}}$ la materia sì come dimanda la deta scienza. Et cominciamo ad honore$_{\text{onore}}$ e riverenzia$_{\text{reverentia}}$ del nostro signore Gieso$_{\text{Geso}}$ Cristo et de la sua madre vergine =$_{\text{santa}}$ Maria e di tuta la corte celestiale et col'aiuto de nostri predecisori$_{\text{predecessori}}$ et a honore di tuti maestri et scolari di questa sciencia$_{\text{scienza}}$ e di qualunque altra buona persona vedesse e legiesse$_{\text{leggesse}}$ questo trattato divoto e ragionevolemente$_{\text{ragionevolmente}}$.

M.1.3
[F.—]   **Ora** mostreremo$_{\text{mosterremo}}$ la propietade de' sopra deti cinque capitolli$_{\text{capitoli}}$, secondo che dicie$_{\text{dice}}$ Boecio nel'arismetica$_{\text{arismetrica}}$ sua. Lo primo capitolo si è multipricare$_{\text{multiplicare}}$. Lo secondo capitolo =$_{\text{si è}}$ dividere. Lo terzo capitolo sono li numeri rotti. Lo quarto si sono le regole$_{\text{reghole}}$. Lo quinto capitolo si è il general$_{\text{generale}}$ intendimento che si trae de' deti quatro capitoli. E dovete sapere che questi cinque$_{\text{v.}}$ capitoli ànno i'lloro$_{\text{in loro}}$ molte divisioni et menbri. Sicome di multipricare di due o di tre o di quatro o di piue figure. Il dividere si è i$_{\text{di}}$ numeri sanny$_{\text{sani}}$ et roti. I rotti sono multipricare$_{\text{multiplicare}}$, dividere, giungere, sotrare et dire quant'è più l'uno rotto che l'altro, overo quant'è meno e quale, e =$_{\text{è}}$ di conosiergli$_{\text{conosciergli}}$ vegiendogli$_{\text{veggendogli}}$ scriti per figure. Le regole ànno i'lloro$_{\text{in loro}}$ **(fol. 2^{r})** molte maniere et intendimenti et sotilitadi, le quali udirette$_{\text{udirete}}$ ordinatamente seguendo la materia che deto avemo.

M.1.4
[F.—]   **Sicome** in questo tratato lo'nteleto e'l buono ingienio$_{\text{ingegno}}$ sì ci dona a sapere le$_{\text{la}}$ grande sottilitade delle profecie e dele filosofie$_{\text{philosofie}}$ et dele celestiali scriture e dele temporali et sì ci dona ancora a$_{\text{ad}}$ sapere più inanzi che per inteletto e per buono ingiegnio$_{\text{ingegno}}$ e sotile si fano gli omini$_{\text{huomini}}$ molte sperienze et conpilationi$_{\text{compilazion}}$ di tratati, li quali non fuoro$_{\text{furono}}$ ancora fati per altri huomini, e sanno fare nuovi artifici e argomenti di scriture che rafinano le cosse$_{\text{cose}}$ che fuoro$_{\text{furono}}$ fatte per li primay$_{\text{primai}}$ huomini. Dumque (sic), sicome deto avemo di sopra lo nostro trattato si chiama in lingua arabia algorismus$_{\text{alghorismis}}$, perciò dovemo scriver le dette fighure$_{\text{figure}}$ del deto algorismus$_{\text{algorismo}}$ secondo la costumanza deli$_{}$ arabi, perciò che furono trovatori di questa scienzia$_{\text{scientia}}$, cioè che noi dovemo scrivere ad ritroso et legiere$_{\text{leggere}}$ a dirito secondo noi, cioè a dire che ci dovemo cominciare a scrivare$_{\text{scrivere}}$ dal minore numero e legiere$_{\text{leggere}}$ dal magiore numaro$_{\text{numero}}$.

[2. Introduction of the Numerals and the Role of Zero]

M.2.1
[F.—] Queste sono le nostre figure dell'abaco, cho'le_{co'le} quali tu poy_{puoi} scrivere qualunque numaro_{numero} tu voli o di quantunque quantita_{quantitade} fosse. Figure del'arte vechia e de la nuova_{nuova}:^a

^aAs redrawn by Simi [1995: 8], the corresponding shapes in **F** are

M.2.2
[F.—] Ancora scriviamo qui di sotto chome_{come} lievano le dete figure e per cioè che s'intendano meglio et più apertamente si'lle scriviamo per figure et simigliantemente per letere perciò che senza_{sanza} alchuno maestero_{magisterio} huomo per se medesimo le possa intendere. E dovete sapere che'l zevero per se sollo_{solo} non significa nula ma ae_è potentia di fare significare.

[3. Tabulated Writing of Numbers, with Corresponding Roman or Semi-Roman Writings]

(fol. 2v)

9	8	7	6	5	4	3	2	1
viiij	viij	vij	vi	v	iiij°	iij	ij	1

14		13		12		11		10
xiiij°		xiij		xij		xi		x

20	19	18	17	16	15
xx	xviiij°	xviij	xvij	xvi	xv

25	24	23	22	21
xxv	xxviij°	xxiij	xxij	xxi

30	29	28	27	26
xxx	xxviiij°	xxviij	xxvij	xxvi

70	60	50	40
lxx	lx	l	xl

300	200	100	90	80
ccc	cc	c	lxxxx°	lxxx

700	600	500ᵃ	400
$\overset{c}{vij}$	$\overset{c}{vi}$	$\underset{c}{v}$	cccc

3000	2000	1000	900	800
mm	mm	m	$\overset{c}{viiij^{o}}$	$\overset{c}{viij}$

ᵃ **F** has the regular c above v.

(fol. 3ʳ)

7000	6000	5000	4000
$\overset{m}{vij}$	$\overset{m}{vi}$	$\overset{m}{v}$	$\overset{m}{iiij^{o}}$

20000	10000	9000	8000
$\overset{m}{xx}$	$\overset{m}{x}$	$\overset{m}{viiij}$	$\overset{m}{viij}$

60000	50000	40000	30000
$\overset{m}{lx}$	$\overset{m}{l}$	$\overset{m}{xl}$	$\overset{m}{xxx}$

100000	90000	80000	70000
$\overset{m}{c}$	$\overset{m}{lxxxx^{o}}$	$\overset{m}{lxxx}$	$\overset{m}{lxx}$

500000	400000	300000	200000
$\overset{c}{v}$	$\overset{m}{cccc}$	$\overset{m}{ccc}$	$\overset{m}{cc}$

1000000	900000	800000	700000
$\overset{m}{m}$	$\overset{m}{vcccc}$	$\overset{m}{vccc}$	$\overset{m}{vcc}$

135512	51121	3456	234
m c c xxxv v xij	m li cxxi	c mmm iij lvi	ccxxxiiij°

[4. Explanation and Exemplification of the Place-Value Principle]

M.4.1
[F.—] **D**ovete sapere che una solla figura, cioè figura d'uno, quand'e ll'è posta nel primo grado, significa uno, e quand'è posta in secondo grado significa dicie, et in terzo grado, significa ciento, et in quarto grado significa mille, e in quinto grado significa diciemilia, e in sesto grado significa cientomillia **(fol. 3ʳ)** e in settimo grado, significa mille migliata, e così aviene di qualunque figura tu poni in seguente luogho_ nel suo grado et, simigliante-mente, qualunque figura tu poni in seguente luogo significa diecie tanty che'lle figure che sono passate dinanzi secondo figura che fosse.

M.4.2
[F.—] Ancora il mostreremo per altro modo, e diremo, una solla figura, qualche figura fosse, lieva unitade, cioè ad dire da diecie in giuso. E quando fosseno due figure, la prima figura lieva dicine e la seconda unitade, legiendo per drito modo. E quando sono tre figure, la prima lieva cientinaia e la seconda lieva dicine et la terza lieva unitadi =leggendo per dritto modo. E quando sono quatro figure, la prima lieva migliaia e_ la seconda lieva centinaia e la terza dicine et la quarta unitadi. E quando sono cinque figure, la prima lieva dicine di migliaia e la seconda unitadi di migliaia et la terza centinaia e la quarta dicine et la quinta unitadi. E quando sono sey figure, la prima lieva centinaia di migliaia e la seconda lieva dicine di migliaia e la terza e la quarta e la quinta e la sesta lievano come detto avemo di sopra. E quando sono sete figure, la prima lieva migliaia di migliaia e'lla seconda lieva centinaia di migliaia e'lla terza lieva dicine di migliaia e la quarta e la quinta e la sesta e'lla settima lievano nel modo che deto avemo di sopra per regolla.

M.4.3
[F.—] Et dovete sapere che la figura rapresenta nel primo grado tante unitadi quant'è la figura medessima e nel secondo grado rapresenta la figura tante dicine quant'è la figura medessima e nel terzo grado rapresenta la figura tante centinaia quant'è la figura medessima e nel quarto grado rapresenta la figura tante migliaia quant'è la figura medessima e nel quinto grado rapresenta la figura tante dicine di migliaia quant'è **(fol. 4ʳ)** la figura medessima e nel sesto grado rapresenta la figura tante centinaia di migliaia quant'è la figura medessima e nel septimo grado rapresenta la figura tante migliaia di

migliaia quant'è la figura medessima. Et per ogni figura che cresce ef rappresenta dicie tanti che quelle che sono passate inanzi per regolla.

M.4.4
[F.—]   Avemo deto del rapresentamento dele figure da una figura infino_insino in 7. Ora diremo da sete infino in ifinito_infinito.

M.4.4*
[F.—]   E diremo chosie, dovete sapere che'lle figure da una infino_insino in 7_septe rapresentano ciaschuna per sua regolla, ma da sete in ifinito modo rapresentano et lievano tute per una regola. In questo modo scrivi le figure che tu vogli levare per ordine et poi ti comincia_ricomin-_cia da la parte del'unitadi e conta sete figure et poi fae un ponto e queste sete figure rapresentano sicome deto avemo di sopra e poi fae un ponto a ogni quatro figure, e tanti ponti quanti trovaray tante migliaia di migliaia ti rapresentarano. Et questo sarae_farae secondamente che figura fosse e in qua⟨l⟩ luogho fosse posta.

M.4.5
[F.—]   Et in ciò diremo uno asempro. Dimmi quanto lievano queste nove figure: 987654321. Sapie che lievano noveciento otanta sete migliaia di migliaia et seciento cinquantaquatro migliaia et treciento ventuno. Et per questa regola leverieno le più.

M.4.6
[F.—]   =_E Dimi quanto lievano queste quindici_xv figure chi ti mostro: 234567898765432. Sapie che lievano dumilia treciento quarantacinque migliaia di migliaia di migliaia et semilia seteciento otantanove migliaia di migliaia et oto_ottanta milia seteciento sesantacinque miglia et quatrociento trenta due. E per questo modo levereboro quante figure fosseno:

M.4.7
[F.—]   • Dovete sapere che la figura la quale si troverae nel primo **(fol. 4ʼ)** grado rapresenta se medessima, e questo s'intende, se la figura d'unitadi si troverae nel primo grado rapresenta uno, et se fosse figura di due rapresentarebe due et in questo modo rapresentano insino in nove. E la figura la quale si troverae nel secondo grado rapresenta tante dicine quante unitadi nel primo grado e simigliantemente "*la figura la quale si troverae* nel terzo grado rapresentarae tante centinaia quante unitadi la figura rapresenta nel primo grado. Et in simigliante modo, sicome deto avemo di sopra, rapresenta ciascuna figura nel suo grado per la deta regola.

M.4.7A
[F.—]   Et dicina o cientinaia o migliaia non si puote scrivere senza questo segnale .0. lo quale si chiama zevero e'llieva con uno in questo modo .10. di^e^cie e seguita in ifinito per la regola.

[6. Multiplication Tables]

2	2	4	3	10	30	8	10	80
3	3	9	⊏⊐	⊏⊐	⊏⊐	9	10	90
4	4	16	4	5	20	10	10	100
5	5	25	4	6	24	⊏⊐	⊏⊐	⊏⊐
6	6	36	4	7	28	⊏⊐	⊏⊐	⊏⊐
7	7	49	4	8	32	20	20	400
8	8	64	4	9	36	30	30	900
9	9	81	4	10	40	40	40	1600
10	10	100	⊏⊐	⊏⊐	⊏⊐	50	500	2500
⊏⊐	⊏⊐	⊏⊐	5	6	30	60	60	3600
2	3	6	5	7	35	70	70	4900
2	4	8	5	8	40	80	80	6400
2	5	10	5	9	45	90	90	8100

(fol. 5ᵛ)

a						200	700	140000
50	60	3000	90	100	9000	200	800	160000
50	70	3500	100	100	10000	200	900	180000
50	80	4000				200	1000	200000
300	400	120000				200	1000	200000
300	500	150000	500	600	300000			
300	600	180000	500	700	350000	800	900	720000
300	700	210000	500	800	400000	800	1000	800000
300	800	240000	500	900	450000	900	1000	900000
300	900	270000	500	1000	500000			
300	1000	300000						
			600	700	420000	7	8	56
400	500	200000	600	800	480000	7	80	560
400	600	240000	600	900	540000	7	800	5600
400	700	280000	600	1000	600000	7	8000	56000
400	800	320000				7	80000	560000
400	900	360000	700	800	560000	7	90000	630000
400	1000	400000	700	900	630000			

[a] In **F**, empty cases are still filled out by the drawing that was rendered ⊏⊐ in the previous table (actually ⊟, with variations like ⊞ when greater breadth is asked for).

(fol. 6ʳ)

2	11	22	11	100	110	12	80	960
3	11	33				12	90	1080
4	11	44	2	12	24	12	100	120
5	11	55	3	12	36			
6	11	66	4	12	48	2	13	26
7	11	77	5	12	60	3	13	39
8	11	88	6	12	72	4	13	52
9	11	99	7	12	84	5	13	65
10	11	110	8	12	96	6	13	78
			9	12	108	7	13	91
11	20	220	10	12	120	8	13	104
11	30	330				9	13	117
11	40	440	12	20	240	10	13	130
11	50	550	12	30	360			
11	60	660	12	40	480	13	20	260
11	70	770	12	50	600	13	30	390
11	80	880	12	60	720	13	40	520
11	90	990	12	70	840	13	50	650
13	60	780	14	20	280	8	15	120

(fol. 6ᵛ)

13	70	910	14	30	420	9	15	135
13	80	1040	14	40	560	10	15	150
13	90	1170	14	50	700			
13	100	1300	14	60	840	15	20	300
2			14	70	980	15	30	450
2	14	28	14	80	1120	15	40	600
3	14	42	14	90	1260	15	50	750
4	14	56	14	100	1400	15	60	900
5	14	70				15	70	1050
6	14	84	2	15	30	15	80	1200
7	14	98	3	15	45	15	90	1350
8	14	112	4	15	60	15	100	1500
9	14	126	5	15	75			
10	14	140	6	15	90	2	16	32
			7	15	105	3	16	48
4	16	64	2	17	34	17	100	1700
5	16	80	3	17	51			
6	16	96	4	17	68	2	18	36

(fol. 7ʳ)

7	16	112	5	17	85	3	18	54
8	16	128	6	17	102	4	18	72
9	16	144	7	27	119	5	18	90
10	16	160	8	17	136	6	18	108
			9	17	153	7	18	126
16	20	320	10	17	170	8	18	144
16	30	480				9	18	162
16	40	540	17	20	340	10	18	108
16	50	800	17	30	510			
16	60	960	17	40	680	18	20	360
16	70	1120	17	50	850	18	30	540
16	80	1280	17	60	1020	18	40	720
16	90	1440	17	70	1190	18	50	900
16	100	1600	17	80	1360	18	60	1080
			17	90	1530	18	70	1260
18	80	1440	19	60	1140	13	17	221
18	90	1620	19	70	1330	13	18	234

(fol. 7ᵛ)

18	100	1800	19	80	1520	13	19	247
			19	90	1710	13	20	260
2	19	38	19	100	1900			
3	19	57				14	15	210
4	19	76	12	13	146ᵃ	14	16	224
5	19	95	12	14	168	14	17	238
6	19	114	12	15	180	14	18	252
7	19	133	12	16	192	14	19	266
8	19	152	12	17	204	14	20	280
9	19	171	12	18	216			
10	19	190	12	19	228	15	16	240
			12	20	240	15	17	255
19	20	380				15	18	270
19	30	570	13	14	182	15	19	285
19	40	760	13	15	195	15	20	300
19	50	950	13	16	208			

ᵃ 4 is written heavily, replacing a correct 5 which can just be distinguished. **F** has 156.

[7. Tables of Higher Squares and Products]

(fol. 8ʳ)

256	225	196	169	144	121
4 | 16 16	0 | 15 15	7 | 14 14	7 | 13 13	0 | 12 12	4 | 11 11
529	484	441	400	361	324
7 | 23 23	7 | 22 22	0 | 21 21	4 | 20 20	1 | 19 19	0 | 18 18[a]
441[b]	784	729	676	625	576
4 | 29 29	1 | 28 28	0 | 27 27	1 | 26 26	4 | 25 25	0 | 24 24
1225	1156	1089	1024	961	900
1 | 35 35	4 | 34 34	0 | 33 33	7 | 32 32	7 | 31 31	0 | 30 30
1681	1600	1521	1444	1369	1296
7 | 41 41	7 | 40 40	0 | 39 39	4 | 38 38	1 | 37 37	0 | 36 36
2209	2116	2025	1936	1849	1764
4 | 47 47	1 | 46 46	0 | 45 45	1 | 44 44	4 | 43 43	0 | 42 42
2809	2704	2601	2500	2401	2304
1 | 53 53	4 | 52 52	0 | 51 51	7 | 50 50	7 | 49 49	0 | 48 48
3481	3364	3249	3136	3025	2916
7 | 59 59	7 | 58 58	0 | 57 57	4 | 56 56	1 | 55 55	0 | 54 54

[a] As we notice, 17×17 has been left out. Similarly in **F**.
[b] **F** has the correct 841.

(fol. 8ᵛ)

4225	4096	3969	3844	2721[a]	3600
4 · 65/65	1 · 64/64	0 · 63/63	4[b] · 62/62	4 · 61/61	0 · 60/60
5041	4900	4761	4624	4489	4356
1 · 71/71	4 · 70/70	0 · 69/69	7 · 68/68	7 · 67/67	0 · 66/66
5929	5776	5625	5476	5329	5184
7 · 77/77	7 · 76/76	0 · 75/75	4 · 74/74	1 · 73/73	0 · 72/72
6889	6724	6561	6400	6241	6084
4 · 83/83	1 · 82/82	0 · 81/81	1 · 80/80	4 · 79/79	0 · 78/78
7921	7744	7569	7396	7225	7050[c]
1 · 89/89	4 · 88/88	0 · 87/87	7 · 86/86	7 · 85/85	0 · 84/84
9025	8836	8649	8464	8281	8100
7 · 95/95	7 · 94/94	0 · 93/93	4 · 92/92	1 · 91/91	0 · 90/90
276	242	9801	9604	9409	9216
6 · 12/23	8 · 11/22	0 · 99/99	1 · 98/98	4 · 97/97	0 · 96/96
522	476	432	390	350	312
0 · 18/29	8 · 17/28	0 · 16/27	3 · 15/26	8 · 14/25	6 · 13/24

[a] **F** has the correct 3721.
[b] **F** has the correct 1.
[c] The scribe probably misread a 6 with a weak upward stroke in his original. **F** has the correct 7056.

(fol. 9ʳ)

840	782	726	672	620	570
3 24 35	8 23 34	6 22 33	6 21 32	8 20 31	3 19 30
1230	1160	1092	1026	962	900
6 30 41	8 29 40	3 28 39	0 27 38	8 26 37	0 25 36
1692	1610	1530	1452	1376	1302
0 36 47	8 35 46	0 34 45	3 33 44	8 32 43	6 31 42
2226	2132	2040	1950	1862	1776
3 42 53	8 41 52	6 40 51	6 39 50	8 38 49	3 37 48
2832	2726	2622	2520	2420	2322
6 48 59	8 47 58	3 46 57	0 45 56	8 44 55	0 43 54
3510	3392	3276	3162	3050	2940
0 54 65	8 53 64	0 52 63	3 51 62	8 50 61	6 49 60
4260	4130	4002	3876	3752	3630
3 60 71	8 59 70	6 58 69	6 57 68	8 56 67	3 55 66
5082	4940	4800[a]	4662[b]	4526	4392
6 66 77	8 65 76	3 64 75	0 63 74	8 62 73	0 61 72

[a] **F** has an erroneous 4840.
[b] **F** has an erroneous 4602.

(fol.9ᵛ)

5976	5822	5670	5520	5372	5226
72 0 | 83	71 8 | 82	70 0 | 81	69 3 | 80	68 8 | 79	67 6 | 78
6942	6776	6612	6450	6290	6132
78 3 | 89	77 8 | 88	76 6 | 87	75 6 | 86	74 8 | 85	73 3 | 84
7980	7802	7626	7452	7280	7110
84 6 | 95	83 8 | 94	82 3 | 93	81 0 | 92	80 8 | 91	79 0 | 90
51376	50625	8712	8526	8342	8160
76 4 | 676	75 0 | 675	88 0 | 99	87 3 | 98	86 8 | 97	85 6 | 96
64124	63261	62400	53641	52884	52129
82 8 | 782	81 0 | 781	80 3 | 780	79 1 | 679	78 0 | 678	77 1 | 677
69344	68469	67596	66725	65856	64989
88 8 | 788	87 6 | 787	86 6 | 786	85 8 | 785	84 3 | 784	83 0 | 783
84036	83049	82064	81081	80100	70221
94 3 | 894	93 6 | 893	92 2 | 892	91 0 | 891	90 0 | 890	89 3 | 789
98901	89001	88004	87009	86016	85025
99 0 | 999	99 0 | 899	98 2 | 898	97 6 | 897	96 3 | 896	95 2 | 895

(fol. 10ʳ)

(lost, from **F**, fol. 10ᵛ)

209764		208849		207936		207025		206116		205209	
	458		457		456		455		454		453
1	458	4	457	0	456	7	455	7	454	0	453
318096		316969		315844		314721		313600		210681	
	564		563		562		561		560		459
0	564	7	563	7	562	0	561	4	560	0	459
448900		323761		322624		321489		320356		319225	
	670		569		568		567		566		565
7	670	4	569	1	568	0	567	1	566	0ᵃ	565
456976		455625		454276		452929		451584		450241	
	676		675		674		673		672		671
1	676	0	675	1	674	4	673	0	672	7	671
611524		609961		608400		461041		459684		458329	
	782		781		780		679		678		677
1	782	4	781	0	780	0ᵇ	679	0	678	4	677
620944		619369		617796		616225		614656		613089	
	788		787		786		785		784		783
7	788	7	787	0	786	4	785	1	784	0	783
799236		797449		795664		783881		792100		622521	
	894		893		892		891		890		789
0	894	4	893	1	892	0	891	1	890	0	789
998001		808201ᶜ		806404		804609		802816		801025	
	999		899		898		897		896		895
0	999	1	899	4	898	0	897	7	896	7	895

ᵃ Error for 4.
ᵇ Error for 7.
ᶜ Apparently corrected from a mistaken 808291.

(fol. 10ᵛ)

(lost, from **F**, fol. 11ʳ)

155722	154926	154132	153340	107505ᵃ	83062
343; 4 | 454	342; 0 | 453	341; 7 | 452	340; 0ᵇ | 451	239; 0 | 450	238; 1 | 349
195440	159732	159926ᶜ	158122	157320	156520
349; 5 | 560	348; 0 | 459	347; 4 | 458	346; 1 | 457	345; 0 | 456	344; 1 | 455
257530	256510	255492	254476	253462	252450
455; 4 | 566	454; 1 | 565	453; 0 | 564	452; 1 | 563	451; 4 | 562	450; 0 | 561
376992	375760	307530	260602	259576	258552
561; 0 | 672	560; 1 | 671	459; 0 | 670	458; 7 | 569	457; 7 | 568	456; 0 | 567
384426	383182	381940	380700	379462	378226
567; 0 | 678	566; 7 | 677	565; 7 | 676	564; 0 | 675	563; 4 | 674	562; 1 | 673
527632	526176	524722	523270	443820	385672
673; 7 | 784	672; 0 | 783	671; 4 | 782	670; 1 | 781	569; 3 | 780	568; 4 | 679
604310	534942	533476	532012	530550	529090
679; 5 | 890	678; 0 | 789	677; 1 | 788	676; 4 | 787	675; 0 | 786	674; 7 | 785
703360	701680	700002	698326	696652	694980
785; 1 | 896	784; 4 | 895	783; 0 | 894	782; 7 | 893	781; 7 | 892	780; 0 | 891

ᵃ Error for 107550.
ᵇ Error for 7.
ᶜ Error for 158926.

(fol. 11ʳ)

(lost, from **F**, fol. 11ᵛ)

3175760	2602530	3092603	2087576	2082552
560	459	458	457	456
2 5671	0 5670	3 4569	8 4568	6 4567
3206940	3200700	3194462	8188226[a]	3181992
565	564	563	562	561
6 5676	3 5675	2 5674	3 5673	6 5672
4543270	3857820	3225672	3219426	3213182
670	569	568	567	566
7 6781	6 6780	0 5679	0 5678	2 5677
4580550	4573090	4565632	4558176	4550722
675	674	673	672	671
0 6786	1 6785	4 6784	0 6783	7 6782
6154980	5357310	4602942	4595476	4588012
780	679	678	677	676
6 7891	6 7890	0 6789	4 6788	1 6787
6198360	6186680[b]	6181002	6172326	6163652
785	784	783	782	781
6 7896	2 7895	0 7894	0 7893	2 7892
8879112	7991112	6224412	6215726	6207042
888	888	788	787	786
0 9999	3 8999	3 7899	2 7898	3 7897

[a] Error for 3188226.
[b] Error for 6189680.

(fol. 11ᵛ)

(lost, from **F**, fol. 12ʳ)

43151761	43138624	43125489	43112356	43099225
1 6569 / 6569	4 6568 / 6568	0 6567 / 6567	7 6566 / 6566	7 6565 / 6565
58890276	58874929	58859584	58844241	43164900[a]
0 7674 / 7674	7 7673 / 7673	7 7672 / 7672	0 7671 / 7671	0 7570 / 7570
58967041	58951684	58936329	58920976	58905625
4 7679 / 7679	1 7678 / 7678	0 7677 / 7677	1 7676 / 7676	7[b] 7675 / 7675
77158656	77141099[c]	77123524	77105961	58982400
0 8784 / 8784	1 8783 / 8783	4 8782 / 8782	0 8781 / 8781	0 7680 / 7680
51057	51048	41139	41130	41121
0 9 / 5673	0 9 / 5672	0 9 / 4571	0 9 / 4570	0 9 / 4569
51102	51093	51084	51975	51066
0 9 / 5678	0 9 / 5677	0 9 / 5676	0 9 / 5675	0 9 / 5674
611055	611046	611037	611028	611019
0 9 / 67895	0 9 / 67894	0 9 / 67893	0 9 / 67892	0 9 / 67891

[a] This is actually 6570×6570, to which also corresponds the remainder 0 mod 9.
[b] Error for 4.
[c] Error for 77141089.

(fol. 12ʳ)

1681		1369		529	
41	6 —	37	2	23	2
1 · $\frac{2}{3}$ 13	4 · $\frac{1}{3}$ 12	4^a · $\frac{1}{2}$ 11			
41	6	37	2	23	2
$\frac{2}{3}$ 13	$\frac{1}{3}$ 12	$\frac{1}{2}$ 11			
$\frac{7}{9}$ 186	$\frac{1}{9}$ 152	1^b $\frac{1}{4}$ 132			

4489		961		3249	
67	3	31 —	3	57	1
2 · $\frac{3}{4}$ 16	2 · 1^c $\frac{1}{2}$ 15^d	1 · $\frac{1}{4}$ 14^e			
67	3	31	3	57	1
$\frac{3}{4}$ 16	$\frac{1}{2}$ 15	$\frac{1}{4}$ 14			
$\frac{9}{16}$ 280	$\frac{1}{4}$ 240	$\frac{1}{16}$ 203			

9604		8464		7396	
98	0	92	1	86	2
0 · $\frac{3}{5}$ 19	1 · $\frac{2}{5}$ 18	4 · $\frac{1}{5}$ 17			
98	0	92	1	86	2
$\frac{3}{5}$ 19	$\frac{2}{5}$ 18	$\frac{1}{5}$ 17			
$\frac{4}{25}$ 384	$\frac{14}{25}$ 338	$\frac{21}{25}$ 295			

4489		16129		10816	
67	4	127	1	104	6
2 · $\frac{1}{3}$ 22	1 · 1^f $\frac{1}{6}$ 16 21	1 · $\frac{4}{5}$ 20			
67	4	127	1	104	6
$\frac{1}{3}$ 22	1^g $\frac{1}{6}$ 127 21	$\frac{4}{5}$ 20			
$\frac{7}{9}$ 498	$\frac{1}{36}$ 448	$\frac{16}{25}$ 432			

[Explanation: These schemes, unexplained in **F** as well as **M**, show how to multiply mixed numbers and how to cast out sevens. The calculations in the middle of the lower row of ol. 13ʳ may serve as an example. They concern the multiplication $82\frac{1}{3} \times 93\frac{3}{4}$ (we observe that the fractions are written to the left of the integer part, in the "Arabic" way, and with a separation that suggests that the scribe did not understand that mixed numbers were meant; **F** is similar). Since $82\frac{1}{3} = {}^{247}\!/_3$ and $93\frac{3}{4} = {}^{375}\!/_4$, 247 and 375 are written above the mixed numbers, and their remainders modulo 7 (2 respectively 4) in the two small boxes to the right. The product $247 \times 375 = 92625$ is written on top of everything, and its remainder modulo 7 (1) in the upper small box to the left; as requested, $2 \times 4 \equiv 1$ modulo

7. Division of 92625 by 3×4 gives 7718$\frac{9}{12}$, which is shown at bottom, with the fraction reduced to $\frac{3}{4}$ and enclosed in the lower left box.]

[a] Missing in **F**.
[b] Missing in **F**.
[c] Missing in **F**.
[d] Missing in **F**.
[e] By error, **F** has $\frac{1}{4}$.
[f] By error, **F** has 16 instead of $\frac{1}{6}$.
[g] By error, **F** has 16 instead of $\frac{1}{6}$.

(fol. 12ᵛ)

24025			5476			2209		
155		1	74		4	47		5
1	$\frac{5}{6}$	25	2	$\frac{2}{3}$	24	4	$\frac{1}{2}$	23
155		1	74		4	47		5
	$\frac{5}{6}$	25		$\frac{2}{3}$	24		$\frac{1}{2}$	23
$\frac{13}{36}$	667		$\frac{4}{9}$	608		$\frac{1}{2}$[a]	552	
3960[b]			3⟨6⟩481			33489		
199		3	191		2	183		1
2	$\frac{3}{7}$	28	4	$\frac{2}{7}$	27	1	$\frac{1}{7}$	26
199		3	191		2	183		1
	$\frac{3}{7}$	28		$\frac{2}{7}$	27		$\frac{1}{7}$	26
$\frac{9}{49}$	808		$\frac{25}{49}$	744		$\frac{22}{49}$	683	
49729			46225			42849		
223		6	215		5	207		4
1[c]	$\frac{6}{7}$	31	4[d]	$\frac{5}{7}$	30	2	$\frac{4}{7}$	29
223		6	215		5	207		4
	$\frac{6}{7}$	31		$\frac{5}{7}$	30		$\frac{4}{7}$	29
$\frac{43}{49}$	1014		$\frac{18}{49}$	943		$\frac{23}{49}$	874	
75625			17689			66049		
275		2	133		0	257		5
4	$\frac{3}{8}$	34	0	$\frac{1}{4}$	33	4	$\frac{1}{8}$	32
275		2	133		0	257		5
	$\frac{3}{8}$	34		$\frac{1}{4}$	33		$\frac{1}{8}$	32
$\frac{41}{64}$	1181		$\frac{9}{16}$	1105		$\frac{17}{64}$[e]	1032	

[a] Error for $\frac{1}{4}$ (shared with **F**).
[b] Error for 39601 (shared with **F**).
[c] By error, **F** has 2.
[d] Missing in **F**.
[e] Error for $\frac{1}{64}$ (shared with **F**).

(fol. 13ʳ)

126511		74129		36239	
371	[0]	293	[6]	217	[0]
0 · $\frac{1}{5}$ · 74		6 · $\frac{1}{4}$ · 73		0 · $\frac{1}{3}$ · 72	
341	[5]	253	[1]	167	[6]
$\frac{1}{4}$ · 85		$\frac{1}{3}$ · 84		$1\frac{1}{2}$ᵃ · 83	
$11\frac{b}{20}$ · 6325		$\frac{5}{12}$ · 6177		$\frac{5}{6}$ · 6039	

380689		278759		194381	
617	[1]	533	[1]	451	[3]
1 · $\frac{1}{8}$ · 77		5 · $\frac{1}{7}$ · 76		5 · $\frac{1}{6}$ · 75	
617	[1]	523	[5]	431	[4]
$\frac{1}{7}$ · 88		$\frac{1}{6}$ · 87		$\frac{1}{5}$ · 86	
$\frac{1}{56}$ · 6798		$\frac{9}{14}$ · 6636ᶜ		$\frac{11}{30}$ · 6479	

88275		43259		501239	
321	[6]	239	[1]	703	[3]
5 · $\frac{1}{4}$ · 80		6 · $\frac{2}{3}$ · 79		4 · $\frac{1}{9}$ · 78	
275	[2]	181	[6]	713	[6]
$\frac{2}{3}$ · 91		$\frac{1}{2}$ · 90		$\frac{1}{8}$ · 89	
$\frac{1}{4}$ · 7356		$\frac{5}{6}$ · ~~7~~7209		$\frac{47}{72}$ · 6961	

118144		92625		90906	
416	[3]	247	[2]	327	[5]
5 · $\frac{1}{5}$ · 83		1 · $\frac{1}{3}$ · 82		4 · $\frac{3}{4}$ · 81	
284	[4]	375	[4]	278	[5]
$\frac{2}{3}$ · 94		$\frac{3}{4}$ · 93		$\frac{2}{3}$ · 92	
$\frac{4}{15}$ · 7874ᵈ		$\frac{3}{4}$ · 7718		$\frac{1}{2}$ · 7575	

ᵃ By error, **F** has $\frac{1}{3}$.

ᵇ By error, **F** has $\frac{11}{26}$.

ᶜ $6636\frac{9}{14}$ is the outcome of $278739 \div 42$. The result of the division $278759 \div 42$ is $6637\frac{5}{42}$. The error is shared by **F**.

ᵈ Error for 7876 (shared with **F**).

(fol. 13ᵛ)

690561		94864		204930	
831		308		506	
	[5]		[0]		[2]
4 $\frac{7}{8}$ 103	[0]	0 $\frac{2}{3}$ 102	[5] $\frac{1}{5}$	506 101	
831	[5]	308	[0]	405	[6]
$\frac{7}{8}$ 103		$\frac{2}{3}$ 102		$\frac{1}{4}$ 101	
$\frac{1}{64}$ 10790		$\frac{4}{9}$ 10540		$\frac{1}{2}$ 10246ᵃ	

3725030		1912689		1075369	
2719		1383		1037	
	[3]		[4]		[1]
1 $\frac{3}{4}$ 678ᵇ	[2] $\frac{3}{4}$	345	[1] $\frac{2}{5}$	207	
1370	[5]	1383	[4]	1037	[1]
$\frac{2}{3}$ 456		$\frac{3}{4}$ 345		$\frac{2}{5}$ 207	
$\frac{5}{6}$ 227085ᵈ	$\frac{1}{4}$ᶜ	119543	$\frac{19}{25}$	43014	

```
9 7 5 4 6 1 0 5 7 7 8 9 9 7 1 0 4 1


        9   8   7   6   5   4   3   2   1
    0   9   8   7   6   5   4   3   2   1


    8   8   8   8   8   8   8   8   8   9



                                    9
    0   9   8   7   6   5   4   3   2   1
```

ᵃ By erroneous anticipation, **F** has 1075369 already here (and again, correctly, in the next line).
ᵇ In **M** as well as **F**, the subsequent calculations run as if this number had been 679.
ᶜ Error for $\frac{1}{16}$ (shared with **F**).
ᵈ 227085$\frac{5}{6}$ is the outcome of 2725030 ÷12. The error is shared with **F**.

[8. Divisions *a regolo* and *a danda*]

(fol. 14ʳ)

		24816328			5152587	
	0	12408164		2	1030517	
	0	06204082		2	4206103	
	0	03102041		3	4841220	
	1	01551020		0	6968244	
2	0	50775510		4	1393648	5
	0	25387755		3	8278729	
	1	12693877		4	7655745	
	1	56346938		0	9531149	
	ᵃ	78173469		4	1906229	
	1	39086734		4	8381245	
	0	69543367		0	9676249	
		36912154			61829671	
	1	12304051		1	10304945	
	2	37434683		3	18384157	
	1	79144894		1	53064026	
	2	59714964		0	25510671	
3	2	86571654		3	04251778	6
	2	95523884		4	50708629	
	1	98507961		5	75118104	
	1	96169320ᵇ		2	95853017	
	2	55389106ᶜ		1	49308836	
	0	85129702		0	24884806	
	1	28376567		4	04147467	
	0	42792189		1	67357911	
		4816329			7142835	
	1	1204082		0	1020405	
	2	2801020		1	0145772	
	0	5700255		0	1449396	
	3	1425063		4	0207056	
4	3	7856265		1	5743865	7
	1	9464066		4	2249123	
	2	4866016		1ᵈ	6035586	
	0	6216504		0	2290798	
	0	1554126		6	0327256	
	2	0388531		3	8618179	
	3	5097132		5	5516882	

ᵃ The scribe has filled in this entire column first, forgetting the 1 that should be here and leaving no space for it; when filling out afterwards the adjacent column to the right without aligning corresponding numbers from the two columns he has not noticed the error. The subsequent schemes are more carefully made.
ᵇ Error for 66169320.

[c] This (with the remainder 2) is the outcome of the division 166167320÷3. The error is shared with **F**.

[d] Error for 0 (shared with **F**). Since the wrong number is used in the next step, a calculational, not merely a copying error is involved.

(fol. 14ᵛ)

	81632649			132639	
8					13
1	10204081		0	010203	
1	13775510		11	000784	
6	14221938		2	846214	
1ᵃ	76777742		7	218939	
6	22097217		0	555303	
1	77762152		8	042715	
0	22220269		5	618670	
5	02777533		5	432205	
5	62847191		12	417861	
7	70355898		1	955220	

	91827369			173451	
9					17
0	10203041		0	010203	
2	01133671		3	000600	
6	22348185		15	176505	
3	69149798		10	892735	
2	41016644		2	640749	
1	26779627		3	155338	
2	14086625		2	185608	
7	23787402		3	128565	
4	80420822		4	184033	
3	53380091		7	257884ᵇ	
5	39264454		6	426934	

	1122334			193857	
11					19
4	0102030		0	010203	
1	3645639		0	000537	
7	1240512		5	000028	
1	6492939ᶜ		7	263159	
1	1499358		10	382271	
2	1045396		6	546435	
10	1904126ᵈ		3	344548ᵉ	
5	9264011		16	176028	
4	5387637		17	851369	
9	4126148		14	939545	
			6	786281ᶠ	

ᵃ Calculation error for 2 (shared with **F**). Used in the next step.

ᵇ This number (including the remainder) is the outcome of the division 4384035÷17. It is shared with **F** and used in the next step.

ᶜ This result, shared with **F**, is the outcome of the division 71422330÷11; on its part, 1422330, remainder 7, is the outcome of the division 15645637÷11. It thus appears that either the original calculator reviewing his calculations or a copyist coming no later than the compiler of the archetype for **M+F** discovered that result of the division 13645639÷11 was wrong and corrected it, but did not change the subsequent numbers. Most likely is perhaps that this partial correction was made by the compiler who inserted the new higher multiplications and these new divisions *a regolo* in Jacopo's treatise.

ᵈ This number, shared with **F**, is the outcome of the division 20945396/11. If really made *a regolo*, the calculator forgot to carry the remainder of the division $100 \div 11$.

[e] This number including the remainder, shared with **F**, is the outcome of the division $654641/5 \div 19$.

[f] This number including the remainder, shared with **F**, is the outcome of the division $1493345 \div 19$.

(fol. 15^{r})

```
  5                         5          1                      16             26693360
                           ──                                ──       37 │987654321│1
                           23                                29            6 .  .  .
                                                                          38.  .  .
                                                                          14.  .  .
                                                                          247 .  .
                                                                          18  .  .
                                                                          ──
              2   12                                                       67  .  .
                                                                          42  .  .
            2311331                      13 26 16                          256 .  .
  2         16193115                1    31160156                          18  .  .
                                                                          ──
 23         987634321              29    987654321                         76  .  .
                                                                          42 .  .
  6          42941492               1    34057045                          ───
                                                                          345.  .
                                                                          27 .  .
  2                        1          4                      6             ──
                           ──                                ──            75.  .
                           37                                43            63.  .
                                                                          ───
                                                                          124 .
                                                                           9  .
                                                                          ──
            23112                        4233                              34  .
            2677344                      145552                            21  .
  2         34542320                1    12197022                          ───
                                                                          133 .
 37         987654321              43    987654321                          9  .
                                                                          ──
  1          26693360               4    22968705                          43  .
                                                                          21  .
                                                                          ───
  0                        10         4                      34            222.
                           ──                                ──            18  .
                           43                                79            ──
                                                                          42.
                                                                          42.
                                                                          ──
            312543                       3   142                           01
            45336763                     154 7276
  4         45386268                2    29918527

 53         987654321              79    987654321

  0          18634987               2    12501953
```

[Fol. 15^{r-v} shows examples of *danda* division, a method developed from a division algorithm used on the dust board but corresponding to the conditions of paper, where deletion and replacement was not possible. As an example we may discuss the division 987654321÷37 in the middle of the left column. The explanation may be correlated with the "extended long division" shown to the right of the scheme, in which numbers that are actually written during the *danda* division are italicized.

At first we notice that 37 divides 98 twice. In the dust-board algorithm we would therefore first replace 9 by 9–2×3 = 3. Instead of deleting and replacing we now write above; any digit above which another digit is written is thus to be understood as superseded, and we are left with a reduced dividend 387654321. From 38 we still need to subtract 2×7 = 14, which leaves 24 – once more written above and thus superseding 38. We are now left with a reduced dividend 247654321. Since 37 divides 247 six times, we subtract 6×3 = 18 from 24, leaving 06. 0 is omitted, 6 is written above 4, meaning that the dividend is now reduced to 67654321. From 67, 6×7 = 42 still has to be subtracted, leaving 25, which therefore supersedes 67 – leaving a reduced dividend 25654321. 37 again divides 256 6 times, for which reason 25 is first replaced by 25–6×3 = 7, and 76 next by 76–6×7 = 34. Thereby the dividend is reduced to 3454321 – etc. Stepwise, the quotients are written below the original dividend, giving in the end the integer part of the total quotient. The final remainder 1 occurs as $\frac{1}{37}$ in the box in the upper right corner.

The number 2 written in the box above the divisor 37 represents the remainder of the division 37÷7, and the number 1 written in the box to the left of the integer part of the quotient the remainder of this number. The product 2×1 = 1 of these two remainders is written in the upper left box. When added to the remainder (here 1), this should be equivalent modulo 7 to the remainder 3 of the division 987654321÷7, which indeed it is (since the dividend is the same in all examples, this remainder is not written).

The absence of any explanation makes one suspect that the compiler just copied these division schemes from elsewhere without understanding them. This suspicion is corroborated by the right-to-left organization of the preceding schemes and the right-to-left writing of mixed numbers in these, and also by the disagreement between the way mixed numbers are multiplied in fols 12^r–13^v and the rule given for the same matter in **M.11**.14 (a rule which again disagrees with the actual calculations when these go back to Jacopo's original treatise).]

(fol. 15ᵛ)

```
 5 |          54      | 1 |          30ᵃ
   |          83      |   |          97


        11                        1
        787348                    771
        17006598                  018990 5
     6  15428710            6
    83  987654321          97  987654321
     2  11899449            6  10182003


 3 |         126      | 1 |         114
   |         131      |   |         311


        1                         1212
        1    1                    212121
        23122                     222121
        252467                    237212
        7525683                   5383232
     5  27012508           3  05458969
   131  987654321         311  987654321
     2  7539345            5  3175737


 4 |         209      | 8 |        2334ᵇ
   |        1234      |   |        9011


                                  3
        1                         63
        12                        63
        102                       2663
        204                       2664
        820                       2464
        18434                     624643ᶜ
     2  12349518           2  08655760
  1234  987654321        9011  987654321
     2  800368            4  100727ᵈ
```

ᵃ **F** has an erroneous $^{39}/_{97}$.

ᵇ **F** instead has $^{3324}/_{9011}$. Both are of course wrong, the true result of the division is $109605\,^{3666}/_{9011}$. Both could have seen to be wrong if the check by casting out sevens had really been performed (they should leave a remainder 2), since $9011 \equiv 2$, and $100727 \equiv 4$, as correctly stated in both manuscripts.

ᶜ From this point on, things go awry for the calculator. The first digit in the row should be 5.

ᵈ This result, shared with **F**, should have been 109605.

[9. Graphic Schemes Illustrating the Arithmetic of Fractions]

(fol. 16ʳ)

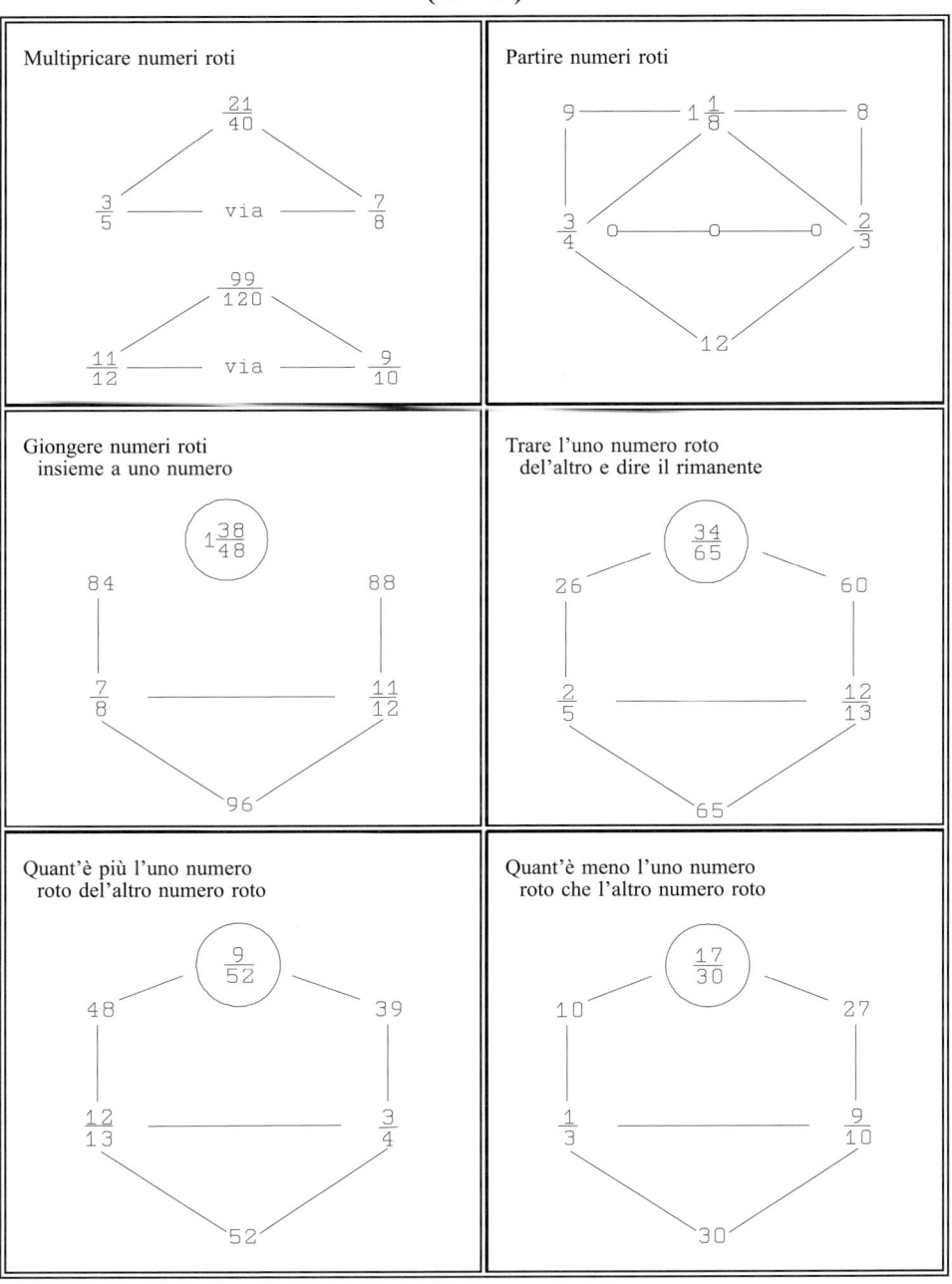

[10. Examples Explaining the Arithmetic of Fractions]

(fol. 16ᵛ)

M.10.1
[F.—] **A**vemo deto dele multipricationi e dele divisioni et di tuto quello che intorno a ciò di necesitade si convene. [F.—] Ora lasciamo questo et diremo per propria et $=_{per}$ legitima regola sopra tute maniere di numeri rotti, sicome proponemo dinanzi nel prolago, per cioe che danno grande arghomento al'altre ragioni e senza essi non si può fare sotilmente.

M.10.2
[F.iii.1] $=_E$ **P**rimamente incominciamo al nome del'altissimo Idio et diremo cosie, dimi quant'è gionto insieme $\frac{1}{2}$ et $\frac{1}{3}$. Fa cosie, die, mezo e terzo si trova in sey, percioe che 2 via 3 fanno 6. E prendi il mezo e'l terzo di 6, che sono 5, et parti 5 per 6, che ne viene 5 sesti$_{5/6}$. E diremo che $\frac{1}{2}$ mezzo e $\frac{1}{3}$ terzo sieno giunti insieme 5_{cinque} sesti. E in questa maniera poi giungere insieme qualunque numaro rotto fosse.

M.10.3
[F.iii.2] **D**imi quant'è gionto insieme $\frac{1}{3}$ et $\frac{1}{4}$. Di' cosie, $\frac{1}{3}$ et $\frac{1}{4}$ si trova in 12. Ora prendi il $\frac{1}{3}$ terzo e'l $\frac{1}{4}$ quarto di 12, ch'è 7, et parti ^7^ per in 12, che ne viene $\frac{7}{12}$. E diremo che $\frac{1}{3}$ et $\frac{1}{4}$ siano gionti insieme $7\,\frac{7}{12}$. E cossì fa di tuti gli altri numari.

M.10.3A
[F.iii.3] **D**imi quant'è giunto insieme $\frac{1}{4}$ et $\frac{1}{5}$. Anchora somigliantemente die, $\frac{1}{4}$ et $\frac{1}{5}$ si trova in 20, e prendi il $\frac{1}{4}$ di 20 e'l $\frac{1}{5}$ di 20, che sono 9. e parti 9 in 20, che ne viene $9\,\frac{9}{20}$, e tanto è gionto insieme $\frac{1}{4}$ e $\frac{1}{5}$, cioè nove ventesimi.

M.10.6
[F.—] **E**t sapie che quarto si scrive uno di sopra e quatro di sotto e una vergha in mezo, e chossì si fae per simiglianza d'altri numari.

M.10.7
[F.iii.4] **D**imi quanto son gionti insieme $\frac{2}{3}$ et $\frac{3}{4}$. Fa cosie, die, $\frac{2}{3}$ et $\frac{3}{4}$ si trova in 12. Ora prendi $\frac{2}{3}$ di 12, che sono 8, e prendi $\frac{3}{4}$ di 12, sono 9, et giungi insieme 8 e 9, sono 17, e parti 17 in 12, che ne viene uno sano et piu $\frac{5}{12}$, e diremo che **(fol. 17ʳ)** {che} $\frac{2}{3}$ et $\frac{3}{4}$ farano giunti insieme j et $\frac{5}{12}$·=

M.10.8
[F.iii.5] **D**imi quanto sono giunti insieme $\frac{4}{5}$ et $\frac{5}{6}$. Di' cosie, $\frac{4}{5}$ et $\frac{5}{6}$ si trova in 30. Ora prendi $\frac{4}{5}$ di 30, che sono 24, et prendi $\frac{5}{6}$ di 30, che sono 25, e giungi insieme 24 e 25, fano 49, et parti 49 per 30, che ne viene uno sano e $\frac{19}{30}$ 19 trentesimi· E diremo che $\frac{4}{5}$ e $\frac{5}{6}$ sono, giunti insieme, j$_{uno}$ sano et piùe $\frac{19}{30}$.

M.10.8A
[F.iii.6] **Dimi** quanto sono giunti insieme $\frac{7}{8}$ et $\frac{11}{12}$. Fa cosie, die, $\frac{7}{8}$ e $\frac{11}{12}$ si trova in 96. Ora prendi $\frac{7}{8}$ di 96, che sono 84, e prendi $\frac{11}{12}$ di 96, che sono 88. Ora giungi insieme 84 e 88, fanno 172, et parti 172 per 96, che ne viene uno e $\frac{38}{48}$, che sono $\frac{19}{24}$, e diremo che $\frac{7}{8}$ e $\frac{11}{12}$ sono, giunti insieme, j$_{uno}$ sano e piue $\frac{19}{24}$. Et in questo modo giungi insieme tuti i$_{-}$ numeri rotty.

M.10.9
[F.iii.7] **Ancora** diremo, giungi$_{giugnimi}$ insieme $\frac{1}{2}$ et $\frac{1}{3}$ e $\frac{1}{4}$. Fa così, die, $\frac{1}{2}$, • $\frac{1}{3}$, • $\frac{1}{4}$ si trova in 12. Ora prendi il mezo$_{1/2}$ di 12 ch'è 6, e prendi il $\frac{1}{3}$ di 12, $^{ch'à}$*ch'è* 4, e prendi il $\frac{1}{4}$ di 12, ch'è 3. Ora giungi insieme 6 e 4 e 3, che$_{-}$ fano 13, e$_{-}$ parti 13 per 12, che ne viene j$_{uno}$ sano e $\frac{1}{12}$. E tanto è giunto insieme $\frac{1}{2}$, • $\frac{1}{3}$, • $\frac{1}{4}$, cioè uno sano et più $\frac{1}{12}$, ed è fata.

M.10.9A(5)
[F.iii.8] **Dimmi** quant'è giunto insieme $\frac{1}{4}$ e $\frac{1}{5}$ e $\frac{1}{6}$ e$\frac{1}{10}$. Fa cosie, die, $\frac{1}{4}$ et $\frac{1}{5}$ et$\frac{1}{6}$ et $\frac{1}{10}$ si trova {si trova} in 60. Ora prendi il $\frac{1}{4}$ di 60, ch'è 15, et prendi il $\frac{1}{5}$ di 60, ch'è 12, e prendi il $\frac{1}{6}$ di 60, ch'è 10, et prendi il $\frac{1}{10}$ di 60, ch'è 6. Ora giungi insieme 15 et 12 et 10 et 6, che sono in tuto 43, e parti 43$_{-}$ per 60, che ne viene $\frac{43}{60}$. E tanto è giunto insieme $\frac{1}{4}$, • $\frac{1}{5}$, • $\frac{1}{6}$, • $\frac{1}{10}$, **(fol. 17ᵛ)** cioè $\frac{43}{60}$. Ed è fata, e in questo modo fae de simiglianti numeri.

M.10.10
[F.iii.9] **Ancora** diremo, giungi insieme $\frac{2}{3}$ et $\frac{3}{4}$ et $\frac{7}{8}$ et $\frac{5}{12}$. Fa così, die, $\frac{2}{3}$ et $\frac{3}{4}$ et$\frac{7}{8}$ et $\frac{5}{12}$ si trova in 24. Ora prendi $\frac{2}{3}$ di 24, sono 16, et prendi $\frac{3}{4}$ di 24, sono 18, et prendi $\frac{7}{8}$ di 24, sono 21, e prendi $\frac{5}{12}$ di 24, sono 10. Ora giungii insieme 16, • 18, • 21 e 10, che sono in tuto 65. Ora parti 65$_{-}$ per 24, che ne vene due sany et $^{\{17/65\}}$ $\frac{17}{24}$. Ed è fata, e tanto sono giunti insieme $\frac{2}{3}$ et $\frac{3}{4}$ et $\frac{7}{8}$ et $\frac{5}{12}$, cioè due sany e piue $^{\{17/65\}}$ $\frac{17}{24}$. E per questo modo e per questa regola giungi insieme tuti numeri roti di qualunque quantitade fosse.

M.10.11
[F.iii.10] **Avemo** deto del giungimento$_{giudicamento}$ di numeri rotti. Ora diremo di trare l'uno numaro rotto del'altro, e sapere quant'è il rimanente. Primamente diremo cosie, trami$\frac{2}{3}$ di $\frac{11}{12}$ e dimi quant'è il rimanente. Fa cosie, die: $\frac{2}{3}$ et $\frac{11}{12}$ si trova in 12. Ora die, $\frac{2}{3}$ di 12 sono 8, et $\frac{11}{12}$ di 12 sono 11. Ora die, da 8 infino$_{insino}$ in 11 si à 3. Dunque traendo$\frac{2}{3}$ di $\frac{11}{12}$ rimane $\frac{3}{12}$, cioè $\frac{1}{4}$, ed è fata. O più apertamente puoi dire, $\frac{2}{3}$ sono $\frac{8}{12}$. Dunque, traendo $\frac{8}{12}$ di $\frac{11}{12}$ rimane $\frac{3}{12}$. È tuto a$_{-}$ uno modo.

M.10.12
[F.iii.11] **Trami** $\frac{2}{7}$ di $\frac{9}{11}$ et dimi quant'è il rimanente. Fa cosie, die: $\frac{2}{7}$ e $\frac{9}{11}$ si trova in 77. Ora prendi $\frac{2}{7}$ di 77$_{-}$, sono 22, e die che $\frac{2}{7}$ sieno $\frac{22}{77}$. Ora simigliantemente prendi $\frac{9}{11}$ di 77, che sono 63, et die che $\frac{9}{11}$ sieno $\frac{63}{77}$. Or ày questi roti recati a setantasetesimi. Ora trai 22$_{-}$ di 63, rimane 41, e questi 41 sono setantasetesimi. Dunque traendo $\frac{2}{7}$ di $\frac{9}{11}$ rimane $\frac{41}{77}$, e per questa regola tray $^{\{i\ roti\ tuti\}}$*tutti i rotti* l'uno del'altro.

M.10.13
[F.iii.12]　Anchora diremo, trami $\frac{1}{13}$ di $\frac{1}{7}$ e dimi quant'è il rimanente. Fa cosie, die, $\frac{1}{13}$
e $\frac{1}{7}$ si trova in 91. Ora prendi il $\frac{1}{13}$ **(fol. 18ʳ)** di 91, sono 7, e prendi il $\frac{1}{7}$ di 91, sono
13. Ora die che j tredecisimo$_{1/13}$ sia $\frac{7}{91}$ e die che j$_{uno}$ setimo sia $\frac{13}{91}$. Ora $^{\{ay\ trati\}}$*trai* 7 di
13, rimane 6, e diremo che traendo $\frac{1}{13}$ di $\frac{1}{7}$ rimane $\frac{6}{91}$ aponto.

M.10.14
[F.iii.13]　Avemo deto del giungimento e del $^{\{so\ rimanente\}}$*sottraimento* de numeri roti. Ora
diremo che part'è l'uno numaro rotto del'altro. Et primamente diremo cosie, dimmi quant'è $\frac{1}{3}$
di $\frac{5}{7}$. Di' cosie, sicome tu vedi $\frac{1}{3}$ si scrive uno di sopra e 3 di soto et una vergha in mezo,
et $\frac{5}{7}$ si scrive cinque di sopra et sete di soto e una vergha in mezo, sicome vedi. Ora
multiprica le parti di sopra le verghe, cio⟨è⟩ uno via cinque, fano 5, e multiprica tre$_3$ via
7, cioè le parte di soto, fano 21, e pony simigliantemente 5$_{cinque}$ sopra 21, in questo modo,
$\frac{5}{21}$, e die che'l terzo di 5 $\frac{5}{7}$ si è $\frac{5}{21}$. E in questo modo fae di tuti altri numuri roty, sempre
multiprica le figure che sono di sopra le verghe l'una contra l'altra, e simigliantemente
multiprica le figure che sono di sotto l'una contra l'altra et poni le figure di sopra, poni
di sopra, e quelle di sotto poni di sotto, cioè la loro multipricatione. E per questa regola
puoi sempre fare di qualunque numaro rotto ti fosse detto.

M.10.14*
[F.iii.14]　Dimi quanto sono i $\frac{3}{5}$ di $\frac{7}{8}$. Multiprica 3 via 7, fano 21, et multiprica 5 via 8,
fano 40, e poni 21 f sopra 40 in questa maniera, $\frac{21}{40}$, e diremo che siano $\frac{21}{40}$ aponto.

M.10.15
[F.iii.15]　Dimi quanto sono i $\frac{3}{4}$ di $\frac{9}{10}$. Multiprica 3$_{tre}$ via 9$_{nove}$, fanno 27, e multiprica 4
via 10$_{diece}$, fano 40, e poni 27 sopra 40 in questo modo, $\frac{27}{40}$, e diremo che $\frac{3}{4}$ di $\frac{9}{10}$ sia
aponto $\frac{27}{40}$.

(fol. 18ᵛ)

M.10.16
[F.iii.16]　Ora mostreremo qual'è più l'uno numaro rotto del'altro e quanto. E primamente
chominciamo chosie et diremo, dimi qual'è piue tra $\frac{2}{3}$ e $\frac{7}{8}$. Fa cosie, die $\frac{2}{3}$ e $\frac{7}{8}$ si trova
in 24. Ora prendi $\frac{2}{3}$ di 24, sono 16, et prendi $\frac{7}{8}$ di 24, sono 21, e die, da 16 fino$_{insino}$ in
21 si ae 5, e parti 5$_{cinque}$ per$_{in}$ 24, che ne viene $\frac{5}{24}$. E diremo che $\frac{7}{8}$ sia più che $\frac{2}{3}$, $\frac{5}{24}$.
E così fae di tuti simiglianti numeri.

M.10.17
[F.iii.17]　Dimmi qual'è più e quanto intra $\frac{5}{6}$ et $\frac{4}{5}$. Fa cosie, die, $\frac{5}{6}$ et $\frac{4}{5}$ si trova in 30. Ora
prendi $\frac{5}{6}$ di 30, sono 25, et prendi $\frac{4}{5}$ di 30, sono 24, e die, da 24 fino in 25 si ae uno,
et parti uno per 30$_{trenta}$, che ne viene $\frac{1}{30}$, e diremo che $\frac{5}{6}$ siae più che $\frac{4}{5}$, uno trentesimi,
ed è fatta aponto.

M.10.17A
[F.iii.18]　Dimi qual'è più et quanto intra $\frac{7}{12}$ ed oto tredecimi. Di' cosie, $\frac{7}{12}$ et $\frac{8}{13}$ si
~~tredecimi~~ trova in 156, per cioe che 12 via 13 fanno 156. Ora prendi $\frac{7}{12}$ di 156, che sono

91, et prendi $\frac{8}{13}$ di 156, che sono 96. Ora die, da 91 infino$_{\text{insino}}$ in 96 si ae 5, e parti 5 per et 156, che ne viene $\frac{5}{156}$, e diremo che $\frac{8}{13}$ sieno piue che $\frac{7}{12}$, $\frac{5}{156}$ aponto. Ed è fata, e per questa regola poi fare di qualunque numero rotto fosse.

M.10.18
[F.iii.19] **D**imi qual'è meno e quanto intra $\frac{7}{8}$ et $\frac{8}{9}$. Fa cosie, die, $\frac{7}{8}$ et $\frac{8}{9}$ si trova in 72. Ora prendi $\frac{7}{8}$ di 72, sono 63, e prendi $\frac{8}{9}$ di 72, sono 64, e die, da 64 $^{\{\text{infino}\}}$ insino in 63 menoma uno, e parti uno in 72, che ne viene $\frac{1}{72}$, e diremo che $\frac{7}{8}$ sono meno di $\frac{8}{9}$, $\frac{1}{72}$. Ed è fata, e in questo modo e per questa regola puoi fare di qualunque numero rotto fosse.

[11. The Rule of Three, with Examples]

M.11.2
[F.iv.a] Se ci fosse deta alchuna ragione nela quale si proponese in tre cosse, sì debiamo multipricare quela cosa **(fol. 19^r)** chc noi volemo sapere contra quella che non è di quela medessima e partire nel'altra, cioè nella terza cossa.

M.11.3
[F.iv.1] Voglioti dare l'asempro ala deta regola e voglio$_{\text{voglioti}}$ dire cosie, 7 tornesi vagliono 9 parixini, dimi quanto varano 20 tornesi. Fa cosie, sicome la regola dicie, la cossa che noi vogliamo sapere si sono li 20$_{\text{venti}}$ tornesi, e quella che nonn'è di quella medesima si sono li 9 parisini, e però dovemo multipricare 20 via 9 parixini, fanno 180 parixini, e parti in 7, ^però^ ch'è la terza cossa, che ne viene 25 e $\frac{5}{7}$, e diremo che vinti tornesi varano parixini 25 et $\frac{5}{7}$.=

M.11.3A
[F.iv.2] Ancora diremo: 7 tornesi vagliono 9 parixini, dimi quanto varano li 30 parixini. Fa cosie, sicome la nostra regola dicie, la cosa che noi vogliamo sapere si sono li 30 parixini e quella che nonn'è di quella medesima si sono li 7 tornexi e però dovemo multipricare 30 via 7 tornexi, fanno 210 tornexi e parti 210 per 9, che ne viene 23 e $\frac{1}{3}$. E diremo che 30 parisini varanno tornesi 23•$\frac{1}{3}$, a ragione che 7 tornesi vagliano 9 parisini.

M.11.4
[F.iv.3] Anchora diremo, 7 tornesi vagliono 9 parisini, dimi quanto varano le 120 de tornesi. Fa cosie, die, poiché 7 tornesi vagliono 9 parisini, dunque ß 7$_{7\text{ soldi}}$ di tornesi vagliono 9 ß di parisini e 7 libre di tornesi vagliono £ 9 di parisini. Dunque multiprica 9 via 120 £ di parisini, fano £ 1080, e parti =$_£$ 1080 per 7, che ne viene £ 154 =$_e$ ß 8 et δ 6 et $\frac{6}{7}$. E diremo che 100 £^{a} di tornesi varano £ 154 • ß 8 • δ 6 et $\frac{6}{7}$ =d'uno danaio, ed è fata aponto.

^{a} Should be 100 £. The error is common to **F** and **M**.

M.11.5
[F.iv.4]   Anchora diremo un altro asempro ala deta regola et **(fol. 19ᵛ)** diremo cosie, 7 tornesi vagliono 9 parisini, dimi per 150 £ et ß 13 et 4 δ_denari 4 di tornesi, quanti parisini n'avemo_avren noi. Di' cosie, la cossa che noi volemo sapere si sono le 150 £ et ß 13 • δ 4 di tornesi, e quella che nonn'è di quella medesima si sono li 9 parisini e però dovemo multipricare 9 via £ 150 • ß 13 • δ4, =_che fano £ 1356, e parti £ 1356 per 7 che ne vene £ 193 • ß 14 • δ 3 et $\frac{3}{7}$. E tanto vagliono le_ 150 £_£ 150 • ß 13 • δ 4 di tornexi, cioè £ 193 • ß 14 • δ 3• $\frac{3}{7}$.

M.11.6
[F.iv.5]   Anchora diremo chosie, se 5 via 5 facesse 26, dimi quanto farebe 7 via 7 in quela medesima proportione. Fa cosie, die, 5 via 5 fano 25. Et io dico che fano 26, et 7 via 7 fanno 49. Dunque diremo: 25 vale 26, quanto varà 49. Dovemo multipricare 26 via 49, che fano 1274, e parti 1274 per 25, che ne viene 50 et $\frac{24}{25}$. E diremo che 7 via 7 facia 50 et $\frac{24}{25}$ a quela medesima rasone.

M.11.7
[F.iv.6]   Anchora diremo, se 3 via 4 faciese 13, dimi quanto farae 7 via 9. Fa cosie, die, 3 via 4 fanno 12, et io dico che fanno 13. Et 7 via 9 fanno 63. Dunque multiprica 13 via 63, che fano 819, e parti per 12, che ne viene $68\frac{1}{4}$, et tanto farae 7 via 9, cioè $68\frac{1}{4}$ 68 et 1/4 a quela medesima rasone.

M.11.8
[F.iv.b]   Se ci fosse deta alchuna rasone nela quale si proponese in tre cosse e, dal'una dele due parti dinanzi avesse rotto, sì dobiamo multipricare ambo le parti dinanzi per tal numero quent'è quel rotto.

M.11.9
[F.iv.7]   Ora diremo asempro ala detta regola e voglio dire **(fol. 20ʳ)** cosie, tornesi 3 et$\frac{1}{3}$ vagliono 4 parixini, dimi quanto varano li 25 tornesi. Fa cosie, sicome la regola dicie_dice la regola, multiprica ambo le parti dinanzi per 3, e die, 3 via tornesi 3• $\frac{1}{3}$, fano 10 tornesi, e die, 3_tre via 4 parisini, fano 12 parisini. E die che 10 tornesi vagliono 12 parisini, e noy volemo sapere quanto varano li 25 tornesi. Multiprica 12 {12} via 25 parisini, fanno 300 parisini, e parti 300 per 10, che ne viene 30, e diremo che 25 tornesi_ varano 30 parisini a quela medesima rasone.

M.11.10
[F.iv.8]   Anchora diremo un altro asempro ala deta regola, e diremo cosie, tornesi 4• $\frac{1}{4}$ vagliono 6 parisini, dimi, per £ 100_100 £ di parisini, quanti =_£ di tornesi avremo noi. Fa cosie, sicome la regola dicie, multiprica ambo le parti dinanzi per 4, per cioe che'l rotto è {quatro}*quarto*, e die, 4 via 4 $\frac{1}{4}$ et uno quarto fano 17, et_or die, 4_quattro via 6_sei fanno 24. Ora diremo che 17 tornexi vagliono 24 parisini, e noi vogliamo sapere quanto varano le 100 £ di parisini. Dunque dovemo multipricare 17 via 100 £ di parisini_, fano £ 1700, e parti £

1700 per 24 per cioe ch'è la terza cossa, che ne viene £ 70 • ß 16 • δ 8, ⁼*et diremo che per £ 100 di parigini avren noi £ 70 et soldi 16 et denari 8 di tornesi* a ragione che tornesi 4• $\frac{1}{4}$ vagliano 6 parisini. E per questo modo puoi fare tute le ragioni dela seconda regola ₀ₑ che vi si posono rechare.

M.11.11
[F.iv.c]
Se ci fosse deta alcuna ragione, la quale si proponesse in tre cosse ed ambo le parti dinanzi avese rotti sì dobbiamo sapere in che numero si trovano que' rotti. Saputo in che numero si trovano que' rotti, sì dobbiamo multipricare ambo le parti dinanzi per tal numero in quente si trovaro que' roti. **(fol. 20ʳ)** Diremo l'asempio_.

M.11.12
[F.iv.9]
Voti dare l'asempro ala deta regola, e voglio dire cosie, tornesi 2 et $\frac{2}{3}$ vagliono parisini 3 et $\frac{3}{4}$. Dimi quanto varano le 200 £ di tornesi. Fa cosie, sicome la regola dicie, die, i roti sono {3} $\frac{2}{3}$ et $\frac{3}{4}$, et $\frac{2}{3}$ due terzi et $\frac{3}{4}$ tre quarti si trova in 12. Et però dovemo multipricare ambo le parti dinanzi per 12 e die_dire, 12 via tornesi 2 et $\frac{2}{3}$ fano 32 tornesi, e die_dire, 12 via parisini 3 et $\frac{3}{4}$ fanno 45 parisini. Dunque diremo che 32 tornesi vagliano 45 parisini, ⁼*et altrettanto è a dire ke 32 tornesi vagliono 45 parigini* quanto tornesi 2 et $\frac{2}{3}$ vagliono parisini 3• $\frac{3}{4}$. E noi vogliamo sapere quanto varano le 200 £ di tornesi. Dunque multiprica 45 via 200 £_£ 200 di parisini, che fano £ 9000 di parisini_, e parti per 32, che ne viene £ 281 • ß5. E diremo che 200 £ di tornesi varano £ 281 • ß 5 di parisini. Ed è fata. =_E Chosì fae tute le simiglianti.

M.11.13
[F.iv.10]
Anchora diremo, se 3$\frac{1}{2}$ tornesi vagliono parisini 4 et $\frac{1}{3}$, quanto_ke varano li 20 parisini. Di' cosie, $\frac{1}{2}$ un mezzo e $\frac{1}{3}$ un terzo si trova in ? 6. Dovemo multipricare ambo le parti dinanzi per 6 et dire, 6 via tornesi 3$\frac{1}{2}$ 3 e mezzo fanno 21 tornesi, e die, ? 6 via parisini 4 et $\frac{1}{3}$ fano 26 parisini. Ed ài che 21 tornesi vagliono 26 parisini, e noi volemo fa sapere quanto varae no li 20 parisini. Multiprica 20 via 21, fano 420. E parti in 26, che ne viene 16 et $\frac{2}{13}$, e diremo che 20 parisini varano tornesi 16 et $\frac{2}{13}$ di tornese, ed è fata.

M.11.14
[F.iv.d]
Se noi avesimo a multipricare numero sano et rotto contra numaro sano et rotto, sì dobbiamo multipricare il minore numero contra tuto l'altro numero e poi **(fol. 21ʳ)** il rotto del minore numero contra tuto l'altro numero e diremo in ciò asempro.

M.11.15
[F.iv.11]
Ora diremo l'asempro ala deta regola et voglio dire cosie, multipricami 3$\frac{1}{3}$ 3 et 1/3 via 3 et $\frac{1}{3}$. Fa cosie, di'_Di' cosie, 3 via 3 et $\frac{1}{3}$ fanno 10, e prendi il $\frac{1}{3}$ terzo di 3 et $\frac{1}{3}$, sicome dice la regola, ch'è j₁ e $\frac{1}{9}$, et giungi 1 et $\frac{1}{9}$ sopra 10, fano_sono 11 et $\frac{1}{9}$, e diremo che 3 et $\frac{1}{3}$ via 3 et $\frac{1}{3}$ facia 11 et $\frac{1}{9}$. Ed è fata, et per questo modo poi multipricare qualunque numero roto tuo vuoli.

[12. Computations of Non-Compound Interest]

M.12.1
[F.iv.e] Se ci fosse deta alcuna ragione di merito, cioè la livra è prestata a cotanti δ lo mese, e noi volessimo sapere le cotante £ in cotanto tempo, sì dobbiamo sapere quanto vale la lira in tuto il termine, e saputo quanto vale la livra in tuto il termine sì dobbiamo multipricare contra la soma dele £.

M.12.2
[F.iv.12] Asempro ala deta regola, e voglio dire cosie, la livra guadagna il mese 3 denari, dimi quanto guadagnarano le 100 £ in sey₆ mesi. Fa cosie, sapia quanto vale la livra in tuto il termine, cioè in 6 mesi. Multiprica sei via 3 δ, fano 18 δ, e tanto valle la livra in tuto il termine, cioè 18 δ. E noi volemo sapere quanto varano le 100 £$_{£\,100}$. Die, poichè la livra vale 18 δ, le 100 £$_{£\,100}$ varano ß 150, cioè £ 7 • ß 10, e tanto guadagnarano le 100 £ in sei mesi, cioè ß 150, che sono £ 7 • ß 10. E per questo modo puoi fare di quantunque livre fosseno et a quantunque fosse prestata, overo guadagnase la livra il mese et quantunque fosse il tempo per la deta regola.

(fol. 21ᵛ)

M.12.3
[F.iv.f] Se ci fosse deta alcuna ragione di merito, cioè la livra è prestata overo guadagnia cotanti δ lo mese, et noi volessimo sapere le quante £ guadagniarano il die un dinaio, sì dobbiamo partire livre 30 per tante parti quanti denari guadagnia overo è prestata li livra il mese.

M.12.4
[F.iv.13] Asempro ala deta regola. E diremo cossie, la livra è prestata il mese a 3 δ, dimi le quante £ guadagnarano il die un danaio. Sicome la regola dicie, parti £ 30 per 3, che ne viene £ 10$_{10\,£}$. Et le 10 £ guadagnerano il die un danaio.

M.12.5
[F.iv.g] Se ci fosse deta alcuna rasone di merito, cioè il centenaio mi guadagnia l'anno cotante £ et noi volessimo sapere le quante £ guadagnano il die un danaio, sì dobbiamo partire 150 £ per ^tante^ parti quante lire guadagnial'anno il centinaio.

M.12.6
[F.iv.14] Asempro ala deta regola. E diremo cosie, il centinaio mi guadagnia l'ano £ 12, dimi le quante £ =$_{mi}$ guadagnerano il die un danaio. Fa cosie, sicome dicie la regola$_{la\ regola}$ $_{dice}$, parti £ 150 per 12, che ne vene £ 12 • ß 10, et le 12 £$_{£\,xij}$ et 10 ß$_{ß\,x}$ guadagnerano il die un₁ danaio. Ed è fatta. • Così fae tuti le simiglianti.

M.12.7
/[F.iv.h] Se ci fosse deta alcuna rasone di merito, cioè la livra è prestata a cotanti δ lo mese

et noi volessimo sapere le cotante £ in quanto tempo sarano doppie, senza fare capo d'anno, sì dobiamo partire 20 anny per tante parti quanti δ è prestata overo guadagnia la livra il mese. Et in cioè diremo asempro.

M.12.8
[F.iv.15] Asempro ala deta regola. Et diremo cosie, la livra è prestata il **(fol. 22ʳ)** mese a 3 denari $=_{lira}$, dimi in quanto tempo sarano dopie le 100 £. Fa cosie, sicome dicie la nostra regola, parti 20 anny per 3, che ne viene 6_{sei} anny e otto mesi, e in 6 anny et 8 mesi sarano dopie le 100 £, a non fare capo d'ano. E in altrettanto tempo sarebono dopie quantunque £ o ß o δ che tu avesi detto.

M.12.9
[F.iv.i] Se ci fosse deta alcuna rasone di merito, cioè il centinaio mi guadagnia l'anno cotante £, e noi volesimo sapere le quante £ guadagnano il die un dinaio, sì dobiamo partire 100 anny per tante parti quante £ guadagnano l'anno il centinaio.

M.12.10
[F.iv.16] Asempio ala deta regola. E diremo cossì, il centinaio mi guadagna l'anno 6 £, dimi in quanto tempo sarano dopie le 1000 £. Fa cosie, sicome la nostra regola dicie, parti 100 anny in 6_{sei}, che ne viene 16 anny et 8 mesi, e in cotanto tempo sarano dopie le 1000_{mille} £, cioè, in 16_6 anni et 8 mesi, e in altratanto tempo sarebero dopie quantunque £ tu avessi deto. In tanto tempo si radopiano le poche $^{quanto}come$ l'asay.

M.12.11
[F.iv.l] Se ci fosse deta alchuna ragione di merito, $ciò_{cioè}$ il centinaio mi guadagna l'anno cotante £, et noi volessimo sapere quanto guadagnia il centinaio il die, sì dobiamo pigliare $\frac{2}{3}$ di tanti danari quante £ guadagna l'anno il centinaio.

M.12.12
[F.iv.17] Asempro ala deta regola. E voglio dire cosie, il centinaio mi guadagnia l'anno 12 £, dimi quanto mi guadagnia il centinaio il die. Piglia i due terzi di 12 $=_{denari}$, che sono 8 δ, et diremo che'l centinaio guadagnia il die 8 δ. E se fosse prestato overo guadagnasse il 100 l'anno 16 £, sì prendi $\frac{2}{3}$ di 16, che sono 10 et $\frac{2}{3}$, et tanty δ guadagnia il centinaio il die, cioè δ 10 et $\frac{2}{3}$.

[13. Problems Involving Metrological Shortcuts]

(fol. 22ᵛ)

M.13.1
[F.iv.m] Se ci fosse deta alchuna ragione in questo modo, e diremo chosie, la charicha del pepe, o di qualunque altra cossa fosse, la qual'è 300 libre, vale $tante_{cotante}$ £, overo cotanti ß, overo cotanti δ, et noi volessimo sapere quanto varae la libra. Sì dovete sapere che

per ogni libra che vale la caricha, valle $\frac{4}{5}$ d'un$_{di}$ danaio la libra, e per ognie soldo che vale la carica, vale $\frac{1}{25}$ d'un danaio $=$*la* libra, e per ognie danaio che valle la carica vale la libra uno trecientesimo d'un danaio.

M.13.2
[F.iv.18]   **Ora** diremo asempro ala deta regola et diremo chosie, la caricha del pepe vale 18 £, dimi quanto varae la libra. Fa cosie, die, però che per ogni livra che valle la caricha vale la libra $\frac{4}{5}$ d'un danaio, sì multiprica 18 via 4, fano 72, et parti per 5, che ne viene 14 et $\frac{2}{5}$. Et tanto varae la libra, cioè δ 14 et $\frac{2}{5}$ d'un danaio. Ed è fatta, e in questo modo di quantunque £ tonde fossero, et non la potrai giamay falire.

M.13.2*
[F.iv.19]   **Et** s'io diciesse, la libra quanto varae, vagliendo la charica soldi 17, sapie che varae $\frac{17}{25}$ d'un danaio, e simigliantemente, s'io dicesse, la caricha valle 20 δ, quanto varae la libra. $=$*Sappie* che varrae $\frac{20}{300}$ d'un danaio, che sono $\frac{1}{15}$.

M.13.3
[F.iv.n]   **Anchora** mostreremo questa regola per altro modo e diremo chosie, sapie che quantunque danari valle la livra, mete vi suso il quarto, et tanti δ quanto ti verae, tante £ varae la caricha. Et questo s'intende la caricha che fosse 300 libre. [F.iv.20] E diremo chosie, la libra vale δ 20, dimi quanto varae la caricha. Fa cosie, sicome dicie la nostra regola, **fol. 23^r** die, il $\frac{1}{4}$ di 20 si è 5, e poni 5 sopra 20, sono 25. Et tante £ varae la caricha, cioè £ 25, a ragione di δ 20 la libra.

M.13.4
[F.iv.21]   **Anchora** diremo, la libra del pepe vale ß 10 • δ 8, dimi quanto ~~val~~ varae la caricha. Fa cosie, die, 10 ß et 8 δ sono 128 δ, e die, il $\frac{1}{4}$ di 128 si è 32. Ora poni 32 sopra 128, sono 160, et tante £ varae la caricha, cioè £ 160. Et in questo modo fa sempre, recha ~~a~~ £ a δ e metevi suso il $\frac{1}{4}$, e non potrà falire.

M.13.5
[F.iv.22]   **Anchora** diremo, la libra del pepe vale denari 17, dimi quanto varae la caricha. Prendi il $\frac{1}{4}$ di 17, ch'è 4 et $\frac{1}{4}$, et poni 4 et $\frac{1}{4}$ sopra 17, sono 21 et $\frac{1}{4}$, e diremo che la caricha varae £ 21 e $=_{\text{quarto di livra, cioè}}$ ß 5$_{\text{v}}$. • tanto varrae la caricha, cioè £ 21 et soldi 5. Et dovete sapere che secondo questa regola ogni quarto si contò per 5 soldi e'l mezo per 10 soldi.

M.13.6
[F.iv.o]   **Ora** diremo sopra questa regola il contrario, e diremo chosie, sapie che quantunque £ vale la caricha, la qual'è 300 libra, trai il quinto$_{1/5}$ di quelle £ e tante £$_=$ quante ti rimarà, tanti δ varae la libra. Et in ciò diremo l'asempro. [F.iv.23] La carica vale 40 £, dimi quanto varae la libra. Di' cosie, il $\frac{1}{5}$ di 40 £$_{£\,40}$ sono 8 £. Ora tray 8$_=$ di 40, rimane 32, e tanti δ varae la libra, cioè 32 δ. E per questo modo fae di quante £ fossero.

M.13.7
[F.iv.24] Anchora diremo, la caricha di £ 300 vale 57 £, dimi quanto varae la libra. Di' cosie, il $\frac{1}{5}$ quinto di 57 si è 11 et $\frac{2}{5}$. Ora trai di 57, 11 et $\frac{2}{5}$, rimane 45 et $\frac{3}{5}$, et tanti δ varae la libra, cioè δ 45 et $\frac{3}{5}$. Et sapie che in questa regola s'intende cinque$_{ogni\ 5}$ **(fol. 23ʳ)** ß uno quinto [*sic*, read "quarto"] di danaio, secondo che deto avemo.

M.13.8
[F.iv.p] Avemo detto dela caricha la qual'è 300 libre. Ora, per cioe che in alcuna parte la caricha si conta piue o meno di 300 libre$_{libre\ 300}$, e simigliantemente il quintale si conta meno $^{\{e\}}o$ piue di 100 libre, sì diremo una gienerale regola dela caricha e del quintale di quantunque libre fossero l'uno e l'altro.

M.13.9 Et cominciamo in questo modo, uno quintale solamente pogniamo che si facia in alchuna parte di £ 104 e la carica si è$_{-}$3$_{tre}$ quintali, domque la caricha $^{\{farae\}}$*sarae* libre 312. E pogniamo che sapie quanto valle la deta caricha, e $^{\{h\ volli\}}$*che vuoli* sapere quanto varae la libra, sì ne daremo questa regolla.

M.13.10
[F.—] Dovete sapere che per ogni soldo che valle la caricha, valle uno grano la libra, e per ogni soldo che vale il quintalle di £ 104, sì vale tre grani la libra. Et sapie che secondo questa regola il danaio si è 26 grany, perchè'l $\frac{1}{4}$ quarto di 104 si è 26. E sapie che di quantunque libre è il quintale, prendi il $\frac{1}{4}$ quarto, e tanti grani varae$_{verrà}$ il danaio. [F.—] Et se noi dicessimo che la carica fosse di 324 libre, dumque varebe il danaio grany 27, però che se la carica è libre 324, dumque è il quintale il $\frac{1}{3}$ terzo di 324, cioè 108. Ora sicome deto avemo, prendi il $\frac{1}{4}$ quarto di 108, ch'è 27, e tanti grany varae$_{verrae}$ il danaio, cioè grany 27. E, sicome avemo fato il quintale di libre 104 e la caricha di £ 312, a quella medessima ragione, possiamo fare di quantunque libre'ffosse la caricha e'l quintale. E, poichè sai che ogni 26 grani si contano uno danaio secondo il proponimento ch'avemo fatto, si poi rechare i grani a δ secondamente **(fol. 24ʳ)** che mostraremo inanzi per più asempri, per meglio intendere.

M.13.11
[F.iv.25] Ora diremo uno asempro ala deta caricha, e diremo chosie, la charicha vale £ 13 et ß 8, dimi quanto varae la libra. Di' cosie, 13 £ et ß 8$_{8\ soldi}$ sono 268$_{dugentosesantotto}$ ß, e secondo che deto avemo ogni soldo vale un grano. Dunque varae la libra 268 grani. Ora se voli recarli a δ, parti 268 per 26, che ne viene δ 10 et $\frac{4}{13}$ di danaio, e tanto varae la libra, cioè δ 10 e $\frac{4}{13}$ di δ. E se noi avessimo deto che'l quintale fosse di £ 108, sì avrey contato ogni danaio 27 grani, però che'l $\frac{1}{4}$ quarto di 108 si è 27. E s'io avesi deto che'l quintale fosse di £ 102, sì conterei ogni danaio grani 25$\frac{1}{2}$, però che'l $\frac{1}{4}$ di 102 si è 25$\frac{1}{2}$. Sempre prendi il $\frac{1}{3}$ di quantunque pesa la caricha, e tanto sarae il quintale. E prendi il$\frac{1}{4}$ di quantunque pesa il quintale, et tanto quanto ti verae, tanto varae il danaio, cioè tanti

grani. E questa regola de grani è fatta per le libre ispezate, che sono piue o meno di 100 e per li ß spezati piue di libre.

M.13.12
[F.iv.26] Anchora diremo cosie, il quintale, lo qual'è libre 104, valle £ 4 et ß 12, dimi quanto varae la libra. Or sapie, sicome deto avemo, che per ogni ß che vale il quintale, si valle 3 grani la libra. Dunque die, £ 4 • ß 12 sono 92 ß, e multiprica 3 via 92, fano 276. Et tanti grani varae la libra, cioè grany 276. Ora parti 26 276 per 26, però che 26 grani valle il danaio, che ne viene 10 δ et $\frac{8}{13}$. E tanto varae la libra, cioè δ 10 et $\frac{8}{13}$ di danaio.

M.13.13
[F.iv.27] La charicha vale £ 3 ß 5, dimi quanto varae la libra. Die, £ 3 e ß 5 sono ß 65. Dunque varae 65 grani la libra. E pongho il quintale 108 libre. Dunque parti 65 per 27, che ne viene 2 et $\frac{11}{27}$. Et tanti δ varae la libra, entendesi la carica di £ 324 e'l quintale di £ 108.

[14. Mixed Problems, Including Partnership, Exchange and Genuine "Recreational" Problems]

(fol. 24ᵛ)

M.14.1
[F.v.1] Sono 3 compagni che fano conpagnia insieme, et l'uno compagno mete in corpo di conpagnia £ 150, et el secondo compagnio mete in corpo di conpagnia £ 230, e'l terzo compagnio mete in corpo di conpagnia £ 420. Or viene a chapo d'un tempo, c'ànno guadagniato £ 100 e vogliole partire. Dimi quanto ne verae a ciascuno per sua parte, rimagniendo fermo il capitale di ciascuno di questi 3 conpagny. Fa cosie, primamente giongi insieme tuto quelo ch'àno messo in corpo di conpagnia, cioè £ 150 et £ 230 et £ 420, che sono in tuto £ 800. Ora parti quello ch'àno guadagniato, cioè £ 100, per 800, che ne viene₍nn'aviene₎ ß 2 • δ 6, e cotanto ne viene per livra, cioè ß 2ᵢⱼ et δ 6. Ora multiprica 150 via ß 2 δ 6, che fano £ 18 et ß 15, e tanto de' avere lo primo conpagnio che mise in corpo di conpagnia £ 150, cioè £ 18 ß 15. Ora multiprica 230 via ß 2 δ 6, che fanno £ 28 ß 15, e tanto de' avere lo secondo conpagnio che misse in corpo di conpagnia £ 230, cioè £ 28 • ß 15. Ora multiprica 420 via ß 2 • δ 6, che fano £ 52 et ß 10, et tanto de' avere lo terzo conpagno che misse in corpo di conpagnia £ 420, cioè £ 52 ß 10. Ed è fata. Ora giungi insieme tute queste parti, cioè £ 18 • ß15 et £ 28 • ß 15 et £ 52 • ß 10, che sono in tuto £ 100. Dunque avemo ben partito, e per questo modo e per questa regola fae di quantunque conpagni fossero e di quantunque avesse messo ciascuno

conpagnio in corpo di conpagnia, et vedi quanto ne viene per livra.

M.14.2
[F.v.2] Uno mercatante de' dare ad un altro £ 200 di qui a due **fol. 25ʳ** mesi e mezo. Dicie quelo chi de' avere le dete 200 £, dameli ogi et scontati i danari tuoi a ragione di δ 2 per lira il mese. Dimi quanto gli de' dare inanzi per le dete 200 £. Fa cosie, die, in due mesi e mezo, a 2 δ_{denari 2} per livra vale la livra denari 5. Fa cosie, apponti ale 195 £_{lire 195}, e sapia quanto vagliono di merito le dete 195 £, che vagliono £ 4 ß 1 δ 3. Ed è per tuto £ 199 ß 1 δ 3, manchavi_{campavi} ß 18 • δ 9, che vagliono di merito δ 5_. Ora trai di ß 18 • δ 9, δ 5, campano ß 18 δ 4. Ora giungi ß 18 • δ 4 sopra 195 libre, ed ài per tutto £ 195, ß 18 δ 4, ed è fata. E diremo che gli dee pagare inanzi per le dete £ 200, £ 195 ß 18 δ 4. E in questo modo fa tute le simiglianti.

M.14.3
[F.v.3] I'ò a'ffare in Bolognia uno pagamento di £ 100 di bolognini picioli. E a Bologna vale il bolognino grosso δ 13•$\frac{1}{3}$ di bolognini picioli. E in Firenze vale il deto bolognino δ 15•$\frac{1}{4}$. E a Bologna vale il fiorino d'oro_{dell'oro} ß 31_{13} et δ 6 di bolognini picioli. E in Fiorenze vale il deto fiorino d'oro ß 39 δ 6 dela moneta di Firenze. Dimi quale mi meterà meglio aportare a Bolognia, partendo da Firenze per fare il deto pagamento, intra fiorini d'oro o bolognini grossi, e quanto mi meterae meglio ale dete 100 £. Fa cosie, sapia primamente quanto in bolognini grossi gli conviene portare, e_ multiprica 100 via 15 et $\frac{1}{4}$, fano 1525_{millecinquecento venti}, e parti per 13$\frac{1}{3}$, che ne viene £ 123 ß 12 δ 11 et $\frac{25}{37}$ di bolognino, et tanto gli conviene portare in bolognini grossi, cioè £ 123 ß 12 δ 11 et$\frac{25}{37}$ di bolognino. Ora sapiamo quanto gli conviene portare in fiorini d'oro, e multiprica 100 via 39 et $\frac{1}{2}$, fanno 3950, **(fol. 25ᵛ)** e parti 3950 per 31 et $\frac{1}{2}$, che ne viene £ 123 • ß 7 δ 9, e tanto gli convien portare in *f* d'oro, cioè £ 123 ß 7 δ 9. Dunque sarae meglio aportare *f* d'oro che bolognini, tanto quant'è_{quant'à} da 123 £ et 7 ß et δ 9_{9 denari} infino in 123 £ et ß 12 δ 11_{12 soldi et 11 denari} et $\frac{25}{37}$, che v'ae ß 5 et_ δ 2 et $\frac{25}{37}$ d'un danàio. Dumque diremo che gli meterae meglio aportare fiorini d'oro che bolognini grossi, e meteragli meglio_ a tuto il pagamento dele dete 100 £ • soldi 5_v • δ 2 et $\frac{25}{37}$ aponto.

M.14.4
[F.v.4] Il so^l^do de' provenigini valle δ 40 di pisani, e'l soldo degl'inperiali valo 32 pisani. Dimi quanto avrò io di queste due monete mischiate insieme per 200 £ di pisany. Fa cosie, giungi insieme 40 et 32, fanno 72, che sono 6 ß, e parti 200 £ per 6, che ne viene 33 £ e 6 ß et δ 8_{8 denari}. E cotanto avray di ciaschuna di queste due monete, cioè £ 33 ß 6 δ 8, per le dete 200 £ di pisani. Ed è fatta. Se la voli provare, sapia quanto vagliono £ 33 • ß 6 • δ 8 di provinigini a 40 δ il soldo, e sapia quanto vagliono altretanti inperiali a δ 32 il soldo, che vagliono in tuto £ 200. Dunque avemo ben fatto. In questo modo fa le similiglianti.

M.14.5
[F.v.5] l'oe fiorini d'oro nuovi e vechi, e'l fiorino d'oro vechio valle ß 35, e'l novo valle ß 37. E i'ò chambiati 100 fiorini d'oro intra novi e vechi et ò ne avuto £ 178. Dimi quanty fiorini v'ebe de' vechi et quanti n'ebe de' nuovi. Fa cosie, poni primamente che fossono tuti vechi e sapia quanto vagliono 100_c fioriny d'oro tuti vecchi, a 35 $ß_{soldi\ 35}$ l'uno, che vagliono £ 175. Et die, insino in ^{78}*lire 178* si ae 3 **(fol. 26^r)** £. Ora recha queste 3 £ a soldi, che sono 60 soldi, e parti nela diferenzia ch'à da 35 infino$_{insino}$ in 37, cioè per 2, questi 60 ß, che ne viene 30 ß, e die che 30 fiorini d'oro vogliono essere de' novi e lo rimanente vechi, cioè 70$_{settanta}$. E diremo che n'ebe$_{verrebbe}$ 30 f novi et 70 vechi. E simigliantemente ti sarebe venuto se tu avessi posto che fossono tuty novi. Or pogniamo che fossono tuti novi, e sapia quanto vagliono 100 f d'oro a ß 37 l'uno, che vagliono £ 185. $^=$*E die, simigliantemente da 178 infino in 185 si à 7 £*, che sono 140 ß, e parti per quella medesima differentia, cioè per 2, che ne viene 70. Et die che 70 sarano vechi e lo rimanente cioè fino$_{insino}$ in 100, cioè 30, sarano novi. Ed è fata. Se la voli provare, sapia quanto vagliono 70 f d'oro vechi a ß 35 l'uno, che vagliono £ 122 ß 10, e sapia quanto vagliono 30 f d'oro novi a ß 37 l'uno, che vagliono £ 55 ß 10. Ora giungi insieme quelo che vagliono i novi e quelo che vagliono i vechi, cioè £ 122 ß 10 e £ 55 • ß 10, che sono in tuto £ 178. Dunque avemo ben fatto, e per questa reghola poi fare di quanta moneta o valuta fosse.

M.14.6
[F.v.6] Il fiorino del'oro vale a Gienova ß 14 di gienovini e l'aghulino valle a Gienova δ 12 ğ, e l'aghulino vale in Firenze ~~dela moneta di Firenze~~ δ 33. Dimi quanto varae il f d'oro in Firenze dela moneta di Firenze, a quela medesima ragione. Fa cosie, multiprica 14 via 33, fano 462 soldi, e parti per 12, che ne viene ß 38 • δ 6, e tanto varae$_{verrae}$ il f d'oro in Firenze dela moneta di Firenze, cioè ß 38 • δ 6. Ed è fata, et in questo modo et per questa reghola poi fare di qualunque moneta fosse e di qualunque valuta fosse la moneta, sicome deto avemo.

(fol. 26^v)

M.14.7
[F.v.7] Uno merchatante prestoe a un suo amicho una libra d'oro, la quale tenea δ^z due di di rame. Quando venne a$_{in}$ chapo d'un tempo e'l meratante gli ridomandoe la deta libra d'oro e'l buono homo dicie, io non oe di quelo cotale oro che tu mi prestasti, ma i'ò d'un altro oro, lo quale tiene onzie 3 di rame per libra, dimi quanto oro gli converae rendere di quelo che tiene δ^z 3 per libra per quelo che tenea δ^z 2 per libra. Fa cosie, abatti del primayo oro che tenea δ^z 2 per libra di rame, ^δ^z 2^, rimane δ^z 10 d'oro fine et puro et neto, e abati del secondo oro δ^z 3 per libra, rimane δ^z 9 di fine et puro. Ora die, 10 δ^z vagliono 9 oncie, quanto varano le 12 δ^z _. Multiprica 10 via 12, fano 120, e parti per 9, che ne viene δ^z 13 et $\frac{1}{3}$. Et dè fata.^{a} E diremo che gli de' rendere δ^z 13 et $\frac{1}{3}$ del'oro che tenea δ^z 9 per libra, per quela libra che tenea δ^z 10 per libra.

ᵃ The orthography "et de", repeated in M.14.25, might reveal that the recurrent phrase "e de fata" was *not* intended by the scribe as "ed è fata" (though that is how I interpret it throughout, in agreement with its obvious origin) but as "e de fata" or perhaps "è de fata", cf. "e di fatto". Indeed, in *Trattato di tutta l'arte*, T**f**, fol. 66ᵛ, l. 7 we also find "et de fatta", and again on fol. 70ᵛ, l. 15 "et defatta appunto".

M.14.8
[F.v.8] Una coppa, la quale pessa ỡ^z 14 in questo modo che'l nappo è d'oro e pesa oncie 7 e'l gambo è d'argiento e pesa ỡ^z 4 e'l piede è di rame et pesa ỡ^z 3. Or viene ch'io fo fondere tutta questa copa insieme, mischiato insieme l'oro e l'argiento e'l rame. E quand'è fonduto et io ne prendo un pezo, così mischiato insieme, lo quale pesa ỡ^z 6, dimi in queste 6 ỡ^z quanto v'ae del'oro e quanto del'argiento e quanto del rame. Fa cosie, Primamente giungi insieme

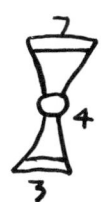

l'oro et l'argiento e'l rame, che sono ỡ^z 14, tutto mischiato. Ora, però che'l pezzo che tu prendesti pesava ỡ^z 6, sì multiprica 6 via 7 onzie d'oro. fano ỡ^z 42 e parti in 14, che ne viene ỡ^z 3, e tanto v'ebbe del'oro, cioè ỡ^z 3. Ora multiprica 6 via 4 ỡ^z d'ariento, fano ỡ^z 24$_{xxiiij}$ e parti per 14, che ne viene ỡ^z 1 et $\frac{5}{7}$. Et tanto v'ebbe del'argento, **(fol. 27ʳ)** cioè oncie 1 et $\frac{5}{7}$. Ora multiprica 6 via 3 ỡ^z di rame, fanno 18 ỡ^z, et parti per 14, che ne viene ỡ^z 1 et $\frac{2}{7}$, e tanto v'ebe del rame, cioè ỡ^z 1 et $\frac{2}{7}$. Ed è fata. Se'lla voli provare, giongi insieme l'oro e l'argiento e'l rame, cioè ỡ^z 3? e ỡ^z 1•$\frac{5}{7}$ e ỡ^z 1•$\frac{2}{7}$, che sono in tuto ỡ^z 6. Dunque sta bene.ᵃ

ᵃ The illustration is absent from **F**.

M.14.8A
[F.v.9] **Il** marcho del'argento, lo qual'è 8 oncie, mi costa ß 66 di tornesi. Or viene ch'io$_{che}$ foe fondere e afinare$_{farae}$ il detto argento. E, quando l'ò trato da fuocho, il pesò$_{pesolo}$ e truovolo che ogni marcho minoma $\frac{3}{4}$ d'oncia, cioè che ogni marcho mi torna ỡ^z 7 et $\frac{1}{4}$. Dimi quanto mi conviene vendere il marcho a ciò ch'io ne rifacia mio capitale. Fa cosie, die, 8 ỡ^z d'argento vagliono ß 66, quanto varano ỡ^z 7 et $\frac{1}{4}$. Multiprica 8 via 66 ß, fanno £ 26$_{26 £}$ • ß 8$_{8\ soldi}$, et parti per 7 et $\frac{1}{4}$ in questo modo, die, 4 via 7 et $\frac{1}{4}$ fano 29, e die, 4 via 26 £ et ß 8$_{8\ soldi}$ fano £ 105 • ß 12, et parti £$_{in}$ 105 • ß 12 per 29, che ne viene ß 72 • δ 9 e $\frac{27}{29}$ di danaio. E tanto gli conviene vendere il marcho del'argento a rifarne suo capitale, cioè ß 72 • δ 9 e $\frac{27}{29}$ di danaio. Ed è fata. Così fa tute$_-$ le simiglianti.

M.14.9–10
[F.v.10] {Uno}*Un uomo* sta gravemente et vuole fare suo giudicamento. Ed à una sua dona, la qual'è pregnia. E il buon omo si giudica in questo modo, e dicie ala dona, se tu auray uno$_-$ figliolo maschio, li$_{sì}$ lascio a lui le due parti del mio avere, et a te lascio il terzo, e s'adviene che tu abi figliola femina, sì lascio a ley lo terzo di tuto lo mio avere, e a te lascio le due parti. E'l bono homo passò di questa vita, e quando vene a capo d'un

tempo • la dona aparturiò et fecie uno figliolo maschio **(fol. 27ᵛ)** et una femina. Dimi
come si de' partire questo avere, però che non si puote partire chosì come il padre lascioe
ala dona e a' figlioli. Fa cosie, e questa è la sua regola, fae primamente posicione d'uno
e die, quando la figlia femina avese uno, la dona avrebbe due, e quando la madre avesse
due, il figliolo maschio avrebe 4. Dunque, di quantunque avere si dovesse partire fra loro,
d'ogni 7 doverebe avere il figliolo maschio 4 e la donna due e la fanciula femina uno.
Dumque avemo rechata questa ragione ala conpagnia, e diremo chosie, sono 3 tre compagni
ch'ànno e ànno fato conpagnia insieme, e l'uno compagnio mise 4 e l'altro mise 2 e'l terzo
mise uno₁, e ànno guadagnato tanto quanto il giudicamento fosse, quanto ne verae a
ciaschaduno. E questo si fae per lo modo dela compagnia che mostrata avemo adietro.
Or pogniamo che questo giudicamento fosse 1400 fiorini d'oro. Dimi quanto ne de' avere
la madre e quanto il figliolo maschio e quanto la femina. Fa cosie, giongi insieme 4 et
2 et l, sono 7, e questo è il partitore. Ora multiprica 4 via 1400 fiorini d'oro, fanno 5600
ƒ d'oro, e parti 5600 in 7, che ne viene 800 ƒ d'oro, e tanto ne de' avere il figliolo
maschio, cioè 4800 ƒ d'oro. Ora multiprica 2 via 1400 millequattrocento, fanno 2800 et parti
in 7, che ne viene 400, e tanto de' avere la madre, cioè 400 ƒ d'oro. Ora multiprica uno
via 1400, fanno 1400, e parti per 7, che ne viene 200. E tanti ƒ d'oro de' avere la figlia
femina, cioè 200. Ed è fatta, e in questo modo et per questa regola poi partire quantunque
fosse il giudicamento ch'avese lasciato.

(fol. 28ʳ)

M.14.10A(17)
[F.v.11] Uno maestro toglie a murare uno difitio in 30 die. E'l die che'l maestro lavora
si à dal segnore ß 5, e'l die che'l maestro de non lavora rende al segnore ß 7. Ora il deto
maistro à tanto lavorato e tanto stato che non lavoroe che non de' avere nula dal signore
nè'l maestro non de' rendere nulla a luy. Dimi quanto stete il maestro che non lavoroe
et quanto lavoroe, cioè quanti die. Fa cosie, primamente giungii insieme 7 et 5, fanno
12 ß, e quest'è il partitore. Ora multiprica 30 via 5 die, fano 150 die, e parti in 12, che
ne viene die $12\frac{1}{2}$, e tanto stete che non lavoroe, cioè die $12\frac{1}{2}$. E simigliantemente
multipricha 30 via 7 die, fano 210 die, e parti per 12, che ne viene die $17\frac{1}{2}$, e tanto
lavoroe, cioè die $17\frac{1}{2}$. Ora diremo che'l maestro lavoroe die $17\frac{1}{2}$ et stete che non lavoroe
die 12 e $\frac{1}{2}$, che sono in tuto 30 die. Dunque fecie il deto difitio in 30 die. Se la vuoli
provare, di' così, il maestro lavoroe die $17\frac{1}{2}$ e prese il die 5 ß soldi 5, che prese in tuto £
4 ß 7 δ 6. E tanto prese dal signore, cioè £ 8 ß 7 δ 6. E die, il deto maestro stete che
non lavoroe die $12\frac{1}{2}$ e rendeo al segnore ß 7 il die il die soldi 7, che sono in tuto £ 4 • ß 7
δ 6. Dunque tanto prese prese tanto dal signore quanto gli rendeo. Ed è ben fata.

M.14.11 Uno pescie, lo quale pesa la testa il terzo di tuto il pescie e la coda pesa il $\frac{1}{4}$
[F.v.12]
di tuto il pescie e'l corpo del mezo pesa oncie 8. Dimi quanto pesa la testa per se sola

et quanto pesa la coda e quanto pesa tuto il pescie. Fa cosie, die, $\frac{1}{3}$ et $\frac{1}{4}$ si trova in 12, piglia il $\frac{1}{3}$ ₜₑᵣᵤₒ e il $\frac{1}{4}$ quarto di 12, sono 7, e die, da 7 infino_fino in 12 si à 5, e questo è il partitore. Ora, però che 8 oncie pesa il corpo del mezo, sì multiprica 8 **(fol. 28ᵛ)** via 12 oncie, ⁽ᶠᵒⁿᵒ⁾*fanno* 96, e parti per 5, che ne viene ♂ 19 e $\frac{1}{5}$, ct tanto pesa tuto il pescie, cioè ♂ $19\frac{1}{5}$. Se voli sapere quanto pessa la testa per se sola, sì prendi il $\frac{1}{3}$ ₜₑᵣᵤₒ di 19 e $\frac{1}{5}$, ch'è 6 et ⁽³/⁵⁾ $\frac{2}{5}$, et tanto pesa la testa, cioè ♂ 6 et_ ⁽³/⁵⁾ $\frac{2}{5}$ d'oncia. Se vuoli sapere quanto pesa la coda, sì prendi il $\frac{1}{4}$ di 19 e $\frac{1}{5}$, ch'è 4 e $\frac{4}{5}$, e tanto pesa la coda, cioè 4 onzie et $\frac{4}{5}$ d'oncia. Ed è fata. Se la vuoli provare, giongi insieme quelo che pesa la testa, cioè ♂ 6 et $\frac{2}{5}$, et quello che pesa la coda, cioè ♂ 4₄ ₒₙ𝒸ᵢₑ et $\frac{4}{5}$ =_d'oncia, et quelo che pesa il corpo del mezo, cioè 8 oncie, che sono in tuto ♂ 19 et $\frac{1}{5}$. Dunque avemo ben fato. Così fae tute le simiglianti. =_Amen.

M.14.12
[F.v.13]   Uno homo è a Roma et vuole venire a Monpuleri e verebe in 11 die né piue né meno, e un altro homo è a Monpuleri et vuole andare a Roma et andarebe in 9 die né più né meno. Ora si partino a un'ora l'uno da Roma e l'altro da Monpuleri. Dimi in quanti die si trovarano insieme al camino. Fa cosie, die, però che l'uno viene in 9 die e l'altro vae in 11 die, sì multiprica 9 in_via 11, fano 99, e parti 99 per 20, però che 11 et 9 fano 20, che ne vene 5 meno uno ventesimo₁/₂₀, e in cotanto si trovarano insieme i deti homini, cioè in 5 die meno uno ventesimo₁/₂₀ di die.

M.14.13
[F.v.14]   Anchora diremo una simigliate ragione per mostrarla più apertamente. E diremo chosie, uno coriere è a Vignione et vuole andare a Tolosa et andarebe in 5 die, e un altro coriere è a Tolosa et vuole andare a Vignione et andarebe in **(fol. 29ʳ)** quatro₄ die. Or viene che i deti corieri =*si partono* a un'ora, l'uno da Tolosa per andare a Vignione et l'altro si parte da Vignone per andare a Tolosa. Dimi in quanti die si trovarano insieme. Fa cosie, però che l'uno vae in 4quattro die e l'altro in 5cinque, si giungi 4 et 5, sono 9, et questo è il partitore. Ora simigliantemente multiprica ? 4 via 5, fano 20, et parti 20 per 9, che ne viene 2 et $\frac{2}{9}$, e diremo che i deti corieri si trovarano insieme in due₂ die et due noni₂/₉ d'un die. Ed è fatta aponto.

M.14.14
[F.v.15]   Uno mercatante è oltramare chon uno suo conpagnio e vogliono passare di qua da mare. E venghono al porto per pasare, e trovano una nave, dove caricha l'uno 20 sacha di lana e l'altro vi =_ne carica 24 sacha. Ora la nave à fato suo viagio ed è passata di qua da mare. Dicie il patrone dela nave, pagatemi il nolo dela deta lana, e i mercatanti dicono, noi non avemo danari, ma prendi da ciaschuno di noi un sacho di lana, e vendila e pagati et rendeci il soperchio. E'l patrone vendeo le sacha e pagasi del nolo, e rendeo al mercatante che n'ave caricate 20 sacha 8 £₈, e al mercatante che n'avea caricate 24 sacha

rendeo 6 $£_{£6}$. Dimi quanto vendeo il sacho dela lana e quanto tolse di nolo a ciaschuno di questi 2 mercatanti. Fa cosie, sapia primamente quante sacha avea l'uno piue che l'altro, che n'avea 4, et quante £ rendeo al'uno più che $^{=}a$ l'altro, £ 2. Ora parti £ 2 per 4, che ne viene 10 $ß_{soldi\ 10}$, et tanto tolse di nolo di ciaschuno sacho, cioè 10 ß. Dunque tolse al mercatante dele 20 sacha 10 £, et 8 gli ne rendeo, sono £ 18. **(fol. 29ʳ)** E $^{=}a$ l'altro mercatante tolse dele 24 sacha 12 £, et 6 £ gli rendeo ^{et}che sono £ 18. Ed è fata, e diremo che 10 ß tolse di nollo del sacho dela lana, e vendeo ciascuno sacco £ 18 nè più né meno. Ed è fata. Così fa le simiglianti.

M.14.15
[F.v.16]   Uno arbore è soto terra il $\frac{1}{4}$ quarto e'l $\frac{1}{5}$ quinto di tuto l'albore, e sopra tera n'ae 20
$^{\{di\}}=$ braccia. Dimi quanto è longo tuto l'arbore. Fa cosie, die, $\frac{1}{4}$ et $\frac{1}{5}$ si trova in 20. Or prendi il $\frac{1}{4}$ quarto e'l $\frac{1}{5}$ quinto di 20, sono 9, e die, da 9 fino in 20 si ae 11, e quest'è il partitore. Ora multiprica 20 $^{=}via$ 20, fanno 400, e parti 400 per 11 che ne viene 36 et $\frac{4}{11}$. E diremo che tuto quelo arbore sia longho bracia 36 et $\frac{4}{11}$ d'un$_{di}$ bracio. Ed è fata.

M.14.16
[F.v.17]   Sono 3 compagni ch'ànno a partire 20 $ß_{soldi\ 20}$. L'uno ne dee avere il $\frac{1}{2}$ mezzo e l'altro ne de' avere il $\frac{1}{3}$ terzo e l'altro ne de' avere il $\frac{1}{4}$ quarto. Quelo$_{Quelli}$ che de' avere il $\frac{1}{2}$ mezzo dicie, datemi 10 ß, però che'l mezo di 20 $ß_=$ sono 10 soldi. E quelo$_{quelli}$ che de' avere il$\frac{1}{3}$ terzo dicie, datemi ß 6 δ 8$_{6\ soldi\ et\ 8\ denari}$, però che'l $\frac{1}{3}$ di 20 soldi sono ß 6 δ 8$_{ch'è\ il\ terzo\ di\ 20\ soldi}$. E quelo$_{quelli}$ che de' avere il $\frac{1}{4}$ quarto dicie, datemi 5 $ß_{soldi\ 5}$, però che'l $\frac{1}{4}$ di 20 ß sono 5 ß$_{ch'è\ il\ quarto\ di\ 20\ soldi}$. Ora se ciaschuno prendesse quello ch'el domanda, non vi si trovarebero tanti danari, però dimi in qual modo si deboro partire questi 20 ß che niuno non rimanga inganato. Fa cosie, die, $\frac{1}{2}$ uno mezzo e $\frac{1}{3}$ uno terzo et $\frac{1}{4}$ uno quarto si trova in 12. Ora prendi il$\frac{1}{2}$ mezzo e'l $\frac{1}{3}$ terzo e'l $\frac{1}{4}$ quarto di 12, sono 13 e quest'è il partitore. Ora prendi il $\frac{1}{2}$ mezzo di 20 $ß_=$, che sono 10 ß, et multiprica 10 via 12 $ß_=$, fano 120 ß, e parti ß 120$_{120\ soldi}$ per 13, che ne viene ß 9 δ 2$_{9\ soldi\ et\ 2\ denari}$ et $\frac{10}{13}$ d'un danaio, e tanto de' avere quelo$_{quegli}$ che domanda il $\frac{1}{2}$ mezzo, cioè ß 9 δ 2$_{9\ soldi\ et\ 2\ denari}$ et $\frac{10}{13}$. Ora prendi il $\frac{1}{3}$ terzo di 20 ß, che sono ß 6 δ 8$_6$ soldi et 8 denari, et multiprica **(fol. 30ʳ)** 12 via ß 6$_{6\ soldi}$ et δ 8$_{8\ denari}$, che fano soldi 80, e parti 80 ß$_{partigli}$ per 13, che ne viene ß 6$_{6\ soldi}$ et δ 1$_{1\ danaio}$ et $\frac{11}{13}$ d'un danaio, e tanto de' avere quelo$_{quelli}$ che domanda il $\frac{1}{3}$ terzo, cioè ß 6$_{6\ soldi}$ et δ 1$_{1\ danaio}$ et $\frac{11}{13}$ d'un danaio. Ora prendi il $\frac{1}{4}$ quarto di 20 $ß_=$, che sono 5 $ß_{soldi\ 5}$, et multiprica 12 via 5 ß, fano 60 ß, e parti per 13, che ne viene ß 4 et δ 7 et $\frac{5}{13}$, et tanto de' avere quelo$_{quelli}$ che domanda il $\frac{1}{4}$, cioè ß 4 • δ 7 et $\frac{5}{13}$ d'un danaio. Ed è fata. Se la vuoli provari, giongi insieme tute le parti, cioè ß 9$_{9\ soldi}$ et δ 2$_{2\ denari}$ et $\frac{10}{13}$ et ß 6 • δ 1$_{1\ danaio}$ et $\frac{11}{13}$ et ß 4 • δ 7 et $\frac{5}{13}$, che sono in tuto ß 20. Dunque avemo bene partito. In questo modo fa tute le simiglianti ragiony.

M.14.17A   Una botte à 3 canele ed è piena di vino. Et s'io ne traese l'una canela los
[F.v.18]
solamente, si votarbe la bote in due die. Et s'io ne traesi l'altra canela, si votarebe la bote
in tre die. E s'io ne traese la terza canela, si votarebe la bote in 5 die. Or viene ch'io ne
tragho tute queste tre cannele ad un'ora. Dimi in quanti die si vodarae la deta botte. Fa
cosie, die, un mezo et $\frac{1}{3}$ un terzo et $\frac{1}{5}$ un quinto si trova in 30. Or prendi il $\frac{1}{2}$ di 30 $\frac{1}{2}$ mezzo
e'l $\frac{1}{3}$ terzo e'l $\frac{1}{5}$ quinto di 30, sono 31. Or parti 30 in 31, che ne viene $\frac{30}{31}$ d'un$_{di}$ die. E in
cotanto$_{tanto}$ si votarae la bote, cioè in $\frac{30}{31}$ d'un$_{di}$ die.

M.14.18   Una galea è$_-$ a Gienoa e vuole andare in Aguamorta. E la deta galea à due velle
[F.v.19]
tali che, col'una vella andarebe in Aguamorta in 7 die, e chol'alltra vela andarebbe in
9 die. Or viene ch'io collo ambidue queste$_{queste due}$ vele a un'ora. Dimi in quanti die la galea
avrae fatto suo viagio da Gienova infino$_{insino}$ ad Aquamorta, operando ciaschuna di queste
velle per sua potenza. Fa **(fol. 31ʼ)** cosie, die, per cioe ch'ela v'andarebe in 7 die col'una
vella e col'altra v'andarebe in 9 die, sì giongi 7 et 9, fano 16. Et simigliantemente
multiprica 7 via 9, fano 63, e parti 63 per 16, che ne viene 4 meno $\frac{1}{16}$, e in cotanti die
sarae gionta la galea in Aguamorta, cioè in 4 die meno $\frac{1}{16}$.

M.14.19   I'oe 400 drapi, e ò ne fato 38 bale, tali di 10 drapi per bala e tali di 11 drappi
[F.v.20]
per bala. Dimi quante sono quelle di 10 drappi per bala et quanti sono quele di 11 drapi
per balla. Fa cosie, multiprica 10 via 38, fano 380. E die, da 380 fino in 400 si ae 20.
Et tanti sono quele di 11 drapi per bala, cioè 20. Et die, da 20 fino$_{insino}$ in 38 si ae 18,
e tante sono quele di 10, cioè 18. Ed è fatta. Se la vuoli provare, die, 20 bale a 11 drapi
per bala sono 220 drapi, et 18 bale a 10 drapi per balla sono 180 drapi. Ora giungi 180
et 220, sono 400. Dunque stae bene. Così fa le simiglianti.

M.14.20   Uno prestoe a uno suo amicho una archa piena di biada, e questa archa è per ogni
[F.v.21]
verso 4 bracia, cioè longa e ampia e alta. Quando vene a capo d'un tempo • questo$_{questi}$
che avea prestata la biada il dimandò a quelo suo amicho e quelo disse, io non oe archa
così fata come quella che tu mi prestasti, ma io n'oe due che ciascuna per ogni verso
è$_{è per ogni verso}$ 2 braccia, cioè 2 bracia per alto e due per ambio ampio {e 2 per alto} e 2$_{due}$
per longho. Dimi se gli è pagato per queste due arche picole per la sua grande o$_{e}$ quante
volte gli de' dare piene. Fa cosie, primamente recha a bracia quadre la grande archa, e
multipricha$_{multiplicata}$ per la lungheza e per l'ampieza 4 via 4, **(fol. 31ʳ)** fano 16, e per l'alteza
multiprica 4 via 16, fanno 64, e cotante bracia quadre è la grande archa, cioè 64 bracia.
Ora rechiamo a bracia quadre la grande archa, cioè che diremo che sia bracia 64 quadre.
Et, simigliantemente, rechiamo a bracia quadre la picola archa. Et multiprichiamo per
l'ampiezza e per la lungheza 2 via 2, fano 4, e per l'altezza multiprica 2 via 4, fano 8,

e cotante bracia quadre è la pichola archa, cioè 8 bracia. Ora parti 64 per 8, che ne viene 8, e diremo che li debia rendere 8 arche picole piene di biada per una di quelle grandi. Ed è fata.

M.14.21
[F.v.22] Uno si vole vestire et trova drapo che n'à asay in una roba di bracia 11, e'l deto drapo è ampio palmi 3 et $\frac{1}{2}$. Trova un altro drapo, lo qual'è ampio palmi 5 et $\frac{1}{2}$. Dimi di quanto drapo avrae asay a farne una roba di questo ch'è ampio palmi $5\frac{1}{2}$ a quella medesima ragione. Fa cosie, multiprica 11 via 3 e $\frac{1}{2}$, fanno $38\frac{1}{2}$, e parti $38\frac{1}{2}$ per 5 et $\frac{1}{2}$, che ne viene 7, e diremo che 7 braccia di drappo avrae asay per fare la roba.

M.14.22
[F.v.23] Una dona mi manda ad uno giardino a choglire pomeranzie$_{melarance}$ et diciemi, cogline quante a te pare$_{piace}$. Et il$_{al}$ deto giardino si à 3 porte, e a ciaschuna porta si è uno portiere che guardino il giardino. E diciemi la dona, chogline tante che al primo portiere che trovaray al'uscire del giardino gli doni la metade dele poma $=_{o\,vero}$ che coglieray e una più. E al secondo portiere che trovaray, doni la metade dele pome$_{mele}$ che ti sono rimase et una più$_{pie}$. E simigliantemente al terzo portiere doni la metade dele pome$_{mele}$ che ti sono rimase e una $=_{mela}$ più. **(fol. 31ᵛ)** E fatto tuto questo, voglio che ti rimanghino 3 pome$_{mele}$ solamente né piu né meno. Dimi quante pome$_{mele}$ gli conviene a$_{}$ choglire a ciò che no' gli ne soperchino e no' gli ne vegniano meno. Fa cosie, però che la dona vuole che ^3^ pome$_{mele}$ gli soperchino, sì radopia 3, sono 6, e poni suso 2, sono 8. Ora radopia 8, sono 16, et poni suso 2, sono$_{et\,ài}$ 18. Et questo$_{qui}$ è il secondo portiere. Ora radopia 18, sono 36, poni$_{ponvi}$ suso 2, sono 38, e $^{\{questa\}}$quest'è il terzo portiere. E tante ne$_{mele}$ gli conviene a$_{}$ cogliere, cioè 38. Ed è fata. Se la vuoli provare, die, la metade di 38 si è 19, et una, sono 20. Tray 20 di 38, sono$_{rimane}$ 18 et quest'è l'una porta. Ora die, la metade di 18 sono$_{si\,è}$ 9, e una più sono 10, tray 10 di 18, rimane 8, ed ày due porte. Et anche die, il mezo di 8 si è 4, et una piue sono 5, tray 5 di 8, si rimane 3, ed ày tre porte, et son ti rimase 3 pome$_{mele}$ né più né ꝑ meno. Dunque avemo ben fatto. In questo modo fae di quantunque porte fossero: sempre radopia, et poni$_{giugni}$ suso due.

M.14.23
[F.v.24] L'oncia del'oro fine, lo qual'è di 24 carati, valle £ 9 • ß 7 e δ 6. Dimi quanto varano le 125 oncie e 13 teri e 14 grany d'oro, che sia di carati $22\frac{1}{2}$ per oncia. E sapia che 30 teri sono una onzia et 20 grani sono uno terie. Fa cosie, sapia primamente quanto valle l'oncia del'oro di 24 carati, che vale £ 9 • ß 7 δ 6. Ora multiprica 22 et $\frac{1}{2}$ via £ 9 • ß 7 • δ 6, che fano £ 210 • ß 18 • δ 9, e questo parti per 24, però che l'oro fino di prima si è di 24 carati, che ne viene £ 8 15 δ 9• $\frac{3}{8}$, e cotanto valle l'oncia del'oro di carati $22\frac{1}{2}$, cioè £ 8 • ß 15 δ 9 et $\frac{3}{8}$ di danaio, cioè £ 8 ß 15 δ $9\frac{3}{8}$ $_{}$. Ora sapia quanto **(fol. 32ʳ)** varano le 125 oncie. Multiprica 125 via 8 £ e 15 ß$_{soldi\,15}$ e 9 δ et $\frac{3}{8}$, che fano £ 1098

• ß 12 • δ 9. Ora sappiamo quanto vagliono li 13 teri. Multiprica 13 via 8 £ • ß_ 15 • δ 9₉ denari e $\frac{3}{8}$, che fano £ 114 • ß 5₅ soldi • δ 2, e questo parti per 30, cioè £ 114 ß 5 δ 2_, però che 30 teri sono una oncia, che ne viene £ 3 • ß 16 • δ 2, e tanto vagliono li 13 teri_ d'oro {cioè £ 3 ß 16 δ 2 d'oro} = di carati di 22$\frac{1}{2}$. Ora sapiamo quanto vagliono li 14 grani. Fa così, parti quello che valle l'oncia del'oro di carati 22$\frac{1}{2}$, cioè £ 8 ß 15 δ 9 et_ $\frac{3}{8}$, per 600_, cioè per 20 e per 30, perchè 30 teri sono una onzia e 20 grany sono uno₁ teri, che ne viene δ 3 et_ $\frac{1}{2}$. E cotanto valle il grano, cioè 3 et_ $\frac{1}{2}$. Dunque vagliono li 14 grani ß 4 δ 1. Ora giongi insieme quelo che valiono le 125 oncie e quello che vagliono li 13 teri e quello che vagliono li 14 grani, che sono in tuto £ 1102 • ß 13. Ed è fata. E diremo cha δᶻ 125 et 13 teri et 14 grany d'oro di carati 22$\frac{1}{2}$ varae £ 1102 • 13ₓᵢᵢⱼ. Così fa tute le simiglianti.

M.14.24
[F.v.25]   Una borssa è di tre colori, cioè di seta biancha e di setta verde et di setta vermiglia. E la seta biancha pesa il $\frac{1}{4}$ quarto di tuta la borssa. E la seta verde pesa il $\frac{1}{7}$ settimo di tuta la borssa. Ed à vi di sete velmiglia 2 onzie né più né meno. Dimi quanto pesa tuta la borsa. Fa cosie, die, $\frac{1}{4}$ et $\frac{1}{7}$ si trova in 28. Piglia il $\frac{1}{4}$ quarto et il $\frac{1}{7}$ settimo di 28, ch'è 11, et die, da 11 fino insino in 28 sì à 17, e quest'è il partitore. Ora multiprica 28 via 2, fano 56, e parti 56 per 17, che ne viene 3 et $\frac{5}{17}$. Et tante cotante oncie pesa tuta la borsa, cioè δᶻ 3 et $\frac{5}{17}$. Se la vuoli proare, piglia il $\frac{1}{4}$ quarto et $\frac{1}{7}$ 'l septimo di 3 et $\frac{5}{17}$, {che sono} ch'è uno et $\frac{5}{17}$, et giogii {ni}vi δᶻ 2, ed ày per tuto δᶻ 3 et $\frac{5}{17}$. Ed avemo bene trovato. **(fol. 32ᵛ)** E per questo modo fae di quantunque quantitade qualitade fosse.

M.14.24A
[F.v.26]   Uno panno lo qual'è ampio 7 palmi e costami la cana, la qual'è 8 palmi, 70 ß. Un altro panno, lo qual'è ampio 5 palmi, mi costa la deta cana, la qual'è 8 palmi, ß 30. Dimi qual'è migliore mercato e quant'è migliore mercato_ la canna del'uno e che del'altro. Fa cosie, sapia primamente quanti palmi quadri è l'uno panno e l'altro. Multiprica per lo primo panno 7 via 8, fanno 56, e per lo secondo multiprica 5 via 8, fanno 40. E diremo che'l primo panno sia quadro palmi 56, e'l secondo_ sia quadro palmi 40. Ora parti ß 70 per 56, che ne viene δ 15, e cotanto valle il palmo quadro di quelo che costa ß 70 la cana, cioè δ 15. Ora parti ß 30 per 40, che ne viene δ 9, e cotanto valle il palmo quadro di quello che costa soldi 30 la cana, cioè δ 9. Or di' cosie, da 9 fino in 15 si ae 6. Dunque è meglio 6 δ lo palmo di quello che costa ß 30 la cana. Dunque die, per 6 δ il palmo quadro vagliono 56 palmi 28 ß, e per 6 denari lo palmo quadro vagliono li 40 palmi 20 ß. Ora giungi 28 et 20, fanno 48, e parti per mezo, ne viene 24. Et diremo che 24 ß sarae meglio la cana di quello che costava ß 30 la cana che di quelo che costava_ ß 70. Ed è fata. Così fa tute le simiglianti.

M.14.25
[F.v.27]
  Una copa è di tre parti, cioè il gambo e'l nappo e'l piede, e'l napo pesa il $\frac{1}{4}$ quarto di tuta la copa e'l piede pesa il $\frac{1}{6}$ sexto di tuta la coppa $^{\{e\}}$*e'l* gambo pesa 5 oncie. Dimi quanto pesa tuta la copa e quanto pesa il gambo e quanto pesa il piede e quanto pesa il $^{\{gambo\}}$*nappo*. Fa cosie, die, $\frac{1}{4}$ et $\frac{1}{6}$ si trova in 12. Prendi il $\frac{1}{4}$ quarto

e'l $\frac{1}{6}$ sexto di 12, ch'è 5. Ora die, da 5 fino$_{sino}$ in 12 si ae 7, et quest'è
(fol. 33^r) il partitore. Ora per cioe che'l gambo pesa 5 oncie, sì multiprica 5 via 12, fano 60, e parti 60 per 7, che ne viene 8 et $\frac{4}{7}$, e diremo che tuta la copa pesa oncie 8 et$\frac{4}{7}$ d'oncia. Ora se vuoli sapere quanto pesa il nappo per se solo, sì prendi il $\frac{1}{4}$ quarto d'oncie 8 et $\frac{4}{7}$, ch'è 2 et $\frac{1}{7}$, et tanto pesa il napo, cioè ỡ^z 2 et $\frac{1}{7}$ settimo. Ora prendi il $\frac{1}{6}$ d'oncie 8 et $\frac{4}{7}$, ch'è 1 et $\frac{3}{7}$, et tanto pesa il piede, cioè ỡ^z 1 et $\frac{3}{7}$. Et ^{de}è fatta. Se la vuolli provare, giongi insieme quelo che pessa il nappo e'l piede, cioè ỡ^z 2 et $\frac{1}{7}$ et ỡ^z 1 et $\frac{3}{7}$, che sono in tutto ỡ^z 3$_{tre}$ et $\frac{4}{7}$, et metevi suso quelo che pesava il gambo, cioè ỡ^z 5, ed ài per tuto ỡ^z 8 et $\frac{4}{7}$, dunque stae bene.^{a}

^{a} The illustration is absent from **F**.

M.14.26
[F.v.28]
  Una chiesa overo palazo, la qual'è longa bracia 120 ed è ampia bracia 36 né più né meno. E io la voglio lastricare di lastre overo pietre che sieno tute d'una grandeza, e ciaschuna pietra è longha mezo bracio ed è ampia quarto di bracio. Dimi quante pietre $^{\{varae\}}$*vorrae* a lastricare la deta chiesa overo palazzo, né più né meno. Fa cosie, primamente recha a bracia quadre la chiesa overo palazo, e multiprica la longheza contra l'ampieza, $^{=}$*cioè* 120 via 36, $^{=}$$_{che}$ fanno 4320, e cotante bracia quadre è tuto il terreno del palazo, cioè 4320. E simigliantemente recha a bracia quadre la pietra, e multipricha la longheza dela pietra contra l'ampieza, cioè mezo$_{1/2}$ et $^\wedge$via$^\wedge$ $\frac{1}{4}$, fae $\frac{1}{8}$, e diremo che in ogni bracio quadro entrino 8 pietre. E noi volemo sapere quante pietre entranno in bracia 4320. Multiprica 8 via 4320, che fano 34560, e diremo che in tutto quello terreno dela chiesa overo palazo entrarano **(fol. 33^v)** 34560 pietre né più né meno. Ed è fata. Così fa le simiglianti.

M.14.26*
[F.—]
  Se la vuoli provare, die cosie, per la longhezza entrano 240 pietre, e$_=$ per l'ampieza entrano 4 pietre per ogni bracio, dunque multiprica 4 via 240, fanno 960, e tante pietre entrano per ampieza in ogni bracio, cioè 960. E tu vuoli sapere quante entrano$_{n'entrerranno}$ in 36 bracia. Multiprica 36 via 960, fanno 34560. Dunque troviamo bene nostro conto.

M.14.27
[F.v.29]
  In Cicilia e nel Regnio di Puglia si è tut'uno peso e una misura et uno conto. E dovete sapere che in Cicilia et in Puglia si fanno tuti pagamenti a oncie et a teri et a grani. E trenta$_{30}$ teri sono una oncia, et 20 grani sono uno terie. Et questo s'intende a conto, ma non a peso. E'l fiorino d'oro di Firenze si conta teri 6 al conto, e cinque$_5$ fiorini d'oro

sono una oncia a conto, et quatro₄ carlini d'oro sono 1ᵃ_una onzia di conto, e'l carlino d'oro si conta teri 7 et_ $\frac{1}{2}$, e due carlini d'argiento si contano uno teri d'oro di conto et di pagamento.

M.14.27*
[F.—]   Et in cioe diremo uno asempro per meglio intendere, et diremo cosie, il marcho del'argento, lo qual'è 8 oncie, vale teri 36 e grany 13. Dimi quanto varano li 47 marchi e õ_z 6$\frac{1}{2}$ del deto argento. Fa cosie, primamente multiprica 47 via 36 =_teri et die, 47 via una oncia e 6 teri fano 56 oncie et 12 teri. E multiprica 13 via 47 grany, fanno 30 teri et 11 grany. Et ày per tuto õ_z 57 et teri 12 e grani 11. Ora sapia quanto vagliono õ_z 6$\frac{1}{2}$, che vaglion teri 29 e grani 16 et $\frac{9}{16}$ di grano. Ed ày per tuto õ_z 58 et teri 12 et grani 7 et $\frac{9}{16}$ di grano._ Ed è fata, e tanto varano li 47 **(fol. 34ᵛ)** marchi_ e oncie 6$\frac{1}{2}$ d'argento, cioè õ_z 58 e teri 12 • grani 7 et $\frac{9}{16}$ d'un grano.

M.14.27A
[F.v.30]   Nele fiere di Campagnia si compra e vende e fanosi tuti i pagamenti a provinigini forti, e j_provenigini si vendono ad ⁽ᵒⁿᶜⁱᵃ⁾*dozzina*. Et in ciò diremo uno asempro. La dozzina de' forti, cioè £ 12, vale £ 37₃₇ £, ß 10. Dimi quanto varano le 1443 £ de forti. Or sapie che dey così fare, e die cosie, le 1200 £_mille dugento ᵉ*sono una dozzina di centinaia, dunque le 1200 £* di forti varano £ 3750. Ora vi campano £ 243, e'lle 240 £ sono due dozzine di dicina, e ognie 120 £ di provinigini vagliono £ 375, dunque due dozine varano £ 750, ed ày per tuto £ 4500. Et avemo a fare le 3 £, che vagliono il $\frac{1}{4}$ _quarto_ di £ 37 ß 10, cioè £ 9 ß 7 δ 6. Ed ày per tuto £ 4509 • ß 7 δ 6. Ed è fata, e diremo che £ 1443 di forti vagliono £ 4509 ß 7 δ 6 di qualunque moneta tu poni a ragione di £ 37 et $\frac{1}{2}$ la dozina de provinigini.

M.14.27B
[F.v.31]   Et_ dovete sapere, se 12 forti vagliono δ 37$\frac{1}{2}$ a fiorini, dunque 12 ß_soldi 12 di forti varano ß 37 e $\frac{1}{2}$ _mezzo_ di soldo, cioè sey danari. Et 12 £ di forti varano £ 37 • ß 10 di fiorini_a'ffior. Et 120 £ di forti varano £ 370 e mezo di dizina, cioè £ 5, dunque varano £ 375. Et simigliantemente 1200 £ di forti varano £ 3750 a fiorini. E per questa regola dele dozine poi fare la valuta di quantunque forti ti fossero ᵈᵃᵗⁱ*detti*.

M.14.28
[F.v.32]   Uno mercatante conproe il quintale dela lana, lo quall'è 100 libre, £ 10. Or viene a capo d'un tempo che questa lana si trovoe bagnata e'l deto mercatante la puose ad sciugare. E quando fue asciuta, trovoe che ogni quintale era meno_menomato 5 libre, cioè che ogni 100 libre erano tornate £ 95. Dimi **(fol. 34ᵛ)** quanto gli conviene vendere il c°_100, a ciò che =_ne rifacia suo capitale. Fa cosie, die, 100 libre di lana vagliono £ 10, quanto varano libre 95? Fa cosie, multiprica 10 via 100 £, fano 1000 libre, e parti 1000 £_ per 95, che ne viene £ 10 ß 10 • δ 6 et $\frac{6}{19}$ di danaio, e diremo che gli converae vendere lo

quintalle dela lana, quando sarae asciuta, a rifarne suo capitalle, £ 10 ß 10 et_ δ 6 et $\frac{6}{19}$.
Ed è fata, e in questo modo e per questa regola fae di quantunque menomasse la lana
overo qualunque altra cossa fosse, per lo modo dela ^{ron}*regola* dele 3 cosse.

M.14.29
[F.v.33]
I'oe a'lato due borsse, nelle quali oe danari. Nel'una borsa oe il mezo e'l terzo
di tuti i danari che sono in ambidue le borsse. E nel'altra borssa oe 13 danari. Dimi quanti
danari io avea intr'ambidue' le borsse. Fa cosie, die, $\frac{1}{2}$ uno mezzo et $\frac{1}{3}$ uno terzo si trova in 6,
et prendi il $\frac{1}{2}$ mezzo et il $\frac{1}{3}$ terzo di 6, =che sono 5, e die, da 5 fino$_{insino}$ in sey$_6$ si ae 1, et
quest'e il partitore. Ora multiprica 6 via 13, fanno 78, e parti per uno$_1$, che ne viene 78,
e tanti denari avea intr'ambidue le borsse, cioè 78 δ. Ed è fata. Se la vuoli provare, die,
il $\frac{1}{2}$ mezzo e'l $\frac{1}{3}$ terzo di 78 $^{\{si\,à\}}$*si è* 65. Poni$_{Ponvi}$ suso δ 13 ch'arano nel'altra borsa et sono
δ 78. Dunque sta bene aponto.

M.14.30
[F.v.34]
Egli è uno tereno lo qual'è ampio 12 bracia, cio un muro, ed è alto bracia 7 ed
è grosso braccia j$_1$ et $\frac{1}{4}$. E io l'ò tuto murato di quadroni, e i quadroni sono tuti d'una
grandeza e sono così fati che ciascaduno quadrone è longo mezo$_{1/2}$ bracio ed è ampio
terzo$_{1/3}$ di bracio ed è alto quarto di bracio. Dimi quanti quadroni ae in questo muro. Fa
cosie, prima(fol. 35^r)mente recha a bracia quadre tuto il muro. E però ch'egli è alto 7
bracia ed è ampio 12 bracia, sì multiprica 7 via 12, fano 84. E per la groseza multiprica
84 via uno e quarto, fano 105. Et tant'è quadro tuto il muro, cioè bracia 105 =$_{\{et\,tanto\,è\,quadro}$
$_{tutto\,il\,muro\}}$. Ora rechiamo a bracia quadre =$_{tutto}$ il quadrone, e die cosie, per l'ampieza e per
la longheza multiprica $\frac{1}{2}$ via $\frac{1}{3}$, fae $\frac{1}{6}$, et per l'alteza multiprica $\frac{1}{6}$ via $\frac{1}{4}$, fae $\frac{1}{24}$. E
diremo che in ognie bracio quadro entrino 24 quadroni. Et noi volemo sapere quanti
quadroni entrino in 105 bracia. Multiprica 24 via 105, che fano 2520, e diremo che nel
deto muro entrino 2520 quadroni, né più né meno. Ed è fata. Et in questo modo et per
questa regola fae di quantunque fosse la grandeza del muro e del quadrone.

M.14.31
[F.v.35]
La libra de zendadi mi costa a Luca £ 6 • ß 5 a fiorini, $^{\{a\}}$*e* la li bra di Luca si
è 12 oncie né più né meno. Portola a Monpulieri. E ogni libra di Monpulieri si è δ^z 15•$\frac{1}{4}$,
cioè che ogni libra di Lucha torna a Monpuleri oncie 15 et $\frac{1}{4}$. E'l fiorino del'oro di Firenze
vale a Monpuslieri ß 13 • δ 4 di tornesi, e in Luccha vale il fiorino d'oro$_{dell'oro}$ ß 29 a
fiorini. Dimi per quanto potroe dare la libra al peso di Lucha in Monpuslieri, cioè per
quanti ß di tornesi, a ciò ch'io ^ne^ rifacia$_{faccia}$ mio capitale. Fa cosie, die, 12 δ^z vagliono
£ 6 • ß 5, quanto varano oncie 15 et_ $\frac{1}{4}$. Fa cosie, multiprica 12 via 6 £ et 5 ß, che fanno
£ 75$_{lxxv}$, e parti 75 =$_£$ per 15 et $\frac{1}{4}$ in questo modo, die, 4 via 15 et $\frac{1}{4}$ fano 61, e die, 4
via 75 =$_£$ fano £ 300$_{300\,£}$, e altretal'è a partire £ 300 per 61$_{16}$, come £ 75 per 15 et $\frac{1}{4}$, che
une viene £ 5 • ß 18 δ 4 et $\frac{30}{61}$ di danaio. Et per cotanto **(fol. 35^v)** puote dare la libra di

Lucha, cioè ð^z 12, al peso di Monpuslieri, cioè per £ 5 • ß 18 δ 4 et $\frac{30}{61}$ di danaio a fiorini. Ora gli rechamo a tornesi, che £ 5 • ß 16 sono 4 fiorini d'oro. Campanvi ß 2, et_δ 4 a fiorini, che vagliono a tornesi δ 13. Dunque varae la libra in Monpuslieri fiorini d'oro 4 et δ 13 tornesi,_ che vagliono a tornesi ß 54 δ 5, e tanto gli converae dà [sic, read "la"] libra di Lucha al peso di Monpuslieri vendere, cioè che 12 oncie di Luca varano al peso di Monpeslieri ß 54 • δ 5 di tornesi. Ed è fata. Così fa tute le simiglianti.

M.14.31A
[F.v.36]   La cana del panno la qual'è 4 braccia in Firenze et è palmi 8 et $\frac{3}{4}$ ala misura di Nimisi, =_{si} mi costa soldi 43 a fiorini, dimi per quanti ß di tornesi potroe dare la cana di Nimisi, la qual'è 8 palmi. E'l fiorino d'oro_{dell'oro} vale in Firenze ß 29 a fiorini e in Nimisi =_{vale} ß 13 et_δ 4 di tornesi. Fa cosie, die, palmi 8 e $\frac{3}{4}$ di Firenze vagliono ß 43 a fiorini, quanto varanno 8 palmi di Nimisi? Multiprica soldi 43 via 8 et $\frac{3}{4}$, che fanno £ 18 ß 16 δ 3, e parti £ 18 ß 16 δ 3 per 8, che ne viene ß 47, • $\frac{3}{8}$ d'uno danaio, e cotanto li converae vendere la cana di Nimisi, cioè ß 47 et $\frac{3}{8}$ d'uno danaio a fiorini. Ora gli rechiamo a tornesi, et die, soldi 43$\frac{1}{2}$, cioè uno fiorino d'oro e mezo, vagliono ß xx₂₀ di tornesi, campanvi ß 3 δ 6 a fiorini, che vagliono di tornesi δ 21, ed ày per tuto ß 21 δ 9, e cotanto gli converae vendere la chana del panno in Nimisi, cioè ß 21 e δ 9 di tornesi. Ed è fata.

M.14.32
[F.v.37]   Sono 3 compagni, che ciascuno ae δ in borsa. Dicie l'uno agli altri due, i'oe in borssa il $\frac{1}{4}$ di tuti i_δ che noi avemo intra noi. Dicie l'altro conpagnio: i'oe l'otavo di tuti i δ **(fol. 36^r)** che noi avemo intra noi. Dicie l'altro, et i'oe un danaio solamente e non più. Dimi quanti danari aveano intra tuti e tre conpagni, e quanti n'avea ciascuno per se solo. Fa cosie, die, $\frac{1}{4}$ _{un quarto} et $\frac{1}{8}$ _{uno ottavo} si trova in 8. Prendi il $\frac{1}{4}$ di 8 e l'otavo_{lo 1/8} di 8_{otto}, sono 3_{tre}, e die, da_{di} 3 fino_{insino} in 8 si ae cinque, et quest'è il partitore. Ora multiprica 8 via 1, fano 8, e parti per 5, che ne viene 1 et $\frac{3}{5}$, et tanto aveano intra tuti e tre conpagni, cioè 1 et $\frac{3}{5}$. Ora prendi il $\frac{1}{4}$ _{quarto} di 1_{uno} et $\frac{3}{5}$ _{tre quinti}, ch'è_{che nne viene} $\frac{2}{5}$. E tanto avea lo primo conpagnio, cioè $\frac{2}{5}$ d'un danaio. Ora piglia l'otavo di 1_{uno} e $\frac{3}{5}$ _{tre quinti}, ch'è $\frac{1}{5}$ _{uno quinto}, et tanto avea l'altro conpagno che dicea ch'avea l'otavo. Ed è fata. Se la vuoi provare, giongi insieme tutte e tre le parti, cioè quelo_{quelli} che dimanda il $\frac{1}{4}$, ch'ae $\frac{2}{5}$ d'un danaio, e quelo_{quelli} che dimanda l'otavo, cioè $\frac{1}{5}$ d'un danaio, et per quello_{quelli} che dice ch'ae un danaio. Ed ày per tutto danari 1 et $\frac{3}{5}$. Dunque avemo trovato bene_{bene trovato} nostro conto. E in questo modo et per questa regola fae di quanti fossoro li conpagni e di qualunque parte eglino domandassero.

[15. Practical Geometry, with Approximate Computation of Square Roots]

M.15.1
[F.—] **Al** nome di Dio amen. Qui cominciamo ad dire di tute
maniere di misure, e primamente diremo del tondo a conpasso.
Et in cioe diremo_daremo asempro et mostreremo per propria regola.

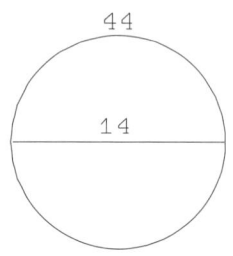

M.15.2
[F.vi.1] **Egli** è uno tereno, lo qual'è tuto ritondo a conpasso, ed è
la sua circunferentia, cioe {el}*che* gira d'intorno, bracia 44_quarantaquattro·
Dimi quant'è il suo diamitro, cioè per lo drito di mezo. Questa
è la sua propria e legitima regola. Senpre quando tu sapi la
circonferentia d'un tondo et tu voli sapere quant'è per lo drito di mezo, sì parti la
circonferentia per 3 et_ $\frac{1}{7}$, et quelo che'nne verae, tanto sarae il suo diamitro, cioè il drito
di mezo. E, simigliantemente, **(fol. 36ʼ)** quando tu say il drito di mezo d'una circonferenzia
et tu vuoli sapere quanto gira d'intorno, sì multiprica il drito del mezo per 3 e $\frac{1}{7}$, e tanto

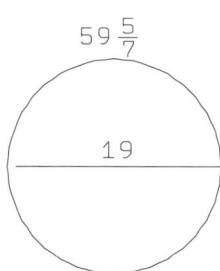

quanto farae, tanto girarae d'intorno il deto ritondo aponto.
Dunque, sicome dice la nostra regola, dovemo partire la circon-
ferencia del tondo, cioè 44, per 3 et_ $\frac{1}{7}$. Et die, 7 via 3 et $\frac{1}{7}$ fano
22, e die, 7 via 44 fano 308. Et tanto è a partire 308 per 22 quanto
44 per 3 et_ $\frac{1}{7}$, che ne viene 14. Et diremo che 14 bracia sarae
la deta circonferentia per lo drito di mezzo. Ed è fatta, e in questo
modo e per questa regola fae di tute le circonferentie, quando voli
sapere il diamitro, sichome io ti mostro ˮ*la forma* disegnata qui
apresso.

M.15.3
[F.vi.2] **Ora** diremo un altro asempro del tondo. Avemo deto del
retondo*tondo* quant'è il suo diamitro, ora diremo il drito di mezo et
mostreremo quante sarae_è tuta la circonferentia. Et diremo cosie,
egli è uno tondo a conpasso, lo qual'è per lo drito bracia 19, dimi
quanto gira tuto d'intorno. Fa cosie, sicome dicie la nostra reghola,
multiprica 19 via 3 et $\frac{1}{7}$, fanno 59 et $\frac{5}{7}$, e diremo che la sua circonferentia sia bracia 59 et $\frac{5}{7}$.

M.15.4
[F.vi.3] **Uno** terreno, lo quale è tuto ritondo a compasso et gira tuto intorno bracia 22, dimi
quanto sarae tuto questo terreno quadro dentro dal cierchio. Fa cosie, parti il cerchio per
3 et $\frac{1}{7}$, che ne viene 7, e tant'è il diamitro del deto cerchio. Ora multiprica 7 via 22, fanno
154, e parti 154 per 4, che ne viene 38$\frac{1}{2}$, e diremo che tuto questo terreno sia quadro

bracia $38\frac{1}{2}$, sicom'io$_{\text{siccome}}$ ti mostro la forma.

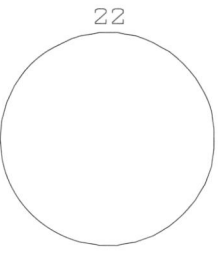

(fol. 37^r)

M.15.5 $_{\text{[F.vi.4]}}$ Uno terreno con tre cantoni$_{\text{cantora}}$, i due lati diriti e gl'altro canto $=_{\text{tutti}}$ isquadrando, di questa grandeza, che l'uno lado, cioè lato drito si è 30 bracia e l'altro lato si è 40 bracia. Dimi quanto sarae lo squadrante del terreno, cioè dala$_{\text{la}}$ ponta del'uno lato del terreno al'altro.

Fa cosie, multiprica 30 via 30, fanno 900, e multiprica 40 via 40, fa 1600. Ora giungi insieme 900 e 1600, fano 2500. Ora trova radice di 2500, cioè 50, ed è fata. Et diremo che 50 bracia abia dal'una ponta del terreno al'altra, sicom'io ti mostro quie la sua forma.

M.15.6 $_{\text{[F.vi 5]}}$ Uno terreno quadro, lo qual'è per ogni facia 10 bracia, dimi quant'à dal'uno canto del terreno al'altro. Fa cosie, multiprica 10 via 10, fanno 100, e$_{\text{Ora}}$ radoppia 100, sono 200. Ora trova radice di questo numero, cioè di 200, la qual'è 14 et $\frac{1}{7}$. Ed è fata. Et diremo che quello terreno sia, al quadrare per lo canto, bracia 14 et $\frac{1}{7}$, sicom'io ti mostro per forma qui apresso. Et in questo modo fae tute le simiglianti $=_{\text{ragioni}}$.

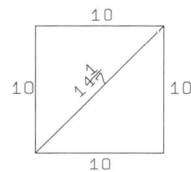

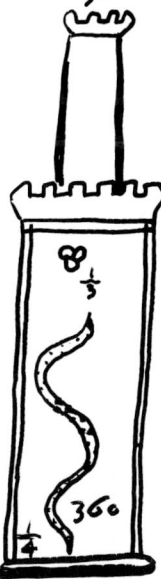

M.15.7 $_{\text{[F.vi.6]}}$ Una serpe è al piede d'una tore, la quale tore è alta 30 bracia, e la deta serpe vuole montare suso la tore. Et ogni die monta terzo$_{1/3}$ d'un$_{\text{di}}$ bracio, e la note discende quarto$_{1/4}$ d'un bracio. Dimi in quanti die montarae la deta serpe in cima dela tore. Fa cosie, die, $\frac{1}{3}$ et $\frac{1}{4}$ si trova in 12, e multiprica 12 via 30, fano 360. Ora piglia il $\frac{1}{3}$ di 12, ch'è 4, e piglia il $\frac{1}{4}$ di 12, ch'è 3. Ora tray 3 di 4, rimane uno$_1$, e parti 360 per j$_1$, che ne viene 360. Et diremo che in 360 die montarae la serpe incima dela tore, sicom'io ti mostro per forma disegnata. Et anchora potresti fare la deta ragione per altro modo et dire, $\frac{1}{3}$ et $\frac{1}{4}$ **(fol. 37^v)** si trova in 12, dunque uno terzo si è $\frac{4}{12}$, e uno quarto si è $\frac{3}{12}$. Dunque è il $\frac{1}{3}$ il terzo è più che'l $\frac{1}{4}$ quarto, $\frac{1}{12}$. E però che $\frac{1}{3}$ è piue $\frac{1}{12}$ d'un quarto$_{\text{che uno quarto}}$, si avanza la serpe ogni die $\frac{1}{12}$ d'un$_{\text{di}}$ bracio. Dunque in 12 die avanzarae uno$_1$ bracio. Et noi volemo ch'avanzi 30 bracia. Multiprica 12 via trenta, fanno 360. Ed è fata. Dunque così vene per l'uno modo come per l'altro.

M.15.8
[F.vi.7] Uno terreno lo qual'è per le due facie magiori sicome vedi disignato, bracia 60, e per l'altre due facie si è per ciascuna facia 17 bracia, dimi quant'è tuto questo terreno quadro. Fa cosie, però ch'egli è per l'una facia 60 bracia e per l'altra facia 17 bracia, sì multiprica 17 via 60, che fano ? 1020. Et diremo che tuto questo terreno sia quadro 1020 bracia. E sempre, quando voli recare a quadro alcuno terreno che'ssieno çle facie iguali, sicome deto avemo, sì multiprica la longheza contra l'ampiezza.

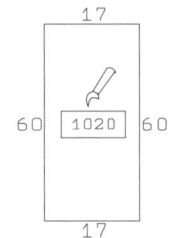

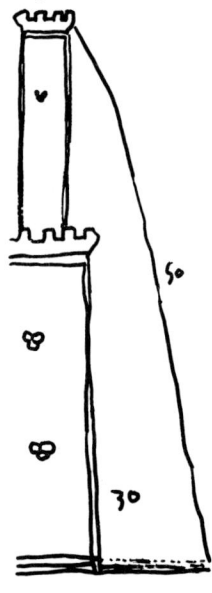

M.15.9
[F.vi.8] Una tore, la qual'è alta 50 bracia. E al piede de questa tore si ae un fosso, lo qual'è ampio 30 bracia. Ora voglio pore una fune overo corda ch'agiongha da l'orlo del fosso infino in cima de la tore. Dimi quanto sarae longha la deta corda. Fa cosie, die, però che 50 bracia è alta la tore, sì multiprica 50 via 50, fanno 2500, e però che 30 bracia è ampio_{alto} il fosso, sì multiprica 30 via 30, fanno 900. Et giongi insieme 2500 e 900, sono 3400. Ora trova radicie di 3400, la qual'è 58 et $\frac{9}{29}$, et tanto vuole es-sere longa la fune ch'agiongha dal'orlo del fosso infino in cima dela tore, cioè bracia 58 et $\frac{9}{29}$ di bracio. Ed è fata. Et io ti mostro qui apresso la forma per' meglio intendere.

(fol. 38ʳ)

M.15.10
[F.vi.9] Una tore la qual'è alta 40 bracia, e al piede dela deta tore si a un fossato. E io pongho una corda la qual'è longha 50 bracia e agiongie dal'orlo del fosso infino in cima della tore. Dimi ~~qua~~ quant'è ampio il deto fossato. Fa cosie, multiprica 50 via 50, fano 2500. E però che 40 bracia è alta la tore, sì multiprica 40 via 40, fano 1600. Ora tray 1600 di 2500, rimane 900, e trova radice di 900, lo qual'è 30. Et diremo che 30 bracia sia ampio il fossato, sicome vedi per forma designato.

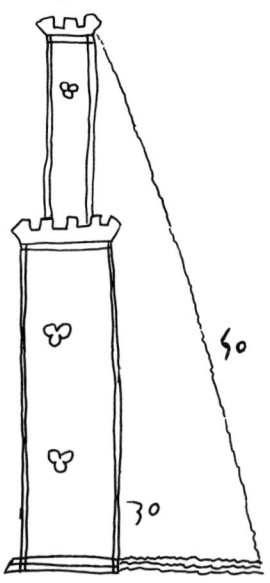

M.15.11
[F.vi.10] Uno ritondo a conpasso, lo quale gira d'intorno 100 bracia, dimi quant'è il suo diamitro, cioè il drito di mezo. Fa cosie, e quest'è la sua propia regola, parti 100 per 3

et_ $\frac{1}{7}$ in questo modo, die, 7 via 3 et_ $\frac{1}{7}$ fano 22, e die, 7 via 100

fano 700. =$_{\text{\{et altrettale è il partitore, cioè a partire 700 per 22 come 100 per 3 e 1/7 in questo modo.}}$

$_{\text{Die, 7 via 3 et 1/7 fano 22. E die: 7 via 100, fano 700\}}}$ Et altratal'è a partire 700 per

22_{ventidue} come 100 per 3 e $\frac{1}{7}$, che ne viene 31 et $\frac{9}{11}$. E tant'è quelo

tondo per la drito_ di mezo, cioè bracia 31 et $\frac{9}{11}$ di bracio, sicom'io

ti mostro la forma disignata.

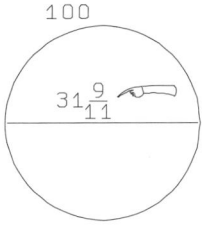

Figure 23

M.15.12
[F.—] Questa è una reghola la quale ci mostra a trovare radice
d'ogni numero del quale si puote trovare radice overo la più pressa radice che si puote
trovare. E = ciò mostreremo per propia regola.

M.15.12* Primamente diciamo chosie, per asempro, la radice di 4 si è 2, però che 2 via
[F.—]
2 fanno 4. E la radice di 9 si è 3, però$_{\text{poni}}$ che 3 via 3 fanno 9. E la radice di 16 si è 4,
però che 4 via 4 fanno 16. E la radice di 100 si è 10, però che 10
via 10 fanno 100. E la radice dy 169 si è 13, però che 13 via 13
fanno 169. E la radice di 10000 si è 100, però che 100 via 100 fa
10000. Et chosì diviene di tuti gli altri numeri che ogni numero lo
quale tu_ multiprichi **(fol. 38$^{\{v\}}$)** in se medesimo, quelo medesimo
{medesimo} numero è la radice dela sua multipricatione, sichom'ày
inteso.

M.15.13 Ora diremo in qual$_{\text{qualunque}}$ modo si trova radice_ a ogni
[F.—]
numero del quale si puote trovare, overo la più presso radice. Sappie
che dey così fare. Tu dey trovare un numero che multipricandolo
$^{\{in\}}$per se medessimo sia più presso a quelo numero honde voi
trovare ¨radice, che niun altro numero. E poi parti il rimanente per
lo doppo di quelo cotale numero che tu multiprichi. Et in questo
modo si trova radice vera o più pressa.

M.15.14 Et in ciò diremo l'ap?ntesempro, e diremo così, trovami radice di 10. Fa cosie,
[F.vi.11]
die, 3 via 3 fano 9, ¨e die, da 9 fino_ in 10 si à j$_1$. Ora parti 1 per lo dopio di 3, cioè per
6, che ne viene $\frac{1}{6}$, e giongi $\frac{1}{6}$ sopra 3, sono 3 et $\frac{1}{6}$. E diremo che la radice di 10 sia
3 et $\frac{1}{6}$, cioè la più presso radice che si puote trovare. E in questo modo et per questa
regola poi trovare radice a ciascuno numero, overo la più pressa radice che si puote trovare,
per la sopra deta regola.

M.15.15 Anchora diremo un altro asempro di trovare radice, e diremo cosie, trovami la
[F.vi.12]

radice di 67. Fa cosie, die, 8 è quello numero che, multipricandolo in se medessimo, è più presso a 67 che niuno altro numero. E però die, 8 via 8 fanno 64, e simigliantemente die, 8 e 8 fano 16. E die, da 64 fino$_\text{sino}$ in 67 si ae 3, e parti 3 per 16, che ne viene $\frac{3}{16}$. E diremo che la radice di 67 si è 8 et $\frac{3}{16}$. Ed è fata per la reghola.

M.15.16
[F.vi.13] Anchora diremo cosie, dimi quant'è la radice di 82. Fa cosie, et die, 9 è quelo numero che, multipricandollo **(fol. 39^r)** in sè medessimo_ è più presso a 82 che niuno altro numero. Et però die, 9 via 9 fano 81, e die, da 81 infino$_\text{insino}$ in 82 si à 1. Ora quelo che tu multipricasti, cioè 9, sì la radopia, che sono 18, e parti 1 per 18, che ne viene $\frac{1}{18}$. E diremo che la radice di 82 sia 9 et $\frac{1}{18}$. Et in questo modo fae di tuti gli altri numeri onde tu voli trovare radice. Sempre parti il soperchio che ti viene il multipricato per lo dopo di quello che multiprichi in sè.

M.15.17
[F.vi.14] Uno terreno lo qual'è longho bracia 567 ed è ampio bracia 31, sicom'io ti mostro quì disignato per forma. Et io lo voglio tuto acasare. Die, lo voglio tuto fare acasare di case che sia longha l'una bracia 11 e sia ampia bracia_ 7, né più né meno. Dimi quante case v'alogarai tu, a cioe che tue empi tuto quelo terreno. Fa cosie, primamente recha a bracia quadre tuto il terreno, e multiprica la longheza contra l'ampieza, cioè 31 via 567, che fano 17577. Et =$_\text{le}$ cotante$_\text{tante}$ bracia quadre è tuto il terreno, cioè bracia 17577. Et simigliantemente recha la casa a bracia quadre, e multiprica la longheza contra l'ampieza, cioè 7 via 11, fano 77. Et cotante bracia quadre è la casa, cioè 77 bracia. Hora, se vuoli sapere quante case v'alogharay, sì parti 17577 per 77, che ne viene 228 et $\frac{3}{11}$. Ed è fatta. E diremo che in tuto quelo terreno vi si farebero 228 case et $\frac{3}{11}$ di casa, né più né meno. In questo mo do fa le simiglianti.

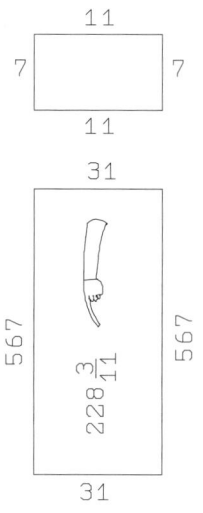

M.15.18
[F.vi.15] Uno pozzo quadrato, lo qual'è per ogni facia 2 bracia, ed è profondo bracia 50, ed è pieno e raso d'aqua. Or viene **(fol. 39^v)** che vi chade entro una colona quadrata, la qual'è per ogni facia un bracio ed è longha bracia 25. Dimi quant'acqua uscirae fuori del detto pozo per questa colona che vi chade. Fa cosie, primamente racha a bracia quadre il pozo, e multiprica 2 via 2, fanno 4. E per la profunditade multiprica 4 via 50, fano 200. Et cotante bracia quadre è tuto il pozo, cioè 200 ¬*braccia*. Ora, simigliantemente recha a bracia quadre la colona, e multiprica j_1 via j_1, fa j_1. E per la longheza, j multiprica j_1 via 25, fae 25. Et cotante bracia quadre è la colona, cioè bracia 25. Ora parti 200 per 25, che ne viene 8. E diremo che 8 bracia quadre d'aqua uscirano fuori del pozzo per

questa colona che vi chade entro, sicom'io ti mostro la forma del pozzo
et dela cholona.

M.15.19
[F.vi.16] Uno terreno con cinque facie iguali,
sicome vedy qui disignato, lo quale si chiama il
pentagone ed è per ogni facia 8 bracia. Dimi
quant'è tuto quello terreno quadro. Questa è la sua
reghola, multiprica l'una dele sue_ facie in se
medessimo, cioè 8 via 8, fanno 64, e multiprica

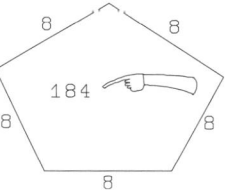

3 via 64, fano 192. E di 192 tray l'una dele facie, cioè 8, rimane 184. Ed
è fata, e diremo_ che quello terreno sia tuto quadro bracia 184. E in questo
modo et per questa regola fae di quantunque fosse il terreno per facia,
esendo le facie iguali ed_ esendo 5 facia, sempre multiprica l'una dele facie
in se medessima e poi fae tre via quela multipricagione, e dela soma tray l'una dele facie,
e'l rimanente sarae tuto quello terreno quadro sicome deto avemo.

(fol. 40ʳ)

M.15.21
[F.vi.17] Uno padiglione, lo qual'è longo il feristo del mezo 40 bracia
e'l panno è longho dala cima del feristo di sopra in fino al'orlo del
padiglione di sotto bracia 50. Dimi quant'è tuto questo panno et
quanto terra posede soto sè il deto padiglione. Fa cosie, die, però che'l
panno è longo 50 bracia sì multiprica 50 via 50, fano 2500. Et, però
che'l feristo è longho 40 bracia, sì multiprica 40 via 40, fanno 1600 mille
secento· Ora tray di 2500 1600, rimane 900, e trova radice di 900, ch'è

30. E radopia 30, fano 60, e tanto è cotant'è ampio il padiglione per lo
drito di mezo, cioè 60 bracia. Ora multiprica 60 per 3•$\frac{1}{7}$, che fano
188 $\frac{4}{7}$· cento ottantotto et quattro settimi, et tanto è tuto il cerchio del padiglione d'intorno. Ora, se
voli f sapere quanta terra posiede sotto sè, sì parti il drito del mezzo de padiglio(n)e per
mezo_, cioè 60, che ne viene 30. E simigliantemente parti il cerchio del padiglione, cioè
188•$\frac{4}{7}$ per mezo, che ne viene 94 et $\frac{2}{7}$. Ora multiprica 30 via 94 et $\frac{2}{7}$7, che fano 2828
et $\frac{4}{7}$. Et cotanta terra posiede soto sè il deto_ padiglione, cioè bracia 2828•$\frac{4}{7}$ quadre. Ora
se voli sapere quant'è tuto il panno, parti il diamitro, cioè 60, per $\frac{1}{2}$, che ne viene 30,
e multiprica 30 via 50, fanno 1500, et cotante bracia quadre è tuto il panno del deto
padiglione, cioè 1500 bracia. Ed è fata, sicome vedi la forma designata.

M.15.21A(20)
[F.vi.18] Uno schudo, cioè uno trianghola, lo qual'è per lo drito di mezo bracia 5, dimi
{dimi} quanto sarae il deto diritto trianghola per ciaschuna facia. Fa cosie, multiprica 5 via
5, fanno 25, e parti 25 per 3, che ne viene 8 et $\frac{1}{3}$, et giongi **(fol. 40ᵛ)** sopra 25 8 et $\frac{1}{3}$,

sono 33 et $\frac{1}{3}$, e$_{or}$ sapia che la radice di 33 et $\frac{1}{3}$ sarae il deto triangolo

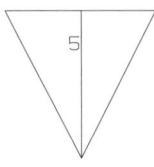

per facia. La radice trova secondo la reghola che deto avemo, la quale radice diciamo che sia 5 et $\frac{5}{6}$ meno $\frac{17}{54}$ non aponto, e tanto sarae lo scudo per la faccia. Io ti mostro_ la forma per meglio intendere. Così fa tuti i simiglianti. Et questo s'intende d'uno scudo che abia le facie iguali di misura.

M.15.22
[F.vi.19] **A**nchora diremo un altro asempro del triangolo per mostrarlo più apertamente. E diremo cosie, uno triangholo lo qual'è iguali per faccia,

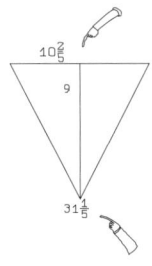

cioè tanto per l'una facia quanto per l'altra, ed à dala ponta di soto infino$_{insino}$ ala facia di sopra, cioè per lo drito di mezo, 9 bracia. Dimi quant'è $^{\{tute\ e\ tre\}}$*tutte'lle* facie, cioè quant'è per ciascuna facia. Fa cosie, multiprica 9 via 9, fanno 81, e parti 81$_{ottantuno}$ per 3, che ne viene 27. Et giongi 27 sopra 81, sono 108, e di questo numero trova radice, cioè di 108, ch'è 10 et $\frac{2}{5}$. Et tanto sarae per facia il deto terreno, cioè bracia 10 et $\frac{2}{5}$ di bracio. Se voli sapere quanto è tuto il triangolo d'intorno, multiprica l'ampieza d'una dele facie, cioè 10 et $\frac{2}{5}$, per 3, fano 31 et $\frac{1}{5}$. Et diremo che tuto il triangholo giri d'intorno bracia 31 et $\frac{1}{5}$. $^{\{Et\}}$= In questo modo fa tute le simiglianti.

M.15.23
[F.vi.20] **S**ono 2$_{due}$ lancie, le quali sono fitte in uno piano, e l'una lancia è longha 10 bracia e l'altra è longa bracia 17, e da l'una lancia a l'altra si ae 20 bracia. Dimi quante bracia avrae dal'una cima dela lancia al'altra. Fa così, tray 10 di 17, rimane 7, e multiprica 7 via 7, fanno 49. Et simi**(fol. 41$^{\{r\}}$)**gliantemente multiprica 20 via 20, fanno 400, e giongi insieme questi due numeri, cioè 49 et 400, che sonno in tuto 449. E trova radice di 449, ch'è 21 et $\frac{4}{21}$. $^{=}$*Et diremo che, da l'una punta della lancia a l'altra, abbia braccia 21 et* $\frac{4}{21}$

di bracio. Ed è_ fata. Io ti mostro la forma.

M.15.24
[F.vi.21] **S**ono 2$_{due}$ tori in uno piano, sicom'io ti mostro disignato. E l'una tore è alta 20 bracia e l'altra è alta 25 bracia, et nel mezo di queste due tori si à una coppa, sicome vedi disignato. E dal'una tore al'altra si ae 100 bracia. Et in $^{=}$*su* ciascuna di queste 2 tori si ae una colonba, le quali vogliono ire a bere in questa copa, et muovansi ad un'ora, e volano $^{\{di\}}$*d'uno* volare così l'una come l'altra. Dimi quanto sarae più tosto l'una colonba che l'altra a bere nela copa. Fa cosie, die, però che dal'una tore al'altra si è 100 bracia,_ sì parti 100 per mezo$_{1/2}$, che ne viene 50, e multiprica 50 via 50, fano 2500. Et però che l'una tore è alta 20 bracia, sì multiprica 20 via 20, fanno 400. Et giongi 400 sopra 2500,

ed ày 2900. Ora trova radice di 2900, ch'è 54 meno $\frac{4}{67}$ 4/77. Et in cotanto sarae a bere $=_{\text{nela coppa}}$ la colomba ch'è su la tore ch'è alta 20 bracia, cioè in 54 bracia meno $\frac{4}{67}$ 4/77 di bracio. Se voli sapere in quanto vi sarae l'altra colonba, sì multiprica 25 via 25, fano 625. Et simigliantemente giongi sopra 2500, sono 3125. Et di questo trova radice, cioè di 3125, lo qual'è 56 et $\frac{1}{112}$, e in cotanto sarae l'altra colomba a bere ala coppa, 1cioè in bracia 56 et $\frac{1}{112}$ di bracio. Ora tray di 56 et $\frac{1}{112}$ $=_{\text{di}}$ 54 e $\frac{4}{67}$ 4/77, rimane 1 et $\frac{73}{77}$, et in cotanto sarae a bere ala copa $=$*prima* l'una colomba che l'altra, cioè bracia 1 et $\frac{73}{77}$ di bracio, cioè quela ch'è sula tore dele 20 bracia.

(fol. 41ᵛ)

M.15.25
[F.vi.22]
Uno palazzo lo qual'è quadro ed è tuto iguali il muro di sopra, ed è ampio 40 bracia ed è longho altre 40 bracia et io vi_ voglio porre suso un teto a due piovitoi, e voglio che sia alto il detto teto nel colmigniolo bracia 13. Dimi quante bracia vorano essere i corenti che giongano dal colmigniolo fino sule mura. Fa cosie, però che'l palacio è ampio 40 bracia, sì dividi 40 per mezzo, $=_{\text{che}}$ ne viene 20, e multiprica 20 via 20, fanno 400. E però che'l teto è alto nel colmignio 13 bracia, sì multiprica 13 via 13, fano 169, e giongi insieme 400 et 169, sonno 569. E trova radice di 569, la qual'è 23 et $\frac{20}{23}$, e diremo che

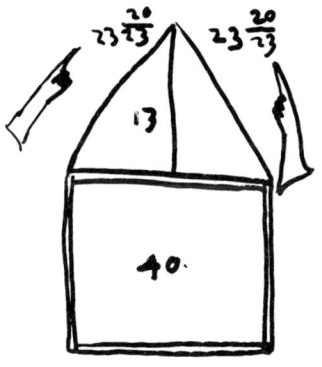

i corenti che si riposano sopra il colmigni e pogiano suso il muro vogliono esser longhi_ bracia 23 et $\frac{20}{23}$ di bracio l'uno, sicom'io ti mostro disignato qui apresso per meglio intendere.

M.15.25A
[F.vi.23]
⟨d⟩imi quant'è la radice di 101. Fa cosie, multiprica 10 via 10, fanno 100, e simigliantemente radopia quelo che tue multipricasti, cioè 10, sono 20. Et die, da 100 fino in 101 si ae uno1. Ora parti 1 per 20, che ne viene $\frac{1}{20}$, e giongi 10, sonosopra 10

et $\frac{1}{20}$. Et tant'è radice di 101.

[20. Tabulated Degrees of Fineness of Coins]

M.20.1
[F.—] Al nome di Dio amen. Qui apresso sarano scrite tute maniere di leghe di monete, e simigliantemente tuti alegamenti$_{i\ legamenti}$ d'oro et d'argento et di rame, come s'aleghano l'una moneta overo bolzone $=_o$ d'oro in verghe, o d'argiento di tutte ragiony.

M.20.2
[F.—] Et cominciamo cosie. Dovete intendere che $=_{in}$ una oncia d'oro fine si è$_à$ 24 carati. E quanto l'oro è pegliore, meno carati ae nel'oncia. E quanto l'oro è migliore, ae nell'oncia più carati$_{più\ karati\ ae\ nell'oncia}$. E simigliantemente aviene del'argento, ma l'argento s'alega a δ^z overo a danari pesi. E l'argento che tiene 12 oncie per libra s'intende che'l$_{che}$ sia argento fine e bono aponto.

(fol. 42^r)

M.20.3 [F.—] Fiorini d'oro di Firenze sono a carati	24	per oncia
Agostany d'oro sono a carati	$20\frac{1}{2}$	per oncia
Perperi paglialocati sono a carati	15	per oncia
Dobble dela mira sono a carati	$23\frac{1}{2}$	per oncia
Dobble del rascietto sono a carati	$23\frac{1}{4}$	per oncia
Castelany d'oro sono a carati	$23\frac{1}{2}$	per oncia
Anfosini d'oro sono a carati	$20\frac{1}{2}$	per oncia
Tornesi d'oro sono a carati	$23\frac{3}{4}$	per oncia
Bisanti vecchi d'oro sono ^a^ carati	24	per oncia
Perperi vecchi comunali e mezany sono a carati	17	per oncia
Bisanti saracinati d'oro, che ne vano dodici per oncia sono a charati	15	per oncia
Luchesi d'oro a cavalo sono a carati	18	per oncia
Luchesi d'oro a piede sono a carati	23	per oncia
Perperi novi sono a charati	14	per oncia
Gianovini d'oro a chavalo sono a carati	24 meno $\frac{1}{15}$	per oncia
Gianovini d'oro a piede sono a carati	$23\frac{1}{4}$	per oncia
Carlini d'oro sono a carati	24	per oncia
Pezzeti di bisanti a carati	12 meno $\frac{1}{4}$	per oncia
Romany d'oro a carati	24 meno $\frac{1}{18}$	per oncia

Parigini d'oro coll'agnus dei a carati	24 meno $\frac{1}{4}$	per oncia
Ducati d'oro di Vinegia a carati	24 iscarsi	per oncia
Ragonesi d'oro sono a carati	24 men $\frac{1}{4}$	per oncia
Bisanti d'Acri colla crocie sono a carati	$16\frac{1}{3}$	per oncia
Santolene fini sono a carati	24	per oncia
Marrabortini d'oro sono a carati	21	per oncia

(fol. 42ᵛ)

Medaglie Massamutine sono a carati 2	24	per oncia

Oro di pagliola, ˢⁱᶜʰᵒᵐᵉ*secondo chome* s tiene, il migliore si è a charati

22 e'l comunale si è a carati	20 sino in 21	
Bisanti vechi d'Alesandria a carati	24	per oncia

30 teri sono una oncia et $=_i$ venti grany sono un teri d'oro

M.20.3A   Franchi di Francia a pie sono a carati 23 g̃ʳ 10

Franchi di Francia a cavalo a carati $23\frac{5}{8}$

Nobili d'Inghilterra sono a carati $23\frac{5}{8}$

Fiorini romany cioè ducati romany sono a carati $23\frac{1}{4}$

Franchi del Ducha d'Angio sono a carati 22

Fiorini de Aragonia sono a carati 17

Fiorini di papa e di reyna e di Vignone sono a carati $23\frac{1}{8}$

Scudi del Re di Francia sono a carati $23\frac{3}{4}$

Fiorini ungari del gilio e del'Amagnia a carati $23\frac{3}{4}$

(fol. 43ʳ)

M.20.4 Qui sono scritte tute tenute di monete d'argento
[F.—]

Tornesi grossi sono a oncie 11 et $\frac{1}{2}$ per libra, entendesi che la libra sia oncie 12 d'argento
       fine in tuti allegamenti.

Medaglie di tornesi primiere sono a oncie $11\frac{1}{2}$ per libra

Medaglie ⁽ᵈⁱ⁾= terzeriole sono a oncie 11 per libra

Carlini et Merghuzesi et Barzalonesi sono a oncie 11•$\frac{1}{4}$ per libra.

Sterlini sono a oncie 11 • δ 2 per libra

Viniciani di Vinegia sono a oncie 11•$\frac{3}{4}$ per libra

Popolini di Firenze e da Siena e di Pisa sono comunamente
       a oncie 11 et δ 15 per libra

Aghulini vechi da Pisa sono a oncie 11 per libra

Bolognini grossy sono a oncie 9 et δ 21 per libra

Astigiany sono a oncie 8 et δ 18 per libra

Inperiali et Piacentini sono a oncie 9 per libra

Romany di peso di tornese sono a oncie 11 δ 8 per libra

Gianovini sono a oncie 11 δ 12 per libra

Baldachini coll'aghuglia sono a oncie 11 et_ δ 8 per libra

Fregiachiesi d'Aghulea col'aghuglia et dela tore e del giglio e de-
      la luna sono a oncie 8 et δ 8 $10\frac{1}{2}$ per libra

Agontany grossi sono a oncie 11 et δ 15 per libra

Ravingniany grossi sono a oncie 10 • δ 12 per libra

Sanesi vechi sono a oncie 11 • δ 6 per libra

Volterany grossi sono a oncie 9 per libra

Et intendesi oncie 12 la libra, et 24 danari pesi sono 1 oncia.

M.20.5
[F.—]   **Q**ui sono scrite leghe di monete piciole

Parigini primieri sono a δ 5 e grani 18 di legha

Parigini sicondi sono a δ 4 • grani 16 di lega

(fol. 43ᵛ)

Parisini terzi sono a δ 3 et grani 14 di legha

Tolosani vechi ala crocie sono a δ 6 • grani 18 di legha

Tolosany ala fiore sono a δ 7 • grani 4 di legha

Murlany sono a δ 7 • grani 7 di legha

Reali primeri sono a δ 4 • grani 18 di legha

Reali secondi sono a δ 3 • grani 18 di legha

Reali terzi sono a δ 3 di legha

Ternali sono a δ 3 • grany 14 di legha

Medaglie ternali sono a δ 3 • grana 3 di legha

Coronati del Re Carlo primeri sono a δ 4 di legha

Coronati sicondi sono a δ 3 • grani 18 di legha

Coronati terzi sono a δ 3 di legha

Rinforzati sono a δ 3 • grani 15 di legha

Reali di Marsilia a δ 3 • grani 18 di legha

Merchuzesi et Valenzany e chapo di Re a δ 3 et $\frac{1}{2}$ di legha

Coronati vechi a δ 2 • grany 18 di_ legha

Caorsini sono a δ 3 di legha

Vaselamento di Parigii et di Torso et di Monpuslieri
      sono a oncie 11 et $\frac{3}{4}$ per libra

Vaselamento di Marsilia a oncie $11\frac{1}{2}$ per libra

M.20.5A   Tenute di monete lombarde come qui di sotto.[a]

Otini di Milano sono a δ 4 ğʳ 10, per marcho õᶻ 2 δ 22 ğʳ 16

Quatrini di Milano e Pavia a δ 3 ğʳ 15, per m° õᶻ 2 δ 10 ğʳ 0

Quatrini da Cremona da leone a δ 3 ğʳ 15, per m° õᶻ 2 δ 10 ğʳ 0

Quatrini vegi da Milano a δ 3 ğʳ 22, per mᵘ õᶻ 2 δ 14 ğʳ 16

Quatrini da Crema δ ğʳ , per m° õᶻ δ ğʳ

Sexini vegi da Milano a δ 5 ğʳ 22, per m° õᶻ 3 δ 22 ğʳ 16

(fol. 44ʳ)

Sexini novi da Milano e Pavia a δ 5 ğʳ 21, per m° õᶻ 2 δ 14 ğʳ 0

Pigioni dela † e galeazo a δ 7 ğʳ 10, per m° õᶻ 4 δ 22 ğʳ 16

Pigioni vegi vegi di più sorti a δ 7 ğʳ 22, per m° õᶻ 5 δ 6 ğʳ 16

Dodexini da Milano a δ 7 ğʳ 10, per m° õᶻ 4 δ 22 ğʳ 16

Grossi vegi de Milano a δ 10 ğʳ 18, per m° õᶻ 7 δ 4 ğʳ 16

Inperiali del galeazo da Milano a δ 1 ğʳ 10, per m° õᶻ δ 22 ğʳ 16

Inperiali dele lettere da Milano a δ 1 ğʳ 22, per m° õᶻ 1 δ 6 ğʳ 16

Bisuoli da Milano e Pavia a δ ğʳ 22†, per m° õᶻ δ 15 ğʳ 0

Bisuoli da Monza e da Cantù a δ ğʳ 22, per m° õᶻ δ 14 ğʳ 16

Bisuoli da Como a δ ğʳ 16, per m° õᶻ δ 10 ğʳ 16

Inperiali da Brescia a δ 1 ğʳ 1, per m° õᶻ δ 16 ğʳ 16

Inperiali da Cremona a δ 1 ğʳ , per m° õᶻ δ 16 ğʳ 0

Inperiali da Crema a δ ğʳ 22, per m° õᶻ δ 14 ğʳ 16

Inperiali da due di Milano a δ 2 ğʳ 4, per m° õᶻ 1 δ 10 ğʳ 16

Inperiali da due da Monza a δ 1 ğʳ 16, per m° õᶻ 1 δ 4 ğʳ 16

Pigioni novi da Milano e Pavia a δ ğʳ , per m° õᶻ 4 δ 4 ğʳ 16

Pigioni da Como a δ ğʳ , per m° õᶻ δ ğʳ

Sexini da Monza a δ 5 ğʳ 22, per m° õᶻ 3 δ 22 ğʳ 16

Grosi di Gienova vegi a δ 11 ğʳ 10, per m° õᶻ 7 δ 14 ğʳ 16

Grossi di Gienova novi a δ ğʳ , per m° õᶻ δ ğʳ

Otini di Gienova a δ 5 ğʳ 22, per m° õᶻ 3 δ 22 ğʳ 16

Denari da due di Gienoa a δ ğʳ , per m° õᶻ δ ğʳ

Inperiali da Gienova a δ ğʳ , per m° õᶻ δ ğʳ

Grossi di Vinegia a δ 11 ğʳ 14, per m° õᶻ 7 δ 18 ğʳ 16

Soldini di Vinegia a δ 11 ğʳ 14, per m° õᶻ 7 δ 18 ğʳ 16

Grossi da Firenze a δ 11 ğʳ 13, per m° õᶻ 7 δ 16 ğʳ 16

Grossi di Pisa a δ 11 ğʳ 11, per m° õᶻ 7 δ 15 ğʳ 8

(fol. 44ᵛ)

Quatrini da Brescia a δ 3 ğʳ 22, per m° õᶻ 2 δ 14 ğʳ 16

[a] In this section, each piece weighs ¹⁄₁₆ mark. One ounce is 24 weight *denari*, and each of these

24 grani.

  This additional list is probably due to a different hand (though similar to the one that wrote the rest). Firstly, it has the orthography *vegi* while **M** uses *vechi* consistently; secondly, "5" is written in one place (in the line "Bisuoli da Milano e Pavia") with a shape that is used nowhere else in **M** but only in **V**.

[21. Alloying Problems]

(fol. 45^r)

M.21.2
[F.vii.1] I'oe 60 once d'oro, lo qual'è a carati 16 per oncia, e vogliolo$_{voglio}$ metere a fuocho e afinarlo tanto che tor⟨ni⟩mi a carati$_{liar}$ 21 per oncia. Dimi quanto deboro tornare queste 60 oncie a peso, t⟨r⟩atolo dal fuocho e sia di carati 21, né più né meno. Fa cosie, sapia quanti carati d'oro ae nelle dete 60 oncie d'oro$_—$ che tu metesti a fuoco di prima, et multiprica 16 via 60, fano 960, e tanti carati era l'oro che tu metesti a fuoco di prima, cioè carati 960. Ora se voli sapere quanto torna a peso, sì parti carati 960 per 21 $^=$*però che tu vuo' che'tti torni di carati 21*, che ne viene 45 et $\frac{5}{7}$, e sono oncie. E diremo che le dete 60 oncie che tu metesti a fuocho $^{\{e\}}$*a carati 16 per oncia, tornerae, tratto del fuocho,* oncie 45 et $\frac{5}{7}$ $=_{d'oncia}$ e sarae a carati 21 per oncia. Ed è fata$_—$.

M.21.2A
[F.vii.2] I'oe 18 oncie d'oro, lo qual'è a charati $20\frac{1}{2}$ per oncia. Et io lo voglio mettere a fuocho et afinarlo tanto che torni oro fino, cioè a 24 carati $=_{per\,oncia}$. Dimi quanto torneranno le dete 18 oncie a peso, trato dal fuocho e sia oro fino di 24 carati. Fa cosie, sapia primamente quanti carati d'oro ae nele dete 18 oncie che tu $=_e$ meti a fuocho, e multiprica 18 via $20\frac{1}{2}$, fanno 369, e tanti carati d'oro ae nele dete 18 oncie, cioè carati 369. Hora se voli f sapere quanto torna a peso, sì parti 369 per 24, però che tu vuoli che ^{ritorni}*torni* di 24 carati per oncia, che ne viene 15 et $\frac{3}{8}$. Ed è fata. E diremo che le dete 18 once, lo quale tu mettesti a fuocho a carati $20\frac{1}{2}$ per oncia, tornerae a peso, trattolo da fuocho, oncie 15 et $\frac{3}{8}$ a carati 24 per oncia.

M.21.3
[F.vii.3] I'oe oncie 7 d'oro, lo qual'è a carati 19•$\frac{1}{2}$ per oncia, et honne once 9, lo qual'è a carati 20 {20} et $\frac{1}{4}$ per oncia, et honne once 16, lo qual'è a carati 21 et $\frac{2}{3}$ per oncia, e anche n'oe once 20, lo qual'è a carati 23 et$_—$ $\frac{3}{4}$ per oncia. Ora voglio **(fol. 45^v)** tuti questi quatro ori fare fondere insieme et farne una vergha di tuto così mischiato insieme. Dimi quanto sarae tuta questa vergha a pesso e di quanti carati tornerae per oncia apponto. Fa cosie, primamente sapia quanti carati d'oro ae nele 7 once d'oro primere$_{prime\,7\,once}$, lo qual'è a carati $19\frac{1}{2}$ per oncia, e multiprica 7 via oncie $19\frac{1}{2}$ $_—$, fano oncie 136 et$_—$ $\frac{1}{2}$, cioè che sono carati. E diremo che nele dete 7 oncie siano carati $136\frac{1}{2}$. Ora sapiamo ~~sapia~~

quanti carati d'oro ae nelle 9 oncie, lo qual'è a carati 20 et $\frac{1}{4}$ per oncia. Multiprica 9 via 20 et $\frac{1}{4}$, fanno 182 et $\frac{1}{4}$, e tanti carati ae nele dete 9 once, cioè caratti 182 et $\frac{1}{4}$. ⸗*Et tanti carati ae nelle dette 9 once*. Et sapia quanti carati ae nele dete 16 oncie, lo qual'è a carati 21 et $\frac{2}{3}$ per oncia, e multiprica 16 via 21 et $\frac{2}{3}$, ⸗$_{che}$ fano 346 et $\frac{2}{3}$, e tanti carati ae nele dete 16 oncie, cioè 346 et $\frac{2}{3}$ ⸗. E sapia quanti carati ae nele dete 20 oncie, lo qual'è a carati 23 et $\frac{3}{4}$ per oncia. Multiprica 20 via 23 et $\frac{3}{4}$, fanno 475, e tanti carati ae nele dete 20 once, cioè carati 475⸗. Ora giongi insieme tuti questi carati, cioè carati $136\frac{1}{2}$ e carati 182•$\frac{1}{4}$ et carati 346•$\frac{2}{3}$ e carati 475, che sono in tuto carati 1140 et $\frac{5}{12}$ di carato. Ora simigliantemente giongi insieme tuto l'oro, cioè oncie 7 et oncie 9 et oncie 16 et oncie 20, che sono in tuto oncie 52. Ora parti tuti questi carati, cioè 1140 et $\frac{5}{12}$, ⸗*per 52*, che ne viene carati 21 et $\frac{581}{624}$, che sono questi roti bonamente $\frac{3}{4}$. Et diremo che tuta questa verga sarae oncie 52, e sarae a carati 21 et $\frac{3}{4}$ di carato. Ed è fata.

M.21.4
[F.vii.4] I'oe biglione lo qual'è a δ 11 di legha e oe biglione lo qual'è a δ 4 di lega. E io voglio fare una moneta, la **(fol. 46ʳ)** quale sia a δ 7 di legha, né più né meno, e voglione alegare 100 marche. Dimi in questi 100 marchi quanto meteroe di ciaschuno di questi due biglioni, e'ssia 100 marchi a δ 7 di legha. Fa cosie, die, la lega ch'io voglio fare si è a δ 7, e il più alto biglione ch'i'oe⸗ si è a δ 11. Dunque dovemo dire, da 7 fino$_{insino}$ in 11 si à 4. Et prendi marchi 4 del contradio biglione, cioè di quello ch'è a δ 4 di lega. Et simigliantemente die, da 7 fino$_{insino}$ in 4 menoma 3, et prendi marche 3 del contrario biglione, ehe cioè di quello ch'è à δ 11 di legha. Ora ày alegati marchi 7 a δ 7 di lega. Ed à vi messo marchi 4 de^l^$_{di}$ biglione, lo qual'è a δ 4 di lega, ed à vi meso marchi 3 del$_{di}$ biglione, lo qual'è a δ 11 di lega. E noi volemo alegare 100 marchi. Dunque multiprica 3 via 100 marchi, fanno 300 marchi, e parti 300 per 7, che ne viene 42 et $\frac{6}{7}$. Et tanto vuole del biglione, lo qual'è a δ 11 di lega, cioè marchi 42 et $\frac{6}{7}$. Ora multiprica 4 via 100, fanno 400, e parti in 7, che ne viene 57 et $\frac{1}{7}$, et tanto vi⸗vole del biglione lo qual'è a δ 4 di legha, cioè marchi 57 et $\frac{1}{7}$ di marcho. Ora avemo alegato marchi 100 di biglione, lo qual'è⸗a δ 7 di legha. Et avemovi messo marchi 42 et $\frac{6}{7}$ del biglione lo quale è a δ 11 di legha, ed à vi messo marchi 57 et $\frac{1}{7}$ del$_{di}$ biglione lo qual'è a δ 4 di legha. Ora giongi insieme marchi 42 et $\frac{6}{7}$ e marchi 57•$\frac{1}{7}$, che sono 100 marchi. Dunque n'ày alegati 100 marchi. E per questa regola ne poi alegare quantumque quantitade tu vuoli et di qualumque legha tu voli⸗. Ora proviamo s'avemo bene **(fol. 46ᵛ)** alegato, et provasi in questo modo. Et di' cosie, ne' deti 100 marchi che tu ày alegati a δ 7 di legha, si entrò denari 700 di lega. Ora vegiamo se noi ritroviamo i deti 700 δ. Di' cosie, noi avemo alegati et messevi marchi 42 et $\frac{6}{7}$ di marchi a δ 11 di legha, che v'à entro 471 δ et $\frac{3}{7}$ di danaio. Et à vi mesi marchi 57 et $\frac{1}{7}$ di biglione lo qual'è a δ 4 di lega per marco, che v'à entro δ 228 et $\frac{4}{7}$. Ora giongi insieme questi danari, cioè δ 471 et $\frac{3}{7}$ et

δ 228• $\frac{4}{7}$, che sono in tuto denari 700. Dunque avemo bene alegato, però ch'avemo aponto ritrovati i deti 700 δ. S'avessimo trovato piue o meno starebbe male.

M.21.5
[F.vii.5] I'oe marchi 5 d'ariento, lo qual'è a oncie 9 per libra, et ò ne marchi 8, lo qual'è a oncie 10 et $\frac{1}{4}$ per libra. Ed oe marchi 2 di rame $^{p\,roto}$*pretto*. Et io foe fondere tuto questo argiento e'l rame insieme et fo ne un pane. Dimi quanto questo pane pesarae et di che legha tornerae, l'argiento e'l rame mischiato insieme. Fa così, giongi insieme l'argiento e'l rame, che sono marchi 15, et quest'è il partitore. Ora sapia quante oncie ae ne' 5 marchi, che v'ae entro oncie 45. Et sapia quante oncie ae nele 8$_{otto}$ marchi, che v'à entro oncie 82. Et simigliantemente sapia quante oncie ae nele due marchi di rame, che non vi n'ae nula d'argento, però ch'è preto rame. Et giongi$_{raggiugni}$ insieme queste oncie, $^=$*cioè* õ^z 45 e õ^z 82, che sono õ^z 127, e parti õ^z 127 per 15, che ne viene õ^z 8 et $\frac{7}{17}$ d'oncia. E diremo che tuto questo argento e'l rame sì mischiato insieme sia_ marchi 15 e sia a oncie 8 et $\frac{7}{15}$ d'oncia per libra.

(fol. 47^r)

M.21.6
[F.vii.6] I'oe argento fine, lo quale tiene oncie 12 per libra, e ò ne$_{once}$ d'un altro argento basso, lo quale tiene oncie 8• $\frac{1}{2}$ per libra, e io voglio fare una moneta che sia a oncie 9$\frac{1}{2}$,$_=$ et voglione alegare 20 marchi. Dimi, in questi 20 marchi quanto vi meteroe di ciascuno di questi due argenti a ciò che sia alegati a oncie 9$\frac{1}{2}$ per libra. Fa cosie, die, la lega ch'io voglio fare si è a oncie 9$\frac{1}{2}$, e'l più alto argiento che tu ày si è fino, cioè a oncie 12 per libra. Dunque die, da 9• $\frac{1}{2}$ infino$_{insino}$ in 12 si ae 2$\frac{1}{2}$, e prendi marchi 2$\frac{1}{2}$ del contradio, cioè del'ariento lo qual'è a õ^z 8$\frac{1}{2}$ per libra. Et die, da 9 et_ $\frac{1}{2}$ fino in 8$\frac{1}{2}$ si menoma 1, e prendi uno marcho del contrario, cioè del'argiento lo qual'è a oncie 12 per libra. Ora ày alegati marchi 3$_{iij}$ $\frac{1}{2}$ a oncie 9 et $\frac{1}{2}$ per libra, ed à vi messo =$_1$ marchi 2• $\frac{1}{2}$ del'argiento lo quale à oncie 8$\frac{1}{2}$ per libra ed à vi messo uno$_1$ marcho del'argiento fino, lo quale à oncie 12 per libra. Et noi =$_{ne}$ volemo alegare marchi 20. Multiprica 20 via 2$\frac{1}{2}$, fanno 50, e parti in 3$\frac{1}{2}$, che ne viene marchi 14 et $\frac{2}{7}$ di marcho, e tanto v'ae del'argiento lo qual'è a oncie 8$\frac{1}{2}$ per libra, cioè marchi 14 et $\frac{2}{7}$ di marcho. Ora multiprica 20 via uno$_1$, fae 20, e parti 20 per 3$\frac{1}{2}$, che ne viene 5 et $\frac{5}{7}$, e tanti marchi vi meteray del'argiento, lo qual'è a oncie 12 per libra. Ed è fata. Ora giongi insieme questi marchi, cioè marchi 14 et$\frac{2}{7}$ et marchi 5 et $\frac{5}{7}$, che sono in tuto marchi 20. Dunque n'avemo alegati bene 20 marchi. Ora proviamo s'avemo bene alegato. Provasi in questo modo, die, noi avemo alegati 20 marchi d'argiento a oncie 9$\frac{1}{2}$ per libra **(fol. 47^v)** che v'à entro oncie 190, però che 20 via 9$\frac{1}{2}$ fa 190. Ora sapiamo se ritroviamo 190 oncie. Die, noi $^{n'avemo}$*v'avemo* messi marchi 14 et $\frac{2}{7}$ a oncie 8$\frac{1}{2}$ di lega, che dey multipricare 14• $\frac{2}{7}$ via 8$\frac{1}{2}$, che fano 121 et $\frac{3}{7}$, e tante oncie ae ne' 14 marchi et $\frac{2}{7}$, cioè oncie 121 et $\frac{3}{7}$. Ora sapiamo quante oncie ae ne' 5 marchi et $\frac{5}{7}$ a ragione di 12 oncie di lega. Multiprica 12 via 5 et $\frac{5}{7}$, fano 68 et_ $\frac{4}{7}$, et

tante oncie ae neli deti marchi 5 et $\frac{5}{7}$, cioè oncie 68 et $\frac{4}{7}$. Ora giongi insieme tute queste oncie, cioè oncie 121•$\frac{3}{7}$ et 68$\frac{4}{7}$, fano oncie 190. Dunque avemo bene alegato. Et dovete sapere che in questi alegamenti tanto porta a dire libre quanto marche. Et cholae dov'ài detto 20 marchi potresti dire 20 libre, e tanto ti verebbe l'uno modo quanto l'altro, però che in quela posicione che prendi de' marchi, sì prenderesti dele libre. Ma contoti qui marchi, però ch'ariento si pesa a marchi. Et in questo modo e per questa regola poi alegare et provare di qualunque legha ti fosse deta.

M.21.7 Questo è uno generale alegamento di quatro biglioni e per lo deto modo possemo
[F.vii.7] alegare oro e argiento e rame di qualonque tenuta fossero et di quantunque tu volesi fare la lega. E per questo modo poi aleghare di quantunque biglioni overo monete fossero. Et ciò scriviamo qui apresso et simigliantemente il mostriamo materialmente per figure come si fae lo deto alegamento et come si prendono i biglioni.

M.21.8 Primamente die, {diamo} i'ò di quatro maniere di biglione. Lo primo si è biglione
[F.—] basso ed e a δ 3 di lega, e'l secondo si è a δ 4 di lega, e'l terzo ⁻si è a δ 9 di legha, e'l quarto si è a δ 12 di legha. Et io voglio **(fol. ⁴⁸r)** fare una moneta la quale sia a δ 7 di legha, né più né meno, e voglione alegare 30 marchi. Dimi in questi 30 marchi quanto meteroe di ciaschuno di questi biglioni, a cioe che i deti 30 marchi sieno alegati a δ 7. Fa cosie, die, la lega ch'io voglio fare si è_ a δ 7, e'l magiore biglione ch'i'oe si è a δ 12, dunque die, da 7 fino_insino in 12 si à 5. Et prendi marchi 5 del contradio, cioè del più basso biglione, ch'è a δ 3 di legha. Et simigliantemente die, da 7 infino_insino in 3 menoma 4, et prendi marchi 4 del contradio biglione, cioè del più alto, ch'è a δ 12 di legha. Et anchora die, da 7 infino in 9 si ae due, e prendi marchi due₂ del biglione lo qual'è a δ 4 di lega. Et simigliantemente die, da 7 fino_insino in 4 menoma 3, e_or prendi marchi 3 del_di biglione lo qual'è a δ 9 di lega. Ora avemo alegato marchi 14 di biglione a δ 7 di lega, ed à vi messo marchi 4 lo del biglione lo qual'è a δ 12 di lega, ed à vi messo marchi 5 del biglione_ lo qual è a δ 3 di legha, ed à vi messo marchi 3 del_di biglione lo qual'è a δ 9 di legha, ed à vi messo marchi 2 del_di biglione lo qual'è a δ 4 di lega. Or ày saputo in questi 14 marchi quanto vi vole di ciascuno di questy 4_quatro biglioni. Et noi ne volemo alegare 30 marchi. Fa cosie, giongi insieme tuti questi marchi, cioè 4 et 5 et 3 et 2, sono in tuto marchi 14, e quest'è il partitore. Ora, però che tu ne vuoli alegare 30 marchi, sì multiprica 30 via 4, fanno 120, e parti 120 per 14, che ne viene 8 et $\frac{4}{7}$, et tanti marchi d'argiento fine, entrarae ne' deti 30 marchi. Ora multiprica 3 via 30, fano 90, e parti 90 per 14, che ne viene **(fol. 48ᵛ)** 6 et $\frac{3}{7}$, e tanti marchi vi vole del'argiento lo qual'è a δ 9 di lega, cioè marchi 6 et $\frac{3}{7}$ di marcho. Ora multiprica 30 via 2, fanno 60, e parti per 14, che ne viene 4 et $\frac{2}{7}$, e tanti marchi vi vole del'argiento lo qual'è a δ 4 di legha, cioè

marchi 4 et $\frac{2}{7}$. Ora multiprica 30 via 5, fanno 150, e parti per 14, che ne viene 10 et $\frac{5}{7}$, e tanti marchi vi vorae del'argiento overo $=_{del}$ biglione lo qual'è a δ 3 di lega, cioè marchi 10 et $\frac{5}{7}$. Ed è fata. Ora giongi insieme tuti questi marchi, li quali tu ày messi insieme, e sapia se sono 30 marchi, cioè marchi 8 et $\frac{4}{7}$ et marchi 6 et $\frac{3}{7}$ $_{2/7}$ $^{=}$*et marchi 4 et $\frac{2}{7}$* et marchi 10 et $\frac{5}{7}$, che sono in tuto marchi 30. Dunque n'avemo aligati 30 marchi. Et in questo modo puoi fare tuti alegamenti.

M.21.8A Explicit liber₌ tractatus algorismi. Deo Gratias.$=_{Amen}$
[**F.**—]

Sigla

A: *Trattato dell'alcibra amuchabile*, Florence, Biblioteca Riccardiana, Ricc. 2263, ed. [Simi 1994].

C: *Libro di molte ragioni d'abaco*, "le reghole della chosa con asenpri", Biblioteca Statale di Lucca, MS 1754, fols 50^r–52^r, ed. [Arrighi 1973: 108–114].

D: Dardi, *Aliabraa argibra*

D₁: Dardi, *Aliabraa argibra*, Vatican Library, Chigi M.VIII.170.

D₂: Dardi, *Aliabraa argibra*, Siena, Biblioteca Comunale, I.VII.17, ed. [Franci 2001].

D₃: Dardi, *Aliabraa argibra*, Arizona State University Tempe, personal transcription by Warren Van Egmond.

F: Jacopo da Firenze, *Tractatus algorismi*, Florence, Biblioteca Riccardiana, Ricc. 2236, ed. [Simi 1995].

G: Paolo Gherardi, *Libro di ragioni*, Florence, Biblioteca nazionale, Magl. Cl. XI, 87, ed. [Arrighi 1987].

L: *Libro di molte ragioni d'abaco*, "le reghole dell'aligibra amichabile", Biblioteca Statale di Lucca, MS 1754, fols 80^v–81^v, ed. [Arrighi 1973: 194–197].

M: Jacopo da Firenze, *Tractatus algorismi*, Milan, Biblioteca Trivulziana MS 90.

P: *Libro di conti e mercatanzie*, Parma, Biblioteca Palatina, MS 312, ed. [Gregori & Grugnetti 1998]

T_F: *Trattato di tutta l'arte dell'abacho*, Florence, Biblioteca Nazionale Centrale, fond. prin. II,IX.57.

T_r: *Trattato di tutta l'arte dell'abacho*, Roma, Accademia Nazionale dei Lincei, Cors. 1875.

V: Jacopo da Firenze, *Tractatus algorismi*, Vatican Library, Vat. Lat. 4826.

Z: Giovanni di Davizzo, fragment on algebra in *Alchune ragione*, Vatican Library, Vat. lat. 10488, fols 28^v–31^r.

Bibliography

Abdeljaouad, Mahdi, 2002. "Le manuscrit mathématique de Jerba: Une pratique des symboles algébriques maghrébins en pleine maturité". Présenté au Septième Colloque sur l'histoire des mathématiques arabes (Marrakech, 30–31 mai et 1er juin 2002).

Abdeljaouad, Mahdi, 2005. "12th Century Algebra in an Arabic Poem: Ibn Al-Yāsamīn's *Urjūza fi'l-jabr wa'l-muqābala*". *Manuscript*, January 2005.

Aguiló y Fuster, Marian (ed.), 1905. Johanot Martorell, *Libre del valeros e strenu caualler Tirant lo Blanch.* 4 vols. Barcelona: Alvar Verdaguer.

Allard, André (ed., trans.), 1981. Maxime Planude, *Le Grand calcul selon les Indiens.*(Travaux de la Faculté de Philosophie et Lettres de l'Université Catholique de Louvain, 27. Centre d'Histoire des Sciences et des Techniques, sources et travaux, 1). Louvain-la-Neuve: [Faculté de Philosophie et Lettres].

Allard, André (ed.), 1992. Muḥammad ibn Mūsā al-Khwārizmī, *"Le Calcul indien" (Algorismus).* Histoire des textes, édition critique, traduction et commentaire des plus anciennes versions latines remaniées du XIIe siècle. (Collection Science dans l'histoire: Collection d'Études classiques). Paris: Blanchard / Namur: Société des Études Classiques.

Arrighi, Gino (ed.), 1964. Paolo Dell'Abaco, *Trattato d'aritmetica.* Pisa: Domus Galilæana.

Arrighi, Gino (ed.), 1966. Paolo dell'Abbaco, *Regoluzze.* Secondo la lezione del codice 2511 della Biblioteca Riccardiana di Firenze. Prato: Azienda autonoma di Turismo di Prato, 1966.

Arrighi, Gino (ed.), 1967a. Antonio de' Mazzinghi, *Trattato di Fioretti* nella trascelta a cura di M° Benedetto secondo la lezione del Codice L.IV.21 (sec. XV) della Biblioteca degl'Intronati di Siena. Siena: Domus Galilaeana.

Arrighi, Gino, 1967b. "Un «programma» di didattica di matematica nella prima metà del Quattrocento (dal Codice 2186 della Biblioteca Riccardiana di Firenze)". *Atti e memorie dell'Accademia Petrarca di Lettere, Arti e Scienze di Arezzo*, Nuova Serie **38** (1965–67), 117–128.

Arrighi, Gino (ed.), 1969. Filippo Calandri, *Aritmetica.* Secondo la lexione del codice 2669 (sec. XV) della Biblioteca Riccardiana di Firenze. 2 vols. Firenze: Cassa di Risparmio di Firenze.

Arrighi, Gino (ed.), 1970a. Piero della Francesca, *Trattato d'abaco.* Dal codice ashburnhamiano 280 (359*–291*) della Biblioteca Medicea Laurenziana di Firenze. A cura e con introduzione di Gino Arrighi. (Testimonianze di storia della scienza, 6). Pisa: Domus Galilæana.

Arrighi, Gino (ed.), 1970b. Francesco di Giorgio Martini, *La praticha di gieometria.* Dal codice Ashburnam 361 della Biblioteca Laurenziana di Firenze. Firenze: Giunti.

Arrighi, Gino (ed.), 1973. *Libro d'abaco.* Dal Codice 1754 (sec. XIV) della Biblioteca Statale di Lucca. Lucca: Cassa di Risparmio di Lucca.

Arrighi, Gino (ed.), 1974. Pier Maria Calandri, *Tractato d'abbacho.* Dal codice Acq. e doni 154 (sec. XV) della Biblioteca Medicea Laurenziana di Firenze. Pisa: Domus Galiaeana.

Arrighi, Gino (ed.), 1977. *L'Arithmetica* di fra' Leonardo da Pistoia o.p. (secc. XIII–XIV). Dal codice 116 della Biblioteca Riccardiana di Firenze. Firenze: Fondazione Giorgio Ronchi.

Arrighi, Gino, & Cinzia Nanni (eds), 1982. Tommaso della Gazzaia, fl. 1387–1415, *Praticha di geometria e tutte misure di terre* dal ms. C. III. 23 della Biblioteca comunale di Siena. Trascrizione di Cinzia Nanni. Introduzione di Gino Arrighi. (Quaderni del Centro Studi della Matematica Medioevale, 1). Siena: Servizio editoriale dell'Università di Siena.

Arrighi, Gino (ed.), 1985. Giovanni de' Danti Aretino, *Tractato de l'algorisimo.* Dal Cod. Plut. 30. 26 (sec. XIV) della Biblioteca Medicea Laurenziana di Firenze. *Atti e Memorie della Accademia Petrarca di Lettere, Arti e Scienze*, nuova serie **47**, 3–91.

Arrighi, Gino, 1986. "Libri, maestri e botteghe d'abaco in Toscana nel medioevo", pp. 25–42 *in Quarant'anni.* Lucca.

Arrighi, Gino (ed.), 1987. Paolo Gherardi, *Opera mathematica: Libro di ragioni – Liber habaci.* Codici

Magliabechiani Classe XI, nn. 87 e 88 (sec. XIV) della Biblioteca Nazionale di Firenze. Lucca: Pacini-Fazzi.

Arrighi, Gino (ed.), 1989. "Maestro Umbro (sec. XIII), *Livero de l'abbecho*. (Cod. 2404 della Biblioteca Riccardiana di Firenze)". *Bollettino della Deputazione di Storia Patria per l'Umbria* **86**, 5–140.

Bagheri, Muhammad, 1997. "Recreational Problems from Ḥāsib Ṭabarī's *Miftāḥ al-muʿāmilāt*". Paper Presented to the RiP Workshop "Transmission of Problems", Oberwolfach, July 27–August 1, 1997.

Baron, Roger, 1956. "Hugonis de Sancto Victore *Practica geometriae*". *Osiris* **12**, 176–224.

Bernhard, Wilhelm (ed.), 1887. Die *Werke* des Trobadors n'At de Mons. (Altfranzösische Bibliothek, 11). Heilbronn: Gebr. Henninger.

Bonaini, F., 1858. "Memoria unica sincrona di Leonardo Fibonacci, novamente scoperta". *Giornale Storico degli Archivi toscani* **1**, 3–10.

Boncompagni, Baldassare, 1854. *Intorno ad alcune opere di Leonardo Pisano*. Roma: Tipografia delle Belle Arti.

Boncompagni, Baldassare (ed.), 1857a. *Scritti* di Leonardo Pisano matematico del secolo decimoterzo. I. Il *Liber abbaci* di Leonardo Pisano. Roma: Tipografia delle Scienze Matematiche e Fisiche.

Boncompagni, Baldassare (ed.), 1857b. *Trattati d'aritmetica*. I. *Algoritmo de numero indorum*. II. Ioannis Hispalensis *Liber algorismi de pratica arismetrice*. Roma: Tipografia delle scienze fisiche e matematiche.

Boncompagni, Baldassare (ed.), 1862. *Scritti* di Leonardo Pisano matematico del secolo decimoterzo. II. *Practica geometriae* et *Opusculi*. Roma: Tipografia delle Scienze Matematiche e Fisiche.

Bruins, E. M., & M. Rutten, 1961. *Textes mathématiques de Suse*. (Mémoires de la Mission Archéologique en Iran, XXXIV). Paris: Paul Geuthner.

Bubnov, Nicolaus (ed.), 1899. Gerberti postea Silvestri II papae *Opera mathematica* (972 – 1003). Berlin: Friedländer. Available at http://gallica.bnf.fr.

Bukharin, Nikolay I., 1931/1971. "Theory and Practice from the Standpoint of Dialectical materialism", pp. 11–33 *in Science at the Cross Roads*. Papers Presented to the International Congress of the History of Science and Technology [...] 1931. With a New Foreword by Joseph Needham and a New Introduction by P. G. Werskey. London: Frank Cass, 1971.

Burnett, Charles, 2002. "Indian Numerals in the Mediterranean Basin in the Twelfth Century, with Special Reference to the 'Eastern Forms'", pp. 237–288 *in* Yvonne Dold-Samplonius et al (eds), *From China to Paris: 2000 Years Transmission of Mathematical Ideas*. (Boethius, 46). Stuttgart: Steiner.

Busard, H. L. L., 1968. "L'algèbre au moyen âge: Le «Liber mensurationum» d'Abû Bekr". *Journal des Savants*, Avril-Juin 1968, 65–125.

Busard, H. L. L. (ed.), 1998. Johannes de Muris, *De arte mensurandi*. A Geometrical Handbook of the Fourteenth Century. (Boethius, 41). Stuttgart: Franz Steiner.

Cantor, Moritz, 1875. *Die römischen Agrimensoren und ihre Stellung in der Geschichte der Feldmesskunst. Eine historisch-mathematische Untersuchung*. Leipzig: Teubner.

Cantor, Moritz, 1900. *Vorlesungen über Geschichte der Mathematik*. Zweiter Band, *von 1200–1668*. Zweite Auflage. Leipzig: Teubner.

Cardano, Girolamo, 1539. *Practica arithmetice, et mensurandi singularis*. Milano: Bernardini Calusco.

Cardano, Girolamo, 1663. Hieronymo Cardani Mediolanensis Philosophi ac Medici celeberrimi *Operum* tomus quartus; quo continentur *Arithmetica, Geometrica, Musica*. Lyon: Jean Antoine Huguetan & Marc Antoine Ragaud.

Cassinet, Jean, 2001. "Une arithmétique toscane en 1334 en Avignon dans la cité des papes et de leurs banquiers florentins", pp. 105–128 *in Commerce et mathématiques du moyen âge à la renaissance, autour de la Méditerranée*. Actes du Colloque International du Centre International d'Histoire des Sciences Occitanes (Beaumont de Lomagne, 13–16 mai 1999). Toulouse: Éditions du C.I.H.S.O.

Caunedo del Potro, Betsabé, & Ricardo Córdoba de la Llave (eds), 2000. *El arte del alguarismo*. Un libro castellano de aritmética comercial y de ensayo de moneda del siglo XIV. (Ms. 46 de la Real Colegiato de San Isidoro de León). Salamanca: Junta de Castilla y León, Consejeria de Educación y Cultura.

Caunedo del Potro, Betsabé, 2004. "*De Arismetica*. Un manual de aritmética para mercaderes". *Cuadernos de Historia de España* **78** (2003–04), 35–46.

Chalhoub, Sami (ed., trans.), 2004. *Die Algebra, Kitab al-Gabr wal-muqabala des Aby Kamil Soga Ibn Aslam*. (Quellen und Studien über die Geschichte der Arabischen Mathematik, 7). Aleppo: University of Aleppo,

Institute for the History of Arabic Science.

Chemla, Karine, & GUO Shuchun (eds), 2004. *Les neuf chapitres. Le Classique mathématique de la Chine ancienne et ses commentaires*. Paris: Dunod.

Chiarini, Giorgio, et al (eds), 1972. [Pietro Paolo Muscharello], *Algorismus. Trattato di aritmetica pratica e mercantile del secolo XV*. 2 vols. Verona: Banca Commerciale Italiana.

Ciano, Cesare (ed.), 1964. *La «Pratica di mercatura» datiniana (secolo XIV)*. (Biblioteca della Rivista «Economia e Storia», 9) Milano: A. Giuffrè.

Clagett, Marshall, 1964. *Archimedes in the Middle Ages*. Vol. I. *The Arabo-Latin Tradition*. Madison, Wisconsin: University of Wisconsin Press.

Clagett, Marshall, 1984. *Archimedes in the Middle Ages*. Volume V. *Quasi-Archimedean Geometry in the Thirteenth Century*. (Memoirs of the American Philosophical Society, 157 A+B). Philadelphia: The American Philosophical Society.

Colebrooke, H. T. (ed., trans.), 1817. *Algebra, with Arithmetic and Mensuration from the Sanscrit of Brahmagupta and Bhascara*. London: John Murray.

Cullen, Christopher, 2004. The *Suàn shù shū*, "Writings on Reckoning": A Translation of a Chinese Mathematical Collection of the Second Century BC, with Explanatory Commentary. (Needham Research Institute Working Papers, 1). Cambridge: Needham Research Institute, 2004. Web Edition http://www.nri.org.uk/ suanshushu.html.

Curtze, Maximilian (ed.), 1902. *Urkunden zur Geschichte der Mathematik im Mittelalter und der Renaissance*. (Abhandlungen zur Geschichte der mathematischen Wissenschaften, vol. 12–13). Leipzig: Teubner.

Datta, Bibhutibhusan, & Avadhesh Narayan Singh, 1962. *History of Hindu Mathematics. A Source Book*. Parts I and II. Bombay: Asia Publishing House. [1]Lahore: Motilal Banarsidass, 1935–38.

Davis, Margaret Daly, 1977. *Piero della Francesca's Mathematical Treatises. The 'Trattato d'abaco' and 'Libellus de quinque corporibus regularibus'*. Ravenna: Longo Editore.

Dictionnaire universel françois et latin, 8 vols. Paris: Compagnie des Libraires associés, 1771.

Diderot, Denis, & Jean le Rond d'Alembert (eds), 1751. *Encyclopédie, ou Dictionnaire raisonné des sciences, des arts, et des métiers*. 17 vols. Paris: Chez Briasson, David l'aîné, Le Breton, & Durand, 1751–1772.

Djebbar, Ahmed, 1981. *Enseignement et recherche mathématiques dans le Maghreb des XIIIᵉ-XIVᵉ siècles (étude partielle)*. (Publications mathématiques d'Orsay, 81–02). Orsay: Université de Paris-Sud. *Preprint*, Centro de Matematica da Universidade do Porto, April 27, 2006.

Djebbar, Ahmed, 1988. "Quelques aspects de l'algèbre dans la tradition mathématique arabe de l'Occident musulman", pp. 101–123 *in Histoire des Mathématiques Arabes*. Premier colloque international sur l'histoire des mathématiques arabes, Alger, 1.2.3 décembre 1986. Actes. Alger: La Maison des Livres.

Edler, Florence, 1934. *Glossary of Mediaeval Terms of Business*. Cambridge, Mass.: The Mediaeval Academy of America.

Elfering, Kurt (ed., trans.), 1975. *Die Mathematik des Aryabhata I: Text, Übersetzung aus dem Sanskrit und Kommentar*. München: Wilhelm Fink.

Eneström, Georg, 1906. "Über zwei angebliche mathematische Schulen im christlichen Mittelalter". *Bibliotheca Mathematica*, 3. Folge **7** (1906–07), 252–262.

Epstein, Isidore (general editor), 1960. *The Babylonian Talmud*. 35 vols. London: Soncino Press, 1960–1976. Single referenced volumes under the particular editors.

Evans, Allan (ed.), 1936. Francesco Balducci Pegolotti, *La pratica della mercatura*. Cambridge, Mass.: The Medieval Academy of America.

Folkerts, Menso, 1970. *"Boethius" Geometrie II. Ein mathematisches Lehrbuch des Mittelalters*. Wiesbaden: Franz Steiner.

Folkerts, Menso, 1978. "Die älteste mathematische Aufgabensammlung in lateinischer Sprache: Die Alkuin zugeschriebenen *Propositiones ad acuendos iuvenes*. Überlieferung, Inhalt, Kritische Edition". *Österreichische Akademie der Wissenschaften, Mathematisch-Naturwissenschaftliche Klasse. Denkschriften*, 116. Band, 6. Abhandlung (Wien).

Folkerts, Menso (ed.), 1997. *Die älteste lateinische Schrift über das indische Rechnen nach al-Ḫwārizmī.* Edition, Übersetzung und Kommentar. Unter Mitarbeit von Paul Kunitzsch. (Bayerische Akademie der Wissenschaften. Philosophisch-historische Klasse. Abhandlungen, neue Folge, 113). München: Verlag der Bayerischen Akademie der Wissenschaften.

Folkerts, Menso, 2003. *Essays on Early Medieval Mathematics. The Latin Tradition.* (Variorum). Aldershot, Hampshire, & Burlington: Ashgate.

Franci, Raffaella, & Laura Toti Rigatelli, 1985. "Towards a History of Algebra from Leonardo of Pisa to Luca Pacioli". *Janus* **72**, 17–82.

Franci, Raffaella, & Marisa Pancanti (eds), 1988. Anonimo (sec. XIV), *Il trattato d'algibra* dal manoscritto Fond. Prin. II. V. 152 della Biblioteca Nazionale di Firenze. (Quaderni del Centro Studi della Matematica Medioevale, 18). Siena: Servizio Editoriale dell'Università di Siena.

Franci, Raffaella (ed.), 2001. Maestro Dardi, *Aliabraa argibra*, dal manoscritto I.VII.17 della Biblioteca Comunale di Siena. (Quaderni del Centro Studi della Matematica Medioevale, 26). Siena: Università degli Studi di Siena.

Franci, Raffaella, 2002a. "Jealous Husbands Crossing the River", pp. 289–306 *in* Yvonne Dold-Samplonius et al (eds), *From China to Paris: 2000 Years Transmission of Mathematical Ideas.* (Boethius, 46). Stuttgart: Steiner.

Franci, Raffaella, 2002b. "Trends in Fourteenth-Century Italian Algebra". *Oriens–Occidens* 2002 n° 4, 81–105.

Franci, Raffaella, 2003. "Leonardo Pisano e la trattatistica dell'abaco in Italia nei secoli XIV e XV". *Bollettino di Storia delle Scienze Matematiche* **23**, 33–54.

Freudenthal, Gad, 1993. "Les sciences dans les communautés juives médiévales de Provence: leur appropriation, leur rôle". *Revue des études juives* **152**, 29–136.

Giusti, Enrico, 1993. "Fonti medievali dell'*Algebra* di Piero della Francesca". *Bollettino di Storia delle Scienze matematiche* **13**, 199–250.

Giusti, Enrico, 2002. "Matematica e commercio nel Liber Abaci", pp. 59–120 *in* Enrico Giusti (ed.), *Un ponte sul mediterraneo: Leonardo Pisano, la scienza araba e la rinascita della matematica in Occidente.* Firenze: Edizioni Polistampa.

Goldthwaite, Richard A., 1972. "Schools and Teachers of Commercial Arithmetic in Renaissance Florence". *Journal of European Economic History* **1**, 418–433.

Graham, Loren R., 1993. *Science in Russia and the Soviet Union. A Short History.* (Cambridge History of Science). Cambridge: Cambridge University Press.

Grayson, Cecil (ed.), 1973. Leon Battista Alberti, *Opere volgari.* Volume terzo. *Trattati d'arte, Ludi rerum mathematicarum, grammatica della lingua toscana, Opuscoli amatori, Lettere.* (Scrittori d'Italia, N. 254). Bari: Laterza.

Gregori, Silvano, & Lucia Grugnetti (eds), 1998. Anonimo (sec. XV), *Libro di conti e mercatanzie.* Parma: Università degli Studi di Parma, Facoltà di Scienze Matematiche, Fisiche e Naturali, Dipartimento di Matematica.

Grendler, Paul F., 1989. *Schooling in Renaissance Italy: Literacy and Learning, 1300–1600.* Baltimore & London: Johns Hopkins University Press.

Grimm, Richard E., 1976. "The Autobiography of Leonardo Pisano". *Fibonacci Quarterly* **21**, 99–104.

Guttmann, Miquel, (ed.) & J. Millàs i Vallicrosa (trans.), 1931. Abraam bar Hiia. *Llibre de geometria. Hibbur hameixihà uebatixbòret.* (Biblioteca Hebraico-Catalana, 3). Barcelona: Alpha.

Hahn, Nan L. (ed), 1982. *Medieval Mensuration: QUADRANS VETUS and GEOMETRIE DUE SUNT PARTES PRINCIPALES ...* (Transactions of the American Philosophical Society, vol. 72, Part 8). Philadelphia: The American Philosophical Society.

Heiberg, J. L. (ed., trans.), 1912. Heronis *Definitiones* cum variis collectionibus. Heronis quae feruntur *Geometrica.* (Heronis Alexandrini Opera quae supersunt omnia, IV). Leipzig: Teubner.

Heiberg, J. L. (ed., trans.), 1914. Heronis quae feruntur *Stereometrica* et *De mensuris.* (Heronis Alexandrini Opera quae supersunt omnia, V). Leipzig: Teubner.

Hermelink, Heinrich, 1978. "Arabic Recreational Mathematics as a Mirror of Age-Old Cultural Relations Between Eastern and Western Civilizations", pp. 44–52 *in* A. Y. Hassan et al (eds), *Proceedings of the First International Symposium for the History of Arabic Science, April 5–12, 1976.* Vol. II, Papers in European Languages. Aleppo: Institute for the History of Arabic Science, Aleppo University.

Hinz, Walther, 1970. *Islamische Masse und Gewichte umgerechnet ins metrische System.* Photomechanischer Nachdruck mit Zusätsen und Berichtigungen. (Handbuch der Orientalistik, Erg. Bd. 1, Heft 1). Leiden: Brill.

Hochheim, Adolph (ed., trans.), 1878. *Kâfî fîl Hisâb (Genügendes über Arithmetik)* des Abu Bekr Muhammed ben Alhusein Alkarkhi. I-III. Halle: Louis Nebert.

Høyrup, Jens, 1988. "Jordanus de Nemore, 13th Century Mathematical Innovator: an Essay on Intellectual Context, Achievement, and Failure". *Archive for History of Exact Sciences* **38**, 307–363.

Høyrup, Jens, 1997. "Mathematics, Practical and Recreational", pp. 660–663 *in* Helaine Selin (ed.), *Encyclopaedia of the History of Science, Technology, and Medicine in Non-Western Cultures*. Dordrecht etc.: Kluwers.

Høyrup, Jens, 1998a. "'Oxford´ and 'Gherardo da Cremona´: on the Relation between Two Versions of al-Khwārizmī's Algebra", pp. 159–178 *in Actes du 3me Colloque Maghrébin sur l'Histoire des Mathématiques Arabes, Tipaza (Alger, Algérie), 1–3 Décembre 1990*, vol. II. Alger: Association Algérienne d'Histoire des Mathématiques.

Høyrup, Jens, 1998b. "A New Art in Ancient Clothes. Itineraries Chosen between Scholasticism and Baroque in Order to Make *Algebra* Appear Legitimate, and Their Impact on the Substance of the Discipline". *Physis*, n.s. **35**, 11–50.

Høyrup, Jens, 2001. "On a Collection of Geometrical Riddles and Their Role in the Shaping of Four to Six 'Algebras'". *Science in Context* **14**, 85–131.

Høyrup, Jens, 2002a. *Lengths, Widths, Surfaces: A Portrait of Old Babylonian Algebra and Its Kin.* (Studies and Sources in the History of Mathematics and Physical Sciences). New York: Springer.

Høyrup, Jens, 2002b. "Pedro Nuñez: Innovateur bloqué, et dernier témoin d'une tradition millénaire". *Gazeta de Matemática* nº 143 (Julho 2002), 52–59.

Høyrup, Jens, 2004. "Mahāvīra's Geometrical Problems: Traces of Unknown Links between Jaina and Mediterranean Mathematics in the Classical Ages", pp. 83–95 *in* Ivor Grattan-Guinness & B. S. Yadav (eds), *History of the Mathematical Sciences*. New Delhi: Hindustan Book Agency.

Høyrup, Jens, 2006. "Generosity: no doubt, but at times excessive and delusive". Contribution to the meeting "The Generosity of Artificial Languages in an Asian Perspective", Amsterdam, 18–20 May 2006. *Mimeo.* To appear in *Journal of Indian Philosophy*.

Høyrup, Jens, 2007. "The 'Unknown Heritage´: Trace of a Forgotten Locus of Mathematical Sophistication". *Filosofi og Videnskabsteori på Roskilde Universitetscenter.* 3. Række: *Preprints og Reprints* 2007 Nr. 1. To be published in *Archive for History of Exact Sciences*.

Hughes, Barnabas B. (ed., trans.), 1981. Jordanus de Nemore, *De numeris datis*. A Critical Edition and Translation. (Publications of the Center for Medieval and Renaissance Studies, UCLA, 14). University of California Press.

Hughes, Barnabas B. (ed.), 1986. "Gerard of Cremona's Translation of al-Khwārizmī's *Al-Jabr*: A Critical Edition". *Mediaeval Studies* **48**, 211–263.

Hughes, Barnabas B., 1987. "An Early 15th-Century Algebra Codex: A Description". *Historia Mathematica* **14**, 167–172.

Hughes, Barnabas B. (ed.), 1989. Robert of Chester's Latin translation of al-Khwārizmī's *Al-jabr*. A New Critical Edition. (Boethius. Texte und Abhandlungen zur Geschichte der exakten Naturwissenschaften, 14). Wiesbaden: Franz Steiner.

Hughes, Barnabas B., 2001. "A Treatise on Problem Solving from Early Medieval Latin Europe". *Mediaeval Studies* **63**, 107–141.

Hunger, Herbert, & Kurt Vogel, 1963. *Ein byzantinisches Rechenbuch des 15. Jahrhunderts. 100 Aufgaben aus dem Codex Vindobonensis Phil. Gr. 65*. (Österr. Akad. d. Wiss., phil.-hist. Kl., Denkschriften, Bd. 78, Abh. 2). Wien.

Ius Romanum. http://www.thelatinlibrary.com/ius.html (30.6.2004).

Jayawardene, S. A., 1970. "Bombelli, Rafael", pp. 279–281 *in Dictionary of Scientific Biography*, vol. II. New York: Scribner.

Karpinski, Louis C., 1910. "An Italian Algebra of the Fifteenth Century". *Bibliotheca Mathematica*, 3. Folge **11** (1910–11), 209–219.

Karpinski, Louis C., 1929a. "The Italian Arithmetic and Algebra of Master Jacob of Florence, 1307". *Archeion* **11**, 170–177.

Karpinski, Louis C., & Frank Egleston Robbins, 1929b. "Michigan Papyrus 620: The Introduction of Algebraic Equations in Greece". *Science* **70**, 311–314.

Kaunzner, Wolfgang, 1985. "Über eine frühe lateinische Bearbeitung der Algebra al-Khwārizmīs in MS Lyell 52 der Bodleian Library Oxford". *Archive for History of Exact Sciences* **32**, 1–16.

Kaunzner, Wolfgang, 1986. "Die lateinische Algebra in MS Lyell 52 der Bodleian Library, Oxford, früher MS Admont 612", pp. 47–89 in G. Hamann (ed.), *Aufsätze zur Geschichte der Naturwissenschaften und Geographie*. (Österreichische Akademie der Wissenschaften, Phil.-Hist. Klasse, Sitzungsberichte, Bd. 475). Wien: Österreichische Akademie der Wissenschaften.

Kirzner, E. W. (ed., trans.), revised by M. Ginsberg, 1964. *Baba Ḵamma*. (Hebrew-Engish Translation of the Babylonian Talmud). London: Soncino Press.

Knorr, Wilbur R., 1990. "John of Tynemouth alias John of London: Emerging Portrait of a Singular Medieval Mathematician". *British Journal for the History of Science* **23**, 293–330.

Kokian, P. Sahak (ed., trans.), 1919. "Des Anania von Schirak arithmetische Aufgaben". *Zeitschrift für die deutsch-österreichischen Gymnasien* **69** (1919–20), 112–117.

Kunitzsch, Paul, 2005. "Zur Geschichte der 'arabischen' Ziffern." *Bayerische Akademie der Wissenschaften, Philosophisch-Historische Klasse. Sitzungsberichte*, 2005 no. 3, 1–39.

Lafont, R., & G. Tournerie (eds), 1967. Francès Pellos, *Compendion de l'abaco*. Montpellier: Édition de la Revue des Langues Romanes.

LAM Lay Yong, 1977. *A Critical Study of the Yang Hui Suan Fa, a Thirteenth-Century Mathematical Treatise*. Singapore: Singapore University Press.

Langeli, Atilio Bartoli, 2000. "I notai e i numeri (con un caso perugino, 1184–1206)", pp. 225–254 in Paolo Freguglia, Luigi Pellegrini & Robeto Paciocco (eds), *Scienze matematiche e insegnamento in epoca medioevale*. Atti del Convegno Internazionale di Studio, Chieti, 2–4 maggio 1996. Napoli: Edizioni Scientifiche Italiane.

Levey, Martin (ed., trans.), 1966. The *Algebra* of Abū Kāmil, *Kitab fi'al-jābr* (sic) *wa'l-muqābala*, in a Commentary by Mordechai Finzi. Hebrew Text, Translation, and Commentary with Special Reference to the Arabic Text. Madison etc: University of Wisconsin Press.

Lévy, Tony, 2003. "L'Algèbre arabe dans les textes hébraïques (I): Un ouvrage inédite d'Isaac ben Salomon al-Ahdab (XIVᵉ siècle)". *Arabic Sciences and Philosophy* **13**, 269–301.

Lévy, Tony, 2007. "L'Algèbre arabe dans les textes hébraïques (II): Dans l'italie des XVᵉ et XVIᵉ siècles, sources arabes et sources verniaculaires". *Arabic Sciences and Philosophy* **17**, 81–107.

Libri, Guillaume, 1838. *Histoire des mathématiques en Italie*. 4 vols. Paris, 1838–1841. Available at http://gallica.bnf.fr.

Luckey, Paul, 1941. "Ṯābit b. Qurra über den geometrischen Richtigkeitsnachweis der Auflösung der quadratischen Gleichungen". *Sächsischen Akademie der Wissenschaften zu Leipzig. Mathematisch-physische Klasse. Berichte* **93**, 93–114.

Malet, Antoni (ed.), 1998. Francesc Santcliment, *Summa de l'art d'Aritmètica*. Vic: Eumo Editorial.

Marre, Aristide (ed.), 1880. "Le Triparty en la science des nombres par Maistre Nicolas Chuquet Parisien". *Bullettino di Bibliografia e di Storia delle Scienze Matematiche e Fisiche* **13**, 593–659, 693–814.

Marre, Aristide (ed.), 1881. "Appendice au *Triparty en la science des nombres* de Nicolas Chuquet parisien". *Bullettino di Bibliografia e di Storia delle Scienze matematiche e fisiche* **14**, 413–435.

Michel, Cécile, 1998. "Les marchands et les nombres: l'exemple des Assyriens à Kaniš", pp. 249–267 in J. Prosecky (ed.), *Intellectual Life of the Ancient Near East*. Papers Presented at the 43rd Rencontre Assyriologique Internationale, Prague, July 1–5, 1996. Prague: Academy of Sciences of the Czech Republic, Oriental Institute.

Miura, Nobuo, 1981. "The Algebra in the *Liber abaci* of Leonardo Pisano". *Historia Scientiarum* **21**, 57–65.

Mušarrafa, ʿAlī Muṣṭafā, & Muḥammad Mursī Ahmad (eds), 1939. Al-Khwārizmī, *Kitāb al-mukhtaṣar fi hisāb al-jabr wa'l-muqābalah*. Caïro.

Needham, Joseph, 1969. *The Grand Titration. Science and Society in East and West*. London: Allen and Unwin.

Nesselmann, G. H. F., 1842. *Versuch einer kritischen Geschichte der Algebra*. Nach den Quellen bearbeitet. Erster Theil, *Die Algebra der Griechen*. Berlin: G. Reimer.

Nesselmann, G. H. F. (ed., trans.), 1843. Mohammed Beha-eddin ben Alhossain aus Amul, *Essenz der Rechenkunst*, arabisch und deutsch. Berlin: Reimer.

Neugebauer, Otto, 1935. *Mathematische Keilschrift-Texte*. I-III. (Quellen und Studien zur Geschichte der Mathematik, Astronomie und Physik. Abteilung A: Quellen. 3. Band, erster-dritter Teil). Berlin: Julius Springer, 1935–1937.

Oaks, Jeffrey A., & Haitham M. Alkhateeb, 2005. "*Māl*, enunciations, and the Prehistory of Arabic Algebra".

Historia Mathematica **32**, 400–425.

Pacioli, Luca, 1494. *Summa de Arithmetica Geometria Proportioni et Proportionalita.* Venezia: Paganino de Paganini, 1494. [All references refer to the first, arithmetical part.]

Parker, Richard A., 1972. *Demotic Mathematical Papyri.* Providence & London: Brown University Press.

Paton, W. R. (ed., trans.), 1979. *The Greek Anthology.* In Five Volumes. (Loeb Classical Library) Cambridge, Mass.: Harvard University Press / London: Heinemann. [1]1918.

Pedersen, Fritz Saaby (ed), 1983. Petri Philomena de Dacia et Petri de S. Audomaro *Opera quadrivialia.* Pars I. *Opera* Petri Philomenae. (Corpus philosophorum danicorum medii aevi X.i). København: Gad.

Pieraccini, Lucia (ed.), 1983. M° Biagio, *Chasi exenplari all regola dell'algibra nella trascelta a cura di M° Benedetto* dal Codice L. VII. 2Q della Biblioteca Comunale di Siena. (Quaderni del Centro Studi della Matematica Medioevale, 5). Siena: Servizio Editoriale dell'Università di Siena.

Procissi, Angiolo (ed.), 1954. "I Ragionamenti d'Algebra di R. Canacci". *Bollettino Unione Matematica Italiana*, serie III, **9**, 300–326, 420–451.

Proust, Christine, 2000. "La multiplication babylonienne: la part non écrite du calcul". *Revue d'Histoire des Mathématiques* **6**, 293–303.

Raṅgācārya, M. (ed., trans.), 1912. The *Gaṇita-sāra-sangraha* of Mahāvīrācārya with English Translation and Notes. Madras: Government Press.

Rashed, Roshdi, & Ahmed Djebbar (eds), 1981. *L'oeuvre algébrique d'al-Khayyām.* Établie, traduite et analysée. (Sources and Studies in the History of Arabic Mathematics, 3). Aleppo: University of Aleppo, Institute for the History of Arabic Science.

Rebstock, Ulrich, 1992. *Rechnen im islamischen Orient. Die literarischen Spuren der praktischen Rechenkunst.* Darmstadt: Wissenschaftliche Buchgesellschaft.

Rebstock, Ulrich, 1993. *Die Reichtümer der Rechner (Ġunyat al-Ḥussāb) von Aḥmad b. Ṯabāt (gest. 631/1234). Die Araber – Vorläufer der Rechenkunst.* (Beiträge zur Sprach- und Kulturgeschichte des Orients, 32). Walldorf-Hessen: Verlag für Orientkunde Dr. H. Vorndran.

Rebstock, Ulrich (ed., trans.), 2001. ʿAlī Ibn al-Ḥidr al-Qurašī (st. 459/1067), *At-taḏkira bi-uṣūl al-hisāb wa l'farāʾiḍ (Buch über die Grundlagen der Arithmetik und der Erbteilung).* (Islamic Mathematics and Astronomy, 107). Frankfurt a.M.: Institute for the History of Arabic-Islamic Science at the Johann Wolfgang Goethe University.

Rivolo, Maria Teresa (ed.), 1983. Anonimo Maestro Lombardo, *Arte giamata aresmetica.* Un'antologia dal Codice N. III. 53 della Biblioteca nazionale di Torino. (Quaderni del Centro Studi della Matematica Medioevale, 8). Siena: Servizio Editoriale dell'Università di Siena.

Robbins, Frank Egleston, 1929. "P. Mich. 620: A Series of Arithmetical Problems". *Classical Philology* **24**, 321–329.

Rohlfs, Gerhard, 1966. *Grammatica storica della lingua italiana e i suoi dialette.* I. *Fonetica.* II. *Morfologia.* III. *Sintassi e formazione delle parole.* Torino: Einaudi: 1966–1969.

Rosen, Frederic (ed., trans.), 1831. The *Algebra* of Muhammad ben Musa, Edited and Translated. London: The Oriental Translation Fund. Available at http://gallica.bnf.fr.

Saliba, George A., 1972. "The Meaning of al-jabr wa'l-muqābalah". *Centaurus* **17** (1972-73), 189–204.

Salomone, Lucia (ed.), 1982. M° Benedetto da Firenze, *La reghola de algebra amuchabale* dal Codice L.IV.21 della Biblioteca Comunale de Siena. (Quaderni del Centro Studi della Matematica Medioevale, 2). Siena: Servizio Editoriale dell'Università di Siena.

Sánchez Pérez, José A. (ed., trans.), 1916. *Compendio de algebra de Abenbéder.* Texto arabo, traduccion y estudio. (Junta para Ampliación de Estudios e Investigaciones Científicas, 47). Madrid: Centro de Estudios Históricos.

Santini, Daniela (ed.), 1982. Filippo Calandri, *Una racolta di ragioni* dal Codice L.VI.45 della Biblioteca Comunale di Siena. (Quaderni del Centro Studi della Matematica Medioevale, 4). Siena: Servizio Editoriale dell'Università di Siena.

Sapori, Armando, 1955. *Studi di storia economica: (secoli XIII-XIV-XV).* 3 vols. Firenze: Sansoni, 1955-67.

Sarma, Sreeramula Rajeswara, 2002. "Rule of Three and Its Variations in India". pp. 133–156 in Yvonne Dold-Samplonius et al (eds), *From China to Paris: 2000 Years Transmission of Mathematical Ideas.* (Boethius, 46). Stuttgart: Steiner.

Sarton, George, 1931. *Introduction to the History of Science.* II. *From Rabbi ben Ezra to Roger Bacon.* In two parts. (Carnegie Institution of Washington, Publication 376). Baltimore: William & Wilkins.

Sayılı, Aydın, 1962. *Abdülhamid Ibn Türk'ün katışık denklemlerde mantıkî zaruretler adlı yazısı ve zamanın cebri (Logical Necessities in Mixed Equations by ʿAbd al Ḥamîd Ibn Turk and the Algebra of his Time)*. (Publications of the Turkish Historical Society, Series VII, N° 41). Ankara: Türk Tarih Kurumu Basımevi.

Scholz, Cordula, 2001. "In fremden Landen Handel treiben – ausländische Händler in Byzans". *Das Mittelalter* **6**, 91–107.

Sela, Shlomo, 2001. "Abraham Ibn Ezra's Scientific Corpus: Basic Constituents and General Characterization". *Arabic Sciences and Philosophy* **11**, 91–149.

Sesiano, Jacques, 1984a. "Une arithmétique médiévale en langue provençale". *Centaurus* **27**, 26–75.

Sesiano, Jacques, 1984b. "Les problèmes mathématiques du *Memoriale* de F. Bartoli". *Physis* **26**, 129–150.

Sesiano, Jacques, 1988. "Le Liber Mahameleth, un traité mathématique latin composé au XIIᵉ siècle en Espagne", pp. 69–98 *in Histoire des Mathématiques Arabes*. Premier colloque international sur l'histoire des mathématiques arabes, Alger, 1.2.3 décembre 1986. Actes. Alger: La Maison des Livres.

Sesiano, Jacques (ed.), 1993. "La version latine médiévale de l'Algèbre d'Abū Kāmil", pp. 315–452 *in* M. Folkerts & J. P. Hogendijk (eds), *Vestigia Mathematica. Studies in Medieval and* Early Modern Mathematics in Honour of H. L. L. Busard. Amsterdam & Atlanta: Rodopi.

Sethe, Kurt, 1916. *Von Zahlen und Zahlworten bei den Alten Ägyptern, und was für andere Völker und Sprachen daraus zu lernen ist*. (Schriften der Wissenschaftlichen Gesellschaft in Straßburg, 25. Heft). Straßburg: Karl J. Trübner.

Silberberg, M. (ed., trans.), 1895. Abraham ben Ezra, *Sefer ha-mispar. Das Buch der Zahl*. Zum ersten Male herausgegeben, ins Deutsche übersetzt und erläutert. Frankfurt a.M.: J. Kaufmann.

Silva, M. Céu, 2006. "The Algebraic Contents of Bento Fernandes' *Tratado da arte de arismetica* (1555)".

Simi, Annalisa (ed.), 1991. Orbetano da Montepulciano, *Regole di geometria pratica* dal manoscritto Moreni 130 (sec. XV) della Biblioteca Riccardiana di Firenze. (Quaderni del Centro Studi della Matematica Medioevale, 19). Siena: Servizio Editoriale dell' Università di Siena.

Simi, Annalisa (ed.), 1993. Anonimo fiorentino, *Trattato di geometria pratica* dal Codice L.IV.18 (sec. XV) della Biblioteca Comunale di Siena. (Quaderni del Centro Studi della Matematica Medioevale, 21). Siena: Servizio Editoriale dell' Università di Siena.

Simi, Annalisa (ed.), 1994. Anonimo (sec. XIV), *Trattato dell'alcibra amuchabile* dal Codice Ricc. 2263 della Biblioteca Riccardiana di Firenze. (Quaderni del Centro Studi della Matematica Medioevale, 22). Siena: Servizio Editoriale dell' Università di Siena.

Simi, Annalisa, 1995. "Trascrizione ed analisi del manoscritto Ricc. 2236 della Biblioteca Riccardiana di Firenze". *Università degli Studi di Siena, Dipartimento di Matematica. Rapporto Matematico* N° 287.

Simonson, Shai (ed.), 2000. "The Missing Problems of Gersonides – a Critical Edition". *Historia Mathematica* **27**, 243–302, 384–431.

Slotki, Israel W. (ed., trans.), 1964. *Yebamoth*. (Hebrew-Engish Translation of the Babylonian Talmud). London: Soncino Press.

Souissi, Mohamed (ed., trans.), 1969. Ibn al-Bannāʾ, *Talkhīṣ aʿmāl al-ḥisāb*. Texte établi, annoté et traduit. Tunis: L'Université de Tunis.

Souissi, Mohamed (ed., trans.), 1988. Qalaṣādī, *Kašf al-asrār ʿan ʿilm ḥurūf al-ǧubār*. Carthage: Maison Arabe du Livre.

Spiesser, Maryvonne (ed.), 2003. *Une arithmétique commerciale du XVᵉ siècle*. Le *Compendy de la praticque des nombres* de Barthélemy de Romans. (De Diversis artibus, 70) Turnhout: Brepols.

Steele, Robert (ed.), 1922. *The Earliest Algorisms in English*. (Early English Text Society, Extra Series, 118). London: Humphrey Milford, Oxford University Press.

Stifel, Michael, 1615. *Die Coss* Christoffs Rudolffs. Mit schönen Exemplen der Coss gebessert und gemehrt. Amsterdam: Wilhelm Janson, 1615.

Swetz, Frank J., 1987. *Capitalism and Arithmetic: The New Math of the 15th Century*, Including the Full Text of the *Treviso Arithmetic* of 1478, Translated by David Eugene Smith. La Salle, Illinois: Open Court.

Tannery, Paul (ed., trans.), 1893. Diophanti Alexandrini *Opera omnia* cum graecis commentariis. 2 vols. Leipzig: Teubner, 1893–1895.

Tobler-Lommatsch: *Altfranzösisches Wörterbuch*. Adolf Toblers nachgelassene Materialien, bearbeitet und herausgegeben von Erhard Lommatsch. Wiesbaden: Franz Steiner, 1925–(2002).

Travaini, Lucia, 2003. *Monete, mercanti e matematica. Le monete medievali nei trattati di aritmetica e nei libri di mercatura.* Roma: Jouvence.

Tropfke, J./Vogel, Kurt, et al, 1980. *Geschichte der Elementarmathematik.* 4. Auflage. Band 1: *Arithmetik und Algebra.* Vollständig neu bearbeitet von Kurt Vogel, Karin Reich, Helmuth Gericke. Berlin & New York: W. de Gruyter.

Ulivi, Elisabetta, 2002a. "Scuole e maestri d'abaco", pp. 121–159 *in* Enrico Giusti (ed.), *Un ponte sul mediterraneo: Leonardo Pisano, la scienza araba e la rinascita della matematica in Occidente.* Firenze: Edizioni Polistampa.

Ulivi, Elisabetta, 2002b. "Benedetto da Firenze (1429–1479), un maestro d'abbaco del XV secolo. Con documenti inediti e con un'Appendice su abacisti e scuole d'abaco a Firenze nei secoli XIII–XVI". *Bollettino di Storia delle Scienze Matematiche* **22**:1, 3–243.

Van Egmond, Warren, 1976. *The Commercial Revolution and the Beginnings of Western Mathematics in Renaissance Florence.* Dissertation, Indiana University.

Van Egmond, Warren, 1977. "New Light on Paolo dell'Abbaco". *Annali dell'Istituto e Museo di Storia della Scienza di Firenze* **2**:2, 3–21.

Van Egmond, Warren, 1978. "The Earliest Vernacular Treatment of Algebra: The *Libro di ragioni* of Paolo Gerardi (1328)". *Physis* **20**, 155–189.

Van Egmond, Warren, 1980. *Practical Mathematics in the Italian Renaissance: A Catalog of Italian Abbacus Manuscripts and Printed Books to 1600.* (Istituto e Museo di Storia della Scienza, Firenze. Monografia N. 4). Firenze: Istituto e Museo di Storia della Scienza.

Van Egmond, Warren, 1983. "The Algebra of Master Dardi of Pisa". *Historia Mathematica* **10**, 399–421.

Van Egmond, Warren, 1986. "The Contributions of the Italian Renaissance to European Mathematics". *Symposia Mathematica* **27**, 51–67.

Victor, Stephen K., 1979. *Practical Geometry in the Middle Ages. Artis cuiuslibet consummatio* and the *Pratike de geometrie.* Edited with Translations and Commentary. (Memoirs of the American Philosophical Society, vol. 134). Philadelphia; The American Philosophical Society.

Villani, Giovanni, 1823. *Cronica.* 8 vols. Firenze.

Vocabolario degli accademici della Crusca. Venezia: Giovanni Alberti, 1612. Web edition http://vocabolario. biblio.cribecu.sns.it/vocabolario (25.12.2004).

Vogel, Kurt, 1930. "Die algebraischen Probleme des P. Mich. 620". *Classical Philology* **25**, 373–375.

Vogel, Kurt (ed., trans.), 1968a. *Ein byzantinisches Rechenbuch des frühen 14. Jahrhunderts.* (Wiener Byzantinische Studien, 6). Wien: Institut für Byzantinistik der Universität Wien/Hermann Bohlau.

Vogel, Kurt (ed., trans.), 1968b. *Chiu chang suan shu. Neun Bücher arithmetischer Technik. Ein chinesisches Rechenbuch für den praktischen Gebrauch aus der frühen Hanzeit (202 v. Chr. bis 9 n. Chr.).* (Ostwalds Klassiker der Exakten Wissenschaften. Neue Folge, Band 4). Braunschweig: Friedrich Vieweg & Sohn.

Vogel, Kurt, 1977. *Ein italienisches Rechenbuch aus dem 14. Jahrhundert (Columbia X 511 A13).* (Veröffentlichungen des Deutschen Museums für die Geschichte der Wissenschaften und der Technik. Reihe C, Quellentexte und Übersetzungen, Nr. 33). München.

Wertheim, Gustav (ed., trans.), 1896. Die *Arithmetik* des Elia Misrachi. *Ein Beitrag zur Geschichte der Mathematik.* 2. verb. Auflage. Braunschweig: Friedrich Vieweg und Sohn.

Woepcke, Franz, 1853. *Extrait du Fakhrî, traité d'algèbre* par Aboû Bekr Mohammed ben Alhaçan Alkarkhî; *précédé d'un mémoire sur l'algèbre indéterminé chez les Arabes.* Paris: L'Imprimerie Impériale.

Wolff, Christian, 1716. *Mathematisches Lexicon.* Leipzig: Joh. Friedrich Gleditschens seel. Sohn. Reprint Hildesheim 1965 (Gesammelte Werke. I. Abteilung: Deutsche Schriften, Band 11).

Wyckoff, Dorothy (ed., trans.), 1967. Albertus Magnus, *Book of Minerals.* Oxford: Clarendon Press.

Zupko, Ronald Edward, 1978. *French Weights and Measures before the Revolution: A Dictionary of Provincial and Local Units.* Bloomington: Indiana University Press.

Zupko, Ronald Edward, 1981. *Italian Weights and Measures from the Middle Ages to the Nineteenth Century.* Philadelphia: American Philosophical Society.

Source Index

NB: references to all versions of Jacopo's *Tractatus* are omitted from this index. References to footnotes point to the page where the note begins.

Index of Personal and Geographical Names

This index in split into three parts. The first lists references to authors respectively editors appearing in the bibliography, the second other personal names, the third geographical locations. Locations of manuscripts and places mentioned solely in the coint lists are left out.

References to footnotes point to the page where the note begins.

Authors/Editors Appearing in the Bibliography

Other Personal Names

Geographical Locations

Subject Index

References to footnotes point to the page where the note begins.

Science Networks – Historical Studies (SNHS)

Edited by
Eberhard Knobloch, Technische Universität Berlin, Germany
Erhard Scholz, Bergische Universität Wuppertal, Germany

In cooperation with an international editorial board

The publications in this series are limited to the fields of mathematics, physics, astronomy, and their applications. The publication language is preferentially English. The series is primarily designed to publish monographs. Annotated sources and exceptional biographies might be accepted in rare cases. The series is aimed primarily at historians of science and libraries; it should also appeal to interested specialists, students, and diploma and doctoral candidates. In cooperation with their international editorial board, the editors hope to place a unique publication at the disposal of science historians throughout the world.

SNHS 33: De Risi, V.
Geometry and Monadology. Leibniz's Analysis Situs and Philosophy of Space (2007)
ISBN 978-3-7643-7985-8

SNHS 32: Krömer, R.
Tool and Object. A History and Philosophy of Category Theory (2007)
ISBN 978-3-7643-7523-2

The book is first of all a history of category theory from the beginnings to A. Grothendieck and F.W. Lawvere. Category theory was an important conceptual tool in 20th century mathematics whose influence on some mathematical subdisciplines (above all algebraic topology and algebraic geometry) is analyzed. Category theory also has an important philosophical aspect: on the one hand its set-theoretical foundation is less obvious than for other mathematical theories, and on the other hand it unifies conceptually a large part of modern mathematics and may therefore be considered as somewhat fundamental itself. The role of this philosophical aspect in the historical development is the second focus of the book. Relying on the historical analysis, the author develops a philosophical interpretation of the theory of his own, intending to get closer to how mathematicians conceive the significance of their activity than traditional schools of philosophy of science.

SNHS 31: Keller, A.
Expounding the Mathematical Seed. Vol. 2: The Supplements (2006). ISBN 978-3-7643-7292-7

SNHS 30: Keller, A.
Expounding the Mathematical Seed. Vol. 1: The Translation (2006). ISBN 978-3-7643-7291-0

SNHS 30/31 Set: ISBN 978-3-7643-7299-6

In the 5th century the Indian mathematician Āryabhaṭa (476–499) wrote a small but famous work on astronomy, the Āryabhaṭīya. This treatise, written in 118 verses, gives in its second chapter a summary of Hindu mathematics up to that time. Two hundred years later, an Indian astronomer called Bhāskara glossed this mathematial chapter of the Āryabhaṭīya. Subjects treated in Bhāskara's commentary range from computing the volume of an equilateral tetrahedron to the interest on a loaned capital, from computations on series to an elaborate process to solve a Diophantine equation.
The first volume contains an introduction and the literal translation. The second volume contains explanations for each verse commentary translated in Volume 1. These supplements discuss the linguistic and mathematical matters exposed by the commentator. Particularly helpful for readers are an appendix on Indian astronomy, elaborate glossaries, and an extensive bibliography.

SNHS 29: Guerraggio, A. / Nastasi, P.
Italian Mathematics Between the Two World Wars (2005). ISBN 3-7643-6555-2

SNHS 28: Hesseling, D.
Gnomes in the Fog. The Reception of Brouwer's Intuitionism in the 1920s (2003)
ISBN 3-7643-6536-6

SNHS 27: Dauben, J.W. / Scriba, C.J.
Writing the History of Mathematics – Its Historical Development (2002)
ISBN 3-7643-6166-2 (Hardcover)
ISBN 3-7643-6167-0 (Softcover)

Science Networks – Historical Studies (SNHS)

Edited by
Eberhard Knobloch, Technische Universität Berlin, Germany
Erhard Scholz, Bergische Universität Wuppertal, Germany

In cooperation with an international editorial board